Fundamentals of Springs Mechanics

Vladimir Kobelev

Fundamentals of Springs Mechanics

Third Edition

Springer

Vladimir Kobelev
Department of Natural Sciences
Faculty IV: School of Science
and Technology
University of Siegen
Siegen, Nordrhein-Westfalen, Germany

ISBN 978-3-031-58583-8 ISBN 978-3-031-58584-5 (eBook)
https://doi.org/10.1007/978-3-031-58584-5

This Springer imprint is published by the registered company Springer Nature Switzerland AG
The registered company address is: Gewerbestrasse 11, 6330 Cham, Switzerland

Foreword

Technical springs are well-known machine components that can be reversibly deformed under load and cyclic or oscillating forces. Springs convert kinetic energy into potential energy, store energy, and return it to the system almost without loss when relieved. In order to take advantage of these properties for optimized applications, two essential aspects must be considered: the characteristics of the material used and a spring design that is adapted and optimized for the application. Spring steel alloys have the appropriate material characteristics for springs, which are usually highly stressed. In addition, the well-designed shape of the spring allows the technical requirements and characteristics such as stiffness, fatigue life, etc. to be met. Various spring designs are classified according to their shape and type of loading, which in most cases also provides the basic understanding for their technical calculation. Due to their various technical characteristics and functions, springs are still almost irreplaceable components in every new and modern machine concept, in airplanes, ships, buildings, trains, or automobiles. In order to meet all these high requirements, standards, and specifications, accurate calculation methods are needed, which take into account all important physical effects of springs.

The purpose of this script is to explain the mechanical and physical properties of specific steel alloy springs and to present complementary analytical calculation methods based on existing and summarized calculation models. Approaches for characteristic spring data such as weight and package, life and crack growth, creep and relaxation rate, and lateral vibrations and natural frequencies are presented for specific spring shapes. The script includes calculations for coil springs, disk springs, wave springs, and thin-walled rods with semi-open cross-section. Due to the analytical approach of all calculation models, ambitious development and design engineers get a helpful review and overview of existing and complementary calculation methods for springs.

Professor Vladimir Kobelev was born in Rostow-on-Don, Russian Federation. He studied Physical Engineering at the Moscow Institute of Physics and Technology. After his Ph.D. at the Department of Aerophysics and Space Research (FAKI), he habilitated at the University of Siegen, Scientific-Technical Faculty.

Today, Prof. Kobelev is a lecturer and APL professor at the University of Siegen in the subject area of Mechanical Engineering.

In his industrial career, Prof. Kobelev is an employee at Mubea, a successful automotive supplier located near Cologne/Germany. In the Corporate Engineering Department, Prof. Kobelev is responsible for the development of calculation methods and physical modeling of Mubea components.

Dr.-Ing., Dr. h. c. Thomas Muhr
Managing Partner Muhr and Bender
KG, Mubea
Attendorn, Germany

The original version of the book has been revised: Formulas were corrected in Appendix B.

Preface

This book focuses on the mechanics of elastic elements made of steel alloys, with an emphasis on metal springs for the automotive industry. Industry and scientific organizations are intensively studying the fundamentals of spring element design and constantly improving the mechanical properties of spring materials. The development tasks of spring manufacturing companies include the optimal application of existing material types. The task is therefore to evaluate the mechanical properties in a targeted manner and then to design springs that make full use of the achievable material properties. The first edition of this book (Kobelev, 2018) was revised to reflect the current trends and the state of the art in the modeling of springs. The second edition includes several corrections and improvements of the theory from the first edition. Most of the methods for fatigue assessment of spring materials have been revised. Experimental data for creep and relaxation under temperature have been added. The theory of simultaneous creep and plastic flow has been added to describe the time-delayed presetting of coiled springs. This theory is essential for the fine tuning of presetting processes in machines.

The third edition was considerably extended compared to the second edition (Kobelev, 2021). Chapters 5 and 6 deal with the new theory of disk springs with fixed inner and outer edges. In Chap. 5, disk springs with constant material thickness are studied using the models of thin and moderately thick isotropic shells. The deformation behavior of the disk spring is treated as a one-dimensional inversion of a circular ring of rectangular section about certain inversion centers. The calculation of the disk springs examines the free-gliding edges and the edges with constrained radial motion. The formulations are used to derive load-displacement formulae for the disk springs with different constraints on radial movement on the inner and outer surfaces.

Based on this model, the disk spring with non-constant thickness is calculated in Chap. 6. The disk spring with radially or tangentially variable material is studied. The equations provided are sufficiently accurate to evaluate relatively flat disk springs with smooth thickness variations along the latitude and longitude of the conical central surface. The derived formulae are applicable for design purposes. With these two new chapters, the formulae cover most conceivable applications of disk springs in heavy industry, mechanical engineering and automotive engineering.

Another reorganized topic deals with the calculation of damage during fatigue loading. There are several independently invented criteria for calculating damage during loading with different amplitudes and an average force in the cycle. The engineer must occasionally recalculate the damage from one damage model to another. At first sight, the recalculation appears to be difficult. The new Appendix D provides methods for recalculating the most common damage criteria. This is an important task for the correct application of finite-element methods to damage assessment. A correct description of fatigue processes is important for machinery where potential damage can become a hazard to people. This applies to the aerospace, energy, and automotive industries, among others. In particular, the fatigue life of springs has not yet been fully evaluated. The merging of the multiaxial stress tensor to the equivalent scalar value is one open question. The mean stress sensitivity is the other. Various strength hypotheses have been proposed to answer these questions. The strength hypotheses lead to relationships for a reasonable curve expressed in stress or strain. From this curve, the final allowable amplitudes of stress or strain are obtained. The final allowable amplitudes correspond to a certain number of cycles to failure. The dominant stresses in leaf and disk springs are normal stresses. For shear compression or helical tension springs, the shear stresses should be used instead of the normal stresses. Fatigue characteristics are determined by analyzing the failure parameters and fatigue properties of the materials. The appendix discusses in detail the mean stress sensitivity of the common spring material and compares the different approaches to evaluating the fatigue characteristics of the spring.

The motivation of the previous editions was mainly the fatigue phenomena of the spring elements. The scope of the manuscript was considerably extended in the 2nd and 3rd editions of the book, entitled *Durability of Springs*. Two-thirds of the text of the current volume deal with design and manufacturing aspects of spring technology. This status quo argues for an extension of the title. Since the 3rd edition, the text has been given the new title *Fundamentals of Spring Mechanics*. The logical structure of the text remains largely unchanged. Readers familiar with the old editions will find the updated, corrected, and expanded chapters in the new edition as well.

Siegen, Germany Vladimir Kobelev
Vladimir.kobelev@uni-siegen.de
https://www.mb.uni-siegen.de/kobelev/

Introduction

Aims and Methods of the Book

The integral parts of many mechanical systems are elastic elements or springs (Juvinall & Marshek, 2017, Chap. 12). The springs make it possible to maintain a tension or force in a mechanical system, to absorb the shocks and to reduce the vibrations. Highly stressed spring elements in modern industrial and transport equipment must survive a very high number of cycles with high average stress and high amplitude stress. These springs are manufactured from high-quality wires and by means of special mechanical and heat treatment processes. The standard designations of various spring steels are summarized in (ASTM DS67A, 2002).

The spring is the widespread resilient element, which is used in the industrial machinery and automotive systems:

- coil tension springs and/or torsion springs in disk or drum brakes, locks or locking or blocking systems,
- torsion or bending springs for belt tensioners, safety belts, and for load compensation in clutch pedals,
- disk springs with or without slots for use in clutches, bearings, and for load pre-tensioning,
- leaf springs for chassis suspension,
- helical, or coil, spring for reducing impact events in passenger cars, some heavy trucks, and railroad cars
- coil springs for nozzle holders, in transmissions, as valve springs, injection regulators, or as vibration dampers in clutches and brake cylinders, as diesel fuel pumps, valvetrains, brakes, seats, doors, and control elements.

Coil, of helical, springs are formed by winding a wire or rod of uniform cross-section around a cylinder. A fixed distance is maintained between successive coils of a spring so that the axis of the wire forms a helix. The standard design procedures for helical springs are (SAE, 1996) and (DIN EN, 2012, 2013, 2015). Efficient design procedures for spring elements are based on modern simulation and optimization methods. Achieved by these methods, the reduction of the weight of suspension springs causes the reduction of the unsprung mass of the axle. This reduction has a positive effect on the comfort, traction, and steering characteristics of the car. The development of modern passenger cars has also highlighted the trend toward reduced packaging of suspension components to maximize space for occupants and cargo. Such requirements lead to a reduction in spring dimensions and wire cross-section. Springs can also be found in high-precision test equipment, where springs play the role of energy harvesters.

One of the most important applications for heavy-duty springs is in the valve train of internal combustion engines. Valve springs in internal combustion engines provide forced contact between all moving valve train components during the valve lift up to maximum engine speed. Assuming annual production of 100 million cars having roughly 20 valve springs per engine, one gets a rough number of 2000 million valve springs produced. In Europe and North America, the valve spring is produced mainly from high-tensile wire alloyed with the elements chrome and vanadium. The extremely high oscillating stresses on the surface of the wire achieve the peak values up to 2000 MPa. The requirements on failure rate must be below 1.5% for engine operation test. Hereby, they are subjected to extreme vibration stresses and must endure up to 3×10^8 cycles without failure (Muhr, 1992).

There are some common characteristics of high-quality and properly designed springs:

- the high homogeneity of stresses on the surface of the spring; thus, the absence of stress-concentrators;
- the considerable residual stresses, which, if properly induced, significantly prolong the operation life;
- high sensitivity to imperfections, flaws, inclusions, and corrosion;
- high amount of specific stored elastic energy.

The spring industry has developed specialized materials and sophisticated manufacturing processes to meet the above requirements.

The springs are mostly made from oil-quenched and tempered steel wire, which is wire formed by drawing hot-rolled steel rod through a drawing die and oil-quenching the resulting wire. Oil quenching is a term of art that identifies a process that generally involves heating the wire to austenitizing temperatures, quenching it in oil, quenching it by heating, and recoiling it. This sequence of manufacturing steps increases the tensile strength of the material. However, the ductility of the austenitized material is reduced. The material behaves almost elastically until the moment of fracture. The influence of these two effects, the requirement to increase the ultimate stress and the reduction in ductility, on the fatigue life of the material is contradictory.

The oil-quenched and tempered low-alloy chrome-silicon spring steel wire achieves strengths in excess of 2600 MPa. Even higher strengths can be achieved with chrome-silicon-vanadium alloy steels for valve applications. The wire is peeled or ground before drawing and subjected to non-destructive crack testing after annealing. At the same time, high strength requires high purity and surface finish of the semi-finished product, good formability in cold-forming steels, and corrosion resistance.

High purity also improves ductility while maintaining static strength, which is particularly important for cold-formed springs. The latest phase of materials development for spring steels has therefore been strongly characterized by efforts to achieve very good purity and surface qualities on the one hand, and high-strength steels with the best possible ductility on the other. A metallurgy specially adapted to high-strength spring steel wires allows a very good degree of purity (Hagiwara et al., 1991; Kawahara et al., 1992; Wiemer, 1998). Several publications are devoted to the improvement of wire properties and surface conditions (O'Malley & Hayes, 1990; Postma, 1993). Also included are developments in the direction of thermo-mechanical forming, improvement of toughness properties for a given strength, considerations of the most suitable heat treatment equipment, and the special manufacturing and treatment processes for stainless spring semi-finished products (Lehnert, 1997; Illgner, 1995; Schmitz-Cohnen, 1994).

This setting reduces relaxation and improves creep behavior of springs at operating temperature. It is well known that curing affects the static residual stresses in the spring and changes the cyclic fatigue properties of the spring. In modern spring manufacturing, both cold and hot settings are used. Heat setting refers to the process of applying a time-dependent load to the spring at an elevated temperature. The main physical process during spring setting is material creep.

Shot peening is another mechanical surface treatment used to improve spring performance. Local plastic deformation occurs on the wire surface, resulting in an improvement and strengthening of the properly machined surface. Shot peening significantly increases the fatigue life of springs.

As a result, springs are a highly sophisticated element of modern machines with the highest fatigue life for relatively intensive cyclical stresses. To achieve the required fatigue life, the production of springs involves several specific procedures. The production of modern springs includes several special techniques, which make springs stand out from ordinary machine components. It is worth mentioning that occasionally occurring fatigue failure of springs can cause damage to the entire machine component. In this case, the failure provokes costs that are incompatible even with the price of the highest quality spring.

Structure of the Book

The book reviews the advanced theory of elastic elements from the point of view of structural and material mechanics. The book examines the principal problems that are essential for clarifying the manufacturing and performance of the metallic spring elements. The elements of creep, plasticity, and fatigue serve as the building blocks of the physical background. What all the problems considered have in common is that they are solved in closed form.

The content of the book is divided into three parts. Part I studies springs from the design point of view. The spring elements of machines and vehicles can be roughly divided into helical, leaf, disk, and twist beams. The models of each family of elastic elements and their optimization are discussed in Part I of the book, which covers Chaps. 1–7. Part II explains the basic processes of spring manufacturing. Plastic cold working is the main manufacturing process. Coiling and plastic presetting are described in Chaps. 8–10. Part II examines the creep and relaxation of springs during presetting. Part III of the manuscript estimates the service life of the springs. This part deals with the fatigue phenomena that usually limit the spring life. This part consists of Chaps. 11–14. The appendix includes the mathematical formulary for fatigue estimation of springs and different presentations of fatigue diagrams.

Part I begins with the determination and evaluation of the spring design. The optimization of springs is studied in Chap. 1. The helical springs are the typical energy-storing elements of the valve train in automotive engines, in the suspensions of passenger cars and railway wagons, and in mechanical engineering. The design formulas for linear helical springs with inconstant wire diameter and variable mean spring diameter are presented. These formulas are used to optimize the spring for a given spring rate and wire strength. The basic principles of leaf spring design are also briefly discoursed.

Chapter 2 presents analytical solutions for the torsion problem of an incomplete torus with circular and non-circular cross-sections. The hollow cross-sections of the shape demonstrate a closed form of the analytical solution. The solution is useful for the analysis and design of helical springs with non-circular wires. The torsion problems for straight cylinders with circular and elliptical cross-sections allow for the well-known closed-form solutions that will be needed in the next chapters. Two main load applications, axial force and axial moment, are analyzed.

Chapter 3 explains a powerful method for simplifying the helical spring equations. Instead of a full treatment of the helical wire, the deformation of the virtual center line is studied. The virtual centerline has effective strain, torsion, and bending stiffness. It behaves like an initially straight elastic rod or column. This simplification allows a straightforward solution of several practically important problems. These include the load dependence of the transversal vibrations of helical springs and the transformation of transversal vibrations into buckling mode. Lateral buckling of the spring is considered in the context of dynamic stability as a limit case for vibration analysis.

For proper consideration of dynamic effects, models of flexible springs with solid wire are required. In some cases, such as when the spring is uniform, analytical models for dynamics and buckling can be developed. However, in typical springs, only the central coils are uniform; the ends are often not (e.g., they have a varying helix angle or cross-section). Thus, obtaining analytical models in this case can be very difficult, if possible. A variety of theories can be found in the literature to describe the dynamic behavior of helical springs, which includes the interaction of bending (flexural), torsion, and longitudinal waves. In addition, various approximate methods are used to determine the fundamental frequencies of spring vibrations. The methods used to determine the fundamental frequencies can be roughly divided into three groups:

- analysis methods, based on the concept of an equivalent column;

- exact analysis methods, based on the theory of spatially curved bars;
- numerical methods, based on finite-element formulation for spatially curved bars.

The governing equations for the transversal vibrations of the axially loaded linear helical springs are developed. The method is based on the traditional concept of an equivalent column. The effect of the axial load on the fundamental frequency of the transversal vibrations is shown. The natural frequency of the transverse vibrations of the spring depends on the variable length of the spring. If the number of active coils remains constant, the frequency gradually decreases as the length of the spring is shortened. Finally, when the frequency becomes zero, the lateral buckling of the spring occurs by the divergence mode. Note that progressive springs have the opposite behavior. The number of active coils in compressed progressive springs decreases and the transverse frequency increases. Therefore, the progressive springs rarely became transversely unstable.

The combination of tension, compression, and torque loads often causes deformation of spring elements. These loads sometimes lead to spatial buckling of the elements. The stability of helical springs under the combined tension, compression, and torsion is elucidated.

Disk springs (also known as Belleville washers) are studied in Chap. 4. Disk springs are examples of highly stressed machine elements. The workpiece for disk springs is in the form of a sheet or roll. The disk springs are usually produced by stamping. Other processes to produce disk springs are blanking, stamping, perforating, cutting, drawing, notching, lancing, and bending. Multiple punches can be used together to produce a part in one step. The "Conical Disk Spring," "Belleville Spring," or "Belleville Washer" is typically used as a spring or to apply a preload or flexible quality to a bolted joint or bearing. These springs are the typical energy-storing elements of the valve train in clutches and automatic transmissions of automobiles. Belleville springs are generally made of spring steel and can be subjected to static loads, rarely alternating loads, and dynamic loads. Belleville springs must meet stringent fatigue life and creep requirements. Because the basic characteristics of disk springs include high fatigue life, better space utilization, low creep tendency, high load capacity with small spring deflection. From a mechanical point of view, Belleville washers are flat conical rings subjected to axial load. Normally, the thickness of the ring is constant, and the applied load is evenly distributed between the upper inner edge and the lower outer edge. In this chapter, the equilibrium equations of thin and moderately variable thickness disk springs are obtained. Variational principles for conical shells are used for the derivation. Simplification is based on common deformation hypotheses. Closed-form analytical solutions of thin and thick truncated cone shells are obtained.

Disk wave springs are also analyzed in Chap. 4. Wave springs are coiled, resilient parts made of flat material. The special feature of these springs is the reduction of their spring height under the conditions of spring force and travel. This feature makes wave springs suitable for compact installation. The production of wave springs from wire, as opposed to the punching of disk springs from flat or rolled sheet metal.

This allows a considerable amount of scrap to be saved, which is typical of stamping operations. Both linear and nonlinear disk wave springs are deliberated.

Chapter 5 examines fixed-edge disk springs using the thin and moderately thick isotropic shell models. The thickness of the material is kept constant in this chapter. The calculation of the disk springs examines the free-gliding edges and the edges with constrained radial motion. The variation formulas are used to derive load-displacement formulas for the disk springs with different radial constraints on the inner and outer surfaces. The kinematic hypothesis is used for the shell models of conical shells. The motivating feature of the presented theory is its ability to calculate the cup springs with free-gliding edges and the edges with constrained radial motion. The equations developed here are based on general assumptions and are suitable for disk springs made of isotropic materials, such as spring steel and light metal alloys. The advantage of the methodology is the derivation of closed-form solutions for several common restrictions on the radial motion of the inner and outer edges. The developed formulas are recommended for industrial calculations of free and restricted disk springs and Belleville washers.

Chapter 6 observes the disk spring with variable thickness. The thickness of the material is variable along the meridional and parallel coordinates of the conical coordinate system. The calculation of disk springs includes the cases of free-gliding edges and edges on cylindrical curbs, which restrict the radial movement. The equations developed here are based on common assumptions and are simple enough to be applied to industrial calculations.

The analysis of thin-walled, semi-open-section beams is performed in Chap. 7. An essential characteristic of this class of thin-walled beam-like structures is their closed but flattened profile. In this book, an intermediate class of thin-walled beam sections is studied. The cross-section of the beam is closed, but the shape of the cross-section is elongated and curved. The walls forming the section are nearly equidistant. The unusual shape of semi-open thin-walled beams allows for efficient optimization due to the large variability of shapes. The automotive application of semi-open thin-walled beams is conversed later in Chap. 7. The main application of the theory of semi-open thin-walled beams is the twist beam of the semi-solid trailing arm axle. The analytical expressions for the effective torsional stiffness and effective bending stiffness of the twist beam are derived in terms of the section properties of the semi-open section twist beam. Based on the stiffness coefficients of the twist beam, the roll rate, chamber, and lateral stiffness of the suspension are derived.

Mechanical problems encountered in the manufacture of helical springs are examined in Part II, which includes Chaps. 8 and 9. In Chap. 8, we analyze the coiling of helical springs. For this purpose, we study the plastic flow and the appearance of residual stresses. It is well known that the excessive stresses during the coiling of helical springs can cause the rod to break. In addition, the high level of residual stress in the formed coil spring significantly reduces its fatigue life. For practical estimation of the residual and coiling stresses in the helical springs, the analytical formulas are required. In this chapter, the analytical solution of the problem of elastic-plastic deformation of cylindrical bar under combined bending and torsional moments is found for a special nonlinear stress-strain law. The obtained solution allows the analysis of the active stresses during the combined bending and torsion. In addition, the residual stresses in the beam after spring-back are also derived in a closed analytical form. The obtained results agree with the reported measured values. The developed method does not require numerical simulation and is perfectly suitable for the programming of coiling machines, the estimation of loads during the production of cold-wound helical springs and for the dimensioning and wear calculation of coiling tools.

In Chap. 9, the prestressing calculation for helical springs is developed. The method is based on the deformation formulation of the plasticity theory and common kinematic hypotheses. From a mathematical point of view, the governing equations of prestressing are somewhat analogous to the equations of the coiling process. Two main types of helical springs are studied—compression springs and torsion springs. For the first type (axial compression or tension springs), the spring wire is twisted. The basic approach neglects the pitch and curvature of the coil and replaces the helical wire with a straight cylindrical rod. The elastic-plastic torsion of the straight bar with circular cross-section is studied. The analysis is based on the hypothesis of St. Venant. In the second type (torsion helical springs), the helical wire is in the state of bending. The model analyzes the delayed prestressing, which is accompanied by a significant creep.

The last, Part III, includes the life cycle of the elastic elements, the high static stresses lead to the residual deformation. The loss of sag leads to the gradual reduction of spring forces with the resulting failure if the spring's working length continues. Spring breakage due to static creep is the extremely rare event. Moderate cyclic loading is accompanied by some creep and cyclic sag loss. Severe cyclic loading sooner or later leads to fatigue failure of the spring. These two sources of possible damage to elastic elements are summarized in Chaps. 10–14.

Understanding the long-term behavior of springs under high static loading is essential to their proper design. Creep and relaxation of springs is the subject of Chap. 10. Stress analysis for creep has a long history in engineering mechanics, driven by the need to design for elevated temperatures. In solid mechanics, creep is the tendency of materials to deform gradually or continuously under the action of external mechanical stresses. At stresses below the yield strength of the material, slow inelastic deformation occurs. In the spring industry, this is called creep when a spring under constant load loses length, and it is called relaxation when a spring under constant compression loses load. Creep and relaxation rates depend on temperature, stress in the metal, yield strength, and time. Increased temperature, stress, and time significantly increase the creep and relaxation rates. In particular, temperature and stress have the greatest influence. An accurate description of creep is essential for proper spring design. Finally, Chap. 10 demonstrates the evaluation of creep constants in a wire twisting experiment. In addition, the exact analytical expressions for the torsional and bending creep of bars are derived using the common constitutive models. One of the common creep constitutive models is the Norton-Bailey law, which gives a power law relationship between minimum creep rate and (constant) stress. The power law can be found in high-temperature design and creep numerical codes.

Other common creep laws are the exponential and Garofalo laws, which more adequately describe the stress dependence over a wide range of working stresses. For all these laws, we derive the analytical formulas for creep caused by steady or oscillating loading. In Chap. 10, the generalized expression for the creep law is studied. The new expression is based on experimental data and unifies the primary, secondary, and tertiary regions of the creep curve. The relaxation functions for bending and torsion depend only on the maximum stress in the cross-section, which occurs at the outer surface of the coil. Finally, we explain the temperature dependence for the creep of spring materials.

Chapters 11 and 12 briefly consider the fatigue effects of springs. We begin with the deterministic approach. The results presented show the average fatigue characteristics and evaluate the stress levels at which the majority of springs fail. In this case, we speak about mean S-N lines. The durability of springs under oscillating loads is the subject of Chap. 11. Traditional fatigue design methods are based on collecting a large amount of experimental data in cyclic tests, structuring the data, and extracting empirical formulas. The method of analyzing crack growth under repeated loading is reviewed in Chap. 11. The expressions for the spring length over the number of cycles are derived in terms of a higher transcendental function. The proposed method starts from the micromechanically inspired effects of crack propagation, explains the history of crack propagation, and finally provides the stress-life curves.

Several effects on fatigue life, mainly the effects of stress ratio and multiaxiality, are deliberated in Chap. 12. The solutions presented are used to estimate the fatigue life of springs for asymmetric harmonic loading with substantial mean stress. An attempt is made to unify various traditional methods into a unified Bergmann-Walker formula. Different settings of two fitting parameters in the unified criterion result in the common fatigue criteria.

It is noteworthy that the high-quality springs differ from the low-quality products mainly by the scattering ranges. The evaluation of scatter requires the statistical methods conversed in the next chapters. The statistical effects on fatigue life are further discussed in Chaps. 13 and 14. We study the probability descriptions for the fatigue limit of heterogeneously stressed members. The proposed approach for stress gradient sensitivity of fatigue life is based on the "weakest link" concept. This method is applicable to the exceptionally brittle materials that fail immediately after the failure of the first constituent element. The weakest link approach is used to calculate the number of cycles to complete failure under different probability levels. The effect of fluctuating stresses on the fatigue life of springs is combined with the influence of heterogeneous stress distribution (stress gradient) over the wire cross-section and time-varying stresses. The stress field is inhomogeneous over the cross-section of the spring wire. The stress distribution is uniquely defined by the ratio of the diameter of the wire to the diameter of the spring body. The calculated lifetimes are compared with the lifetimes of helical springs subjected to cyclic loading.

Chapter 14 examines the stochastic effects on the fatigue life of springs. Stochastic crack propagation is typical of the low-stress and high-cycle fatigue regions. The deviation and branching of the crack are caused by the high inhomogeneity of the polycrystalline structure at the micro level. For the low amplitude of stress, the crack extension per cycle is less than the typical size of inhomogeneities. The stochastic differential equation for the traveling crack is derived. The stochastic equation is similar to the forced Brownian motion equation. The methods are based on the unified fatigue laws. These laws lead to analytical solutions for the crack length at the mean value and the range of cyclic variation of the stress intensity factor. In this chapter, we demonstrate the closed-form expressions for the number of cycles to failure as a function of the initial crack size.

Target Audience of the Book

This book was written as a script for the courses "Applied Mechanics of the Automobile," "Automotive Engineering, Chassis, II, III," "Structural Optimization in Automotive Engineering," and "Powertrain Modeling and Optimization," which the author has been teaching at the University of Siegen, North Rhine-Westphalia, Germany, since 2001.

This book is primarily intended for engineers involved in the design and development of springs who have graduated from automotive or mechanical engineering programs at technical colleges or other engineering schools. Researchers working on elastic elements and energy harvesting devices will also find a broad overview of the fundamentals of spring methodology.

The current book presents powerful methods for the analysis of elastic elements made of steel alloys. The focus is on metal springs for the automotive industry.

It is well known that the industry researches the design of spring elements and constantly develops the quality of spring materials. New materials are developed in the factories of material suppliers. The task of the spring manufacturing companies is the optimal application of the existing and newly developed types of materials. The technologically advanced method consists in the target-oriented evaluation of the mechanical properties and the subsequent design of the springs, which makes full use of the measured material properties. Thus, the full development of the improved materials is only possible if their essential properties are rigorously acquired. Design and manufacturing must fully exploit all available capabilities of the semi-finished product.

An enormous number of papers have been written on this and related subjects. This does not mean that the science of spring mechanics and strength has become completely useless. Rather, it is necessary to have a thorough understanding of mechanics and metallic disciplines in order to elucidate the wide range of possibilities. Therefore, much of this book is devoted to these topics.

The precise methods for the design of different types of springs are summarized in the relevant standards. Industrial research has developed reliable methods for estimating fatigue life and creep effects. The purpose of this book is not to replace the established methods of design and pragmatic methods of life assessment of springs. The aim of this book is to qualitatively explain the mechanical behavior of the spring as a unique elastic element, which has some very specific properties. We try to survey the vast and fragmented landscape of springs from a single point of view of classical mechanics. We try to compare the different methods of service life evaluation and to point out the most effective and general methods. There are many experimental values that are currently the subject of speculation. It is possible that these values will be obtained in future studies. Most of the methods presented are recognized and applicable to other heavily loaded structural elements.

A few words about solution methods. There are several recognized commercial finite-element codes, e.g., (ANSYS, 2020; ABAQUS, 2020). However, for the modeling of technical systems, analytical solutions often offer important advantages. First, the transparency of closed-form solutions. Because analytical solutions are represented as mathematical expressions, they provide an understandable view of how variables and relationships between variables affect the result. Second, performance: algorithms and models expressed in terms of analytical solutions are often more effective than the corresponding numerical applications. For example, to compute the solution of an ordinary differential equation for different values of its parametric inputs, it is often faster, more accurate, and more appropriate to evaluate an analytical solution than to integrate numerically. Third, numerical solutions are sometimes extremely abundant. The main reason is that sometimes we either don't have an analytical approach, or the analytical solution is too slow, and instead of computing for hours and getting an exact solution, we rather compute for seconds and get a good approximation. Finally, numerical solutions can seldom contribute to the invention of new ideas. For this reason, the treatment of the material in this book resolves the problems studied to closed-form solutions in the form of mathematical expressions.

The content of this book is logically related to the work *Design and Analysis of Composite Structures for Automotive Applications* by the same author. The latter book is an extension of the present book and covers the subject of composite materials. The manuscript (Kobelev, 2019a) examines the special properties of composite materials, such as their anisotropy, inhomogeneity, load direction dependence, stress coupling, and stacking capabilities.

Contents

Part II Manufacturing of Springs

Symbols[1]

Symbol	Description	Unit[a]
Q_a	Activation energy of diffusion	kJ/mol
χ	Angle between meridian and principal material axis	rad
ϕ_Q	Angle of inclination of the bent axis	rad
α	Angle of pitch of helical spring	rad
ϑ	Angle of twist per unit length	rad
$\bar{\vartheta}$	Angle of twist per unit length after spring-back	rad
$\psi = H/(r_e - r_i)$	Angle, of slope, deformed middle surface of disk spring	rad
$\alpha = h/(r_e - r_i)$	Angle, of slope, free middle surface of disk spring	rad
ϕ	Angle, rotation of the middle surface disk spring	rad
A	Area of the material part of the cross-section	m^2
A_m	Area, enclosed by the curve L_m	m^2
$\phi_1(r), \phi_2(z)$	Auxiliary functions	1
k_1, k_2	Auxiliary functions for survival probabilities and ratio of cycles to failure of spring to straight rod	1
σ_0, N_0	Auxiliary scaling constants, $\sigma'_f = \sigma_0(2N_0)^{-b_0}$	1
T, B	Axes of wire cross-section (thickness and width)	m
$\bar{R} = 1/\bar{\kappa}$	Bending radius after unloading (after spring-back)	m
$R = 1/\kappa$	Bending radius in during active coiling	m
$\theta_a = 2\pi n_a$	Circumferential angle along wire length	rad

(continued)

[1] For the sake of continuous consistency in the manuscript, the designations may be different from those used in the standards.

(continued)

$x_e \leq x \leq x_i$	Coordinate on the meridian of free conical shell	m
$x_c,\ y_c$	Coordinates of the center of mass of the cross-section	m
$\alpha_x,\ \alpha_y$	Coordinates of the twist center of the cross-section	m
$x = \rho \cos \phi,$ $y = \rho \sin \phi$	Coordinates, Cartesian of the circular cross-section	m
$0 \leq \rho \leq r,$ $0 \leq \phi \leq 2\pi$	Coordinates, polar of the circular cross-section	m rad
$k_i = k_i(w)$	Correction factors for stress, $i = 1, .., 4$	1
c_τ	Creep constant, for shear strain	$\frac{s^{-\varsigma}}{Pa^{\xi+1}}$
c_σ	Creep constant, for uniaxial strain	$\frac{s^{-\varsigma}}{Pa^{\xi+1}}$
t_c	Creep, average time	s
$\bar{t}$	Creep, Norton-Bailey constant	s
$\bar{\varepsilon},\ \bar{\gamma}$	Creep, strain rate constants	1/s
ξ	Creep, stress exponent	1
$\bar{\sigma},\ \bar{\tau}$	Creep, stress scaling constants	Pa
ς	Creep, time exponent	1
κ	Curvature in moment of plastic deformation	1/m
κ_i	Curvature principal, $i = 1,\ 2$	1/m
N_L	Cycles number to the failure for a given stress amplitude, highest (failure event of the last homogeneously stressed specimen)	1
N_F	Cycles number to the failure for a given stress amplitude, lowest (failure event of the first homogeneously stressed specimen)	1
$\mu_+^*,\ \mu_-^*$	Deflection at loading and unloading, critical	1
ρ_m	Density of material	kg/m^3
$D^{\widehat{\alpha}}$	Derivative, fractional of order $\widehat{\alpha}$	$1/s^{\widehat{\alpha}}$
$\mathrm{d} = 2r$	Diameter of circular wire or bar	m
D_e	Diameter of middle surface of free spring, external (outside)	m
D_i	Diameter of middle surface of free spring, internal (inside)	m
d_{opt}	Diameter of wire, optimal	m
μ^*	Dimensionless length, character	1
$\tilde{s}$	Displacement, axial, measured from upper inside edge to lower outside edge	m
s_b	Displacement, caused by bending moment	m
s_s	Displacement, caused by shearing force	m
s_Q	Displacement, lateral	m
τ_e	Endurance limit for completely reversed stress	Pa
K_{th}	Endurance threshold limit	$N\sqrt{m}$

(continued)

(continued)

$U_e,\ U_1 \ldots U_5$	Energy, elastic strain	J
U_f	Energy, potential of applied forces	J
Π	Energy, total potential	J
$\langle EI_B \rangle$	Equivalent bending stiffness	Pa m^4
$\langle GS \rangle$	Equivalent shear stiffness	Pa m^2
F^*	Euler's critical load for compression	N
$m_2 > 1$	Exponent, at short-term limit	1
$m_1 > 1$	Exponent, endurance limit	1
$p > 1$	Exponent, fatigue	1
$c(R_\sigma)$	Exponent, fatigue ductility	1
p_S	Exponent, fatigue in high-cycle range	1
p_L	Exponent, fatigue in moderate cycle range	1
$b_0(R_\sigma) = -1/p_\sigma$	Exponent, of fatigue strength	1
k	Exponent, secant	1
k_f	Extent constant of failure region	–
S_f	Factor for safety	1
$\gamma'_f(R_\sigma)$	Fatigue ductility coefficient, shear	1
$\varepsilon'_f(R_\sigma)$	Fatigue ductility coefficient, uniaxial	1
$\lambda \equiv \frac{1}{2}\left(\sigma'_f\right)^{-\frac{1}{b_\sigma}}$	Fatigue equation, constant	1
$\sigma'_f\ (R_\sigma)$	Fatigue strength coefficient for normal stress	Pa
$\tau'_f(R_\sigma)$	Fatigue strength coefficient for shear stress	Pa
F_R	Force acting on the upper middle surface, radial	N
$F_\tau(z,\ s)$	Force tangential, pro unit length	N/m
F	Force, axial on the spring	N
F_θ	Force, Circumferential in the wire direction	N
$\tilde{F}_{1Z}$	Force, corrected total axial	N
$F_\sigma(z,\ s)$	Force, normal, pro unit length	N/m
$F_z(t)$	Force, of spring as the function of time	N
Q	Force, shear	N
F_z^0	Force, Spring at the moment $t = 0$	N
F_{min}	Force, spring loads at lengths L_{comp}	N
F_{max}	Force, spring loads at lengths L_{ref}	N
F_z	Force, total axial acting on the upper middle surface	N
F_{AL}	Force, total axial due to Almen and Laszlo	N
F_{zDIN}	Force, total axial, DIN standard	N
$N_1,\ N_2,\ N_{12}$	Forces, meridional, circumferential and shear direct	N

(continued)

(continued)

ω_N	Frequency of compressed spring, natural	rad/s
ω_N^0	Frequency of free spring, natural	rad/s
Ω_N	Frequency, fundamental, dimensionless, relative, $\Omega_N(\mu) = \omega_N/\omega_N^0$	1
R	Gas constant	$\frac{\text{J}}{\text{mol K}^\circ}$
Y	Geometry parameter, dimensionless	1
M_T^*	Greenhill's critical twist buckling moment	N/m
h	Height of middle surface of disk spring, free state	m
H	Height, of middle surface of disk spring in the deformed state	m
z_i	Height, of the inside edge of the middle surface	m
z_e	Height, of the outer edge of the middle surface	m
$\tilde{h}$	Height, total between utmost edges of the conical spring in its free state	m
$\tilde{H}$	Height, total between utmost edges of the deformed conical spring	m
$w = D/\text{d}$	Index of spring	1
$\mathbf{II}(\boldsymbol{\varepsilon})$	Invariant of deviatory strain tensor, second	Pa^2
$\mathbf{II}(s)$	Invariant of deviatory stress tensor, second	Pa^2
$\mathbf{I}(\boldsymbol{\varepsilon})$	Invariant of strain tensor, first	Pa
$\mathbf{I}(s)$	Invariant of stress tensor, first	Pa
c_i	Inversion center point for cross-section	m
a	Length of crack	m
L	Length of the spring, actual under action of load	m
L_0	Length of the spring, Free	m
$\eta = \varepsilon_p/r$	Length parameter, inverse	1
$\lambda_w = a/l$	Length ratio of disk wave spring	1
${\lambda_c}^2 = \frac{GI_T}{EI_\omega}$	Length, characteristic	m
L_{comp}	Length, compression	m
$\mu = L/L_0$	Length, dimensionless relative	1
L_T	Length, of trailing arm	m
L_B	Length, of twist beam length	m
L_W	Length, of wire	m
L_{rel}	Length, released	m
$n_{I,i}$	Limit cases for number of cycles to fault (type I), $i = 1, 2, 3$	1
$n_{II,i}$	Limit cases for number of cycles to fault (type II), $i = 1, 2, 3$	1

(continued)

(continued)

$F_{I,i}$	Limit cases for propagation function of type I, $i = 1, 2, 3$	1
$F_{II,i}$	Limit cases for propagation function of type II, $i = 1, 2, 3$	1
f_Q	Load, external in the transverse direction	N
B_s	Magnitude of the Burgers vector, screw dislocation	–
m_{opt}	Mass, absolute lowest	kg
m	Mass, of the spring material	kg
$c_f(R_\sigma)$	Material constant for a given stress ratio R_σ	m
K_{max}	Maximum stress intensity factor per cycle	$N\sqrt{m}$
σ_{max}	Maximum stress per cycle	Pa
$D = 2R$	Mean coil diameter	m
$D_m = \frac{D_o+D_i}{2}$	Mean diameter of ring, disk wave spring	m
$K_m = \frac{K_{max}+K_{min}}{2}$	Mean value of stress intensity factor	$N\sqrt{m}$
M_r	Mean stress sensitivity for Haigh diagram	1
μ_r	Mean stress sensitivity for Goodman diagram	1
ξ^*	Minimal slenderness ratio for instability	1
$K = \frac{2(1+\nu)}{3(1-2\nu)}G$	Modulus, bulk	Pa
G_p	Modulus, secant	Pa
W_t	Modulus, sectional for twist of wire	m^3
G	Modulus, shear	Pa
$E = 2(1+\nu)G$	Modulus, young	Pa
I_x	Moment of inertia of the cross-section with respect to the x-axis	m^4
I_y	Moment of inertia of the cross-section with respect to the y-axis	m^4
I_ω	Moment of inertia of the section, Sectorial	m^5
I	Moment of inertia of wire, area (for helical springs: about the helix axis)	m^4
I_r	Moment of inertia of wire, area (for helical springs: about the radius of helix)	m^4
$S_{\omega p}$	Moment of the cross-section, polar	m^4
M_B	Moment, bending	Nm
M_B^0	Moment, bending at the moment $t = 0$	Nm
m_B	Moment, bending the helical spring wire	Nm
M	Moment, circumferential	Nm
M_{AL}	Moment, circumferential due to Almen and Laszlo	Nm
M_S	Moment, due to constrained torsion	Nm
M_H	Moment, due to pure torsion	Nm
W_B	Moment, of resistance for bending load of wire	N

(continued)

(continued)

W_T	Moment, of Resistance for twist of wire	m^3
M_T	Moment, torque	Nm
M_T^0	Moment, torque at the moment $t = 0$	Nm
m_T	Moment, twisting the helical spring wire	Nm
$M_1,\ M_2,\ M_{12}$	Moments, meridional, circumferential and twist	Nm
n_a	Number of active coils, helical spring	1
$n_f(\sigma_a,\ \sigma_m)$	Number of cycles to failure as function of stress amplitude and mean stress	1
$n_f(a,\ \delta,\ \Delta\sigma)$	Number of cycles to failure as function of stress range	1
n_S	Number of cycles to failure of helical spring	1
n_R	Number of cycles to failure of straight rod	1
$2n_f$	Number of reversals to failure	1
$\lambda,\ \mu$	Parameters, dimensionless	1
p	Pitch	m
α	Pitch angle	1
$r_m = \sqrt{\frac{r_e^2+r_i^2}{2}}$	Pivot radius at maximum height of ovate wire	m
ν	Poisson coefficient	1
$\varphi(r,\ z)$	Prandtl stress function	1
P_F	Probability, failure	1
P_F^*	Probability, prescribed survival of the structural element, helical spring	1
P_S	Probability, survival	1
$P_{S,i}$	Probability, survival, of the spring with defects	1
$P_{R,i}$	Probability, survival, of the straight rod with defects	1
Y_p	Projected length of zigzag crack crack, y- component	m
X_p	Projected length of zigzag crack, x- component	m
$X^* = \frac{1}{\pi}\left(\frac{\Delta K^*}{\Delta\sigma}\right)^2$	Projected length, critical	–
Q_p	Quality parameter, spring	–
R	Radius during plastic coiling	m
$r_e = D_e/2$	Radius of free spring, external (outside)	m
$r_i = D_i/2$	Radius of free spring, internal (inside)	m
r	Radius of wire or bar	m
$\bar{R}$	Radius, after spring-back unloaded	m
$\rho = \sqrt{x^2 + y^2}$	Radius, polar	m
$\mu = T/r_i$	Ratio of material thickness to inside radius	1
$\Delta = r_e/r_i$	Ratio of outside radius to inside radius	1

(continued)

(continued)

α_i	Ratio of survival probabilities of spring and rod	1
$C_R(\gamma)$	Ratio stress amplitude to Walker effective stress	–
$i_t = L/L_T$	Ratio, geometrical transmission	1
$\tilde{A}_i$	Ratio, of amplitude for principal strains	1
A_i	Ratio, of amplitude for principal stresses	1
$p_\sigma = -1/b(R_\sigma) > 0$	Reciprocal strength exponent	1
θ_F, θ_M	Reduction of the spring angle due to compression force and torque respectively	rad
$\Lambda\left[R_\sigma,\ R_\sigma{}^*\right]$	Relation between fatigue coefficients	1
$\Phi(t),\ \Psi(t)$	Relaxation function for coiled and disk springs	1
$\tau_r,\ \sigma_r$	Residual stresses after spring-back	Pa
Λ	Root of eigenvalue problem	1
W_b	Section modulus, for bending of wire (for helical springs: with respect to helix axis)	m^3
W_{br}	Section modulus, for bending of wire (for helical springs: with respect radius of helix)	m^3
$z_+(r),\ z_-(r)$	Sections of inner curve, upper and lower	m
$Z_+(r),\ Z_-(r)$	Sections of outer curve, upper and lower	m
S_x	Sectorial moment of the cross-section with respect to the x-axis	m^3
S_y	Sectorial moment of the cross-section with respect to the y-axis	m^3
$\tau_{r\theta}^{(2)}, \tau_{z\theta}^{(2)}$	Shear stresses, second degree Taylor polynomials of	Pa
K_{IC}	Short-term threshold limit	$N\sqrt{m}$
$\zeta = L_o/D$	Slenderness ratio	1
$p_{SWT.\tau}$	Smith-Watson-Topper parameter for shear stress	Pa
$K_\tau,\ K_\sigma$	Solidifying coefficients for shear and tension states	Pa
$n,\ l$	Solidifying exponents for shear and tension states	1
β_i	Solutions of characteristic equation	1
$\tilde{c}_{1Z}$	Spring rate of disk spring, corrected	N/m
c_w	Spring rate of the wave spring, total initial	N/m
c	Spring rate, compression (or extension)	N/m
$c_{\theta F}$	Spring rate, compression-twist	N
c^*	Spring rate, design compression	N/m
c_{zDIN}	Spring rate, DIN standard	N/m
c_f	Spring rate, for axial loading	N/m
c_θ	Spring rate, for twist loading	*Nm*
c_Z	Spring rate, of disk spring	N/m
c_w	Spring rate, of one complete coil	N/m

(continued)

(continued)

$K(\lambda_c L)$	Stiffening factor due to bi-moment	1
r_c	Stiffness camber, axle	Nm/rad
r_a	Stiffness of axle, roll, $r_a = {i_t}^2 r_t$	Nm/rad
r_z	Stiffness, bending of the twist beam	Nm/rad
EI	Stiffness, bending of wire (for helical springs: about the helix axis)	Pa m^4
EI_r	Stiffness, bending of wire (for helical springs: with about radius of helix)	Pa m^4
r_l	Stiffness, lateral of the axle	N/m
r_t	Stiffness, torsion of the twist beam	Nm/rad
$\bar{r_t}$	Stiffness, torsion of the twisted rod without the influence of bi-moment	Nm/rad
$A_\sigma,\ B_\sigma$	Strain coefficients of Johnson-Cook and Ludwik–Hollomon equations	1
$\dot{\varepsilon}$	Strain rate	1/s
$\dot{e_{ij}}$	Strain rate, deviatoric components	1/s
$D^{\widehat{\alpha}} e_{ij}$	Strain rate, fractional, deviatoric components	$1/s^{\widehat{\alpha}}$
$\dot{\varepsilon}_{ref}$	Strain rate, reference (quasi-static)	1
C	Strain rate, sensitivity parameter	1
$\boldsymbol{\varepsilon}$	Strain tensor	1
ε_{ij}	Strain tensor, components	1
$\boldsymbol{e}$	Strain tensor, deviatoric	1
$e_{ij} = \varepsilon_{ij} - \varepsilon_{oct}\delta_{ij}$	Strain tensor, deviatoric components	1
$\gamma_a = \tau_a / G$	Strain, amplitude of shear	1
$\Delta\varepsilon$	Strain, decrements of maximal normal	1
$\Delta\gamma$	Strain, decrements of maximal shear	1
ε_i	Strain, mid-surface, principal, $i = 1,\ 2,\ 3$	1
ε_{oct}	Strain, octahedral normal	1
$\gamma_{oct} = 2\varepsilon_{oct}$	Strain, octahedral shear	1
$\gamma_P,\ \varepsilon_P$	Strain, on the contour of the cross-section during active load (shear and normal)	1
$\mathbf{E}_i$	Strain, principal, $i = 1,\ 2,\ 3$	1
γ_e	Strain, shear component, elastic	1
γ_c	Strain, shear, creep component of	1
γ_w	Strain, ultimate shear	1
τ_w	Strength, ultimate shear	Pa
$\sigma_w = R_m$	Strength, ultimate tensile	Pa
τ_a	Stress amplitude in operation	Pa

(continued)

(continued)

σ_z	Stress due to bending, normal	Pa
$\sigma_z{}^{(i)}$	Stress due to torsion, normal	Pa
$f_c(\tau_{oct}, t) = h_c(t)g_c(\tau_{oct})$	Stress function, isotropic creep	1/s
$f_{Ic}(\tau_{oct}, t)$	Stress function, isotropic primary creep stage	1/s
$f_{IIc}(\tau_{oct}, t)$	Stress function, isotropic secondary creep stage	1/s
$f_{IIIc}(\tau_{oct}, t)$	Stress function, isotropic tertiary creep stage	1/s
K_{max}	Stress intensity factor per cycle, maximum	$\text{Pa}\sqrt{\text{m}}$
K_{min}	Stress intensity factor per cycle, minimum	$\text{Pa}\sqrt{\text{m}}$
$\Delta K = K_{max} - K_{min}$	Stress intensity factor, range	$\text{Pa}\sqrt{\text{m}}$
ΔK^*	Stress intensity factor, transition range	$\text{Pa}\sqrt{\text{m}}$
σ_{max}	Stress per cycle, maximum	Pa
σ_{min}	Stress per cycle, minimum	Pa
$\Delta\sigma = \sigma_{max} - \sigma_{min}$	Stress range, $\Delta\sigma = 2\sigma_a$	Pa
R_σ	Stress ratio of cyclic load	1
$\boldsymbol{\sigma}$	Stress tensor	Pa
σ_{ij}	Stress tensor, components	Pa
$\boldsymbol{S}$	Stress tensor, deviatory	Pa
$s_{ij} = \sigma_{ij} - \sigma_{oct}\delta_{ij}$	Stress tensor, deviatory components	Pa
σ_R	Stress, "comparative" maximal	Pa
$\frac{1}{2}\lvert\sigma_{Be} + \sigma_{Bi}\rvert$	Stress, average bending	Pa
$\frac{1}{2}(\sigma_{Te} - \sigma_{Ti})$	Stress, average tensile	Pa
σ_b	Stress, basic	Pa
$\lvert\sigma_I\rvert, \lvert\sigma_{II}\rvert, \lvert\sigma_{III}\rvert, \lvert\sigma_{IV}\rvert$	Stress, corner	Pa
τ_c	Stress, corrected	Pa
$\mathbf{S}$	Stress, deviatory tensor	Pa
σ_{Bi}	Stress, due to bending, inside diameter D_i	Pa
σ_{Be}	Stress, due to bending, outside diameter D_e	Pa
σ_{Ti}	Stress, due to circumferential strain, inside diameter D_i	Pa
σ_{Te}	Stress, due to circumferential strain, outside diameter D_e	Pa
$\sigma_{eq} = \frac{3}{\sqrt{2}}\tau_{oct}$	Stress, equivalent Mises	Pa
$\tau(\rho, \phi)$	Stress, for the circular wire	Pa
τ_m	Stress, mean in operation	Pa
σ_{oct}	Stress, octahedral normal	Pa
τ_{oct}	Stress, octahedral shear	Pa
σ_p, τ_p	Stress, on the contour of the cross-section during active load	Pa

(continued)

(continued)

σ_i	Stress, principal, $i = 1, 2, 3$	Pa
$\bar{\sigma}_{zz},\ \bar{\tau}_{\varphi z}$	Stress, residual after spring-back	Pa
τ	Stress, shear	Pa
$\langle\tau\rangle$	Stress, shear average	Pa
τ_S	Stress, shear due to bi-moment	Pa
τ_H	Stress, shear due to pure torsion	Pa
$\sigma_c(\phi) = \tau(\rho = r, \phi)$	Stress, shear on the outer surface of the circular wire	Pa
$\tau_{r\theta},\ \tau_{z\theta}$	Stress, shear, components in the cross-section of wire	Pa
$\tau_{r\theta}^{(1)},\ \tau_{z\theta}^{(1)}$	Stress, shear, first degree Taylor polynomials	Pa
$\sigma_I,\ \sigma_{II},\ \sigma_{III},\ \sigma_{VI}$	Stresses on corner points of disk spring	Pa
A_0	Surface element that contains one critical defect, average	m^2
A	Surface of the whole structural element, spring	m^2
ϖ	Surface, middle in the undeformed state	m^2
Ω	Surface, middle of the shell in the deformed state	m^2
m°	Temperature exponent	1
T°	Temperature of deformation, actual	°K
T°_m	Temperature, melting	°K
T°_{ref}	Temperature, Reference (ambient)	°K
T_e	Thickness of material outside diameter	m
T_i	Thickness of material, inside diameter	m
T	Thickness or height or of the cross-section (for helical springs: in the direction of helix axis)	m
t	Time in seconds	s
$t^* = \int_0^t h_c(p)\mathrm{d}p$	Time, scaled	s
M_θ	Torque of the helical torsion spring	Nm
M_{TW}	Torque of wire during plastic coiling	Nm
m_B	Torque, external per unit length	Nm
I_T	Torsion constant for the section of wire	m^4
$\bar{\chi} = 2\pi\bar{\theta}$	Torsion of helix after spring-back	rad/m
$\chi = 2\pi\theta$	Torsion of helix during plastic coiling, instantaneous	rad/m
GI_T	Torsional rigidity of wire about wire axis	Nm^2
s	Travel or axial displacement of spring	m
$\sigma_{w.n}$	Ultimate nominal stress	Pa
F_I	Unified propagation function of type I	1
F_{II}	Unified propagation function of type II	1
$\psi = \sigma_p/K_\sigma$	Variable, dimensionless in flexure problem	1
$\varphi = \tau_p/K_\tau$	Variable, dimensionless in torque problem	1

(continued)

(continued)

$V = \pi r^2 L$	Volume of active material	m^3
V_1	Volume of one coil	m^3
Ω	Volume of the whole structural element, spring	m^3
V_0	Volume that contains one critical defect, average	m^3
$\tilde{\gamma}$	Walker exponent	1
$\psi(r, z)$	Warping function	1
l_w	Wave length of disk wave spring	m
n_w	Wave spring, number of waves of disk	m
p_h	Weibull "risk of rupture" of the element	1
p_f	Weibull immediate "risk of rupture"	1
S_{sh}, S_W, m_W	Weibull stress shift, stress scale and shape parameters	1
$\tilde{b}$	Width between utmost edges of the conical spring in its free state	m
B	Width of disk spring in deformed state of width of disk wave spring	m
$\tilde{B}$	Width of the deformed conical spring, total	m
b	Width of the middle surface of the unloaded disk spring	m
σ_Y, τ_Y	Yield stress	Pa

[a] Only the units of SI system are referred. In industry, instead of *m* or *Pa* the units *mm* and *MPa*, instead of *rad* the degrees are usually used.

Part I
Design of Mechanical Springs

Chapter 1
Principles of Spring Design

Abstract Calculation formulas for linear coil, or helical springs with inconstant wire diameter and with variable mean diameter of the spring are presented. These formulas are used to optimize the spring for a given spring rate and wire strength. The principles of design of optimal leaf springs are briefly presented. This chapter is the introductory section of the first part, which studies springs from the point of view of design.

Keywords Helical springs · Coil springs · Fatigue estimation · Optimal design of helical springs · Leaf springs

1.1 Compression, Extension, and Torque of Helical Springs

1.1.1 Forces and Moments in Coil Springs

Helical, or coil springs, studied in the beginning of this Chapter, are formed by wrapping wire or rod of the certain cross-section (Leiseder, 1997). Compression springs resist when a squeezing force is applied to the ends. Tension springs possess loops and resist the extension force that pulls them apart. The torsion springs resist twisting, when the end coils are rotated around the axis of the spring in different directions… The mean coil diameter of the spring is D.

All coil springs can be wound in either a left-hand or a right-hand direction. A left-hand wound spring will spiral in the same direction as a left land threaded screw. A right-hand wound spring will spiral in the same direction as a right-hand threaded screw. For certainty, the right-hand wound springs are considered, as being more popular.

V. Kobelev, *Fundamentals of Springs Mechanics*, University of Siegen, Faculty IV: School of Science and Technology. Paul-Bonatz-Str. 9–11, D-57076 Siegen. E-mail: Homepage: https://www.mb.uni-siegen.de/kobelev/

V. Kobelev, *Fundamentals of Springs Mechanics*,
https://doi.org/10.1007/978-3-031-58584-5_1

We take as a reference frame cylindrical polar coordinate system (r, θ, z). The axis z of the cylindrical polar coordinate system is aligned with the axis of the cylinder (Fig. 1.1). The mean coil radius of helix $R = D/2$ and the cross-section can differ along the helix. The angle $\alpha = \alpha(\theta)$ designates the inclination of the helix with any plane perpendicular to the axis of the coil (pitch angle, lead angle).. The pitch angle and pitch can vary along the helix. If pitch is maintained, so that the axis of the wire forms a helix. If the radius of helix is fixed, appears the cylindrical spring. Finally, the standard helical spring outcomes, if the cross-section does not alter along the helical spiral. For this type of the spring, the calculation formulas are standard (SAE, 1997, Chap. 5, Design of Helical Springs), (EN 13906–1:2013–11).

Helical springs store elastic energy by means of torsion and bending of wire (Timoshenko, 1948, Chap. VII, Sect. 54) (Fig. 1.2). The elastic energy of the spring is stored its active coils. The compression spring have fewer active coils, if it has closed and squared ends, closed and ground ends, or double closed ends. Ignoring at first other certain complexities of spring technology, we write basic relations for an analysis of the spring. For briefness, we study only the elastic deformation in the active coils. This simplified analysis considers spring ends as "plain", so that only active coils n_a are considered. The polar angle along the active part of the helical spring is θ:

$$0 \leq \theta \leq \Theta_a, \quad \Theta_a = 2\pi n_a.$$

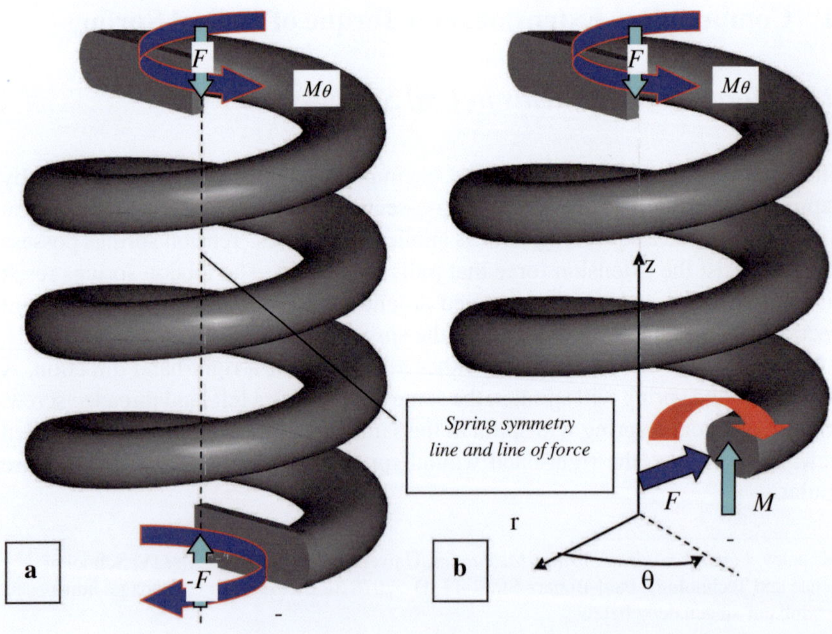

Fig. 1.1 A part of helical spring, subjected to axial loading F and axial torque M_θ

Fig. 1.2 Section of helical cylindrical spring, loaded by force F and torque M_θ

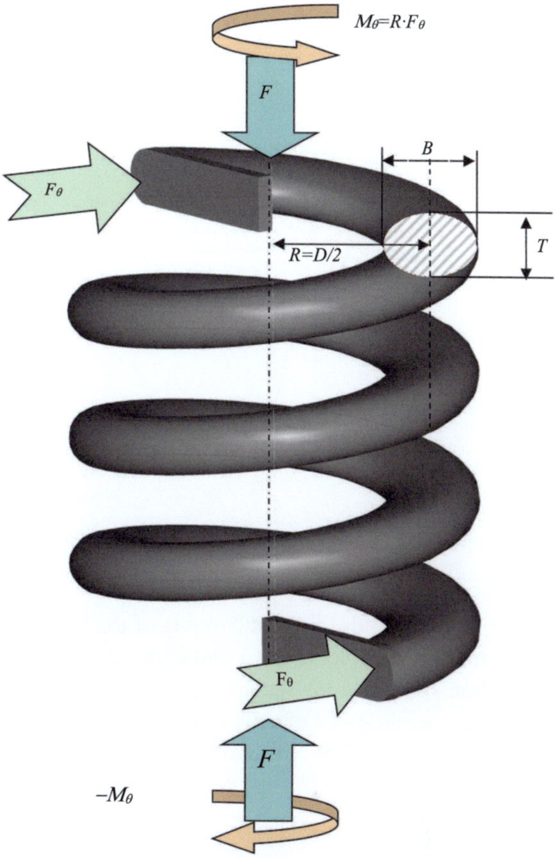

Consider the middle region of the spring, where the transition effects due to the end bars disappear. The transition effects can be neglected. The lower spring end is fixed. Due to the symmetry of the spring, all sections of the spring are deformed identically.

Tension and compression springs store elastic energy principally through torsion of wire. The force F acts ideally along the axis z of the cylindrical coordinate system, such that the line of force coincides with the symmetry axis of the spring. The torsion of helically coiled wire occurs about the axis of the wire by forces applied to the ends of springs. Figure 1.3 shows a view of the spring wire at the cut section. The couple **M** acting in the vertical plane has been shown as a vector. The couple is equal to the product of the axial force **F** with the arm **R** to the force of a couple:

$$\mathbf{M} = \mathbf{F} \times \mathbf{R}. \tag{1.1a}$$

The couple (1.1a) resolves into two components, $M \sin \alpha$ and $M \cos \alpha$, which lie in planes, which are tangential and normal to the helix, respectively (Fig. 1.3a). The

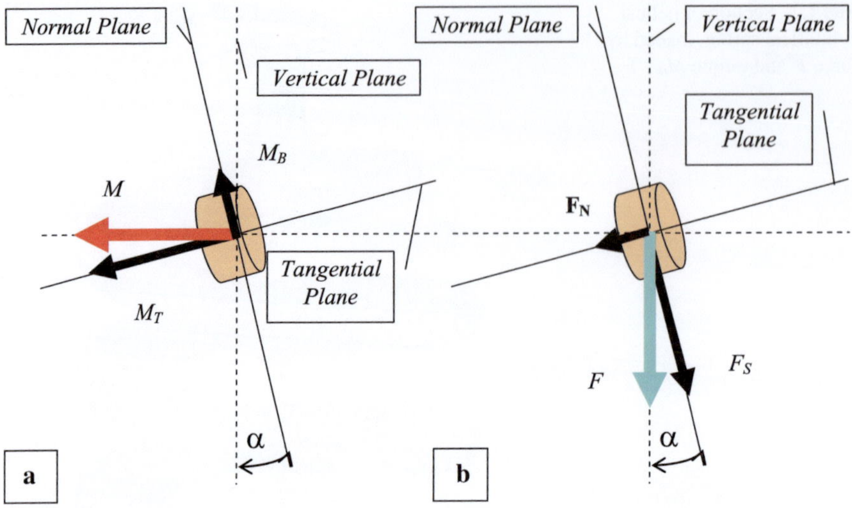

Fig. 1.3 Resolution of moments and forces into normal and tangential components

couple $M \sin \alpha$ tends to cause bending of the spring wire in the plane, normal to the coil plane. The couple $M \cos \alpha$ causes twisting along the tangential to helix curve.

Torsion springs store elastic energy mainly through bending of wire in the coil plane. The bending of helically coiled wire occurs by terminal moments applied to the ends of springs. The terminal moments twist the coil tighter or looser. The torque

$$M_\theta = F_\theta R, \tag{1.1b}$$

on the upper spring acts clockwise, if we look on the spring from above. The circumferential force F_θ pushes towards the wire, applying normal stresses in the wire cross-section. Thus, the torque causes the equal bending moment in each section of the wire.

The springs withstand recurrently to the combination of both loads. In this case, simultaneously act the axial force F and the axial torque M_θ.The magnitudes of the bending and twisting couples in the wire are equal respectively to:

$$m_B = M \sin \alpha - M_\theta \cos \alpha, \quad m_T = M \cos \alpha + M_\theta \sin \alpha. \tag{1.2}$$

Figure 1.3b shows a similar resolution of force F into components, F_N and F_S, which are causing normal and shearing components of stress, respectively, on the cut section:

$$F_N = F \sin \alpha, \quad F_S = F \cos \alpha. \tag{1.3}$$

1.1.2 Elastic Energy of the Coil Spring

As already mentioned, the conventional design formulas (SAE, 1997, Chap. 5, Design of Helical Springs), (EN 13906–1:2013–11) calculate the cylindrical helical springs with the circular wire and the constant pitch. Several applications require the helical springs with the alternating mean diameter of the spring body as well the varying wire diameter along its length. The calculation of such springs needs the generalization of the customary design formulas. Besides this, the formulas are intended to the optimization purposes.

For the element of length, the following expression is valid:

$$\mathrm{d}l = \frac{D \cdot \mathrm{d} \cdot \theta}{2 \cos \alpha}. \tag{1.4}$$

The volume of wire with variable cross-section is given by the integral of the cross-sectional area $A(\theta)$ over the wire length. The total length of the wire L_W and the mass of the spring m depend upon two functions:

$$D = D(\theta), \quad A = A(\theta).$$

With material density is ρ_m, these values are correspondingly:

$$L_W = \int_0^{\Theta_a} \mathrm{d}l, \quad m = \rho_m \int_0^{\Theta_a} A\, \mathrm{d}l. \tag{1.5}$$

To derive the stiffness coefficients, we start with the linear spring loaded from its free state L_0 by simultaneously acting axial force F and axial torque M_θ. The stored elastic energy of the spring reads:

$$U_e = \frac{F^2}{2c} + \frac{F M_\theta}{c_{\theta F}} + \frac{M_\theta^2}{2c_\theta}. \tag{1.6}$$

The coefficients of Eq. (1.6) are the compression spring rate c, the compression-twist springs rate $c_{\theta F}$ and the twist springs rate c_θ.

We derive these coefficients for an arbitrary non-cylindrical helical spring with an arbitrary variable cross-section. The size of the cross-section is assumed much smaller than the radius of coil. For derivation we use Castigliano's method (Teodorescu, 2013). The elastic energy stored in the spring (1.6) is the integral of twist and bending energy of wire over its length L_W:

$$2U_e = \int_{L_W} \left[\frac{m_T^2}{G I_T} + \frac{m_B^2}{E I} \right] \mathrm{d}l. \tag{1.7}$$

We substitute (1.2) for the torsion moment m_T and bending moment m_B in (1.7). The elastic energy density of wire pro unit of its length is:

$$\frac{m_T^2}{GI_T} + \frac{m_B^2}{EI} = M^2\left(\frac{\cos^2\alpha}{GI_T} + \frac{\sin^2\alpha}{EI}\right) + 2MM_\theta \cos\alpha \sin\alpha\left(\frac{1}{GI_T} - \frac{1}{EI}\right) + M_\theta^2\left(\frac{\sin^2\alpha}{GI_T} + \frac{\cos^2\alpha}{EI}\right) \tag{1.8}$$

Using (1.4) in (1.8), reduces the formula for the stored energy to:

$$2U_e = F^2 \int\limits_{L_W} \frac{D^2}{4}\left(\frac{\cos^2\alpha}{GI_T} + \frac{\sin^2\alpha}{EI}\right)\mathrm{d}l + 2FM_\theta \int\limits_{L_W} \frac{D}{2}\cos\alpha \sin\alpha\left(\frac{1}{GI_T} - \frac{1}{EI}\right)\mathrm{d}l + M_\theta^2 \int\limits_{L_W}\left(\frac{\sin^2\alpha}{GI_T} + \frac{\cos^2\alpha}{EI}\right)\mathrm{d}l.$$

According to Castigliano's method, the second derivatives of the stored energy U_e with respect to F and M_θ provide the spring rates. The compression spring rate c, the compression-twist springs rate $c_{\theta F}$ and the twist springs rate c_θ of an arbitrary non-cylindrical helical spring are:

$$\begin{cases} \frac{1}{c} = \frac{\partial^2 U_e}{\partial F^2} = \int\limits_{L_W} \frac{D^2}{4}\left(\frac{\cos^2\alpha}{GI_T} + \frac{\sin^2\alpha}{EI}\right)\mathrm{d}l, \\ \frac{1}{c_{\theta F}} = \frac{\partial^2 U_e}{\partial F \partial M_\theta} = \int\limits_{L_W} \frac{D}{2}\cos\alpha \sin\alpha\left(\frac{1}{GI_T} - \frac{1}{EI}\right)\mathrm{d}l, \\ \frac{1}{c_\theta} = \frac{\partial^2 U_e}{\partial M_\theta^2} = \int\limits_{L_W}\left(\frac{\sin^2\alpha}{GI_T} + \frac{\cos^2\alpha}{EI}\right)\mathrm{d}l, \end{cases} \tag{1.9}$$

for the given design variables:

$$GI_T = GI_T(\theta), \quad EI = EI(\theta), \quad D = D(\theta), \quad \alpha = \alpha(\theta).$$

The twist spring rate c_θ is equal to the moment M_θ, which causes the twist angle of spring of one radiant in the absence of the axial force. The compression-twist springs rate $c_{\theta F}$ describes the moment M_θ, which causes the unit axial travel of the spring in the absence of the axial force. The unit axial travel is one millimeter or one meter in SI system. Notably, that the value $c_{\theta F}$ is equal to the force, which causes the same axial travel in the absence of the twist moment.

1.1.3 Compression and Twist Spring Rates

When the distance between coils is small, the angle α is very small. For small pitch, following relations are valid:

$$\cos\alpha \cong 1, \quad \sin\alpha \cong 0, \quad \mathrm{d}l \cong \frac{1}{2}D \cdot \mathrm{d} \cdot \theta.$$

In this case, the spring is called a closed-coiled spring. With (1.3), the force F_N is also negligible and F_S will approximately equal to F. As follows from (1.2), the torque moment m_T of the wire is nearly equal to M. This means, that for closed-coiled spring, the axial force leads only to the torsion of the wire. Torque of the spring body leads to the bending of wire. The twist and bending moments of wire are thus equal to:

$$m_T = FR, \quad m_B = M_\theta.$$

According to (1.5), the mass of the spring is:

$$m = \frac{1}{2}\rho_m \int_0^{\Theta_a} AD\,\mathrm{d}\theta. \tag{1.10}$$

From (1.9), we get for compression (or extension) and twist spring rates:

$$\frac{1}{c} = \int_0^{\Theta_a} \frac{1}{GI_T}\left(\frac{D}{2}\right)^3 \mathrm{d}\theta, \quad \frac{1}{c_\theta} = \int_0^{\Theta_a} \frac{1}{EI}\frac{D}{2}\mathrm{d}\theta. \tag{1.11}$$

The compression-twist springs rate $c_{\theta F}$ of the closed-coiled spring vanishes.

The above formulas are valid for an arbitrary shape of the wire cross-section. For the springs with elliptic and rectangular shapes of the cross-section the geometric characteristics depend upon the orientation of axes B, T. The height of the cross-section t is measured in the direction of the axis. The equations for spring rates and masses of the springs with circular and non-circular cross-sections of wire are given in Tables 1.1 and 1.2.

Equations (1.10) and (1.11) shorten for the springs with the circular cross-section. The formulas for spring rates of the springs with a circular cross-section and the variable wire diameter $\mathrm{d} = \mathrm{d}(\theta)$ yield after the substitution of the values for GI_T and EI into Eq. (1.11). For the springs with the circular cross-section the expression for mass and for spring rates are:

$$m = \frac{1}{8}\pi\rho_m \int_0^{\Theta_a} \mathrm{d}^2 D\,\mathrm{d}\theta. \tag{1.12}$$

Table 1.1 Spring rates and masses of the linear springs with number of active coils n and a non-circular wire cross-section

	Spring rate	*Mass*
$\begin{cases} A = konst \\ D = konst \end{cases}$	$c = \frac{4GI_T}{\pi n D^3}$	$m = \pi n A \rho D$
$\begin{cases} A = konst \\ D \neq konst \end{cases}$	$c = \frac{8G}{I_T^{-1} \int_0^{2\pi n} D^3(\varphi)\mathrm{d}\varphi}$	$m = \frac{1}{2} A\rho \int_0^{2\pi n} D(\varphi)\mathrm{d}\varphi$
$\begin{cases} A \neq konst \\ D = konst \end{cases}$	$c = \frac{8G}{D^3 \int_0^{2\pi n} I_T^{-1}(\varphi)\mathrm{d}\varphi}$	$m = \frac{1}{2} \rho D \int_0^{2\pi n} A(\varphi)\mathrm{d}\varphi$
$\begin{cases} A \neq konst \\ D \neq konst \end{cases}$	$c = \frac{8G}{\int_0^{2\pi n} D^3(\varphi) I_T^{-1}(\varphi)\mathrm{d}\varphi}$	$m = \frac{1}{2} \rho \int_0^{2\pi n} A(\varphi) D(\varphi)\mathrm{d}\varphi$

Table 1.2 Spring rates and masses of the linear springs with number of active coils n and with the circular wire cross-section

	Spring rate	Mass
$\begin{cases} A = konst \\ D = konst \end{cases}$	$c = \frac{Gd^4}{8nD^3}$	$m = \frac{1}{4}\pi^2 n d^2 \rho D$
$\begin{cases} A = konst \\ D \neq konst \end{cases}$	$c = \frac{\pi G d^4}{4 \int_0^{2\pi n} D^3(\varphi)\mathrm{d}\varphi}$	$m = \frac{1}{8}\pi d^2 \rho \int_0^{2\pi n} D(\varphi)\mathrm{d}\varphi$
$\begin{cases} A \neq konst \\ D = konst \end{cases}$	$c = \frac{\pi G}{4D^3 \int_0^{2\pi n} d^{-4}(\varphi)\mathrm{d}\varphi}$	$m = \frac{1}{8}\pi \rho D \int_0^{2\pi n} d^2(\varphi)\mathrm{d}\varphi$
$\begin{cases} A \neq konst \\ D \neq konst \end{cases}$	$c = \frac{\pi G}{4 \int_0^{2\pi n} D^3(\varphi) d^{-4}(\varphi)\mathrm{d}\varphi}$	$m = \frac{1}{8}\pi \rho \int_0^{2\pi n} d^2(\varphi) D(\varphi)\mathrm{d}\varphi$

$$\begin{aligned} c &= \pi G \left(4 \int_0^{\Theta_a} \frac{D^3 \mathrm{d}\theta}{\mathrm{d}^4} \right)^{-1}, \\ c_\theta &= \pi E \left(32 \int_0^{\Theta_a} \frac{D \mathrm{d}\theta}{\mathrm{d}^4} \right)^{-1}. \end{aligned} \tag{1.13}$$

The study of the optimization problem is based on the formulas for the spring rates Eq. (1.13) and for mass (1.12).

1.1.4 Change in Diameter Due to Simultaneous Compression and Torque Application

The axial compression of the spring (spring travel in the direction of force on the upper spring end) is (Ponomarev et al., 1956):

$$s = s_F + s_M, \tag{1.14}$$

$$\begin{aligned} s_F &= \frac{\pi F D^3 n}{4\cos\alpha}\left(\frac{\cos^2\alpha}{GI_T} + \frac{\sin^2\alpha}{EI}\right), \\ s_M &= \frac{\pi M_\theta D^2 n}{2}\left(\frac{1}{GI_T} - \frac{1}{EI}\right)\sin\alpha. \end{aligned} \tag{1.15}$$

The values s_F and s_M represent the spring travel in cause of compression force and torque respectively. The spring shortens under the action of compression force. Similarly, the length of the right wound spring under the action of clockwise moment on its upper end reduces.

The twist angle of the upper spring end with respect to the axis of the spring reads:

$$\theta = \theta_F + \theta_M, \tag{1.16}$$

$$\begin{aligned} \theta_F &= \frac{\pi F D^2 n}{2}\cdot\left(\frac{1}{GI_T} - \frac{1}{EI}\right)\sin\alpha, \\ \theta_M &= \frac{\pi M_\theta D n}{\cos\alpha}\cdot\left(\frac{\cos^2\alpha}{GI_T} + \frac{\sin^2\alpha}{EI}\right). \end{aligned} \tag{1.17}$$

The values θ_F, θ_M symbolize the reduction of the spring angle in cause of compression force and torque respectively. The positive direction is clockwise. The coil number of spring decreases under the action of compression force. Analogously, the coil number of the right wound spring decreases under the action of clockwise moment on its upper end.

The mean coil diameter of spring increases under the action of axial force and axial moment:

$$\Delta D = \Delta D_F + \Delta D_M, \tag{1.18}$$

$$\Delta D_F = F D^3 \sin\alpha \cdot \left(\frac{1}{2GI_T} - \frac{\cos 2\alpha}{4EI\cos^2\alpha}\right), \tag{1.19}$$

$$\Delta D_M = \frac{M_\theta D^2}{2\cos\alpha}\cdot\left(\frac{2\sin^2\alpha}{GI_T} + \frac{\cos 2\alpha}{EI}\right). \tag{1.20}$$

The diameter of the right wound spring increases under the action of compression force. In other words, the diameter of the compression spring expands when such compression spring has deflected. The diameter of the right wound spring expands under the action of clockwise moment on its upper end. If the moment acts in the opposite direction, such that the circumferential force pulls the wire, the spring's body is being tightened.

1.2 Design Formulas for Compression-Extension Springs

1.2.1 Stiffness and Stored Energy of Cylindrical Helical Springs

Consider the helical spring with a circular wire. In this Chapter, we use for the design purposes the conventional formulas (EN 13906, 2013, 2014; Meissner et al., 2015). The design variables for helical springs are

$$\mathrm{d}, D, n_a, L_0,$$

where n_a is the number of active coils, d is the diameter of wire, D is the mean coil diameter, L_0 is a free length of the spring. The external and internal diameters correspondingly are:

$$D_e = D + \mathrm{d},\ D_i = D - \mathrm{d}.$$

The volume and the mass of the spring material of a cylindrical spring with constant, round cross-section is given by:

$$V = \frac{\pi^2 \mathrm{d}^2 D n_a}{4}, \quad m = \rho_m V. \tag{1.21}$$

where ρ_m is a density of spring material.

Consider for definiteness a compression spring with a free length L_0. Solid length L_s is the height at which the coils of the compressed spring close up (SAE, 1997, Chap. 3, Cold wound helical and spiral springs). The stroke of a compression spring is the spring travel from released length L_{rel} to compressed length L_{comp}:

$$s = L_{rel} - L_{comp}. \tag{1.22}$$

For the compression spring is valid:

$$L_0 > L_{rel} > L_{comp} > L_s > 0.$$

The axial spring stiffness, or spring rate, is the force required to produce a unit deflection. In the simplest case, the force–deflection characteristic is approximately linear and can be calculated from the geometry and shear modulus G of the spring material. In this case, Eq. (1.11) leads to the traditional formula for the cylindrical spring made of the round wire with the constant diameter:

$$c = \frac{Gd^4}{8D^3 n_a}. \tag{1.23}$$

Using Eq. (1.23), the spring loads at lengths L_{comp}, L_{rel} and L_s are correspondingly:

$$F_{\min} = c \cdot (L_0 - L_{rel}), \quad F_{\max} = c \cdot \left(L_0 - L_{comp}\right), \quad F_s = c \cdot (L_0 - L_s).$$

The energy capacity of the linear compression spring is expressed as:

$$U_e = \frac{c}{2}\left[\left(L_{comp} - L_0\right)^2 - (L_{rel} - L_0)^2\right] = \frac{1}{2c}\left(F_{\max}^2 - F_{\min}^2\right). \tag{1.24}$$

The work of applied forces on the spring travel is:

$$U_f = \frac{F_{\min} + F_{\max}}{2} s.$$

The tension spring is handled in the same way. The stroke of a tension spring is the spring travel from released length L_{rel} to extended length L_{ext}:

$$s = L_{ext} - L_{rel}.$$

Clearly, the lengths for tension spring shuffle:

$$0 < L_s < L_0 < L_{rel} < L_{ext}.$$

1.2.2 *Stresses in Spring Wire*

The pitch can be neglected $\alpha = 0$ for the fairly accurate stress evaluation. One turn, or coil, of an undeformed helical spring becomes a torus, generated by rotating the cross-section about the z axis of the cylindrical coordinate system (Fig. 1.4). For the stress calculations, the torus is considered as incomplete. The two ends of the turn are not joined. They carry equal in magnitude and opposite shear stress distributions with resultant F. The line of action of the resultant force F is coincident due to symmetry with the z axis. Any segment of the coil is therefore in equilibrium, because on the ends act two opposite axial forces F with the same magnitude. Nonzero components of the shear stress in cylindrical coordinates are $\tau_{r\theta}$, $\tau_{\theta z}$. Again, due to symmetry, these components are independent of θ.

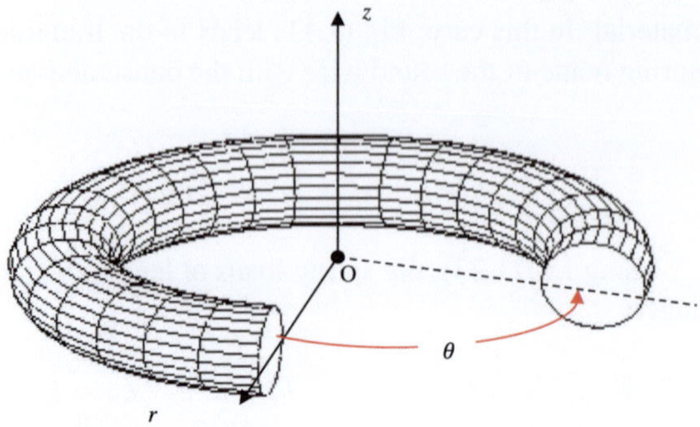

Fig. 1.4 Incomplete torus within the cylindrical coordinate system

One must differ the basic and corrected shear stress in the spring. The basic stress is obtained by dividing the torsion moment in the wire m_T:by its section modulus W_T in torsion (Tables 1.3 and 1.4). Thus, the basic stress is (SAE, 1997, Chap. 5, Design of Helical Springs):

$$\tau_b = \frac{m_T}{W_T} = \frac{8DF}{\pi \mathrm{d}^3}. \tag{1.25}$$

The corrected stress τ_c is calculated by multiplying the basic stress τ_b by the correction factor

$$k_\tau = k_\tau(w), \tag{1.26}$$

Table 1.3 Stiffness of wires with different cross-sections

Cross-section of wire	Area and moments of inertia A, I, I_r	Torsion constant I_T
Circular, diameter d	$\begin{cases} A = \frac{\pi \mathrm{d}^2}{4}, \\ I = I_r = \frac{\pi \mathrm{d}^4}{64} \end{cases}$	$I_T = \frac{\pi \mathrm{d}^4}{32}$
Squared, side size a	$\begin{cases} A = a^2, \\ I = I_r = \frac{a^4}{12} \end{cases}$	$I_T = 0.141..a^4$
Rectangular T-hight B-width	$\begin{cases} A = BT, \\ I = \frac{BT^3}{12}, \\ I_r = \frac{B^3T}{12} \end{cases}$	$I_T = \begin{cases} \text{if} T > B : \xi_1\left(\frac{T}{B}\right)TB^3 \\ \text{if} T < B : \xi_1\left(\frac{B}{T}\right)BT^3 \end{cases}$
Elliptic T-hight B-width	$\begin{cases} A = \frac{\pi}{4}BT \\ I = \frac{\pi BT^3}{64}, \\ I_r = \frac{\pi B^3T}{64} \end{cases}$	$I_T = \frac{\pi B^3T^3}{16(B^2+T^2)}$
$\xi_1(x) = \frac{1}{3}\left(1 - \frac{192}{\pi^5 x}\sum_{k=1,3,5..}^{\infty}\frac{1}{k^5}\tan h\frac{k\pi x}{2}\right) \approx \frac{1}{3}\left(1 - \frac{0.63..}{x} + \frac{0.052..}{x^5}\right)$		

Table 1.4 Section modules of wires with different cross-sections

Cross-section of wire	Bending section modules W_b, W_{br}	Twist section modulus W_t
Circular, diameter d	$W_b = W_{br} = \frac{\pi d^3}{32}$	$W_t = \frac{\pi d^3}{16}$
Squared, side size a	$W_b = W_{br} = \frac{a^4}{6}$	$W_t = 0.208..a^3$
Rectangular T-hight B-width	$W_b = \frac{BT^2}{6}$, $W_{br} = \frac{TB^2}{6}$	$W_t = \begin{cases} \text{if} T > B : \xi_2\left(\frac{T}{B}\right)TB^2 \\ \text{if} T < B : \xi_2\left(\frac{B}{T}\right)BT^2 \end{cases}$
Elliptic T-hight B-width	$W_b = \frac{\pi BT^2}{32}$, $W_{br} = \frac{\pi B^2 T}{32}$	$W_t = \frac{\pi}{16}\min(B^2T, BT^2)$

$\xi_2(x) = 1 - \frac{8}{\pi^2}\sum_{k=1,3,5..}^{\infty}\frac{1}{k^2\cos h\frac{k\pi x}{2}} \approx \frac{1+x^2}{0.35+x^2}\xi_1(x)$

such that

$$\tau_c = k_\tau \tau_b. \tag{1.27}$$

The correction factor k_τ depends upon the ratio of mean coil diameter to wire diameter:

$$w = D/\mathrm{d}. \tag{1.28}$$

The ratio w known as the spring index. A low index indicates a tightly wound spring (a relatively large wire size wound around a relatively small diameter mandrel giving a high rate). The correction factor $k_\tau(w)$ accounts for stress concentration due to curvature of the spring as well as direct shear.

The governing equations for the closed-coiled helical spring were developed using semi-inverse StVenant's method by (Michell, 1899). Unfortunately, the closed form of StVenant's solutions, which is well known in the theory of torsion of straight circular or elliptic rods, does not exist for curved rods in terms of elementary functions. The approximate solutions for rectangular and circular cross-sections were delivered by (Wahl, 1929) and (Göhner, 1931, 1932). The solutions for helical springs with circular cross-sections in terms of series of appropriate Legendre functions in toroidal coordinates were found by Freiberger (1949) and Henrici (1955).

The exact solution according to (Herichi, 1955) expresses the correction factor k_τ as an infinite series:

$$k_1(w) = 1 + \frac{5}{4w} + \frac{7}{8w^2} + \frac{15}{256w^3} + \cdots . \tag{1.29}$$

For springs made of wire with circular cross section, the traditional correction factors for the given coiling ratio were determined by empirically established formulas (SAE, 1997, Table 5.2) or (EN 13906–1: 2013–11, 2013). The following formulas apply to the standard correction factors on the spring index:

$$k_2(w) = \frac{w + 0.5}{w - 0.75}, \text{ (Bergsträsser, EN 13906 - 1 : 2013 - 11)}, \tag{1.30}$$

$$k_3(w) = \frac{4w - 1}{4w - 4} + \frac{0.615}{w}, \text{ (Wahl, 1929)}. \tag{1.31}$$

Shear stresses over the cross-sections and on the outer surfaces of wires in helical springs will be considered more detailed in Chap. 2. In this Chapter, we use for the design purposes the standard formulas (1.30) and (1.31) (Meissner et al., 2015).

An important remark considers the effects of stress concentrations. Properly designed springs possess no notches and stress concentrators, because of very high negative impact on fatigue life. If the cyclic stresses are low, only microscopic plastic deformation occurs in each load cycle. The high cycle fatigue is typical for the most types of springs in automotive and industrial applications of spring elements. The calculations are based on the empirical stress factors. These factors average the stress over the surface of the wire. The factors consider the possibility of defect in the less stressed regions. They estimate the admissible level of stresses more correctly, than the standard stress factors. We explain in Ch. 10 the application of the empirical stress factors for fatigue calculations.

The springs usually have a moderate, rather natural stress concentration.. The explanation is the following. Equation (1.32) estimates the for fatigue relevant stresses based on the ideal stresses. The ideal stresses neglect the escalation of stress on the inside of wire, which faces the midline of helix. This effect could be treated as the stress concentration factor. This specific factor is distinct for helical springs.

There are several further methods to consider the influence of stress concentration on the fatigue of structural elements (see Chaps. 10 and 12). The approximate methods are based on stress–strain conversion rules. The methods were widely applied for the cyclically loaded elements of structures:

1. most simple linear rule, assuming that strain concentration factor equals theoretical stress concentration factor;
2. The rule (Neuber, 1961a, 1961b) relates the product of stress and strain concentration factors to the hypothetical stress concentration factor;
3. Stowell-Hardrath-Ohman expression of stress concentration factor is a function of speculative stress concentration factor and elasticity moduli (Wundt, 1972);
4. "Equivalent Strain Energy Density" approach (Molski & Glinka, 1981). This approach assumes, that the strain energy density in the notch root is related to the energy density due to nominal stress and strain by a certain factor.

The difference between the predictions of the cited approaches could be significant if the significant plastic deformation occurs in each load cycle. Experimental and numerical investigations have been carried out to verify the cited rules. The investigation (Adibi-Asl & Seshadri, 2009) shows that the Neuber rule predicts an upper bound for the maximum local strains, the linear rule provides a lower bound, and the "Equivalent Strain Energy Density" method estimates the fatigue life in the intermediate region. Since the significant plastic deformation is not typical for industrial and automotive spring applications, we follow only the traditional approach.

1.2.3 Fatigue Life and Damage Accumulation Criteria

If the spring is to be deflected a specified number of times, it must be designed so that the material will not fail in fatigue. For design purposes, the simple fatigue criterion for compression spring design is usually assumed to be (Spring Design Manual, 1996):

$$\left(\frac{\tau_{\max}}{\tau_w} + \frac{\tau_a}{\tau_e}\right) S_f \leq 1, \tag{1.32}$$

where

$$\tau_{\min} = k_\tau F_{\min}/W_T,$$

$$\tau_{\max} = k_\tau F_{\max}/W_T,$$

$R_\sigma = \frac{\tau_{\min}}{\tau_{\max}} = \frac{F_{\min}}{F_{\max}} < 1$ is the stress ratio of cyclic load
$\tau_m = \frac{1}{2}(\tau_{\max} + \tau_{\min})$ is the mean stress per cycle,
$\tau_a = \frac{1}{2}(\tau_{\max} - \tau_{\min})$ is the stress amplitude,
τ_w is the ultimate shear strength of material,
τ_e is a endurance limit for completely reversed stress,
S_f is a factor for safety,
The safety factor S_f for simplicity is assumed to be 1.

Both τ_w and τ_e usually vary with wire diameter, as discussed in Chap. 10, Sect. 10.6.1. As we can see, both τ_w and τ_e attain maximum values for a small wire diameter.

The fatigue life of springs is also habitually based on the damage evaluation from the Smith, Watson, Topper rule (Smith et al., 1970; Landgraf, 1973) or Walker rule (Walker, 1970) (See Appendix).

According to the Smith, Watson, Topper rule, the governing parameter for damage characterization is a product of the total strain range and the maximum stress. For a discussion of the applicability of the Smith, Watson, Topper rule to automotive applications, see (Fuchs et al., 1977). During spring deformation, the wire undergoes torsion with shear stresses dominating. Applying this approach to shear deformation, the $\mathbf{P_{SWT}}$ (Smith, Watson, Topper parameter) transforms to:

$$P_{SWT.\tau} = \sqrt{G\gamma_a \tau_{max}}.$$

Here

$$\gamma_a = \tau_a / G$$

is the shear strain amplitude.

The damage parameter is plotted versus number of reversals, so that damage per range between two reversals is a function of damage parameter. The accepted damage for a selected material during fatigue life of the spring is characterized by the condition:

$$P_{SWT.\tau} \equiv \sqrt{\tau_a \tau_{max}} \leq P_{SWT.0}. \tag{1.33}$$

The experimentally acquired constant $P_{SWT.0}$ depends on material properties and accepted damage level for application under consideration (see Chaps. 11 and 12).

The fatigue behavior of the springs is highly dependent on the surface treatment, mainly the shot peened layer on the surface. The highly inhomogeneous stresses in the shot peened layer may be responsible for the cracking despite the compressive stresses. The simulation methods must adequately describe the stress origin and depth variation of the shot peened stress regions. Cold formed springs also retain another type of residual stress due to coiling. The influence of residual stresses on damage accumulation must be considered in fatigue calculations. The fatigue life of springs will be discussed in details in Chaps. 11, 12 and 13. The mechanical properties of spring materials were comprehensively discussed in (Yamada, 2007). The discussion of fatigue life evaluation of springs continues in Chaps. 14, Appendix.(C) and Appendix.(D).

1.3 Helical Springs of Minimal Mass

1.3.1 Restricted Optimization Problem

The spring designer usually seeks the springs of minimum weight for the definite strength and energy capacity limits. In most general case, the mean coil diameter of the spring body D and the wire diameter d depend on the polar angle along the spring wire θ. These two functions (Fig. 1.5) are the design variables:

$$D = D(\theta), \quad \mathrm{d} = \mathrm{d}(\theta).$$

For the analytical treatment, we constrain ourselves to the following optimization problem:

Minimize the mass of the spring:

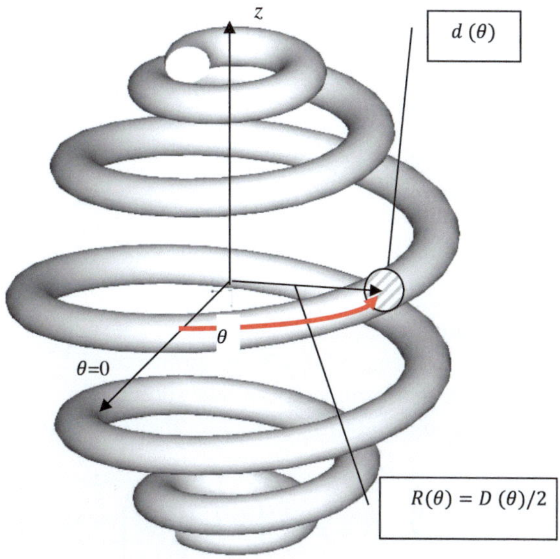

Fig. 1.5 Helical spring with variable wire diameter and non-cylindrical form

$$m \to \min_{D,\mathrm{d}},$$

assuming the spring rate is equal to a given positive constant c^*:

$$c(D, \mathrm{d}) = c^* \tag{1.34}$$

and the forces at installed height $F_{\min}$ and full stroke $F_{\max}$ are prescribed, the fatigue conditions (1.32) and (1.33) fulfilled, and the ideal stress at full stroke is limited:

$$\frac{F_s}{W_T} \leq \tau_w, \quad F_s = c \cdot (L_0 - L_s).$$

1.3.2 Optimization of Helical Springs for Maximal Stress

First, consider the practically important case of non-cylindrical springs with variable circular cross section: the stress at solid height must be less than τ_w to protect the spring from accidental damage. This constraint, applied to the basic shear stress at solid height, is given by:

$$\tau(\mathrm{d}) \equiv \frac{8 F_s D}{\pi \mathrm{d}^3} \leq \tau_w.$$

In terms of wire diameter, this inequality could be expressed as follows:

$$\mathrm{d}(\theta) \geq \mathrm{d}_1(\theta).$$

The optimal diameter of wire:

$$\mathrm{d}_1 \equiv \sqrt[3]{\frac{8F_s D}{\pi \tau_w}} \tag{1.35}$$

is the solution of algebraic equation $\tau = \tau_w$ with respect to $\mathrm{d}(\theta)$.

We use the general formula for spring rate (1.13), which is valid for the changeable diameter of the spring body and the spring wire: $D = D(\theta)$, $\mathrm{d} = \mathrm{d}(\theta)$. Considering that $\mathrm{d}(\theta) \geq \mathrm{d}_1(\theta)$, the inequality for the spring rate reads:

$$c(D, \mathrm{d}) = \pi G \cdot \left(4 \int_0^{\Theta_a} \frac{D^3}{\mathrm{d}^4} \mathrm{d}\theta\right)^{-1} \geq \pi G \cdot \left(4 \int_0^{\Theta_a} \frac{D^3}{\mathrm{d}_1^4} \mathrm{d}\theta\right)^{-1}.$$

We substitute the expression (1.35) for the optimal diameter of the wire into this inequality and express the stiffness requirement (1.39) as follows:

$$c^* = c(D, \mathrm{d}) \geq 4\pi G \cdot \left(\frac{F_s}{\pi \tau_w}\right)^{4/3} \cdot \left(\int_0^{\Theta_a} D^{5/3} \mathrm{d}\theta\right)^{-1}. \tag{1.36}$$

Similarly, we replace the expression (1.35) for the optimal diameter of the wire with the expression (1.33) for the mass of the spring. After this substitution, the second inequality for the mass follows:

$$m \geq \frac{1}{2} \pi \rho_m \cdot \left(\frac{F_s}{\pi \tau_w}\right)^{2/3} \cdot \left(\int_0^{\Theta_a} D^{5/3} \mathrm{d}\theta\right). \tag{1.37}$$

The inequalities with the same sign can be multiplied. The multiplication of inequalities (1.36) and (1.37) gives a final lower limit for the spring mass:

$$m \cdot c^* \geq \frac{2\rho_m G F_s^2}{\tau_w^2}. \tag{1.38}$$

This important inequality establishes the exact lower boundary for the mass of spring of an arbitrary variable shape and an arbitrary variable circular cross-section, which satisfy the stress condition at solid length:

$$m \geq m_1 = \frac{2\rho_m G F_s^2}{\tau_w^2 c^*}. \tag{1.39}$$

The maximal stored elastic energy per unit volume (volume energy density) is equal to $\tau_w^2/2G$. The ultimate stored elastic energy per unit mass (mass energy density) is:

$$\tilde{U}_{e.1} = \frac{\tau_w^2}{2\rho_m G}.$$

The mass in terms of the specific elastic energy density in the spring is given by inequality (1.39):

$$m \geq m_1 = \frac{F_s^2}{\tilde{U}_{e.1} c^*}. \tag{1.40}$$

1.3.3 Design for Fatigue Life

The spring is to operate a definite number of cycles through a deflection s measured as additional compression from L_0. The application of a similar optimization procedure, as applied above, for the fatigue condition (1.32) leads to optimal wire diameter:

$$\mathrm{d}_2 \equiv \sqrt[3]{\frac{8 D F_{\max}}{\pi} \cdot \left(\frac{1}{\tau_w} + \frac{1 - R_\sigma}{2\tau_e} \right)}. \tag{1.41}$$

Using Eq. (1.41), we get in this case the lower boundary for the mass:

$$m \geq m_2 = \frac{2\rho_m G F_{\max}^2}{c^*} \cdot \left(\frac{1}{\tau_w} + \frac{1 - R_\sigma}{2\tau_e} \right)^2. \tag{1.42}$$

Consistently with the Eq. (1.40), the optimal mass expresses in terms of specific energy density:

$$m \geq m_2 = \frac{F_{\max}^2}{\tilde{U}_{e.2} c^*}, \quad \tilde{U}_{e.2} = \frac{1}{2\rho_m G} \cdot \left(\frac{1}{\tau_w} + \frac{1 - R_\sigma}{2\tau_e} \right)^{-2}. \tag{1.43}$$

These expressions determine the optimal spring, acceptable from the viewpoint of fatigue life criterion (1.32).

If the fatigue calculation uses Smith, Watson, Topper rule (1.33), then the optimal wire diameter is:

$$\mathrm{d}_3 \equiv \sqrt[3]{\frac{8DF_{\max}}{\pi \mathrm{P}_{SWT.0}}\sqrt{\frac{1-R_\sigma}{2}}}. \quad (1.44)$$

The optimal mass of spring satisfies with Eq. (1.44) the condition:

$$m \geq m_3 = \frac{2\rho_m G F_{\max}^2}{c^*} \cdot \frac{1-R_\sigma}{2\mathrm{P}_{SWT.0}^2}. \quad (1.45)$$

The optimal mass displays the lower limit for springs, which designed only for fatigue life:

$$m \geq m_3 = \frac{F_{\max}^2}{\tilde{U}_{e.3}c^*}, \quad \tilde{U}_{e.3} = \frac{\mathrm{P}_{SWT.0}^2}{\rho_m G} \cdot \frac{1}{1-R_\sigma}. \quad (1.46)$$

The values $\tilde{U}_{e.1}$, $\tilde{U}_{e.2}$, $\tilde{U}_{e.3}$ express the mass energy density J/kg. These values are significant material constants for spring design. The smallest value signifies the attainable specific energy density of the material, if all three conditions must be simultaneously fulfilled:

$$\tilde{U}_e = \min\left(\tilde{U}_{e.1}, \tilde{U}_{e.2}, \tilde{U}_{e.3}\right). \quad (1.47)$$

1.3.4 Spring Quality Parameter for Helical Springs

Combining the optimization results (1.35), (1.41) and (1.44), we obtain the expression for the optimal wire diameter:

$$\mathrm{d}_{opt} = \sqrt[3]{\frac{8D}{\pi}Q_p}. \quad (1.48)$$

The absolute lowest mass of the spring follows from (1.39), (1.42) and (1.45):

$$m_{opt} = \frac{2\rho_m G}{c^*}Q_p^2. \quad (1.49)$$

Here

$$Q_p = \max\left[\frac{F_s}{\tau_w}, F_{\max} \cdot \left(\frac{1}{\tau_w} + \frac{1-R_\sigma}{2\tau_e}\right), \frac{F_{\max}}{\mathrm{P}_{SWT.0}}\sqrt{\frac{1-R_\sigma}{2}}\right] \quad (1.50)$$

is a spring quality parameter, which accounts for different fatigue and endurance limits.

The optimum wire diameter (1.48) was determined from the stress condition. It was found that the optimum wire diameter guarantees the lowest possible mass of the spring (1.49). This mass depends on the allowable stress, the working loads and the stiffness of the spring.

The usual additional requirements for the spring design include, among others, several practically important requirements:

- the outer diameter of the spring $D + \mathrm{d}$ is to be no greater then D_m;
- the total stored elastic energy in the spring is prescribed;
- the certain natural frequency of the spring ω is limited;
- the assembly volume of the helical spring $\pi D_m^2 L_0$ is limited;
- the volume of the helical spring $\pi D_m^2 L_s$ is limited.

Investigations on minimization of volume and weight of the spring have been carried out, taking into account other technological requirements. The complete mathematical treatment of the design cases is too complex for analytical treatment and requires the application of numerical optimization methods. For details see the review (Kruzelecki, 1990).

The optimization of springs subjected to axial torque is performed in a similar way. Since the wire is in the state of bending, the optimization results are analogous to the discussed above.

1.4 Semi-elliptic Longitudinal and Transverse Leaf Springs of Minimal Mass

1.4.1 Rectangular Cross-Section

Another example of bending-dominated spring is the leaf spring. This type of spring locates, for example, the solid drive axle in the Hotchkiss drive that used in light tracks (Gillespie, 1992) For optimization of this type of springs, the leaf spring is considered. The leaf spring is a simply supported beam of length L_0. For briefness, the deformation of the spring is assumed to be small and we make no difference between the free L_0 and deformed L lengths of the spring. Eyes are provided for attaching the spring to the body. The spring is fixed to the wheel axle by means of a central clamp. Chamber is the amount of bend given to the spring from the central line passing through the eyes. The wheel exerts the force F on the spring and the support reactions at the two eyes of the spring come from the car. The force F results in the bending moment in the cross section of the spring:

$$m_B = m_B(x), \tag{1.51}$$

which depends on the position x along the axis of the spring. For example, the bending moment in the cross-section of the simply supported spring with the force F in its middle section reads:

$$m_B(x) = F \cdot \begin{cases} L/2 + x, & -L/2 \le x < 0, \\ L/2 - x, & 0 \le x \le L/2. \end{cases} \tag{1.52}$$

The wheel-guided transverse leaf spring axle possesses a slightly different moment distribution over the spring length [WO 2008125076 A1]. In this application the distance between the force appliance points is l_f. The bending moment along the spring length reads:

$$m_B(x) = F \cdot \begin{cases} L/2 + x, & -L/2 \le x < -l_f/2, \\ (L - l_f)/2, & -l_f/2 \le x \le l_f/2, \\ L/2 - x, & l_f/2 \le x \le L/2. \end{cases} \tag{1.53}$$

The stored elastic energy with the modulus in tension or bending E is given by:

$$U_e = \frac{1}{2} \cdot \int_L \frac{m_B^2}{EI} \mathrm{d}x = \frac{F^2}{2c}. \tag{1.54}$$

This expression delivers the general expression for spring rate of leaf spring helical, non-cylindrical springs:

$$\frac{1}{c} = \frac{1}{F^2} \cdot \int_L \frac{m_B^2}{EI} \mathrm{d}x. \tag{1.55}$$

The volume of material with variable cross-section is given by the integral of the cross-sectional area $A = A(x)$ over the spring length, so that the mass of the spring is:

$$m = \rho_m \int_L A \mathrm{d}x. \tag{1.56}$$

The stress at solid height must be less than the ultimate tensile strength $\sigma_w = R_m$ to protect the spring from inadvertent damage. This restriction, applied on the bending stress:

$$\sigma \equiv \frac{m_B}{W_B} \le \sigma_w. \tag{1.57}$$

For the springs with a rectangular cross-section with axes of the cross-section

$$B = B(x), \quad T = T(x). \tag{1.58}$$

the geometric characteristics of cross-section are the following (Tables 1.3 and 1.4):

$$W_B = \frac{T^2 B}{6}, \quad I = \frac{BT^3}{12}, \quad A = BT. \tag{1.59}$$

Minimize the mass of the spring (1.56) assuming the spring rate

$$c = c^*. \tag{1.60}$$

and the force F are prescribed and the stress restriction (1.57) is fulfilled.

The following estimation for the height of the cross-section follows from (1.57):

$$T(x) \geq T_o(x), \quad T_o(x) = \sqrt{\frac{6m_B}{B\sigma_w}}. \tag{1.61}$$

The lowest mass possesses the uniform stress beam, which is known as a parabolic spring. For the given moment distribution, the profile height of the uniform stress beam is $T_o(x)$. The substitution of the height from (1.61) into Eq. (1.55) leads to inequality for the spring rate:

$$\frac{1}{c^*} \leq \frac{1}{F^2} \cdot \int\limits_L \frac{M_B^2}{\frac{EB}{12}\left(\frac{6m_B}{b\sigma_w}\right)^{3/2}} \mathrm{d}x = \frac{1}{F^2 \frac{E}{12}\left(\frac{6}{\sigma_w}\right)^{3/2}} \cdot \int\limits_L \sqrt{Bm_B} \mathrm{d}x. \tag{1.62}$$

Similarly, the substitution (1.61) into the expression for the spring mass delivers:

$$\rho_m \cdot \int\limits_l \sqrt{\frac{6Bm}{\sigma_w}} \mathrm{d}x \leq \rho_m \cdot \int\limits_L BT \mathrm{d}x = m. \tag{1.63}$$

The multiplication of inequalities (1.62) and (1.63) leads to an inequality:

$$\frac{\sqrt{6}}{c^* \sqrt{\sigma_w}} \rho_m \int\limits_L \sqrt{Bm_B} \mathrm{d}x \leq \frac{m}{F^2 \frac{E}{12}\left(\frac{6}{\sigma_w}\right)^{3/2}} \int\limits_L \sqrt{Bm_B} \mathrm{d}x. \tag{1.64}$$

The integrals of the positive function $\sqrt{Bm_B}$ are there on the both sides of the inequality (1.64) and could be shortened. From the inequality (1.64) the exact lower mass of the spring with rectangular cross-section reads:

$$m \geq \frac{3E\rho_m}{c^*} \cdot \frac{F^2}{\sigma_w^2}. \tag{1.65}$$

The elastic potential energy per unit volume (specific volume energy density) is equal to $\sigma_w^2/2E$. The elastic potential energy per unit mass (specific mass energy density) is:

$$\tilde{U}_e = \frac{\sigma_w^2}{2\rho_m E}.$$

This value is the material constant at normal stress. The reference mechanical properties for common spring materials are given in Table 1.5. Based on these values, the specific volume energy density and specific mass energy density of common spring materials are shown in Fig. 1.6.

The same method is used for springs made of composite materials. Since the composite materials are anisotropic, the evaluation of the specific mass energy density depends on the orientation of the fibers (Kobelev, 2019).

The inequality (1.65) indicates the lowest mass in terms of stored elastic energy:

$$m \geq \frac{3F^2}{2\tilde{U}_e c^*}.$$

1.4.2 *Circular Cross-Section*

For the springs with a circular cross-section with the diameter $\mathrm{d} = \mathrm{d}(x)$ the geometric characteristics of cross-section are the following (Tables 1.3 and 1.4):

$$W_B = \frac{\pi \mathrm{d}^3}{32}, \quad I = \frac{\pi \mathrm{d}^4}{64}, \quad A = \frac{\pi \mathrm{d}^2}{4}.$$

From (1.57) the estimation for the diameter of the cross-section follows as:

$$\mathrm{d} \geq \sqrt[3]{\frac{32 m_B}{\pi \sigma_w}}. \tag{1.66}$$

The substitution of the wire diameter from (1.66) into Eq. (1.68) leads to inequality for the spring rate:

$$\frac{64\pi^{1/3}\sigma_w^{4/3}}{32^{4/3}E} \cdot \int\limits_L m_B^{2/3} \mathrm{d}x \geq \frac{F^2}{c^*}. \tag{1.67}$$

The substitution (1.66) into the expression for the spring mass reads:

Table 1.5 Selected materials for springs

	Designation	EN	Material characterization	Density	Application temperature	Strength	Moduli		Mass density	Volume density
				ρ_m		σ_W	G	E	$\frac{\sigma_W^2}{2\rho E}$	$\frac{\sigma_W^2}{2E}$
				kg/dm^3	°C	MPa	GPa	GPa	J/kg	J/dm^3
1	1.4301/X5CrNi1810, Spring steel V2A	10088–3	Corrosion resistance	7.9	250	500..700	68	180	126	1000
2	1.4310/X10CrNi18-8, Spring steel V2A	10270–3	High corrosion resistance	7.9	200	500..750	70	185	155	1230
3	1.4401/ X5CrNiMo171-12, Spring steel V4A	10270–3	Corrosion resistant, good relaxation, non-magnetic	8.0	300	500..700	68	180	125	1000
4	1.4436/ X5CrNiMo17133, Spring steel V4A	10088–3	Good corrosion resistance, slightly magnetic	7.98	300	670	68	180	156	1250
5	1.4539/ X1NiCrMoCuN25-20, Spring steel V4A	10088	Heavy corrosion conditions, non-magnetic	8	300	530..730	68	180	137	1100
6	1.4568/X7CrNiAI17-7, Spring steel V4A	10270–3	Low relaxation, high fatigue strength	7.81	350	900..1100	73	195	328	2560
7	1.4571/X6CrNiMoTi17, Spring steel V4A	10270–3	Corrosion resistant, higher strength	7.95	300	500..700	68	185	124	973
8	2.4610/NiMo16Cr16Ti, H astelloy C4	–	In very corrosive atmosphere, non-magnetic	8.64	450	700.900	76	210	176	1520
9	2.4632/NiCr20CO18Ti, Nimonic 90	–	Corrosion resistant against most gases	8.2	500	1200	83	213	412	3380

(continued)

Table 1.5 (continued)

	Designation	EN	Material characterization	Density	Application temperature	Strength	Moduli		Mass density	Volume density
				ρ_m		σ_W	G	E	$\frac{\sigma_W^2}{2\rho E}$	$\frac{\sigma_W^2}{2E}$
				kg/dm^3	°C	MPa	GPa	GPa	J/kg	J/dm^3
10	2.4669/NiCr15Fe7TiAI, Inconel X750	–	High temperature, non-magnetic	8.2	600	980	76	213	274	2250
11	CW101C/CuBe2, Copper beryllium alloy	12,166	Corrosion-resistant, anti-magnetic, spark-free	8.3	−200.80	1150	47	120	663	5510
12	CW452K/CuSn6, Spring bronze	12,166	Nonmagnetic, solderable, weldable, corrosion resistant	8.8	−200.60	340.500	42	115	87	767
13	CW507L/CuZn36, Brass wire	12,166	Non-magnetic	8.45	−200.60	300.560	39	110	198	840
14	CoNiCrFe, Duratherm	–	High temperature	8.45	600	1100.1880	85.95	205.225	619	5230
15	EN 10,270–1, Type DH	10,270–1	Spring steel wire All common springs, high static and medium dynamic stress	7.95	80	2400.3500	81.5	206	2747	21,800
16	EN 10,270–1, Type SH	10,270–1	Spring steel wire All common springs, high static and medium dynamic stress	7.95	80	1900–2940	81.5	206	1908	15,200
17	EN 10,270–2/VDC Valve spring wire	10,270–2	For high continuous vibration stress	7.95	80	1250.2100	79.5	206	989	7860
18	EN 10,270–2/VDSiCr Valve spring wire	10,270–2	High dynamic load over 100C, good relaxation properties	7.95	120	1670.2100	79.5	206	1221	9710

(continued)

Table 1.5 (continued)

	Designation	EN	Material characterization	Density	Application temperature	Strength	Moduli		Mass density	Volume density
				ρ_m		σ_w	G	E	$\frac{\sigma_w^2}{2\rho E}$	$\frac{\sigma_w^2}{2E}$
				kg/dm³	°C	MPa	GPa	GPa	J/kg	J/dm³
19	TiAl6V4, Titanium alloy	–	insensitivity to cold, heat and corrosion	4.43	−200…300	1286	39	111	1794	7950

[1]Material data sheet 1.4301/ X5CrNi1810, HSM Stahl- und Metallhandel GmbH, www.hsm-stahl.de
[2]Material data sheet 1.4310 / X10CrNi18-8, HSM Stahl- und Metallhandel GmbH, www.hsm-stahl.de
[3]Material data sheet 1.4401/ X5CrNiMo171-12–2, HSM Stahl- und Metallhandel GmbH, www.hsm-stahl.de
[4]Material data sheet 1.4436, DEUTSCHE EDELSTAHLWERKE GMBH.
[5]Material data sheet 1.4539 AISI 904 L, Metalcor GmbH, www.metalcor.de
[6]Material data sheet 1.4568 17–7 PH®, Metalcor GmbH, www.metalcor.de
[7]Material data sheet Acidur 4571, DEUTSCHE EDELSTAHLWERKE GMBH.
[8]Material data sheet HASTELLOY® C-4, Zapp Materials Engineering GmbH, specialtymaterials@zapp.com.
[9]Material data sheet 2.4969 ALLOY 90, Metalcor GmbH, www.metalcor.de
[10]Material data sheet 2.4669 ALLOY X-750, Metalcor GmbH, www.metalcor.de
[11]Material data sheet 2.1247 CuBe2, Metalcor GmbH, www.metalcor.de
[12]Material data sheet CW452K, Seeberger GmbH & Co. KG, www.seeberger.net
[13]Material data sheet CuZn36, Deutsches Kupferinstitut.
[14]Material data sheet VACUUMSCHMELZE GmbH & Co. KG, http://www.vacuumschmelze.com
[15]Material data sheet DH, Hartgezogener Federstahldraht, https://metalprice.metalleschmidt.de/de
[16]Material data sheet SH, Hartgezogener Federstahldraht, https://metalprice.metalleschmidt.de/de
[17]Material data sheet VDC, Ölschlussvergüteter Federstahldraht, https://metalprice.metalleschmidt.de/de
[18]Material data sheet VDSiC, Ölschlussvergüteter Federstahldraht, https://metalprice.metalleschmidt.de/de
[19]Material data sheet TiAl6V4, https://www.huh.de/fileadmin/dateien/pdfs/Materialdatenblaetter/huh_Materialdatenblatt_Titan.pdf

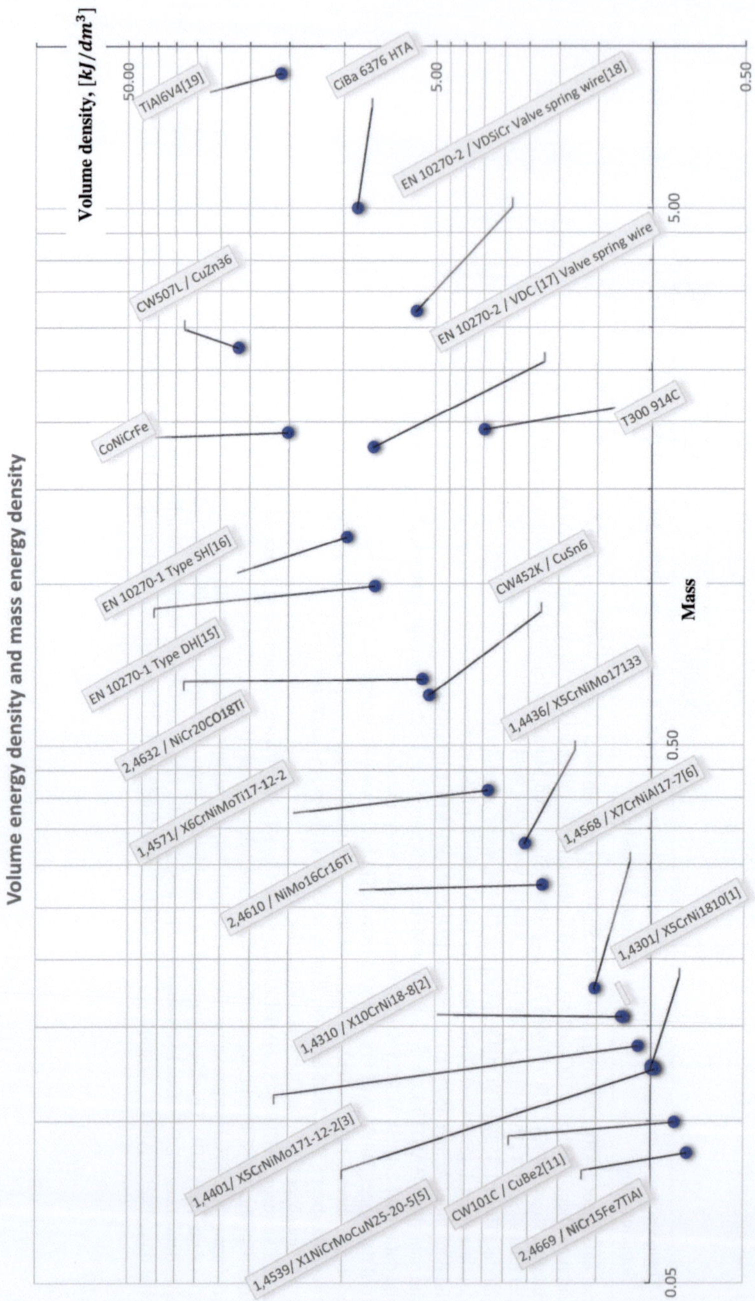

Fig. 1.6 Volume energy density and mass energy density of the common spring materials from Table 1.5

$$m \geq \frac{\rho_m \pi}{4} \cdot \int_L \left(\frac{32 m_B}{\pi \sigma_w} \right)^{2/3} \mathrm{d}x. \tag{1.68}$$

The multiplication of inequalities (1.62) and (1.63) leads to an inequality for the spring mass with the circular cross-section:

$$m \geq \frac{4E\rho_m}{c^*} \cdot \frac{F^2}{\sigma_w^2}. \tag{1.69}$$

The inequality (1.69) specifies the minimal mass in terms of specific elastic energy density in the spring:

$$m \geq \frac{2F^2}{\tilde{U}_e c^*}.$$

The lowest mass of the circular wire is 1/3 more that the lowest mass of the rectangular wire. This relation is valid for all fully-stress-designed springs for an arbitrary moment M_B along the spring axes, if the restrictions (1.56) and (1.57) are satisfied.

The same estimations are valid for other springs, which subjected to bending loads, for example for the twisted helical springs.

1.5 Multi-material Design of Springs

The seminal paper (Ashby & Bréchet, 2003) explored the design of hybrid materials, focusing on the selection of components, their shape and scale. The new design variables were introduced to expand the design space and create the new "hybrid materials" with specific property profiles. The paper (Wargnier et al., 2014), which proposes a multi-material design procedure, can be considered as a continuation of the main effort.

In this section, we study a question of multi-material design for springs. Specifically, we consider a set of helical springs made of different materials, assembled in parallel and in series. Each spring assembly behaves as a single spring. The springs are in series if they are connected at their ends and the forces in each spring are the same. The sum of the reciprocal spring rates equals the spring rate of the series assembly. Similarly, the springs are in parallel when they are connected side by side and each spring has the same travel. The spring rate of the parallel assembly is the sum of the spring rates of the individual springs.

We examine the following question: What is the minimum mass of a serial and a parallel spring assembly, given the total force and the assembly spring rate? For safety, we will examine the design problem for helical springs.

Minimize the mass of the spring assembly:

$$m = \sum_{i=1}^{N} m_i \rightarrow \text{min}, \tag{1.70}$$

assuming the assembly spring rate is equal to a given positive constant c^*, Eq. (1.60), the forces at installed height F_1 and full stroke F_2 are prescribed, the fatigue conditions (1.32) and (1.33) fulfilled, and the basic stress at full stroke for each material

$$\tau_b \leq \tau_{w,i}, \quad i = 1, .., N$$

are limited. Here N is the number of materials with the corresponding admissible working stresses $\tau_{w,i}$, shear modules G_i and densities ρ_i.

For the parallel assemblies the spring travel in each spring is

$$s = \frac{F_i}{c_i} \equiv \frac{F}{c^*}, \quad F = \sum_{i=1}^{N} F_i, \tag{1.71}$$

where F_i is the spring force and c_i is the spring rate of the spring with number i. From these equations follows that the total spring rate of the parallel assembly c^* is the sum of spring rates of all springs c_i:

$$c^* = \sum_{i=1}^{N} c_i.$$

According to Eq. (1.68), the mass of the spring made of material with number i is restricted from below:

$$m_i \geq m_{i.opt} = \frac{2\rho_i G_i F_i^2}{\tau_{w,i}^2 c_i} = \frac{2\rho_i G_i F_i s}{\tau_{w,i}^2}. \tag{1.72}$$

Applying the estimation (1.71) for each spring and calculating the total mass of parallel assembled spring, the following exact boundary for the mass is stated:

$$m = \sum_{i=1}^{N} m_i \geq \sum_{i=1}^{N} m_{i.opt} = \sum_{i=1}^{N} \frac{2\rho_i G_i F_i s}{\tau_{w,i}^2}. \tag{1.73}$$

From Eq. (1.71) and (1.73) follows the inequality:

$$\sum_{i=1}^{N} \frac{2\rho_i G_i F_i s}{\tau_{w,i}^2} \geq Fs \cdot \min_{1 \leq i \leq N} \frac{2\rho_i G_i}{\tau_{w,i}^2} = \frac{F^2}{c^*} \cdot \min_{1 \leq i \leq N} \frac{2\rho_i G_i}{\tau_{w,i}^2}. \tag{1.74}$$

The combination of Eqs. (1.73) and (1.74) yields:

$$m \geq \frac{F^2}{c^*} \cdot \min_{1 \leq i \leq N} \frac{2\rho_i G_i}{\tau_{w,i}^2}. \tag{1.75}$$

The estimation (1.75) proves, that the minimal mass of the springs in parallel possesses the single spring with the uppermost specific elastic energy density of material.

For the springs in series, the force in each spring is equal to F, and the spring travel in each spring is

$$s_i = \frac{F}{c_i}, \quad s = \sum_{i=1}^{N} s_i = \frac{F}{c^*}. \tag{1.76}$$

These equations show that the reciprocal spring rate of the serial assembly c^* is the sum of reciprocal spring rates of each spring in the assembly c_i:

$$\frac{1}{c^*} = \sum_{i=1}^{N} \frac{1}{c_i}.$$

The mass of each spring in the serial assembly is:

$$m_i \geq m_{i.opt} = \frac{2\rho_i G_i F^2}{\tau_{w,i}^2 c_i} = \frac{2\rho_i G_i F s_i}{\tau_{w,i}^2}. \tag{1.77}$$

Applying the estimation (1.77) for each spring, we state the exact lower boundary for the total mass of the serial assembly:

$$m = \sum_{i=1}^{N} m_i \geq \sum_{i=1}^{N} m_{i.opt} = F \cdot \sum_{i=1}^{N} \frac{2\rho_i G_i s_i}{\tau_{w,i}^2}. \tag{1.78}$$

Using Eq. (1.76), from Eq. (1.78) follows:

$$F \cdot \sum_{i=1}^{N} \frac{2\rho_i G_i s_i}{\tau_{w,i}^2} \geq F \cdot s \cdot \min_{1 \leq i \leq N} \frac{2\rho_i G_i}{\tau_{w,i}^2} = \frac{F^2}{c^*} \cdot \min_{1 \leq i \leq N} \frac{2\rho_i G_i}{\tau_{w,i}^2}. \tag{1.79}$$

From Eqs. (1.78) and (1.79) follows the desired inequality:

$$m \geq \frac{F^2}{c^*} \cdot \min_{1 \leq i \leq N} \frac{2\rho_i G_i}{\tau_{w,i}^2}. \tag{1.80}$$

Both lower limits (1.75) and (1.80) guarantee the same minimum mass of the assembly regardless of the type of assembly. The minimum mass of any multi-material assembly is equal to or greater than the mass of the single spring with the highest specific elastic energy density of the material.

1.6 Conclusions

It was proved, that the optimal wire shape, determined from the certain equal stress condition, guarantees the lowest possible mass of the spring. This mass depends only on the ultimate allowable stress for the spring material, the load at full stroke and the spring stiffness. This is an important milestone for comparison of different spring designs and spring materials. As the density and shear module are almost the same for all spring steels, the spring quality parameter can serve as the benchmarking property for spring design.

The minimal mass of the springs in parallel or in series assemblies possesses the single spring with highest specific elastic energy density of material.

The detailed design of springs is performed with numerical methods. The theoretical background and actual methodologies for numerical analysis of springs supplies (Shimoseki et al., 2003). The cited book explains the design examples, calculated by finite-element software and their comparison with experiments on real springs.

Summary of Principal Results
The following expressions are displayed in the current chapter:

- The compression spring rate, the compression-rotational spring rate, and the rotational spring rate of any non-cylindrical helical spring;
- Formulas for spring angle reduction due to compression force and torque;
- The simple expressions for the shear stresses in the cross sections and on the outer surfaces of the wires in helical springs;
- The simple fatigue criterion for the design of compression springs;
- The exact lower limit of the mass of a helical spring of any variable shape and any variable circular cross section that satisfies the stress condition;
- the exact lower mass of the rectangular section leaf spring;
- the exact lower mass of the circular section leaf spring.

It has been demonstrated that.

- The lowest mass of the circular wire is 1/3 more than the lowest mass of the rectangular wire;
- The same minimum mass of the multi-material assembly does not depend on the type of assembly (serial or parallel);
- the minimum mass of each multi-material assembly is equal to or greater than the mass of the single spring with the highest specific elastic energy density of the material.

References

Adibi-Asl, R., & Seshadri, R. (2009). Improved prediction method for estimating notch elastic-plastic strain. *Journal of Pressure Vessel Technology, 135*(4), 041203–041203–9

Ashby, M. F., & Bréchet, Y. J. M. (2003). Designing hybrid materials. *Acta Materialia, 51*(19), 5801–5821.

EN 13906-1:2013-11. (2013). Cylindrical helical springs made from round wire and bar—Calculation and design—Part 1: Compression springs; German version DIN EN 13906-1:2013, Beuth Verlag, Berlin

EN 13906-2:2013-09. (2013). Cylindrical helical springs made from round wire and bar—Calculation and design—Part 2: Extension springs; German version DIN EN 13906-2:2013, Beuth Verlag, Berlin

EN 13906-3:2014-06. (2014). Cylindrical helical springs made from round wire and bar—Calculation and design—Part 3: Torsion springs; German version DIN EN 13906-3:2014, Beuth Verlag, Berlin

Freiberger, W. (1949). The uniform torsion of an incomplete tore. *Australian Journal of Scientific Research, A2*, 354–375. https:doi/org/10.1071/CH9490354

Fuchs, H. O., Nelson, D. V., Burke, M. A., & Toomay, T. L. (1977). Shortcuts in cumulative damage analysis. In: Fatigue under complex loading. Analyses and Experiments. Society of Automotive Engineers, Inc. (Warrendale, PA) pp. 145–161

Gillespie, T. (1992). *Fundamentals of vehicle dynamics*. SAE, Warrendale, PA

Göhner, O. (1931). Schubspannungsverteilung im Querschnitt eines gedrillten Ringstabs mit Anwendung auf Schraubenfedern, *Ing.-Archiv., 2*(1), 1–19.

Göhner, O. (1932). Die Berechnung zylindrischer Schraubenfedern. *Zeitschrift Des VDI, 76*(1932), 269–272.

Henrici, P. (1955). On helical springs of finite thickness. *Quarterly of Appl. Math. XIII, 11*, 106–110.

Kobelev, V. (2019). Design and analysis of composite structures for automotive applications. Chassis and Drivetrain, John Wiley & Sons Ltd, Chichester, UK, ISBN: 9781119513858

Kruzelecki, J. (1990). Optimal design of helical springs. *Mechanika Teoretyczna i Stosowana., 1–2*, 28.

Landgraf, R. (1973). Cumulative fatigue damage under complex strain histories, ASTM STP 519, Cyclic Stress-Strain Behavior, ASTM, pp. 212–227

Leiseder, L. (1997). *Federelemente aus Stahl für die Automobilindustrie*, Bibliothek der Technik, Bd. 140, Verlag Moderne Industrie, ISBN: 3-478-93158-4

Meissner, M., Schorcht, H.-J., & Kletzin, U. (2015). *Metallfedern.* Springer-Verlag 978-3-642-39123-1

Michell, J. H. (1899). The uniform torsion and flexure of incomplete torus, with application to helical springs', Proceedings London. *Mathematics Society 31*, 130–146.

Molski, K., & Glinka, G. (1981). A method of elastic–plastic stress and strain calculation at a notch root. *Material Science and Engineering, 50*, 93–100.

Neuber, H. (1961a). Theory of stress concentration for shear-strained prismatical bodies with arbitrary nonlinear stress–strain law. *Journal of Applied Mechanics, 28*, 544–550.

Neuber, H. (1961b). *Theory of notch stresses: Principles for exact calculation of strength with reference to structural form and material.* Oak Ridge, TN: United States Atomic Energy Commission, US Office of Technical Services, 1961.

Ponomarev, S. D., Biderman, V. L., Likharev, K. K., Makushin, V. M., Malinin, N. N., & Feodos'ev, V. I. (1956). Resistance calculus in construction of machines. *Mashgiz Moscow, I*, 704–835.

Shimoseki, M., Hamano, T., Imaizumi, T. (Eds.) (2003) *FEM for springs.* Springer-Verlag 978-3-540-00046-4

Smith, K. N., Watson P., & Topper T. H., (1970). A stress-strain function for the fatigue of metals. *Journal of Materials, ASTM, 5*(4), 767–778.

SAE. (1996). *Spring design manual.* Part 5, SAE, HS-158 SAE International, Warrendale, PA.

SAE HS 795. (1997). *SAE manual on design and application of helical and spiral springs.* SAE International, Warrendale

Teodorescu, P. P. (2013). *Treatise on classical elasticity*. Springer.

Timoshenko, S. (1948). *Strength of materials.* Van Nostrand.

Wahl, A. M. (1929). Stresses in heavy closely coiled helical springs. APM-51–17, Transactions ASME, 51, s. 185–200

Walker, K. (1970). The effect of stress ratio during crack propagation and fatigue for 2024-T3 and 7075-T6 aluminum. In *Effects of environment and complex load history on fatigue life*, ASTM STP 462, Am. Soc. for Testing and Materials, West Conshohocken, PA, pp. 1–14.

Wargnier, H., Kromm, F. X., Danis, M., & Brechet, Y. (2014). Proposal for a multi-material design procedure. *Materials and Design, 56*, pp. 44–49

Wundt, B. M. (1972). Effect of notches on low-cycle fatigue: A literature survey. *ASTM International*

Yamada, Y. (2007). *Materials for springs.* Springer-Verlag 978-3-540-73811-4

Chapter 2
Stress Distributions Over Wire Cross-Section

Abstract This chapter examines the stress distribution across the wire cross-section of helical springs. For simplicity, the pitch of the helical spring is neglected and the traditional representation of a coil as an incomplete torus is used. This model generalizes StVenant's torsion problem of an elliptical straight rod, considering the curvature of the rod. The closed-form solution for the torsion problem of an incomplete torus is discussed. This chapter is the next section of the first part, which studies the springs from the point of view of design.

Keywords Torsion of wire in helical springs · Non-circular cross-section of wire · Stress in cross-sections · StVenant torsion

2.1 Warping Function

In the present chapter, we consider the helical springs with the elliptical, circular and the ovate cross-sections of the spring wire. The force acts ideally along the axis z of the cylindrical coordinate system, such that the line of force coincides with the symmetry axis of the spring. The radius of the spring helix is $R = D/2$, where the mean diameter of the spring is D. A view of the spring wire at the cut section has been taken from the right and was shown at Fig. 1.3, Chap. 1. The couple $\mathbf{M} = \mathbf{F} \times \mathbf{R}$, acting in the vertical plane, has been shown as a vector. The couple was resolved into two components, which lie in planes which are tangential and normal to the helix, respectively (Fig. 1.3a). The distance between coils is assumed as small. For the closed-coiled spring, the torque moment of the wire is nearly equal to M. The axial force leads only to the torsion of the wire. Thus, both tension and compression helical springs store elastic energy principally through torsion of wire. The component of couple, which tends to cause bending of the isolated portion of the spring wire, is neglected.

V. Kobelev, *Fundamentals of Springs Mechanics*, University of Siegen, Faculty IV: School of Science and Technology. Paul-Bonatz-Str. 9–11, D-57076 Siegen. E-mail: Vladimir.kobelev@uni-siegen.de Homepage: https://www.mb.uni-siegen.de/kobelev/

V. Kobelev, *Fundamentals of Springs Mechanics*,
https://doi.org/10.1007/978-3-031-58584-5_2

The isolated portion of the spring wire is the incomplete torus. For this incomplete torus we search the analytical solution for the torsion problem. If the curvature of the torus vanishes, the isolated portion of the spring wire turns into an ovate straight solid or hollow rod. StVenant examined the torsion problem for an ovate straight solid or hollow rod. The torsion of straight cylinders with circular and elliptical cross-sections was reviewed by Sneddon and Berry (1958) and de Veubecke (1979).

For the helical spring the curvature of the wire could not be neglected (Michell, 1899). For this purpose, consider an elastic torus, generated by rotating the cross-section about the z axis of the cylindrical coordinate system. This torus represents one coil of the helical spring with a negligible pitch ("closed-coiled spring"). The cross-section Ω considered, until explicitly stated otherwise, to be simply connected. The torus is assumed to be incomplete, i.e. the two ends of the turn are not joined.

We examine now the vector displacement field:

$$\langle U(r,\theta,z),\ V(r,\theta,z),\ W(r,\theta,z)\rangle$$

in reference cylindrical coordinate system (r, θ, z).

For the displacement field, which arises in the helical spring, the axial displacement W linearly depends upon the polar angle θ, while the radial displacement U vanishes. The section will not remain planar during the deformation. Each section is deformed, however, in precisely the same way, so we may take the components of displacements of the form:

$$\langle U, V, W\rangle = \langle 0,\ r\psi(r,z),\ \theta \cdot B_s\rangle, \tag{2.1}$$

where $\psi(r, z)$ is the warping function. For the torus, which embodies a single complete coil, $0 < \theta < 2\pi$. The torus turns during the infinitesimal deformation into a helix with the tiny constant pitch. The two ends of coil separate from each other by the relative axial displacement:

$$W|_{\theta=2\pi} - W|_{\theta=0} = 2\pi \cdot B_s. \tag{2.2}$$

Worthwhile mentioned, that the helical spring deforms exactly as an elastic body with a giant screw dislocation. The edge of the dislocation coincides with the axis of the helix. From the physical viewpoint, the constant B_s is equal to the magnitude of the Burgers vector of this giant screw dislocation (Hirth & Lothe, 1991). The magnitude of the Burgers vector B_s is proportional to the displacement of one coil and defines the pitch change due to the axial compression or extension of the helical spring.

According to (2.1), the only non-zero components of the strain tensor in the cylindrical coordinate system in the case of screw dislocation are:

$$\gamma_{r\theta} = r\frac{\partial}{\partial r}\left(\frac{V}{r}\right), \quad \gamma_{\theta z} = \frac{\partial V}{\partial z} + \frac{1}{r}\frac{\partial W}{\partial \theta}. \tag{2.3}$$

After the substitution of representations for displacements from (2.1) into (2.3) the two components of strain tensor reduce to:

$$\gamma_{r\theta} = r\frac{\partial \psi}{\partial r}, \quad \gamma_{\theta z} = r\frac{\partial \psi}{\partial z} + \frac{B_s}{r}. \tag{2.4}$$

The components (2.4) are the same for all cross-sections, being independent of θ. The magnitude of the Burgers vector of screw dislocation B_s in (2.2) depends linearly upon the applied axial force F. The dependency of constant of screw dislocation upon the force will be determined in the next section.

Next, we derive the equilibrium equations in term of the warping. For this purpose, consider equilibrium of the linear-elastic, isotropic body in the absence of any volume forces. With Hook's law and strains in form (2.4), the shear stresses in terms of the warping function $\psi = \psi(r, z)$ will be:

$$\tau_{r\theta} = Gr\frac{\partial \psi}{\partial r}, \tau_{\theta z} = Gr\frac{\partial \psi}{\partial z} + \frac{G}{r}B_s. \tag{2.5}$$

In cylindrical coordinate system, the axial component of the equilibrium equation in the volume of the torus reads:

$$\frac{\partial \tau_{r\theta}}{\partial r} + \frac{\partial \tau_{\theta z}}{\partial z} + 2\frac{\tau_{r\theta}}{r} = 0. \tag{2.6}$$

There are two other components of the equilibrium equation in other directions, but they are satisfied by Eq. (2.5) identically. After the substitution of (2.5) in (2.6) and division by r, the Eq. (2.6) reduces to:

$$\frac{\partial^2 \psi}{\partial z^2} + \frac{\partial^2 \psi}{\partial r^2} + \frac{3}{r}\frac{\partial \psi}{\partial r} = 0. \tag{2.7}$$

In terms of Laplace operator

$$\Delta = \frac{\partial^2}{\partial r^2} + \frac{1}{r}\frac{\partial}{\partial r} + \frac{\partial^2}{\partial z^2},$$

the equilibrium Eq. (2.7) takes the form:

$$\left(\Delta - \frac{1}{r^2}\right)(r\psi) = 0. \tag{2.8}$$

In cylindrical coordinate system, the normal vector to the mantle of the torus is $\langle n_r, 0, n_z \rangle$. The component n_θ disappears for the toroidal surface. The mantle of the torus, i.e. the boundary of torus without bases, is stress-free. This boundary condition requires the vanishing of the shear stress in the normal direction to the surface of the torus:

$$\tau_n = n_r \tau_{r\theta} + n_z \tau_{\theta z} = 0. \tag{2.9}$$

The substitution of Eq. (2.5) in Eq. (2.9) delivers the boundary condition:

$$\tau_n = n_r G r \frac{\partial \psi}{\partial r} + n_z \left(G r \frac{\partial \psi}{\partial z} + \frac{B_s}{r} \right) = G r \frac{\partial \psi}{\partial n} + n_z \frac{B_s}{r} = 0. \tag{2.10}$$

In Eq. (2.10). we use the formula: $\partial/\partial n = n_r \partial/\partial r + n_z \partial/\partial z$.

2.2 Prandtl Stress Function

Alternatively, the torsion problem could be resolved using the Prandtl stress function $\phi = \phi(r, z)$.

The shear stresses in terms of the stress function are:

$$\tau_{r\theta} = -\frac{1}{r^2} \frac{\partial \phi}{\partial z}, \tag{2.11}$$

$$\tau_{z\theta} = \frac{1}{r^2} \frac{\partial \phi}{\partial r}. \tag{2.12}$$

The equilibrium equation in the volume of the coil (2.6) is satisfied by (2.11) and (2.12) identically. The substitution of Eqs. (2.11) and (2.12) in the boundary condition (2.9) leads to the boundary condition in terms of the stress function:

$$\tau_n = -\frac{n_r}{r^2} \frac{\partial \phi}{\partial z} + \frac{n_z}{r^2} \frac{\partial \phi}{\partial r} = \frac{1}{r^2} \frac{\partial \phi}{\partial s} = 0. \tag{2.13}$$

The cross-section Ω is assumed to be simply-connected with the boundary Γ. The upper and lower sections of curve Γ are $Z_+(r)$, $Z_-(r)$ (Fig. 2.1).

As follows from Eq. (2.13) the stress function is constant on the contour Γ:

$$\phi = \phi_0 \quad \text{on } \Gamma. \tag{2.14}$$

The value of the constant ϕ_0 does not influence the value of stresses. This constant it is assumed to be zero:

$$\phi = 0 \quad \text{on } \Gamma. \tag{2.15}$$

The tangential derivative expresses in terms of the partial derivatives as:

$$\partial/\partial s = -n_r \partial/\partial z + n_z \partial/\partial r.$$

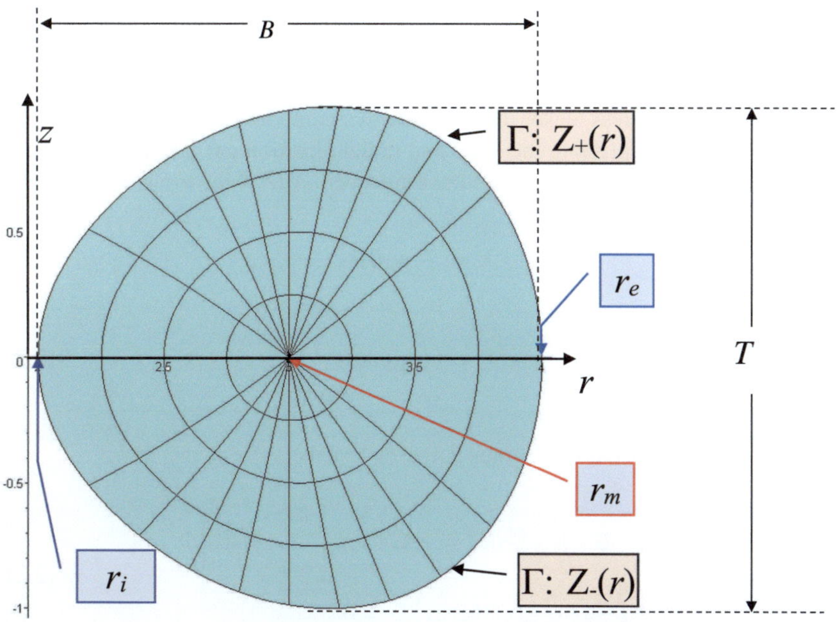

Fig. 2.1 Ovate cross-section. The contour lines of the Prandtl stress function are shown

As mentioned above, the equilibrium equations are identically satisfied by (2.11) and (2.12) via stress function. Notably, that the compatibility conditions should be fulfilled. The easiest way is to use the compatibility conditions in terms of stresses. These conditions are known as are known as Beltrami-Michell equations (Sneddon & Berry, 1958; de Veubecke, 1979). For the stress state, which is given by Eqs. (2.11), (2.12), the Beltrami-Michell equation in cylindrical coordinates assumes the form:

$$\left(\Delta - \frac{4}{r^2}\right)\tau_{r\theta} = 0, \quad \left(\Delta - \frac{4}{r^2}\right)\tau_{z\theta} = 0 \tag{2.16}$$

Substituting (2.11) in the first Beltrami-Michell Eq. (2.16), we obtain equation for the stress function:

$$\frac{\partial}{\partial z}\left(\frac{\partial^2 \phi}{\partial z^2} + \frac{\partial^2 \phi}{\partial r^2} - \frac{3}{r}\frac{\partial \phi}{\partial r}\right) = 0. \tag{2.17}$$

The Eq. (2.17) results in:

$$\left(\Delta - \frac{4}{r^2}\right)\frac{\phi}{r^2} = -\frac{2G}{r^2}B_s. \tag{2.18}$$

The substitution of (2.12) in the second Beltrami-Michell Eq. (2.16) leads to the same governing Eq. (2.18).

Generally saying, the resultant force in the cross-section is a vector. As only two components of the stress tensor remain, only two components of force vector could exist. The resultant forces in the axial and radial direction are given by integration of the shear stresses $\tau_{z\theta}$ and $\tau_{r\theta}$ over the cross-section Ω of the torus, respectively:

$$F = \int_{\Omega} \tau_{z\theta} \mathrm{d}\Omega. \tag{2.19}$$

$$F_r = \int_{\Omega} \tau_{r\theta} \mathrm{d}\Omega. \tag{2.20}$$

The radial component of resultant force F_r vanishes:

$$\begin{aligned} F_r &= -G \cdot \int_{\Omega} \frac{1}{r^2} \frac{\partial \phi}{\partial z} \mathrm{d}\Omega = -G \cdot \int_{r_i}^{r_e} \frac{1}{r^2} \left[\int_{Z_-(r)}^{Z_+(r)} \frac{\partial \phi}{\partial z} \mathrm{d}z \right] \mathrm{d}r = \\ &-G \cdot \int_{r_i}^{r_e} \frac{1}{r^2} \left[\phi(z_+) - \phi(z_-) \right] \mathrm{d}r = 0. \end{aligned} \tag{2.21}$$

The boundary condition (2.5) was applied in Eq. (2.21). The only remaining component of force is the force F in the direction of axis z.

The resultant moments in the cross-section are treated similarly. Consider the radial component of the moment vector with respect to the point O at the origin of the coordinate system. The only two non-zero components of stress tensor $\tau_{z\theta}$, $\tau_{r\theta}$ result in the moment vector. The moment vector is directed normally to the considered cross-section and causes a twist of cross-section. The resultant value of torsional moment about point O is equal to

$$M_r = \int_{\Omega} (\tau_{z\theta} r - \tau_{r\theta} z) \mathrm{d}\Omega = G \cdot \int_{r_i}^{r_e} \frac{1}{r^2} \left[\int_{Z_-(r)}^{Z_+(r)} \left(\frac{\partial \phi}{\partial r} r + \frac{\partial \phi}{\partial z} z \right) \mathrm{d}z \right] \mathrm{d}r. \tag{2.22}$$

The torsional moment M_r vanishes for all cross-sections, which are symmetrical about plane $z = 0$, such that $Z_+(r) = -Z_-(r)$ (Pilgram, 1913). Apparently, two remaining components of the resultant momentum vector with respect to the origin of coordinate system disappear because of the cyclic symmetry of stress state. Thus, the resultant force F_r and resultant value of moment M_r are both equal to zero for the considered stress–strain state. This demonstrates, that the resultant force vector for each cross-section has actually only one non-zero component F and its line of action is aligned with the axis z of the coordinate system.

The elasticity of the axially loaded spring is characterized by means of the spring rate c_w of one complete turn of the helix. The spring rate can be specified through magnitude of the Burgers vector from (2.3). For the linearly elastic spring, the magnitude of the Burgers vector B_s depends linearly upon the axial resultant force F. The spring rate c_w of one coil is, by definition, the force that causes a unit displacement in the axial direction

$$c_w = \frac{F}{W|_{\theta=2\pi} - W|_{\theta=0}} = \frac{F}{2\pi B_s}. \tag{2.23}$$

2.3 Shear Stresses on Surface of Elliptic and Circular Wires

According to Eq. (1.28), the spring index designates the ratio of mean coil diameter to wire diameter

$$w = \frac{R}{r} \equiv \frac{D}{\mathrm{d}}.$$

For the springs with infinite radius of helix R, the spring index infinitely increases and the curvature vanishes. In this case, the stress distribution coincides with the stress field of the twisted straight wire. Thus, the spring with the infinite spring index possesses the constant stress over the surface and this stress is equal to the corresponding basic stress.

The stresses on the surface of spring with the finite spring index vary over the surface of the spring. For the spring with the circular cross-section, the maximum shear stress in a helical spring occurs on the inner face of the spring coils. Thus, the curvature of the helical spring results the shear stresses on the inner faces, that are higher than the basic stress.

However, if the cross-section is elliptical, the maximum of the shearing stress moves to the minor axis of the ellipse. Thus, for the flattened elliptical section the maximum locates in the middle of the upper and lower surfaces of wire. For the moderately flattened elliptical section the maximum of the stress is between the inner and the upper faces of the spring in the upper half of the cross-section. In the lower half, the point of maximum reflects with respect to the symmetry axis of the cross-section. Clear, that the stress distribution depends on the curvature of wire and with higher curvature the point of maximum moves to the inner face.

The easiest way to evaluate the shear stress on the outer surface of the spring wire is to follow the acknowledged results. Göhner (1932) provided the formulas for stresses for the spring with the elliptic wire cross-section. The axes of the elliptic cross-section are T and B (Fig. 2.1). The local coordinate system of the cross-section originates in the center of ellipse. The shear stresses were presented as Taylor series:

$$\tau_{r\theta} = \tau_{r\theta}^{(1)} + \tau_{r\theta}^{(2)} + \cdots, \quad \tau_{z\theta} = \tau_{z\theta}^{(1)} + \tau_{z\theta}^{(2)} + \cdots \tag{2.24}$$

In Cartesian coordinates x, y of ellipse, the first two coefficients of Taylor series in (2.24) are Göhner (1932), Biezeno and Grammel (1971):

$$\tau_{r\theta}^{(1)} = -G\theta \cdot \frac{2T^2 y}{T^2 + B^2}, \tag{2.25}$$

$$\tau_{r\theta}^{(2)} = -G\theta \cdot \frac{2T^2(2T^2 + 3B^2)xy}{(T^2 + B^2)(T^2 + 3B^2)R}, \tag{2.26}$$

$$\tau_{z\theta}^{(1)} = G\theta \cdot \frac{2T^2 x}{T^2 + B^2}, \tag{2.27}$$

$$\tau_{z\theta}^{(2)} = G\theta \cdot \frac{B^2\left[(16T^2 + 12B^2)x^2 - 12T^2y^2 + 3T^2B^2\right]}{4(T^2 + B^2)(T^2 + 3B^2)R}. \tag{2.28}$$

In the above formulas θ is the torsion angle pro length unit. The torsion angle θ relates to the axial displacement W with Eq. (2.1).

The functions $\tau_{r\theta}^{(1)}$ and $\tau_{z\theta}^{(1)}$ are the first degree Taylorpolynomials, and the functions $\tau_{r\theta}^{(2)}$, $\tau_{z\theta}^{(2)}$ are the second degree Taylorpolynomials of Cartesian coordinates x, y.

The octahedral shear stress for the elliptical wire with the same order of precision reads:

$$\tau(\rho, \varphi) \overset{\text{def}}{=} \sqrt{\left(\tau_{r\theta}^{(1)} + \tau_{r\theta}^{(2)}\right)^2 + \left(\tau_{z\theta}^{(1)} + \tau_{z\theta}^{(2)}\right)^2}. \tag{2.29}$$

Equation (2.29) delivers the stress distribution also for the circular wire as its limit case. In this case, the axes of the ellipse turn into the sole wire diameter:

$$T = B = \mathrm{d}.$$

The polar coordinates of the circular cross-section are:

$$x = \rho\cos\varphi,\ y = \rho\sin\varphi, \quad 0 \le \rho \le r = \frac{\mathrm{d}}{2}, \quad 0 \le \varphi < 2\pi.$$

For the circular wire of radius r and for the given spring index w, the octahedral shear stress $\tau(\rho, \varphi)$ yields from Eq. (2.29). After the substitution of expressions for polar coordinates in Eq. (2.29), the octahedral shear stress $\tau(\rho, \varphi)$ reads (Kobelev, 2016):

$$\tau(\rho, \varphi) = \frac{\tau_b}{32r^2 w} \cdot \sqrt{c_0 + c_1\cos\varphi + c_2\cos^2\varphi}, \tag{2.30}$$

$$c_0 = 9R^4 - 72R^2\rho^2 + 1024w^2R^2\rho^2 + 144\rho^4, \tag{2.31}$$

$$c_1 = 192\rho R^3 w + 1792\rho^3 R, \tag{2.32}$$

$$c_2 = 240R^2\rho^2 + 640\rho^4. \tag{2.33}$$

In (2.30), the octahedral shear stress on the surface of the straight circular bar is:

$$\tau_b = \sqrt{\tau_{r\theta}^{(1)2} + \tau_{z\theta}^{(1)2}} = G\theta R. \tag{2.34}$$

Stress τ_b is the "basic", or "uncorrected stress", Eq. (1.25) " (SAE, 1997, Chap. 5, Design of Helical Springs). The basic stress is equal to the shear stress on the surface of the straight circular wire, which is twisted by the same torque. The "basic stress" is obtained by dividing the torsion moment acting on the wire, by the section modulus in torsion the shear stress on the surface of the circular wire. The basic stress represents the mean value of the shear stress on the surface of the wire and does not depend on the spring index w. Specifically, the basic stress is the octahedral stress for the limit case of infinitely high spring index $w = \infty$:

$$\lim_{w \to \infty} \tau(\rho, \varphi) = \tau_b.$$

In this case the stress equalizes over the surface of the wire. Consequently, the basic stress is equal to the shear stress on the surface of the straight cylindrical wire, loaded by the same torsion moment as the wire of the helical spring.

Using the Eq. (2.30), we introduce the stress correction function (Kobelev, 2016):

$$k_0(w, \varphi) = \frac{\sqrt{81 + 1024w^2 + 1984w\cos\varphi + 880\cos^2\varphi}}{32w}. \tag{2.35}$$

The shear stresses on the surface of the wire $\tau_s(w, \varphi) \overset{\text{def}}{=} \tau(\rho = r, \varphi)$ depend solely on polar angle φ and spring index:

$$\tau_s(w, \varphi) = k_0(w, \varphi)\tau_b, \qquad 0 \le \varphi < 2\pi. \tag{2.36}$$

The shear stress $\tau_s(w, \varphi)$ is the function of the spring index w and of the polar angle φ. The maximal value $\tau_c(w)$ is the corrected stress on the surface of the wire, Eq. (1.27):

$$\tau_c(w) = \max_{0 \le \varphi < 2\pi} \tau_s(w, \varphi). \tag{2.37}$$

Consider several springs with the same wire diameter but with different spring indices. Let the torque in the wire remains the same for all springs. The value θ of the torsion angle pro length unit remains equal for all springs with different spring indices. The basic stress does not alter in this case, but the variation of stress over the surface reduces with the increasing spring index w. In the limit case of infinite spring index the variation fades away. Correspondingly the Eq. (2.36) turns into the basic stress σ_b:

$$\begin{aligned}&\lim_{w\to\infty} \tau_s(w,\varphi) = \\ &\lim_{w\to\infty}\left(\frac{\sigma_{id}}{32w}\cdot\sqrt{81+1024w^2+1984w\cos\varphi+880\cos^2\varphi}\right) = \tau_b.\end{aligned} \tag{2.38}$$

The maximum shear stress in a helical spring and is equal in the employed approximation to the value:

$$\tau_c(w) = k_4(w)\tau_b, \tag{2.39}$$

where

$$k_4(w) = \frac{\sqrt{961+1984w+1024w^2}}{32w}. \tag{2.40}$$

The coefficient $k_4(w)$ before τ_b is the stress correction factor. This coefficient is the function of spring index w only. The advantage of newly developed correction function (2.32) is its dependence on the polar angle φ on the surface of the wire. This function allows to account the distribution of defects on the surface of the wire. The distribution of defects on the surface is necessary for the statistical evaluation of failures in course of cyclic loading and evaluation of the failure probabilities.

2.4 Shear Stresses on Surface of Ovate Wire

In this Section we find the solution of the equilibrium equation for an ovate wire. The solution is based on separation of the variables for the stress function. The representation of the stress function is assumed as a sum

$$\phi(r,z) = \phi_1(r) + \phi_2(z) \tag{2.41}$$

of the auxiliary functions $\phi_1(r)$, $\phi_2(z)$, satisfying the ordinary differential equations

$$\frac{\mathrm{d}^2\phi_2}{\mathrm{d}z^2} = C_s, \quad \frac{\mathrm{d}^2\phi_1}{\mathrm{d}r^2} - \frac{3}{r}\frac{\mathrm{d}\phi_1}{\mathrm{d}r} + 2GB_{sep} = -C_s \tag{2.42}$$

with the separation constant C_s.

The general solutions of the auxiliary differential Eqs. (2.42) are:

$$\phi_1(r) = \frac{C_s + 2GB_s}{4}r^2 + \frac{C_{11}}{4}r^4 + C_{12}, \quad \phi_2(z) = \frac{1}{2}C_{sep}z^2 + C_{13}z + C_{14},$$

where $C_{11}, C_{12}, C_{13}, C_{14}$- unknown integration constants. The representation (2.41) delivers the stress function as a polynomial:

$$\phi(r, z) = \frac{C_s + 2GB_s}{4}r^2 + \frac{C_{11}}{4}r^4 + C_{12} + \frac{1}{2}C_s z^2 + C_{13}z + C_{14}. \tag{2.43}$$

Substituting (2.43) in (2.11) and (2.12) provides the following expressions for shear stresses:

$$\tau_{r\theta} = -\frac{C_s z + C_{13}}{r^2}, \quad \tau_{z\theta} = \left(\frac{C_s}{2} + GB_s\right)\frac{1}{r} + C_{11}r. \tag{2.44}$$

Secondly, the solution of the equation for the warping function (2.8) is based on separation of the variables

$$\psi(r, z) = z\psi_1(r), \tag{2.45}$$

where the auxiliary function $\psi_1(r)$ satisfies the differential equation:

$$\frac{\mathrm{d}^2\psi_1}{\mathrm{d}r^2} + \frac{3}{r}\frac{\mathrm{d}\psi_1}{\mathrm{d}r} = 0. \tag{2.46}$$

The solution of the ordinary differential Eq. (2.46) leads to the following representation of the solution for the Eq. (2.8):

$$\psi(r, z) = C_{21}z + C_{22}\frac{z}{r^2}. \tag{2.47}$$

The integration constants C_{21}, C_{22} are not independent and are linked to the integration constants from (2.43). Substitution of (2.47) into the expression for shear stresses (2.5) results in the following representation:

$$\tau_{r\theta} = -\frac{2GC_{22}z}{r^2}, \quad \tau_{z\theta} = G(C_{22} + B_s)\frac{1}{r} + GC_{21}r. \tag{2.48}$$

Equating two different representations of shear stresses (2.44) and (2.48), we obtain the algebraic relations between integration constants from (2.43) and (2.47):

$$C_{21} = \frac{C_{11}}{G}, \quad C_{22} = \frac{C_s}{2G}, \quad C_{13} = 0.$$

The integration constants for a given contour could be obtained from the boundary condition (2.5). The inverse method consists of searching for the shape of the contour, which satisfies the boundary condition (2.5). To find the equation of the unknown contour, which satisfies the boundary Eqs. (2.5), we equate the polynomial (2.43) to zero. Assuming the contour to be symmetrical with respect to the axis $z = 0$ and factorizing the polynomial in (2.43), we obtain the equation of the "quasi-elliptical" contour in a form (Fig. 2.1) (Kobelev, 2002):

$$z^2 = Z_0^2 \cdot \left(r_e^2 - r^2\right) \cdot \left(r^2 - r_i^2\right), \tag{2.49}$$

The integration constants could be expressed through the intensity of screw dislocation B_s and geometrical characteristics of the cross-section. The value

$$T = Z_0 \cdot \left(r_e^2 - r_i^2\right) \tag{2.50a}$$

is the maximum height of the profile at the pivot point

$$r_m = \sqrt{\left(r_e^2 + r_i^2\right)/2}. \tag{2.50b}$$

The solution of torsion problem is also given by the warping function (2.47), satisfying the equilibrium Eq. (2.6) and the boundary conditions (2.8):

$$\psi(r, z) = \frac{B_s z}{r^2} \frac{4T^2 r^2 + (r_e^2 - r_i^2)^2}{(r_e^2 - r_i^2)^2 + 2T^2\left(r_e^2 + r_i^2\right)}. \tag{2.51}$$

The separation constant C_s expresses through the magnitude of the Burgers vector B_s as:

$$C_s = -2B_s G \cdot \frac{(r_i + r_e)^2 (r_e - r_i)^2}{(r_e^2 - r_i^2)^2 + 2T^2\left(r_e^2 + r_i^2\right)}.$$

Substitution of integration constants in the expression (2.43) leads to the final expression of the stress function

$$\phi(r, z) = \frac{B_s G\left(r_e^2 - r_i^2\right)^2}{\left(r_e^2 - r_i^2\right)^2 + 2T^2 \cdot \left(r_e^2 + r_i^2\right)} \cdot \left[z^2 - \frac{T^2}{\left(r_e^2 - r_i^2\right)^2} \cdot \left(r_e^2 - r^2\right) \cdot \left(r^2 - r_i^2\right)\right]. \tag{2.52}$$

2.5 Quasi-elliptical Cross-Section

According to (2.49), the ovate, of "quasi-elliptical" form of cross-section: Ω for incomplete torus is given by the equations:

$$Z_+(r) = Z_0\sqrt{\left(r_e^2 - r^2\right)\left(r^2 - r_i^2\right)},$$
$$Z_-(r) = -Z_0\sqrt{\left(r_e^2 - r^2\right)\left(r^2 - r_i^2\right)}. \tag{2.53}$$

Herein r_i and r_e are respectively the inner and outer radii of torus.

The volume of one coil (volume of the complete torus) with the cross-section (2.53) is:

$$V_1 = 2\pi \cdot \int\limits_{\Omega} r \mathrm{d}\Omega = 2\pi \cdot \int\limits_{r_i}^{r_e} \left[\int\limits_{Z_-(r)}^{Z_+(r)} \mathrm{d}z \right] r \mathrm{d}r = \frac{\pi^2}{4} \cdot T \cdot \left(r_e^2 - r_i^2\right). \tag{2.54}$$

The substitution of (2.51) into (2.4) results in the components of the shear stresses:

$$\tau_{\theta r} = \frac{(r_i + r_e)^2 (r_e - r_i)^2}{(r_e^2 - r_i^2)^2 + 2T^2\left(r_e^2 + r_i^2\right)} \cdot \frac{2B_s G z}{r^2}, \tag{2.55}$$

$$\tau_{z\theta} = \frac{-2r^2 + r_i^2 + r_e^2}{(r_e^2 - r_i^2)^2 + 2T^2\left(r_e^2 + r_i^2\right)} \cdot \frac{2B_s G T^2}{r}. \tag{2.56}$$

The octahedral shear stress reads:

$$\tau_{oct} \stackrel{\text{def}}{=} \sqrt{\tau_{z\theta}^2 + \tau_{\theta r}^2}. \tag{2.57}$$

The stresses on the outer surface of the wire depend on the radii r_i, r_e and the height T of the profile. Figure 2.2 displays the influence of the heights on the octahedral shear stress for the fixed radii. The "stars" mark the points of maximum of the octahedral shear stress. If the shape is flattened along the spring axis, the maximal stress occurs on the inner face of the wire. If the wire has same dimensions in both directions, the maximum region splits in two narrow strips near the inner face. For the gradually flattened along the radius ovate shape, the maximum moves to the top of the cross-section.

Figure 2.3 draws the stresses on the surfaces of the ovate wires with the same profile height and width, but with the different distance of pivot points (2.50b) from the axis of the helix. One again, the maximum region splits with the rising r_m. The stress distributes more evenly over the inner face of the wire. This effect reduces the maximal octahedral stress for the given spring rate and is frequently imposed for the optimization of the fatigue life of the springs.

The force in the axial direction is given by integration of the shear stress $\tau_{z\theta}$ over the cross-section Ω of the torus:

$$F = \int\limits_{\Omega} \tau_{z\theta} \mathrm{d}\Omega = \int\limits_{r_i}^{r_e} \left[\int\limits_{Z_-(r)}^{Z_+(r)} \tau_{z\theta}(r, z) \mathrm{d}z \right] \mathrm{d}r. \tag{2.58}$$

The use of expression (2.56) in (2.58) delivers the relationship between the magnitude of the Burgers vector of the giant screw dislocation and the axial force F:

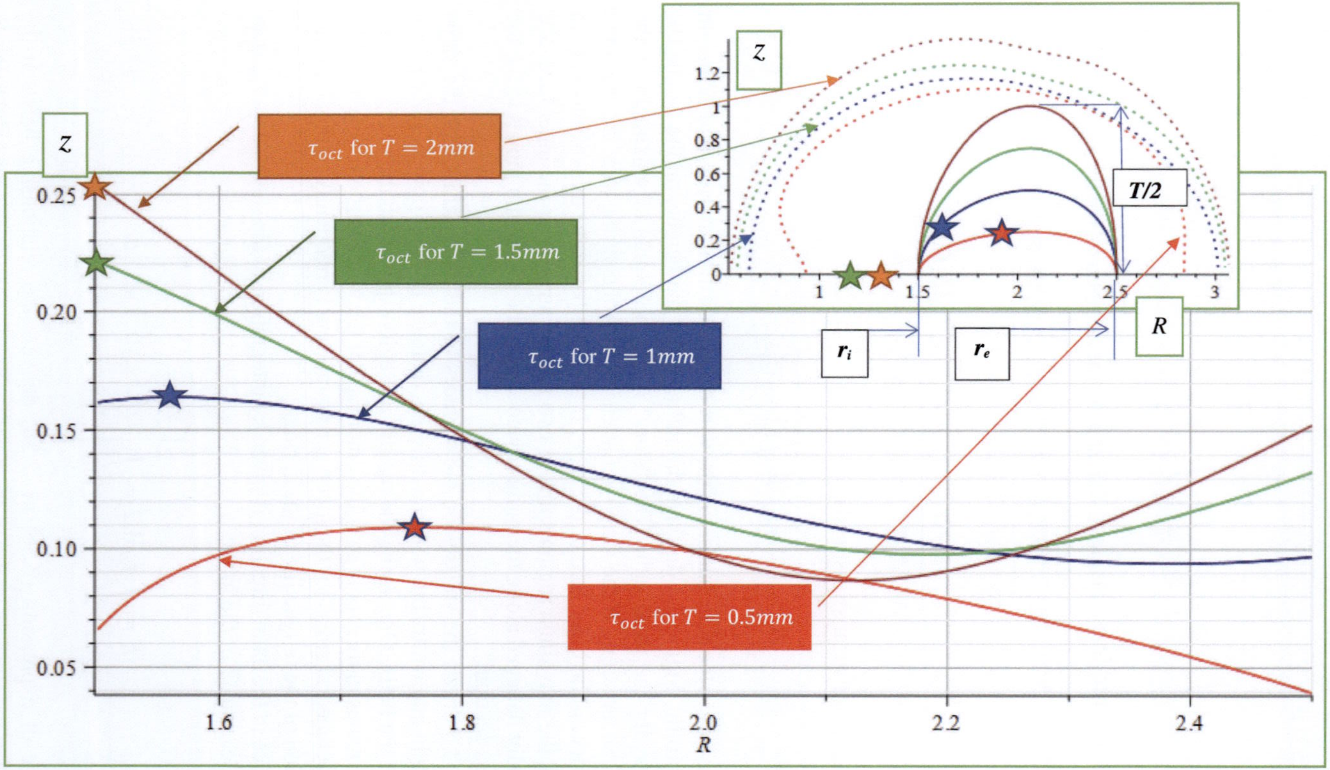

Fig. 2.2 Shearing stress τ_{oct} on the surface of wired with different heights. The contours with equal radii and different heights are shown in the upper sketch together with the polar diagrams for octahedral stresses on the wire surface

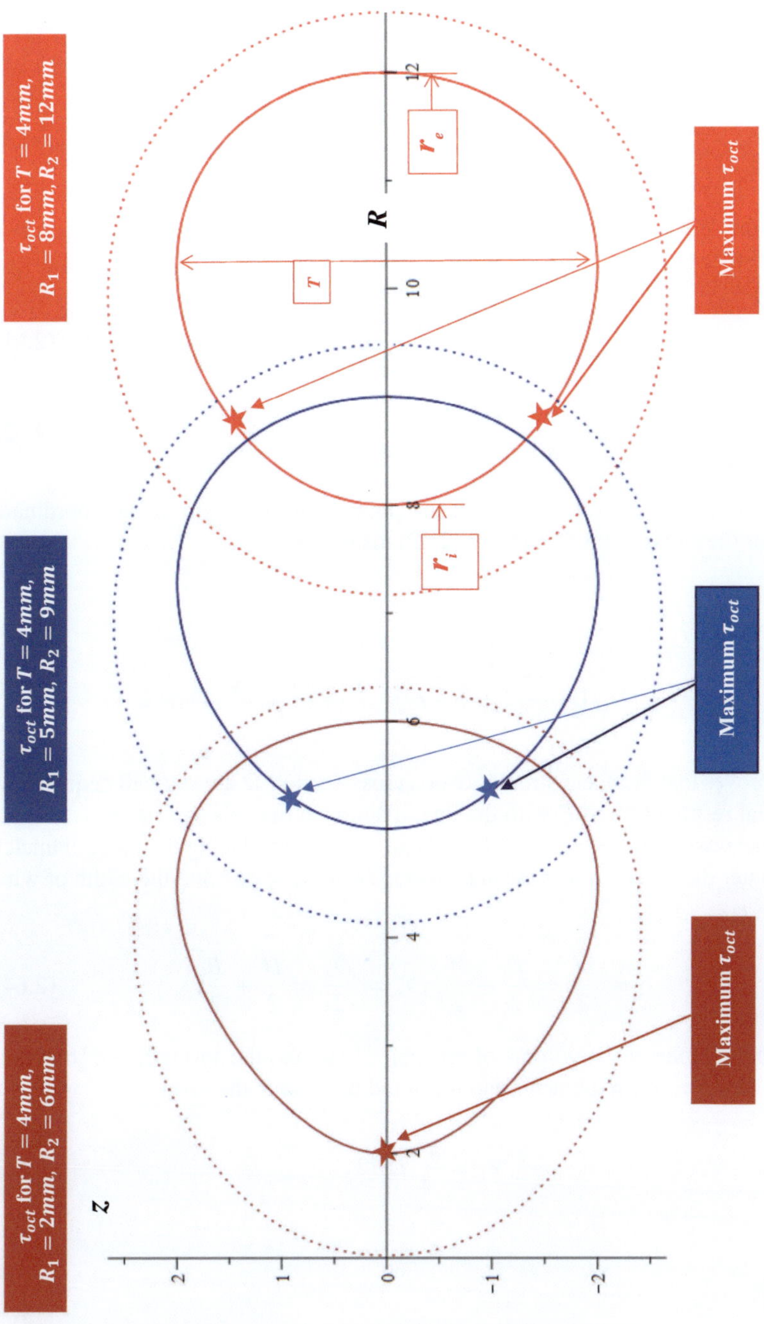

Fig. 2.3 Shearing stress on the surface of wired with different radii and the same height of cross-sections

$$F = \frac{GT^3\pi}{2(r_e + r_i)} \cdot \frac{(r_e - r_i)^3}{(r_e^2 - r_i^2)^2 + 2T^2(r_e^2 + r_i^2)} \cdot B_s. \tag{2.59}$$

The spring rate of one coil, according to (2.13) is given by:

$$c_w = \frac{F}{2\pi B_s} = \frac{GT^3}{4(r_e + r_i)} \cdot \frac{(r_e - r_i)^3}{(r_e^2 - r_i^2)^2 + 2T^2(r_i^2 + r_e^2)}. \tag{2.60}$$

After the replacement of the integration constant through the expression (2.58), the shear stresses from (2.55), (2.56) assume the form:

$$\tau_{r\theta} = \frac{4F}{\pi T^3} \cdot \frac{(r_i + r_e)^3}{r_e - r_i} \cdot \frac{z}{r^2}, \tag{2.61}$$

$$\tau_{z\theta} = \frac{4F}{\pi T} \cdot \frac{(r_i + r_e)}{(r_e - r_i)^3} \cdot \frac{r_i^2 + r_e^2 - 2r^2}{r}. \tag{2.62}$$

The resultant moment (2.12) about the point O at the origin of the coordinate system for the stress fields (2.61) and (2.62) disappears:

$$\begin{aligned} M_r &= \int_{r_i}^{r_e} \left[\int_{z_-(r)}^{z_+(r)} (\tau_{z\theta} r - \tau_{r\theta} z) \mathrm{d}z \right] \mathrm{d}r \\ &= \frac{8F}{3\pi \left(r_e^2 - r_i^2\right)^4} \int_{r_i}^{r_e} \sqrt{\left(r^2 - r_i^2\right)\left(r_e^2 - r^2\right)} \cdot \frac{5r^4 - 2r^2\left(r_e^2 + r_i^2\right) + r_i^2 r_e^2}{r^2} \mathrm{d}r = 0. \end{aligned} \tag{2.63}$$

This proves that the shear stresses over a cross-section Ω are statically equivalent to the axial resultant force F with the line of action at the axis z of helix.

One can rewrite the expressions (2.54) and (2.60), introducing the inner diameter D_i, the outer diameter D_e, the mean diameter D of the spring and the width of wire $B = r_e - r_i$:

$$r_i = \frac{D_i}{2} = \frac{D}{2} - \frac{B}{2}, \quad r_e = \frac{D_e}{2} = \frac{D}{2} + \frac{B}{2}. \tag{2.64}$$

The expressions of the volume of one coil (2.54) and the spring rate of one coil (2.60) in terms of mean diameter and mean radius assume the form

$$V_1 = \frac{\pi^2}{4} DBT, \tag{2.65}$$

$$c_w = \frac{T^3B^3}{T^2D^2 + D^2B^2 + T^2B^2} \cdot \frac{G}{4D}. \tag{2.66}$$

The spring rate of the spring with n_a active coils is c_w/n_a.

The values (2.65), (2.66) are useful in engineering applications for evaluation of weight and consequently, the fundamental frequencies of coil springs. The knowledge of fundamental frequencies is essential for simulation of dynamical behavior of the mechanical systems containing the springs as their parts (Köhler, 1998; Muhr, 1993).

2.6 Hollow Ovate Wire

Helical springs with hollow cross-section are formed by wrapping tube of uniform cross-section around a cylinder. One turn, or coil, of an undeformed helical spring becomes a hollow torus, generated by rotating the voided cross-section about the z axis of the cylindrical coordinate system. We obtain the closed form of analytical solution for the torsion problem of the incomplete torus with the hollow, or tubular, cross-section.

Consider the hollow ovate profile (Fig. 2.4) with the equation of the outer boundary Γ of the cross-section :

$$Z_+(r) = Z_0\sqrt{\left(r_e^2 - r^2\right)\left(r^2 - r_i^2\right)},$$
$$Z_-(r) = -Z_0\sqrt{\left(r_e^2 - r^2\right)\left(r^2 - r_i^2\right)}. \tag{2.66}$$

and the inner boundary γ of the cross-section

$$z_+(r) = \sqrt{Z_0^2\left(r_e^2 - r^2\right)\left(r^2 - r_i^2\right) - \kappa^2},$$
$$z_-(r) = -\sqrt{Z_0^2\left(r_e^2 - r^2\right)\left(r^2 - r_i^2\right) - \kappa^2}. \tag{2.67}$$

The coefficient κ varies in the interval

$$0 \le \kappa \le \frac{1}{2}Z_0\left(r_e^2 - r_i^2\right). \tag{2.68}$$

The shape of the hole can be rewritten in the form similar to the outer curve:

$$z_+(r) = Z_0\sqrt{\left(\rho_2^2 - r^2\right)\left(r^2 - \rho_1^2\right)},$$
$$z_-(r) = -Z_0\sqrt{\left(\rho_2^2 - r^2\right)\left(r^2 - \rho_1^2\right)}, \tag{2.69}$$

where ρ_1, ρ_2 are the real positive roots of the equation:

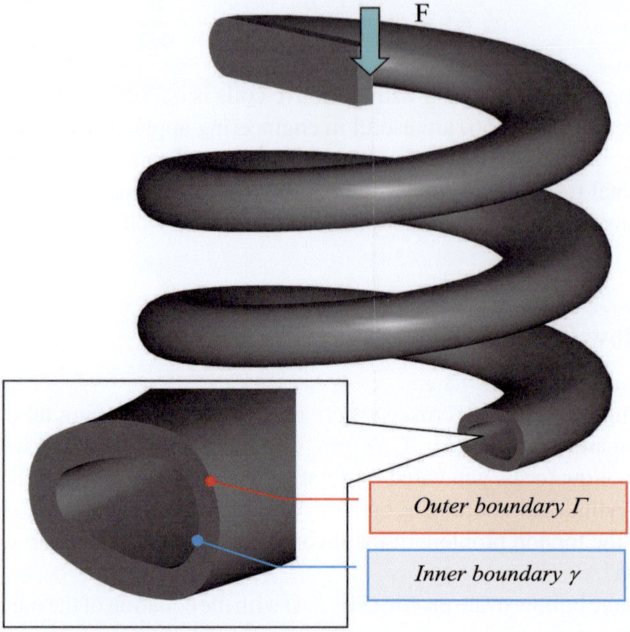

Fig. 2.4 A section of the spring with the hollow profile

$$Z_0^2\left(r_e^2 - r^2\right)\left(r^2 - r_i^2\right) = \kappa^2.$$

These roots are

$$\rho_1 = \frac{1}{\sqrt{2}}\sqrt{r_i^2 + r_e^2 - \sqrt{\left(r_i^2 - r_e^2\right)^2 - 4(\kappa/Z_0)^2}}, \tag{2.70}$$

$$\rho_2 = \frac{1}{\sqrt{2}}\sqrt{r_i^2 + r_e^2 + \sqrt{\left(r_i^2 - r_e^2\right)^2 - 4(\kappa/Z_0)^2}}. \tag{2.71}$$

The stress function for the hollow profile satisfies the boundary conditions on both boundaries of the profile:

$$\phi = 0 \quad \text{on } \Gamma, \tag{2.72}$$

$$\phi = const \quad \text{on } \gamma, \tag{2.73}$$

The stress function $\phi(r, z)$ from Eq. (2.52) satisfies the equilibrium equation over the region Ω. The isolines of stress function $\phi(r, z)$ match with the curves, given by the expression (2.67) with the parameter κ. Therefore, stress function (2.52) is constant along the inner boundary γ. This proves, that the solution of the problem for the hollow cross-section delivers once again the stress function $\phi(r, z)$.

Performing a double integration (2.54) one gets the expression for volume of one hollow coil:

$$V_1 = \frac{\pi^2 D}{4} \cdot (BT - B_i T_i). \tag{2.74}$$

Here

$T_i = Z_0(\rho_2^2 - \rho_1^2)$ is the height of the ovate opening,

$B_i = \rho_1 - \rho_2$ is the width of the ovate opening.

The spring rate for one hollow coil is given by the formula:

$$c_w = \left(\frac{T^3 B^3}{T^2 D^2 + D^2 B^2 + T^2 B^2} - \frac{T_i^3 B_i^3}{T_i^2 D^2 + D^2 B_i^2 + T^2 B_i^2} \right) \cdot \frac{G}{4D}. \tag{2.75}$$

2.7 Helical Spring Deformation Dislocation Character

2.7.1 *Screw and Edge Dislocations*

In the solid-state physics, the screw dislocation is a form of a line defect in which the defect occurs when the planes of atoms in the crystal lattice trace a helical path around the dislocation line (Hull & Bacon, 2011). The Burgers vector of a screw dislocation is parallel to the line of the dislocation, also to the axis of the helix. As it was already mentioned, the axial compression or extension of the helical spring leads to the deformation of the material in form of a screw giant dislocation. The line of the screw dislocation coincides with the axis of the spring. The magnitude of the Burgers vector is equal to the axial displacement of one coil. According to Eq. (2.1), the elastic medium of the spring wire deforms to single surface helicoid (Fig. 2.5). The case of axial compression of the spring, which causes the "giant" screw dislocation in the elastic helix, was already studied in Sect. 2.1. Interesting to note, that if the spring loading causes the plastic deformation of its material, the single giant dislocation of the whole spring splits into a huge number of microscopic edge dislocations of the material. Each microscopic dislocation possesses the Burgers vector of one distance between the atoms in crystal. The sum of all these microscopic Burgers vectors is equal of the giant Burgers vector due to plastic deformation of the helical spring.

In solid-state physics, the edge dislocation formed by inserting one or more extra half-planes of atoms. The Burgers vector of an edge dislocation is normal to the line of the dislocation. Similarly, the torsion M_θ of the helix leads to the deformation of spring material, which is equivalent to the giant edge dislocation. The line of dislocation is once again the helical axis. In this case, the magnitude of the Burgers vector is

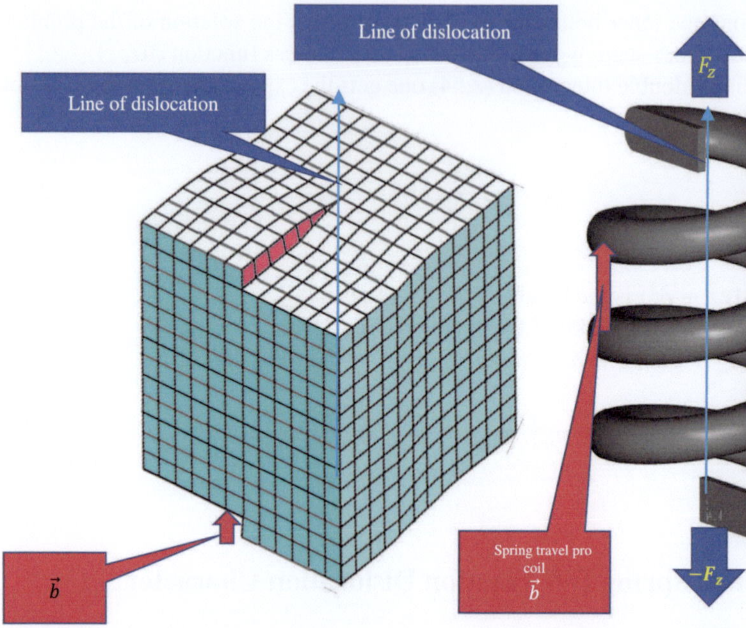

Fig. 2.5 Extension of the helical spring as a screw dislocation

equal to the opening of one coil due to action of torsional moment, which oriented in the direction of the spring axis (Fig. 2.6). The magnitude of Burgers vector is equal to the diameter change of helical spring Two forces, which enforce the increase of the diameter of the incomplete torus, lead to the same deformation of elastic medium, as the insertion of the extra matter in the gap of the torus- We examine below the case of torsion of the spring, which causes the "giant" edge dislocation in the elastic helix. Once again, if the spring undergo the plastic deformation, this giant edge dislocation splits into a massive number of the microscopic edge dislocations. These microscopic edge dislocations accompany the plastic deformation of the material. The sum of the Burgers vectors of microscopic edge dislocations is equal to the Burgers vector of the giant edge dislocation in course of the torsional plastic deformation of the spring.

In general case, both the axial force F and the axial moment M_θ act on the helix simultaneously. Because of axial symmetry of the spring, the dislocation line coincides with the axis of helix. The axial component of the Burgers vector is equal to the axial displacement pro one coil. The radial component of the Burgers vector is equal to the opening of the gap of an incomplete torus due to the twist of helix. Thus, Burgers vector of the giant dislocation lies at an arbitrary angle to the dislocation line. The giant dislocation has a mixed edge and screw character (Nabarro, 1967).

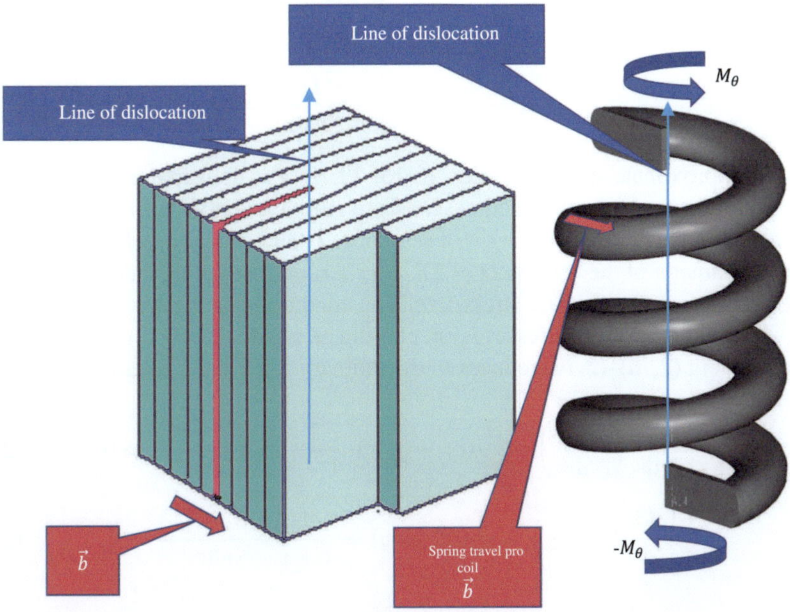

Fig. 2.6 Twist of the helical spring as an edge dislocation

2.7.2 *Torsion of Helical Spring*

We examine in this section the torsion of the helical spring, which causes the "macroscopic" edge dislocation in the elastic helix from the viewpoint of elasticity theory. The torsion was already studied from the perspective of beam theory in Chap. 1. However, the application of the beam theory is insufficient to the correct evaluation of the stress growth on the inner face of the wire. Consequently, for low spring index more accurate solutions are necessary for the precise stress analysis. If pitch vanishes, the torsion of the spring causes the bending of the wire (Michell, 1899). We follow the method of solution (Timoshenko & Goodier, 1951, Chap 13, Art.131) and consider the symmetric elastic stress distribution of an elastic toroidal sector subjected to pure bending under the effect of moments M_θ applied to the free ends. In the case of bending of wire, the first and second Beltrami-Michell Eq. (2.16) are identically satisfied. The third to sixth Beltrami-Michell equations for four appearing stresses $\sigma_r(r, z)$,$\sigma_\theta(r, z)$, $\sigma_z(r, z)$, $\tau_{rz}(r, z)$ read:

$$\begin{aligned} &\Delta\sigma_r - \tfrac{2}{r^2}(\sigma_r - \sigma_\theta) + \tfrac{1}{1+\nu}\tfrac{\partial^2 \mathbf{I}(\sigma)}{\partial r^2} = 0, \\ &\mathbf{I}(\boldsymbol{\sigma}) = \sigma_r + \sigma_\theta + \sigma_z, \end{aligned} \tag{2.76}$$

$$\Delta\sigma_\theta + \frac{2}{r^2}(\sigma_r - \sigma_\theta) + \frac{1}{1+\nu}\frac{\partial \boldsymbol{I}(\boldsymbol{\sigma})}{r\partial r} = 0, \tag{2.77}$$

$$\Delta\sigma_z + \frac{1}{1+\nu}\frac{\partial^2 \boldsymbol{I}(\boldsymbol{\sigma})}{\partial z^2} = 0, \tag{2.78}$$

$$\Delta\tau_{rz} - \frac{\tau_{rz}}{r^2} + \frac{1}{1+\nu}\frac{\partial^2 \boldsymbol{I}(\boldsymbol{\sigma})}{\partial r \partial z} = 0. \tag{2.79}$$

The approximate solution of Eqs. (2.76)–(2.79) was delivered in the above cited works for the circular cross-section. We search the solution for the rectangular wire, as shown on (Fig. 2.7). The height of section is T, and its width is $B = r_e - r_i$. The middle diameter of the helix is $D = 2R = r_e + r_i$ and the spring index is $w = R/B$. The surfaces of the wire are stress-free. The conditions on the free surfaces assure the independence of the stresses upon coordinate z : $\sigma_r(r)$,$\sigma_\theta(r)$, $\tau_{rz}(r)$ and $\sigma_z(r)$. The problem (2.76)–(2.79) reduces to the ordinary differential equations:

$$\frac{1}{r}\frac{\mathrm{d}\sigma_r}{\mathrm{d}r}\left(r\frac{\mathrm{d}\sigma_r}{\mathrm{d}r}\right) - \frac{2}{r^2}(\sigma_r - \sigma_\theta) + \frac{1}{1+\nu}\frac{\partial^2(\sigma_\theta + \sigma_r)}{\partial r^2} = 0, \tag{2.80}$$

$$\frac{1}{r}\frac{\mathrm{d}\sigma_\theta}{\mathrm{d}r}\left(r\frac{\mathrm{d}\sigma_\theta}{\mathrm{d}r}\right) + \frac{2}{r^2}(\sigma_r - \sigma_\theta) + \frac{1}{1+\nu}\frac{\partial(\sigma_\theta + \sigma_r)}{r\partial r} = 0, \tag{2.81}$$

$$\frac{1}{r}\frac{\mathrm{d}\sigma_z}{\mathrm{d}r}\left(r\frac{\mathrm{d}\sigma_z}{\mathrm{d}r}\right) = 0, \tag{2.82}$$

$$\frac{1}{r}\frac{\mathrm{d}\tau_{rz}}{\mathrm{d}r}\left(r\frac{\mathrm{d}\tau_{rz}}{\mathrm{d}r}\right) - \frac{\tau_{rz}}{r^2} = 0. \tag{2.83}$$

The solution of the Eqs. (2.80)–(2.83) reads:

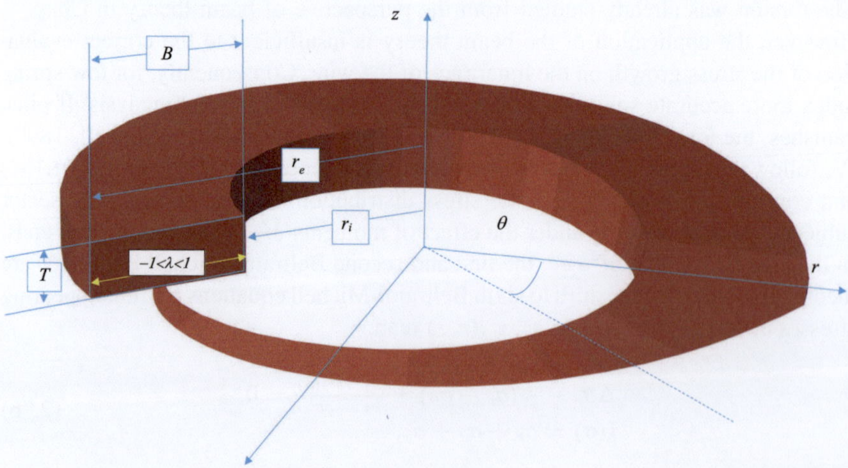

Fig. 2.7 Helical spring with the rectangular wire

$$\sigma_r = \frac{c_1 r^2 + c_2 \ln(r) r^2 + c_3 r^4 + c_4}{r^2},$$
$$\sigma_\theta = c_1 - c_4 r^2 + c_2 \frac{(1+\nu)\ln(r) + 1}{1+\nu} + \frac{c_6}{1+\nu} - \frac{c_3}{r^2}, \tag{2.84}$$
$$\tau_{rz} = c_7 r + \frac{c_8}{r}, \qquad \sigma_z = c_6 \ln(r) + c_5.$$

With the boundary conditions on the free surface the integration constants of (2.84) yield to:

$$c_1 = \frac{\ln(r_i) r_e^2 - \ln(r_e) r_i^2}{\ln(r_e) - \ln(r_i)} c_4 - \frac{\ln(r_i) r_i^2 - \ln(r_e) r_e^2}{r_i^2 r_e^2 (\ln(r_i) - \ln(r_e))} c_3, \tag{2.85}$$
$$c_2 = \frac{r_e^2 - r_i^2}{\ln(r_i) - \ln(r_e)} c_4 + \frac{r_e^2 - r_i^2}{r_i^2 r_e^2 (\ln(r_e) - \ln(r_i))} c_3,$$
$$c_3 = c_4 = c_5 = c_6 = c_7 = c_8 = 0.$$

Two integration constants c_3, c_4 should be determined from the load of the beam sector. The total force over the cross-section F_θ disappears. The bending moment is equal to the given torsion moment M_θ of the helical spring:

$$F_\theta = T \int_{r_i}^{r_e} \sigma_\theta \mathrm{d}r = 0, \quad M_\theta = T \int_{r_i}^{r_e} \sigma_\theta r \mathrm{d}r. \tag{2.86}$$

The linear algebraic Eq. (2.86) determine the constants c_3, c_4. The constants c_1, c_2. result from (2.85). The final formulas are bulky and we provide only the simpler expression the tangential stress in the case $\nu = 0$:

$$\sigma_\theta = \frac{4M_\theta}{Tr^2} \cdot \frac{\ln(r_i) r_i^2 (r^2 + r_e^2) - \ln(r_e) r_e^2 (r^2 + r_i^2) + r^2 (r_e^2 - r_i^2)(1 + \ln(\mathrm{r}))}{4 r_e^2 r_i^2 (\ln(r_e) - \ln(r_i))^2 - (r_e - r_i)^2 (r_e + r_i)^2}. \tag{2.87}$$

The next step is to determine the deformation of the wire and the Burgers constant for the edge dislocation. For derivation we use Castigliano's method (Teodorescu, 2013). The elastic energy stored in the spring 1.6 is the integral of bending energy of wire over its length L_W. With the expression for σ_θ, the elastic strain energy of one coil in the Timoshenko-Michell theory is:

$$U_M = T \int_{r_i}^{r_e} \frac{\sigma_\theta^2}{2E} 2\pi r \mathrm{d}r. \tag{2.88}$$

The spring rate pro one coil according to the Timoshenko-Michell theory is:

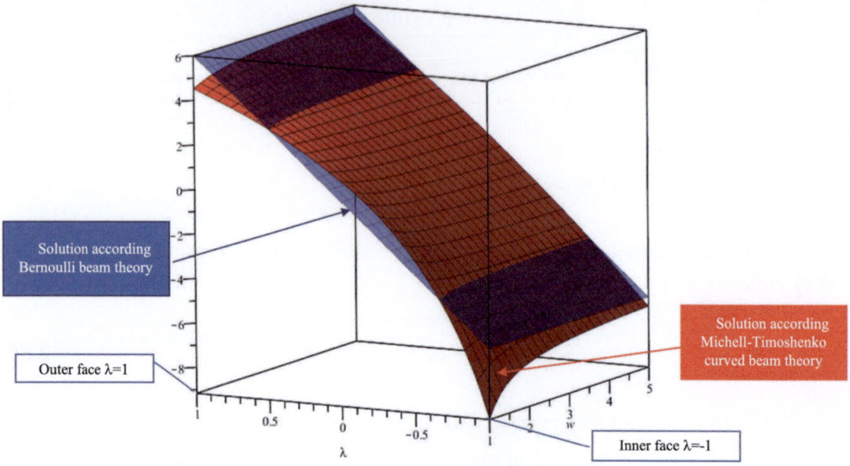

Fig. 2.8 **Tangential stress** $\sigma_\theta(r)$ as the function of spring index w

$$c_M = \left(\frac{\mathrm{d}^2 U_M}{\mathrm{d}M_\theta^2}\right)^{-1}. \tag{2.89}$$

The Bernoulli beam theory provides the expression for the tangential stress:

$$\sigma_\theta = \frac{12 M_\theta}{T\mathrm{B}^3} \cdot (r - R). \tag{2.90}$$

For comparison, on Fig. 2.8 we show the show the stresses in the cross-section of the wire for different spring indices. Two graphs display the Bernoulli function (2.89) in blue color and the Timoshenko- Michell function in red. The parameter λ runs in the interval $-1 \le \lambda \le 1$. This parameter is used instead of radius: $r = r_i + B \cdot (\lambda + 1)/2$. With the increasing spring index both stress distributions become indistinguishable.

With (2.87) the elastic strain energy of one coil and the spring rate in Bernoulli theory read:

$$U_B = \frac{12\pi M_\theta^2 w^3}{R^2 T E}, \quad c_B = \left(\frac{\mathrm{d}^2 U_B}{\mathrm{d}M_\theta^2}\right)^{-1} = \frac{R^2 T E}{24\pi w^3}. \tag{2.91}$$

If the axial moment M_θ applies to the spring, in the elastic material appears the edge dislocation. The magnitudes Burgers vector of the edge dislocation for both approaches will be correspondingly to $M_\theta R/c_M$ and to $M_\theta R/c_B$. These values correspond to the deflections of the cross-sections of one full coil under the action of torque M_θ.

The ratio of stored elastic energies U_M/U_B is shown on Fig. 2.9. Ratios of stored elastic energy according to Timoshenko- Michell and Bernoulli theories for different Poisson ratios are drawn as the functions of spring index $w = R/B$. Evidently, that the Poisson coefficient has no valuable influence on the stored energy. From the stored energies result the spring rates pro coil. Figure 2.10 demonstrates the ratios of the spring rates according to Michell-Timoshenko curved beam theory to Bernoulli-beam theory spring rate. The functions are shown for different Poisson coefficients for rectangular wire as the function of spring index w. One again, the influence of the Poisson ratio is almost negligible. The asymptotic expansions of the ratios of stored elastic energy and spring rates according to Timoshenko- Michell and Bernoulli theories read:

$$\frac{U_M}{U_B} = 1 - \frac{(2+\nu)(2+3\nu)}{60(1+\nu)^2} \cdot \frac{1}{w^2} + o\left(\frac{1}{w^2}\right),$$
$$\frac{c_M}{c_B} = 1 + \frac{(2+\nu)(2+3\nu)}{60(1+\nu)^2} \cdot \frac{1}{w^2} + o\left(\frac{1}{w^2}\right). \tag{2.92}$$

The stress correction factors could be defined in the similar manner, as for the axially compressed springs. For this purpose, we compare the tangential stresses according to the exact theory (2.87) and the stresses in the bending beam (2.90), as shown of Fig. 2.8. The first correction factor is the ratio of the to Michell-Timoshenko stress to the Bernoulli stress on the inner surface $r = r_i$. This stress correction factor is denoted k_{Mi} and it is always greater than 1. The second correction factor is the ratio of both stresses on the outer surface $r = r_e$. This stress correction factor is denoted k_{Me}.This value is less than 1 for all Poisson ratios and spring indices. Instead of providing the bulky formulas for both coefficients, we list only the asymptotic expansions for both factors:

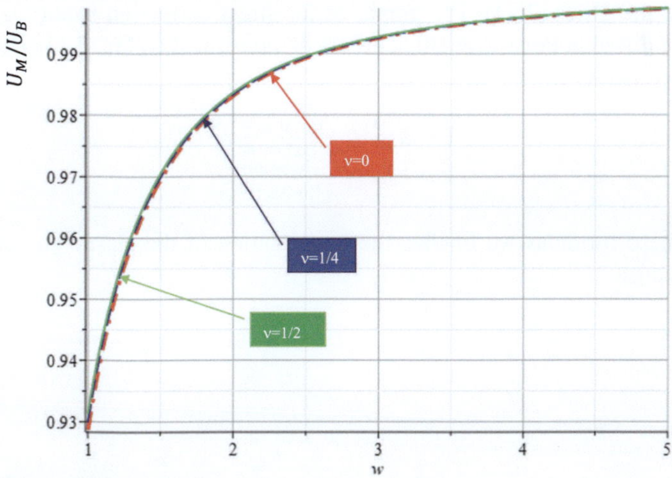

Fig. 2.9 Ratios of stored elastic energy according to Timoshenko- Michell and Bernoulli theories for different Poisson ratios as the functions of spring index $w = R/B$

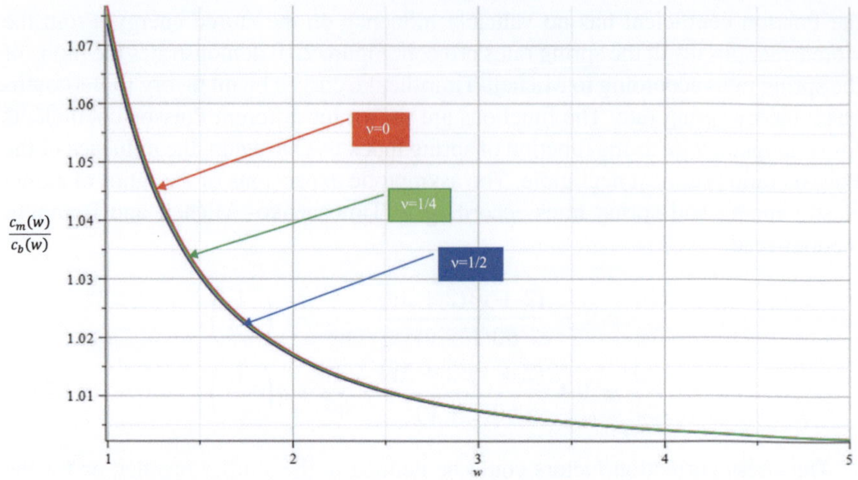

Fig. 2.10 Ratios of the spring rates for different Poisson coefficients according to Michell-Timoshenko curved beam theory to Bernoulli-beam theory spring rate for rectangular wire as the function of spring index w

$$k_{Mi} = 1 + \frac{\nu+2}{6(1+\nu)} \cdot \frac{1}{w} + \frac{4\nu+7}{60(1+\nu)} \cdot \frac{1}{w^2} + \frac{16\nu^2+29\nu+16}{360(1+\nu)^2} \cdot \frac{1}{w^3} + \cdots, \tag{2.93}$$

$$k_{Me} = 1 - \frac{\nu+2}{6(1+\nu)} \cdot \frac{1}{w} + \frac{4\nu+7}{60(1+\nu)} \cdot \frac{1}{w^2} - \frac{16\nu^2+29\nu+16}{360(1+\nu)^2} \cdot \frac{1}{w^3} + \cdots. \tag{2.94}$$

The only difference between the expressions (2.93) and (2.94) is the alternation of signs in the series (2.94). The graphs of the stress correction factor $k_{Mi}(w)$ and $k_{Me}(w)$ for different Poisson coefficients are demonstrated on Fig. 2.11. The bounds of both factors are:

$$\begin{aligned} 1 < k_{Mi} &\leq \frac{16-36\ln 3}{27(\ln 3)^2-48}, \\ 1 > k_{Me} &\geq \frac{-16+4\ln 3}{27(\ln 3)^2-48}. \end{aligned} \tag{2.95}$$

With these formulas we finalize the investigation of the elastic stresses in the helical springs.

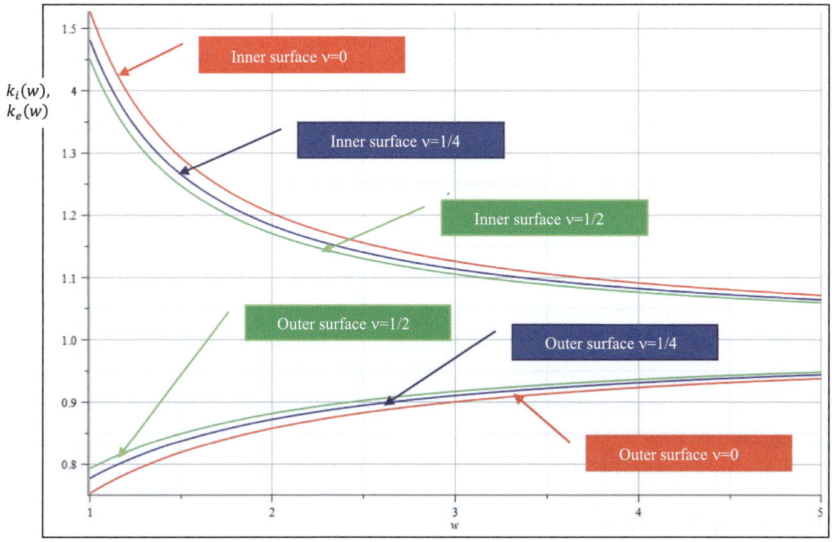

Fig. 2.11 Correction factors for tangential stress $\sigma_\theta(r)$ on the outer and inner surface of wire as the function of spring index w for different Poisson coefficients

2.8 Conclusions

In coil springs with the circular wires, the shear stresses are unevenly distributed over the circumference of the wire. Better utilization of the material is achieved by equalizing the shear stress across the surface of the wire. The idea of coil springs with non-circular, ovate wire cross section with uniform stress over its surface was first mentioned by Fuchs (1959). The initial cross-section of the wire, based on rough approximations, was improved by several studies, Nagaya (1986), Matsumoto (1988). Advanced applications require wires with ovate, optimized cross-sections. The shape optimization of the cross section was based exclusively on finite element methods and numerical optimization algorithms.

An important application of ovate wire valve springs is the valve train of internal combustion engines (Muhr, 1992). Valve springs in internal combustion engines provide a forced connection of all moving valve train components during the valve lift up to the maximum engine speed. Currently, a large proportion of all valve train springs are made from wire with a non-circular cross-section. This type of valve spring is mainly made of high-strength wire alloyed with chromium and vanadium. This environment makes accurate stress analysis of non-circular springs a practically important problem. The stress distributions for arbitrary cross-sections and shape optimization have been studied using finite and boundary element numerical methods (Sato, 1969). The closed-form solution of the torsion problem for torus can serve as a probe for numerical algorithms for calculation and optimization of helical springs, primarily valve springs (Kamiya & Kita, 1990).

As known from solid state physics, line defects are forms of crystallographic defects. There are two forms of these defects, known as edge dislocations and screw dislocations. The difference between screw and edge dislocations is as follows. The screw dislocation occurs when the planes of atoms in the crystal lattice follow a helical path around the dislocation line. The equivalent deformation of the elastic helix is the extension of its compression by the axial force. The edge dislocation occurs when there is an extra half-plane of atoms in the center of the crystal lattice, or when the helical spring or incomplete torus deforms under the action of forces that increase the distance between the cross sections of the torus. The action of these forces is statically equivalent to the moment that causes the bending of the wire. This moment is equal to the torque along the axis of the torus or helix.

When both axial force and axial moment act simultaneously on the helix, the Burgers vector is at an arbitrary angle to the dislocation line and the dislocation is a combination of edge and screw type. The study of dislocations is based on Michell's equations for the axially symmetric problems of elasticity.

Summary of Principal Results

- Stresses on the surface of spring with different shapes of the cross-section of wire
- Stresses for ovate, elliptical, and circular wires of solid and hollow cross-sections.
- Spring rate and volume of wire for the ovate wire.
- For the circular and elliptic wires, the solutions are expressed as the power series of spring index.
- Analogy between screw and edge dislocations and compression and torque deformation of helical spring

References

Biezeno, C. B., & Grammel, R. (1971). *Technische Dynamik* (Vol. 1). Springer-Verlag.

Fraeijs de Veubecke, B. M. (1979). *A Course in Elasticity*. Springer-Verlag.

Fuchs, H. O. (1959). *High efficiency coil springs with equalized stresses*. Metal Improvement Equipment Co.

Göhner, O. (1932). Die Berechnung zylindrischer Schraubenfedern. *Zeitschrift Des VDI, 76*, 269–272.

Hirth, J.P., & Lothe, J. (1991). *Theory of dislocations*. Krieger Publishing Company.

Hull, D., & Bacon, D. J. (2011). *Introduction to dislocations* (5th ed.). Butterworth-Heinemann.

Kamiya, N., & Kita, E. (1990). Boundary element method for quasi-harmonic differential equation with application to stress analysis and shape optimization of helical spring. *Computers and Structures, 37*, 81–86.

Kobelev, V. (2002). An Exact Solution of Torsion Problem for an Incomplete Torus with Application to Helical Springs. *Meccanica, 37*, 269–282.

Kobelev, V. (2016). Weakest link concept for springs fatigue. *Mechanics Based Design of Structures and Machines, 45*(4), 523–543. https://doi.org/10.1080/15397734.2016.1255851

Köhler, E. (1998). *Verbrennungsmotoren: Motorenmechanik, berechnung und Auslegung des Hubkolbenmotors*. Vieweg, Wiesbaden

Matsumoto, Y., Saito, H., & Morita K., (1988). *Wire for coiled spring.* United States Patent Number 4735403

Michell, J. H. (1899). The uniform torsion and flexure of incomplete torus, with application to helical springs', Proceedings London. *Mathematics Society 31*, 130–146.

Muhr, T. H. (1992). *Zur Konstruktion von Ventilfedern in hochbeanspruchten Verbrennungsmotoren.* Ph. D. Thesis, RWTH Aachen, Aachen

Muhr T. H. (1993). New technologies for engine valve springs. In *SAE Paper 930912*, New Engine Design and Engine Component Technology, SAE SP-972 199–209.

Nabarro, F. R. N. (1967). *Theory of crystal dislocations.* Oxford University Press.

Nagaya, K. (1986). Stress analysis of a cylindrical coil spring of arbitrary cross-section. (1st Report). *Bulletin of JSME, 29*, 1664–1678.

Pilgram, M. (1913). Die Berechnung zylindrischer Schraubenfedern. *Artilleristische Monatshefte, 79*, 68–88, *80*(1913) 133–156, *81*(1913) 221–239

Sneddon, I. N., & Berry D. S. (1958). The classical theory of elasticity. In: S. Flügge (Ed.). *Handbuch der Physik* (Vol. VI, pp. 1–126). Springer-Verlag.

SAE AE-22. (1997). *SAE fatigue design handbook*. SAE International, Warrendale.

Sato, M., Matsumoto, Y., Saito, J., & Morita, K. (1969). Stress in a coil spring of arbitrary cross-section. *Transactions Japan Society of Mechanical Engineering, 27*, 86–88.

Teodorescu, P. P. (2013). *Treatise on classical elasticity*. Springer.

Timoshenko, S., & Goodier, J. N. (1951). *Theory of elasticity*, 2nd Ed., McGraw-Hill Book Company, Inc.

Chapter 3
"Equivalent Columns" for Helical Spring

Abstract In this chapter, the coil spring is replaced by a flexible rod located along the axis of the coil. This rod has the same mechanical properties as the spring itself. Its bending, torsion, and compression stiffnesses are equal to the corresponding stiffnesses of the helical spring. This rod is known as the "equivalent column" of the coil spring. The "equivalent column" equations are much easier to handle than the original helical spring equations. The integral properties of the spring, such as axial and lateral stiffness, buckling loads, and natural frequencies, can be determined directly using the equivalent column equations. On the other hand, the local properties, such as stresses in the wire or contact forces, could only be evaluated with the more complicated equations of the helical elastic rod. In this chapter, the stability and transversal vibrations of the spring are studied from the unified point of view based on the "equivalent column" concept. Buckling refers to the loss of stability up to the sudden and violent failure of straight bars or beams under the action of compressive forces whose line of action is the column axis. This concept is applied to the stability of helical springs. An alternative method of approach is based on the dynamic criterion of spring stability. The equations for transverse (lateral) vibrations of compressed helical springs have been derived. This solution expresses the fundamental natural frequency of the lateral vibrations of the column as a function of the axial force, as well as the variable length of the spring. This chapter is the third section of the first part, which studies the springs from the design point of view.

Keywords Equivalent columns of helical springs · Bending of helical spring · Twist of helical springs · Stability of helical springs · Vibrations of springs

3.1 Static Stability Criteria of Helical Springs

A coil spring is a special form of spatially curved pillar. The center of each cross section is located on a helix. The helix is a curve that winds around the surface of a cylinder with a constant slope. In mechanical engineering, the helical spring is typically modeled as a massless, frequency-independent stiffness element. However,

V. Kobelev, *Fundamentals of Springs Mechanics*,
https://doi.org/10.1007/978-3-031-58584-5_3

for a typical suspension spring, these assumptions are only valid in the quasi-static case or at low frequencies. At higher frequencies, the influence of the spring's internal resonances increases and a detailed model is required. In addition, the valve springs of combustion engines are affected by periodic forces with frequencies comparable to the natural frequency of the spring. The massless spring model is not applicable for such excitations.

Accurate vibration and stability analysis based on the theory of spatially curved beams is relatively complicated and difficult for engineering applications. The simulation models provide high accuracy for vibration frequencies and buckling loads. However, the numerical models are not well suited for primary design purposes. Therefore, in most engineering applications, the traditional concept of equivalent column is used for stability and vibration analysis. To simplify the basic equations, the spring is replaced by an equivalent column (Collins et al., 2010, Ch. 14). It is the concept of equivalent column that forms the background of spring calculation in the industry (Encyclopedia of Spring Design, 2013; Helical Springs, 1974). The axis of the equivalent beam coincides with the hypothetical centerline of the spring. Instead of describing the displacement of the points of the coil, the displacement of the point mass at the same height of the center line is used. Such a column must account for axial compressibility and shear effects. The averaged axial stiffness and lateral stiffness of the helical spring are equal to the corresponding axial stiffness and lateral stiffness of the initially straight hypothetical beam that replaces the real helical wire.

The stiffness of the equivalent length of the column and with a circular cross-section was calculated by Grammel (1924), Biezeno and Koch (1925), Haringx (1948) or Dick (1942). There is a considerable discrepancy between the cited above and Ponomariev's solutions (Ponomarev, 1948). This discrepancy results from different estimation of the shear effect. This problem was investigated in detail by Ziegler (1982). He showed that for helical springs a so-called modified approach is the proper way of taking shear, or transverse, force into account, whereas Biezeno and Koch applied Engesser's approach.

The models of equivalent columns were also applied for study of buckling under combined compression and torsion (Ziegler & Huber, 1950; Satoh et al., 1988).

A more exact equivalent column for buckling of helical springs was introduced in Kruzelecki and Zyczkowski (1990). It accounts for the pitch angle and possible buckling in two planes. Non-linear compression rigidity, local bending and shear rigidities as well as lower bounds for the mean values of these rigidities were established.

The concept of an equivalent column was adopted by Skoczeń and Skrzypek (1992) in order to examine the bifurcation buckling of S-shaped bellows. The overall buckling of axially compressed bellows (axial force, internal pressure) in the presence of pre-buckling nonlinearities is investigated. Instability within the class of axisymmetric deformations (snap-through) was also discussed on the basis of the finite deflections and rotations theory. The geometrical limitation of large elastic deflections of the bellows is taken into account. Curves of axial buckling force versus number of segments were presented for various types of bellows segments.

The linearized disturbance equations governing the static buckling behavior of circular-bar helical springs subjected to combined compression and twist were solved in Chassie et al. (1997) numerically using the transfer matrix method to produce buckling design charts. The effects of the number of turns of the spring and the angle of twist were investigated for clamped–clamped ends and the results were compared to existing theories of instability.

In the article Tabarrok and Xiong (1992), the governing equations for the static buckling of spatial rods were derived by considering perturbations about the critical state. These equations involve the curvature and twist of the rod which in general were different from those of the initial, i.e. unloaded, geometry. An incremental procedure is outlined for updating the rod's geometry up to the critical state. Other generalities incorporated into the analysis include the effect of the initial bending moments and transverse forces as well as the axial loads. Based on the theory outlined it is shown how an exact stiffness matrix and an approximate geometric stiffness matrix may be developed for a curved and twisted rod.

By Lee et al. (2009), the dynamic characteristics of the shape memory alloy helical spring proposed to be used in the semi-active suspension platform were considered. The component mounted on the platform generates periodic oscillation due to parametric excitation, such as rotation of an eccentric mass. In this way, the induced vibration of the platform could load the suspension spring in both axial and lateral directions. In the cited paper, the spring constant of the helical spring in transverse deformation was first derived by employing the first theorem of Castigliano's. The derived spring constant was then used to define the equivalent flexural stiffness of the spring.

The equivalent rod was applied for calculations of natural lateral vibrations of springs (Michalczyk, 2015). The model allows one to calculate natural frequencies of the clamped–clamped spring. It was shown that models based on the equivalent beam concept are easier to apply than the models treating the spring as the spatially curved rod and these models provide sufficient accuracy for practical calculations.

3.2 Static "Equivalent Column" Equations

The simplest way to demonstrate the essence of the dynamic stability method consists in its application to the traditional concept of the equivalent column.

On the coil spring and the hypothetical column act the equal force and moment. Under these conditions, the axial s and lateral s_Q displacement of the coil correspond the displacements of the hypothetical beam. Therefore, the hypothetical beam is also referred to as equivalent beam. An equivalent beam deforms exactly like the coil spring under the same load (Fig. 3.1).

Well known, that the Bernoully's hypothesis for the equivalent beams as used by Grammel or Dick leads to considerable overestimation of flexural stiffness of the spring. The Timoshenko-type model is an extension of the Euler–Bernoulli model by considering two additional effects: shearing force effect and rotary motion effect.

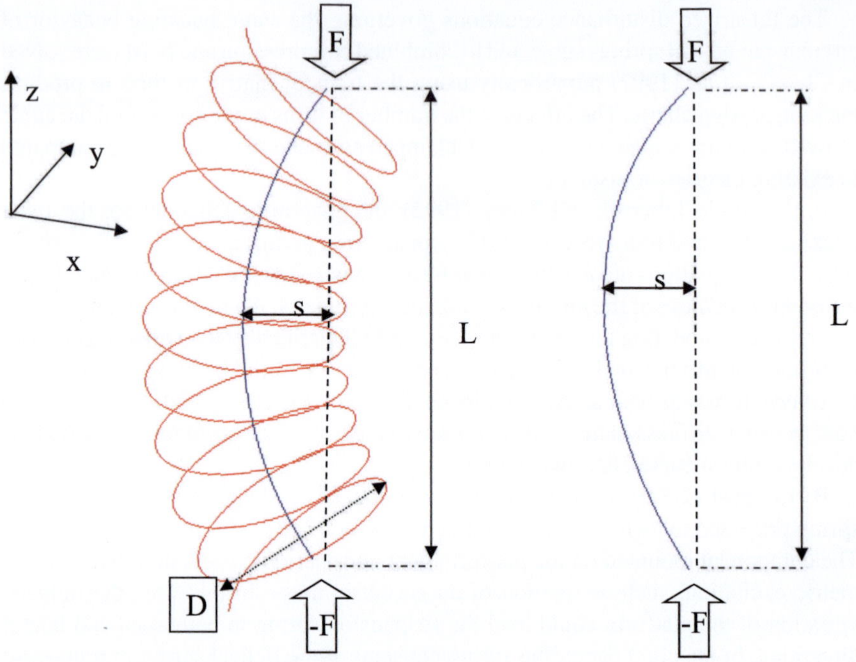

Fig. 3.1 Helical spring (red) and its "equivalent column" (blue)

Consequently, the equations governing the helical spring to the equivalent beam have to be studied in the framework of Timoshenko-type beam theory for the equivalent column.

The centerline of the equivalent to helical spring column is of variable length L and coincides with the z axis.

In static case, the equivalent column is subjected to a static axial load. If the spring is loaded with the lateral force F and the bending moment m_B, the original straight axis of the equivalent column turns to be curved. The angle of inclination of the bent axis can be found by means of the derivation of the lateral movement in a linear approximation:

$$\varphi_Q = \frac{\mathrm{d}s_Q}{\mathrm{d}z}.$$

In any beam except one subject to pure bending only, a displacement due to the shear stress occurs. The solution to the vibration problem requires this displacement to be considered. This total transversal displacement is the sum of two partial displacements:

$$s_Q = s_b + s_s. \tag{3.1}$$

In this equation the following notations for transverse displacements are used:

s_s is the displacement caused by the transverse force,
s_b is the displacement caused by bending moment.

The static equations of the equivalent column are:

$$M_B = \langle EI_B \rangle \frac{\mathrm{d}^2 (s_Q - s_s)}{\mathrm{d}z^2} + F \cdot s_Q, \tag{3.2}$$

$$Q = \langle GS \rangle \frac{\mathrm{d}s_s}{\mathrm{d}z} - F \frac{\mathrm{d}(s_Q - s_s)}{\mathrm{d}z}, \tag{3.3}$$

$$Q + \frac{\mathrm{d}M_B}{\mathrm{d}z} = m_B, \tag{3.4}$$

$$\frac{\mathrm{d}Q}{\mathrm{d}z} = f_Q. \tag{3.5}$$

Here

Q is the shear, or transverse, force
M_B is bending moment
m_B is the external torque per unit length (in y-direction),
f_Q is the external load in the transverse direction per unit length (in the x-direction).

The static deflection of a straight column with the equivalent bending stiffness $\langle EI_B \rangle$, and shear stiffness $\langle GS \rangle$under the action of the torque m_B and external force f_Q is described by Eqs. (3.2)–(3.5).

The first derivation of the equivalent shear, bending and axial compression stiffness was performed in the cited papers of Grammel and Dick. The formulas of Ponomarev and Andreeva (1980) are reasonably general and are valid for an arbitrary wire of noncircular cross-section. The shear, bending and axial compression rigidities of beam, which is equivalent to the helical coil spring, are presented in Table 3.1. The brackets express the fact, that the formulas represent the averaged stiffness of the hypothetical equivalent beam. Mass per unit length $\langle m \rangle$ is a mass related to the unit length of the coil axis of the spring. The stiffness of spring wire in the case of bending in normal direction EI is assumed to be different from the stiffness of wire EI_r by the bending in the direction of binormal. A is the area of the wire cross-section, GI_T is the twist stiffness of wire with respect to its axis, d is the diameter of round wire, D is the mean diameter of coil, n is the number of active coils, ρ_m is the density of the material. The geometric properties of various cross-sections of wires are given in Tables 1.3 and 1.4. Typical deformations of the spring, considered as an "equivalent column", under the action of axial and transverse forces and a bending moment are shown in Table 3.2.

Table 3.1 Effective rigidities and mass of the cylindrical helical spring for shear, bending and compression (I,I_r,I_T in Table 1.3)

	Arbitrary wire	Round wire	Unit
Shear stiffness $\langle GS\rangle$	$\frac{8EI_rL}{\pi n_a D^3}$	$\frac{Ed^4L}{8n_a D^3}$	N
Bending stiffness $\langle EI_B\rangle$	$\frac{2EIL}{\pi n_a D\left(1+\frac{EI}{GI_T}\right)}$	$\frac{Ed^4L}{32(2+\nu)n_a D}$	Nm^2
Axial stiffness $\langle ES\rangle \equiv c_F L$	$\frac{4GI_T}{\pi n_a D^3}L$	$\frac{Gd^4}{8n_a D^3}L$	N
Mass per unit length $\langle m\rangle$	$\frac{\rho_m \pi A n_a D}{L}$	$\frac{\rho_m \pi^2 d^2 n_a D}{4L}$	$\frac{\text{kg}}{\text{m}}$

Table 3.2 Deformation of an "equivalent column" under the action of axial load, transverse force and bending moment

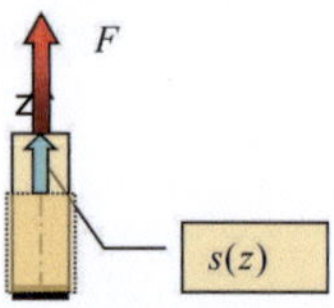	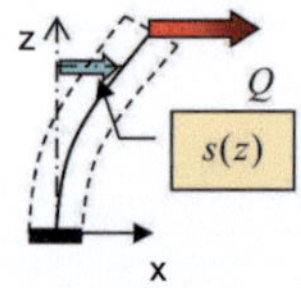	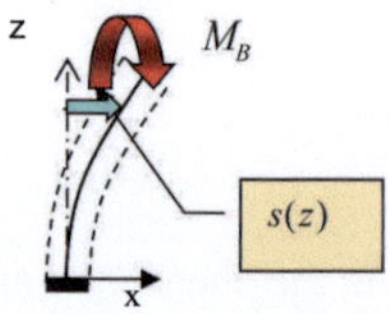
Axial load F	Shear load Q	Bending moment M_B
$s(z) = \frac{F}{\langle ES\rangle}z$	$s(z) = \frac{Q}{6\langle EI_B\rangle}z^2(3L-z)$	$s(z) = \frac{M_B}{2\langle EI_B\rangle}z^2$

3.3 Dynamic "Equivalent Column" Equations

Consider now the equations of the equivalent column in dynamic case. Damping of material is not accounted. The factor that affects the lateral vibration of the beam, neglected in Euler–Bernoulli's model, is the fact that each section of the beam rotates slightly in addition to its lateral motion when the beam deflects. The influence of the beam section rotation is considered through the moment of inertia, which modifies the equation of moments acting on an infinitesimal beam element. The equivalent mass per length element of the hypothetical beam matches the mass spiral spring, distributed over the axis of the spring. The bending moment m_B is the product of static moment of coil

$$\frac{1}{8}\langle m\rangle D^2$$

and the angular acceleration

$$\frac{\partial^3 s_Q}{\partial z \partial t^2},$$

such that the bending moment reads:

$$m_B = \langle m \rangle \frac{D^2}{8} \cdot \frac{\partial^3 s_Q}{\partial z \partial t^2}. \tag{3.6}$$

Accordingly, the transverse force is the mass of the coil multiplied by its acceleration:

$$f_Q = \langle m \rangle \cdot \frac{\partial^2 s_Q}{\partial t^2}. \tag{3.7}$$

Substitution of (3.6) and (3.7) into Eqs. (3.2)–(3.5) delivers the corresponding equations with partial derivatives:

$$M_B = \langle EI_B \rangle \cdot \frac{\partial^2 (s_Q - s_s)}{\partial z^2} + F \cdot s_Q, \tag{3.8}$$

$$Q = \langle GS \rangle \cdot \frac{\partial s_s}{\partial z} - F \cdot \frac{\partial (s_Q - s_s)}{\partial z}, \tag{3.9}$$

$$Q + \frac{\partial M_B}{\partial z} = \langle m \rangle \frac{D^2}{8} \cdot \frac{\partial^3 s_Q}{\partial z \partial t^2}, \tag{3.10}$$

$$\frac{\partial Q}{\partial z} = \langle m \rangle \cdot \frac{\partial^2 s_Q}{\partial t^2}. \tag{3.11}$$

The unknowns of four partial differential Eqs. (3.8) and (3.11) are the functions:

$$M_B, Q, s_Q, s_S.$$

In terms of s_Q, s_s Eqs. (3.8)–(3.11) reduce to partial differential equations:

$$\langle GS \rangle \cdot \tfrac{\partial s_s}{\partial z} - F \cdot \tfrac{\partial (s_Q - s_s)}{\partial z} + \tfrac{\partial}{\partial z}\Big[\langle EI_B \rangle \cdot \tfrac{\partial^2 (s_Q - s_s)}{\partial z^2} + F \cdot s_Q\Big] = \langle m \rangle \tfrac{D^2}{8} \tfrac{\partial^3 s_Q}{\partial z \partial t^2},$$

$$\tfrac{\partial}{\partial z}\Big[\langle GS \rangle \cdot \tfrac{\partial s_s}{\partial z} - F \cdot \tfrac{\partial (s_Q - s_s)}{\partial z}\Big] = \langle m \rangle \tfrac{\partial^2 s_Q}{\partial t^2}.$$

The transverse vibration is represented by two coupled differential equations of second order in time. These equations could be reduced to one differential equation for the entire lateral deformation as an unknown function $s_Q(z, t)$. To account the boundary conditions the expressions for bending moment and transverse force as functions of total lateral deformation of hypothetical axis are required. From Eqs. (3.8) to (3.11) follows:

$$\frac{\partial s_s}{\partial z} = \frac{1}{\langle GS \rangle + F} \cdot \left(Q + F \cdot \frac{\partial s_Q}{\partial z} \right),$$

$$M_B = \frac{\langle EI_B\rangle\langle GS\rangle}{\langle GS\rangle + F} \cdot \frac{\partial^2 s_Q}{\partial z^2} - \frac{\langle EI_B\rangle\langle m\rangle}{\langle GS\rangle + F} \cdot \frac{\partial^2 s_Q}{\partial t^2} + F \cdot s_Q,$$

$$Q = \left[\langle m\rangle\frac{D^2}{8} - \frac{\langle EI_B\rangle\langle m\rangle}{\langle GS\rangle + F}\right] \cdot \frac{\partial^3 s_Q}{\partial z \partial t^2} - \frac{\langle EI_B\rangle\langle GS\rangle}{\langle GS\rangle + F} \cdot \frac{\partial^3 s_Q}{\partial z^3} - F \cdot \frac{\partial s_Q}{\partial z}.$$

These equations also include as required, only the total lateral deformation s_Q as an unknown.

After grouping the terms the basic equation for the transverse vibrations of axially pre-stressed coil spring reads:

$$\frac{\langle EI_B\rangle\langle GS\rangle}{\langle GS\rangle+F} \cdot \frac{\partial^4 s_Q}{\partial z^4} - \langle m\rangle \cdot \left(\frac{\langle EI_B\rangle}{\langle GS\rangle+F} + \frac{D^2}{8}\right) \cdot \frac{\partial^4 s_Q}{\partial z^2 \partial t^2} + \langle m\rangle\frac{\partial^2 s_Q}{\partial t^2} + F\frac{\partial^2 s_Q}{\partial z^2} = 0. \quad (3.12)$$

With these equations, the problem of transverse vibration of helical spring reduces to the problem of vibrations of Timoshenko-type beam. Equation (3.12) is a linear partial differential equation of fourth order in coordinate and second order in time.

The boundary values are determined by the static conditions on the spring ends. There are two boundary conditions at each end (Table 3.3).

In the paper Majkut (2009) was shown, that the solution form of the vibration differential equation of Timoshenko-type beam depends on the examined vibration frequency. The change of the solution form occurs when the frequency crosses a specific value. This value is known from literature as the cut-off frequency (Chan et al., 2002; Stephen & Puchegger, 2006). This phenomenon could be immediately studied for problem under consideration. Namely, displacement, caused by the transverse force, is accounted at the spring ends. For the simplification of the mathematical treatise the boundary conditions for moment-free, hinged ends are investigated:

$$s_Q\big|_{z=0} = 0,$$

Table 3.3 Conditions at the ends of the spring in terms of transverse force, moment and total transversal displacement

Type of clamping		Condition 1		Condition 2	
1. Fixed	Clamped end deflection	Zero deflection	$s_Q = 0$	Zero slope	$s'_Q = 0$
2. Free	Free end transverse force	Zero force	$Q = 0$	Zero moment	$M_B = 0$
3. Hinged	Moment-free end (ball joint)	Zero deflection	$s_Q = 0$	Zero moment	$M_B = 0$
3. Sliding	Lateral force-free, sliding end	Zero slope	$s'_Q = 0$	Zero force	$Q = 0$

$$\left[\frac{\langle EI_B\rangle\langle GS\rangle}{\langle GS\rangle + F}\cdot\frac{\partial^2 s_Q}{\partial z^2} - \frac{\langle EI_B\rangle\langle m\rangle}{\langle GS\rangle + F}\cdot\frac{\partial^2 s_Q}{\partial t^2} + F\cdot s_Q\right]_{z=0} = 0,$$

$$s_Q\big|_{z=L} = 0,$$

$$\left[\frac{\langle EI_B\rangle\langle GS\rangle}{\langle GS\rangle + F}\cdot\frac{\partial^2 s_Q}{\partial z^2} - \frac{\langle EI_B\rangle\langle m\rangle}{\langle GS\rangle + F}\cdot\frac{\partial^2 s_Q}{\partial t^2} + F\cdot s_Q\right]_{z=L} = 0.$$

The following initial values for displacement and velocity are assumed:

$$s_Q(z, t = 0) = s_o(z) \text{ for } 0 < z < L, \tag{3.13a}$$

$$\dot{s}_Q(z, t = 0) = v_o(z) \text{ for } 0 < z < L, \tag{3.13b}$$

$$s_o\big|_{z=0} = s_o\big|_{z=L} = 0, \tag{3.13c}$$

$$v_o\big|_{z=0} = v_o\big|_{z=L} = 0. \tag{3.13d}$$

Here $s'_o(z)$ is the initial transverse deflection and $v_o(z)$ the velocity in the time $t = 0$.

The zero boundary conditions for the displacement simplify the zero boundary conditions for moment at the end section:

$$s_Q\big|_{z=0} = 0, \quad \frac{\langle EI_B\rangle\langle GS\rangle}{\langle GS\rangle + F}s''_Q\big|_{z=0} = 0, \tag{3.14a}$$

$$s_Q\big|_{z=L} = 0, \quad \frac{\langle EI_B\rangle\langle GS\rangle}{\langle GS\rangle + F}s''_Q\big|_{z=L} = 0. \tag{3.14b}$$

Therefore, the zero boundary conditions for moment at the end section require the secondary derivative at the end to vanish:

$$s''_Q\big|_{z=0} = s''_Q\big|_{z=L} = 0.$$

3.4 Natural Frequency of Transverse Vibrations

The transverse vibration is represented by a differential equation of fourth order in place and second order in time. The Fourier method of variable separation is employed to find function $s_Q(z, t)$ satisfying Eq. (3.12). It is assumed that the function can be presented in the form of a product of a function dependent on the spatial coordinate and a function dependent on time. The solution of the model Eq. (3.12) without damping could be obtained by separation of variables:

$$s_Q(z,t) = Z(z)T(t). \tag{3.15}$$

After separation of variables Eq. (3.12) reads:

$$\frac{\langle EI_B\rangle\langle GS\rangle}{\langle GS\rangle+F} Z^{IV}T - \langle m_t\rangle\left(\frac{\langle EI_B\rangle}{\langle GS\rangle+F} + \frac{D^2}{8}\right)Z''\ddot{T} + \langle m\rangle Z\ddot{T} + FZ''T = 0 \tag{3.16}$$

with the common notation for ordinary derivatives:

$$\dot{T} = \frac{\mathrm{d}T}{\mathrm{d}t}, \quad Z' = \frac{\mathrm{d}Z}{\mathrm{d}z}.$$

After separation of variables in (3.16), the following condition established:

$$\frac{1}{\langle m\rangle}\frac{\frac{\langle EI_B\rangle\langle GS\rangle}{\langle GS\rangle+F}Z^{IV} + FZ''}{\left(\frac{\langle EI_B\rangle}{\langle GS\rangle+F} + \frac{D^2}{8}\right)Z'' - Z} = -\frac{\ddot{T}}{T} = \omega^2. \tag{3.17}$$

This condition can be satisfied only for a certain constant ω^2.
Equation (3.17) is separated into two ordinary differential equations:

(a) The eigenvalue problem:

$$\frac{\langle EI_B\rangle\langle GS\rangle}{\langle GS\rangle+F}\cdot Z^{IV} + F\cdot Z'' = \omega^2\cdot\langle m\rangle\cdot\left[\left(\frac{\langle EI_B\rangle}{\langle GS\rangle+F} + \frac{D^2}{8}\right)Z'' - Z\right]; \tag{3.18}$$

(b) The evolution equation:

$$\ddot{T} + \omega^2 T = 0. \tag{3.19}$$

For solution of the eigenvalue problem (a) the representation is used:

$$Z(z) = C\exp(\Lambda z).$$

With this substitution, Eq. (3.18) leads to:

$$\left\{\Lambda^4\frac{\langle EI_B\rangle\langle GS\rangle}{\langle GS\rangle+F} + \Lambda^2 F - \omega^2\langle m\rangle\left[\left(\frac{\langle EI_B\rangle}{\langle GS\rangle+F} + \frac{D^2}{8}\right)\Lambda^2 - 1\right]\right\}C\exp(\Lambda z) = 0.$$

Since

$$\exp(\Lambda z) > 0,$$

This algebraic equation can be satisfied only if:

$$\Lambda^4 \frac{\langle EI_B\rangle\langle GS\rangle}{\langle GS\rangle+F} + \Lambda^2 F = \omega^2\langle m\rangle\left[\left(\frac{\langle EI_B\rangle}{\langle GS\rangle+F} + \frac{D^2}{8}\right)\Lambda^2 - 1\right]. \tag{3.20}$$

The quartic Eq. (3.20) has four solutions:

$$\Lambda = \pm i\beta_1, \quad \Lambda = \pm i\beta_2.$$

The solution to the problem (a) is the linear combination of the partial solutions:

$$Z(z) = c_1 \sin(\beta_1 z) + c_2 \cos(\beta_1 z) + c_3 \sin(\beta_2 z) + c_4 \cos(\beta_2 z). \tag{3.21}$$

On both sides the conditions of hinged ends are used. Displacement, caused by the transverse force, is supposedly omitted at the spring ends. The conditions for bending moment at the hinged ends

$$M_B|_{z=0} = 0, \quad M_B|_{z=L} = 0.$$

in terms of functions $Z(z)$, $T(t)$ transform to:

$$\frac{\langle EI_B\rangle\langle GS\rangle}{\langle GS\rangle+F} T(t)\, Z''\Big|_{z=0} = 0, \quad \frac{\langle EI_B\rangle\langle GS\rangle}{\langle GS\rangle+F} T(t)\, Z''\Big|_{z=L} = 0.$$

The boundary conditions for the hinged end read in this case:

$$Z(0) = Z''(0) = Z(L) = Z''(L) = 0. \tag{3.22}$$

This simplification will be omitted in the final paragraph.

The use of (3.21) into (3.22) gives the following system of homogeneous equations:

$$\begin{bmatrix} 0 & 1 & 0 & 1 \\ 0 & -\beta_1^2 & 0 & -\beta_2^2 \\ \sin\beta_1 L & \cos\beta_1 L & \sin\beta_2 L & \cos\beta_2 L \\ -\beta_1^2 \sin\beta_1 L & -\beta_1^2 \cos\beta_1 L & -\beta_1^2 \sin\beta_2 L & -\beta_1^2 \cos\beta_2 L \end{bmatrix} \begin{bmatrix} c_1 \\ c_2 \\ c_3 \\ c_4 \end{bmatrix} = \begin{bmatrix} 0 \\ 0 \\ 0 \\ 0 \end{bmatrix}. \tag{3.23}$$

The homogeneous system of Eq. (3.23) only leads to non-trivial solutions if its determinant is zero:

$$\left(\beta_1^2 - \beta_2^2\right)^2 \sin(\beta_1 L)\sin(\beta_2 L) = 0. \tag{3.24}$$

The condition (3.18) is satisfied if:

$$\beta_{1,2} = \lambda_N \equiv \pi \frac{N}{L} \text{ with } N = 1, 2, 3... \tag{3.25}$$

The circular natural frequencies resulting to be:

$$\omega_N{}^2 = \frac{\left(\frac{\pi N}{L}\right)^2}{\langle m\rangle} \cdot \frac{\left(\frac{\pi N}{L}\right)^2 \cdot \frac{\langle EI_B\rangle\langle GS\rangle}{\langle GS\rangle+F} - F}{\left(\frac{\langle EI_B\rangle}{\langle GS\rangle+F} + \frac{D^2}{8}\right) \cdot \left(\frac{\pi N}{L}\right)^2 + 1} \quad \text{with } N = 1, 2, 3... \tag{3.26}$$

The corresponding mode shapes:

$$Z_N(z) = c_1 \sin\lambda_N z.$$

Two dimensionless parameters are introduced: the slenderness ratio of the free spring as:

$$\zeta = \frac{L_0}{D}.$$

and relative dimensionless length as

$$\mu = \frac{L}{L_0}.$$

The slenderness ratio of the deformed spring is equal to: $L/D = \zeta\mu$.

The dimensionless deflection μ characterizes the degree of compression of the spring. For the free length of the spring

$$\mu = 1.$$

The flat state of spring corresponds to

$$\mu = 0.$$

In this - although pure imaginary- state the spring is completely flattened. Though, the real springs cannot be compressed to this state due to coils clash.

The axial spring force is a function of the actual spring length L. The axial spring force can be also expressed as the function of dimensionless deflection:

$$F = c_F \cdot (L_0 - L) \equiv \langle ES\rangle \cdot \left(\mu^{-1} - 1\right). \tag{3.27}$$

After inserting the related rigidities in (3.26) the natural frequency ω_n could be obtained as a function of the relative dimensionless length μ and slenderness ratio ζ.

Natural frequency is the function of the deflection dimensionless μ and slenderness ratio ζ (Kobelev, 2014):

$$\omega_N = \frac{\zeta^2 N\mathrm{d}}{n_a L_0^2}\sqrt{\frac{4G\mu}{\rho_m}}\sqrt{\frac{\zeta^2(2+\nu)(\mu-1)(2\mu\nu+\mu+1)+(1+\nu)^2\pi^2N^2}{8\zeta^2(2\mu\nu+\mu+1)\mu^2+\pi^2N^2\left(6\mu+9\mu\nu+2+2\mu\nu^2+\nu\right)}}. \tag{3.28}$$

This equation explicitly demonstrates that the fundamental natural frequency of the transverse vibrations depends on the current compressed length of the spring, which is specified by its relative dimensionless length μ.

The natural frequency ω_N^0 of the free, unloaded spring follows from the formula (3.28), when the dimensionless deflection is $\mu = 1$. For the spring in its free length the natural frequency is equal to (Kobelev, 2014):

$$\omega_N^0 = \omega_N|_{\mu=1} = \frac{\zeta^2 N\mathrm{d}}{n_a L_0^2}\sqrt{\frac{2G}{\rho_m}\frac{(1+\nu)\pi^2N^2}{8\zeta^2+\pi^2N^2(4+\nu)}}.$$

The relative fundamental frequency of transverse oscillations:

$$\Omega_N(\mu) = \frac{\omega_N}{\omega_N^0}$$

is of order N introduced. This formula accounts slenderness as a parameter, fixed for any given spring.

3.5 Stability Conditions and Buckling of Spring

Stability conditions could be derived as the expressions of degeneration of natural frequency of transversal vibrations. Namely, if the relative fundamental frequency $\Omega_N(\mu)$ nullifies, the lateral buckling of the spring with the natural form N occurs. The condition of vanishing of fundamental relative frequency:

$$\Omega_N(\mu) = 0 \tag{3.29}$$

determines the critical spring compression $\mu'^*(N)$. At that moment the degeneration of transversal vibration to buckling mode occurs and the spring loses its stability by divergence mode. The matching mode shape with $N = 1$ corresponds to the buckling of the spring with moment-free, simply supported, hinged ends. The mode $N = 2$matches approximately to the buckling of the spring with clamped ends.

Substitution of the expression for frequency (3.28) into the condition (3.29) leads to the following quadratic equation for critical dimensionless deflection μ as the function of slenderness ratio of the spring ζ:

$$\zeta^2(2+\nu)(\mu-1)(2\mu\nu+\mu+1)+(1+\nu)^2\pi^2N^2 = 0. \tag{3.30}$$

For each mode shape the quadratic Eq. (3.30) has two roots (Kobelev, 2014):

$$\mu_+^*(N) = \frac{\nu}{2\nu+1} + \frac{(\nu+1)\sqrt{(\nu+2)\zeta^2 - (2\nu+1)\pi^2 N^2}}{(2\nu+1)\sqrt{\nu+2}\zeta}, \tag{3.31}$$

$$\mu_-^*(N) = \frac{\nu}{2\nu+1} - \frac{(\nu+1)\sqrt{(\nu+2)\zeta^2 - (2\nu+1)\pi^2 N^2}}{(2\nu+1)\sqrt{\nu+2}\zeta}. \tag{3.32}$$

If the spring is compressed from its free length:

$$\mu = 1,$$

the buckling occurs when μ reaching the critical deflection $\mu_+^*(N)$ Eq. (3.31) from above.

Contrarily, if the spring is unloaded from its flat state:

$$\mu = 0,$$

the buckling occurs at the state of the critical deflection $\mu_-^*(N)$ Eq. (3.32) from below. The first and second axial Euler-Grammel buckling shapes are shown on Fig. 3.2. Figure 3.3 illustrates the first $\mu_\pm^*(1)$ and second $\mu_\pm^*(2)$ critical compressions that correspond to first and second buckling shapes.

There is the unstable region between these two critical states (3.31) and (3.32):

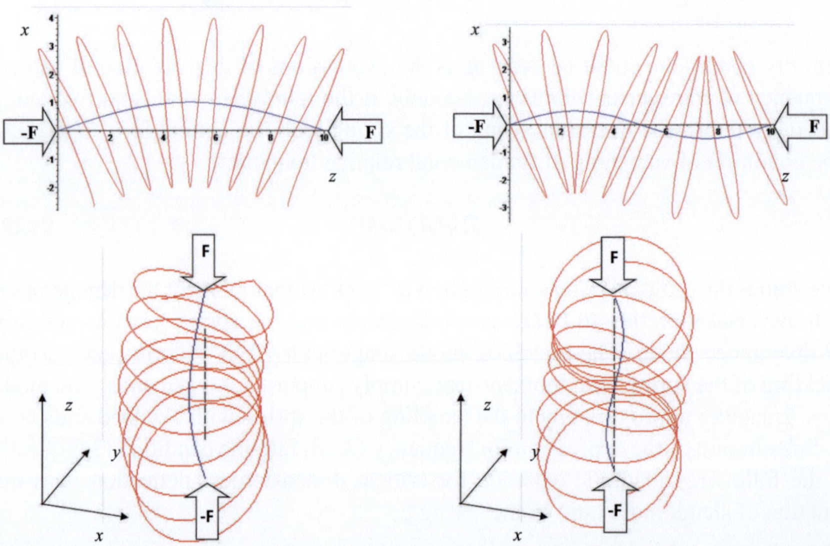

Fig. 3.2 First (left) and second (right) Euler-Grammel axial buckling shapes of simply supported helical spring

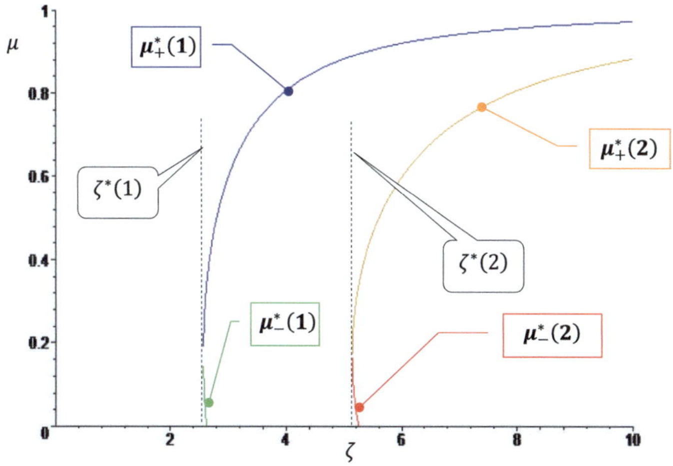

Fig. 3.3 Critical deflection $\mu^*_{\mp}(N)$, $N = 1, 2$ as a function of the degree of slenderness ζ

$$\mu^*_-(N) < \mu < \mu^*_+(N).$$

These two critical states (3.31) and (3.32) exist only if the expression under square root is positive. This happens, if the following condition is satisfied:

$$\zeta \geq \zeta^*(N) = \pi N \sqrt{\frac{2\nu + 1}{\nu + 2}}.$$

Otherwise, if

$$\zeta < \zeta^*(N) = \pi N \sqrt{\frac{2\nu + 1}{\nu + 2}}.$$

the spring is always stable over the compression range.

In this case, no buckling of the spring occurs and the spring remains straight over the full compression range:

$$0 < \mu < 1.$$

The first and second relative natural frequencies of hinged linear spring are shown in Figs. 3.4 and 3.5. The resonant frequency decreases with the degree of compression μ. The shorter the spring, the lower will be usually the first natural, or fundamental, frequency.

On Fig. 3.4 the contours plot of the first fundamental frequency is drawn. The highest relative natural frequency is 1 and this value is achieved at

$$\mu = 1.$$

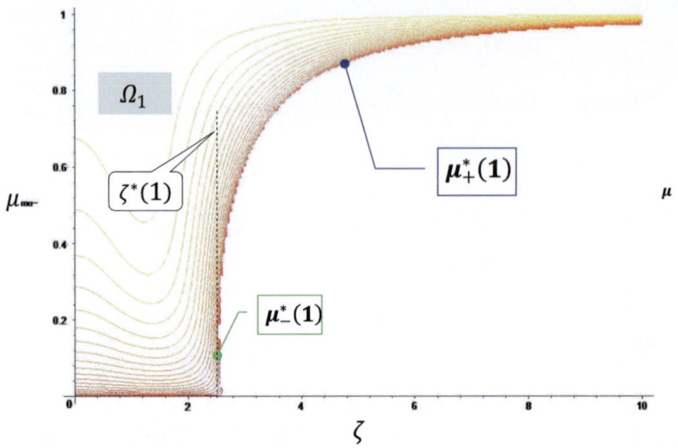

Fig. 3.4 Contour plot of fundamental frequency $\Omega_1 = \omega_1/\omega_1^0$ as a function of deflection μ and slenderness ζ

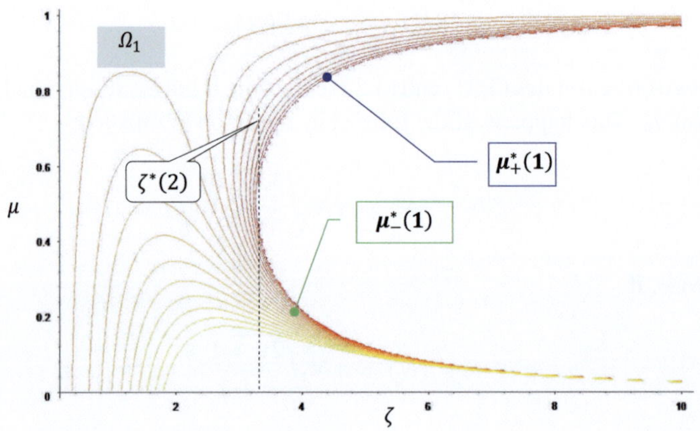

Fig. 3.5 Contour plot of the second natural frequency $\Omega_2 = \omega_2/\omega_2^0$ as a function of deflection μ and slenderness ζ

With the successive compression of spring the relative natural frequency gradually reduces. The shorter the spring, the lower will be the fundamental frequency. In this region, the force action line of the spring drifts from the center line of the spring. On the left region of the plot:

$$\zeta < \zeta^*(1)$$

the relative natural frequency gradually reduces until the spring flattens. On the right region of the plot:

$$\zeta > \zeta^*(1)$$

the relative natural frequency sharply reduces to zero and the spring buckles, when its critical length is achieved:

$$\mu = \mu_+^*(1).$$

Finally, the spring loses its stability on the boundary and lateral buckling of spring occurs. The explanation of data on Fig. 3.5 for the second fundamental frequency is similar.

The spring behaves similarly being unloaded from its flattened shape. Consider the spring in its flattened state. The spring is not stable in this position for all values of the slenderness ratio. Namely, in the tiny region:

$$\zeta^*(1) \leq \zeta < \zeta_{cr}$$

the flattened spring remains stable. Here ξ_{cr} is the solution of the equation

$$\mu_-^*(N) = \frac{\nu}{2\nu + 1} - \frac{(\nu + 1)\sqrt{(\nu + 2)\zeta_{cr}^2 - (2\nu + 1)\pi^2 N^2}}{(2\nu + 1)\sqrt{\nu + 2\zeta_{cr}}} = 0.$$

Stating expanding from its flattened position, the frequency of spring drops rapidly and the spring buckles when:

$$\mu = \mu_-^*(1).$$

For all other values of

$$\zeta \geq \zeta_{cr}$$

the spring is unstable and immediately snaps out being released from its flat state. The flattened form of the spring is also predominantly unstable. The perspective plots of fundamental frequencies are shown respectively on Figs. 3.6 and 3.7. The non-covered regions on these graphs correspond to unstable states of the springs, where the frequencies possess non-vanishing imaginary values.

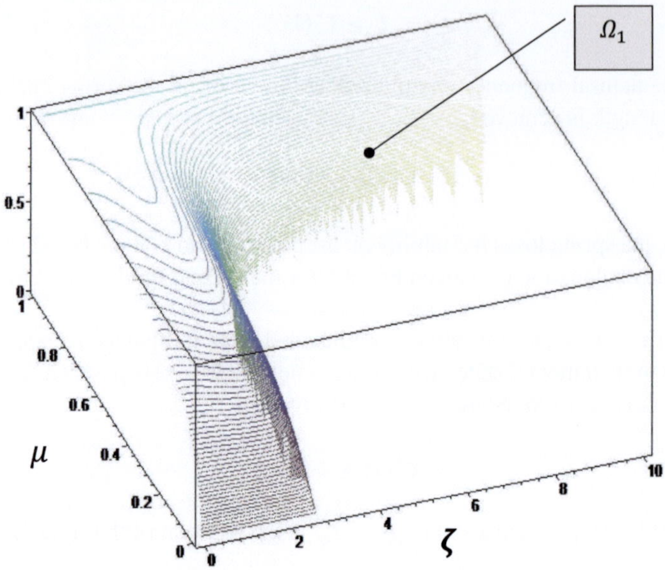

Fig. 3.6 The fundamental frequency $\Omega_1 = \omega_1/\omega_1^0$ as a function of deflection μ and slenderness ζ

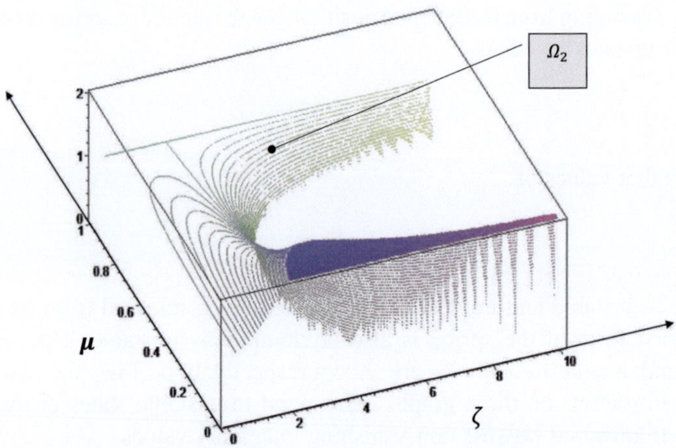

Fig. 3.7 The fundamental frequency $\Omega_2 = \omega_2/\omega_2^0$ as a function of deflection μ and slenderness ζ

3.6 Buckling of Twisted, Compressed, and Tensioned Helical Spring

3.6.1 Instability of Twisted Helical Spring

The counterpart for the Euler buckling problem is the Greenhill's problem, which studies the forming of a loop in an elastic bar under torsion. The torsion spring can wrinkle in a similar way. We study the buckling of the helical spring, twisted by couples M_T applied at its ends alone (Greenhill's problem for helical spring). The ends of the rod are assumed to be attached to the supports by ideal spherical hinges and are free to rotate in any directions.). The twisting couple retains its initial direction during buckling. Introduction of new unknown variables reduces the problem to the exactly solvable auxiliary problem with constant coefficients, such that the critical buckling moment of an equivalent column with an arbitrary shape allows also the exact closed form solution.

Consider an equivalent column of the length L_0, twisted by couples applied at its ends alone (Greenhill's problem for an equivalent column). The twisting couple M_T retains its initial direction during buckling (Bazant & Cedolin, 2010). The effective bending stiffness of the helical spring $\langle EI_B\rangle$ is the same for both directions. For simplicity, we neglect the shear stiffness of the spring.

Consider at first an equivalent column with the constant bending stiffness along the length of the spring (Table 3.1):

$$\langle EI_B\rangle = \frac{E\mathrm{d}^4 L_0}{32(2+\nu)n_a D}.$$

The Greenhill's twist buckling equations are given by:

$$\begin{aligned}\langle EI_B\rangle\frac{\mathrm{d}^2u}{\mathrm{d}z^2} &= M_T\frac{\mathrm{d}v}{\mathrm{d}z},\\ \langle EI_B\rangle\frac{\mathrm{d}^2v}{\mathrm{d}z^2} &= -M_T\frac{\mathrm{d}u}{\mathrm{d}z}.\end{aligned} \tag{3.33}$$

here $u(z)$, $v(z)$ are the deflections of the equivalent column in directions of axes x, y respectively. The ends of the rod $z = 0$ and $z = L_0$ are assumed to be attached to the supports by ideal spherical hinges and are free to rotate in any directions:

$$u(0) = v(0) = u(L_0) = v(L_0) = 0. \tag{3.34}$$

The integration of Eq. (3.33) delivers new equations with two unknown constants c_1, c_2:

$$\langle EI_B\rangle\frac{\mathrm{d}u}{\mathrm{d}z} = M_T v + c_1,$$

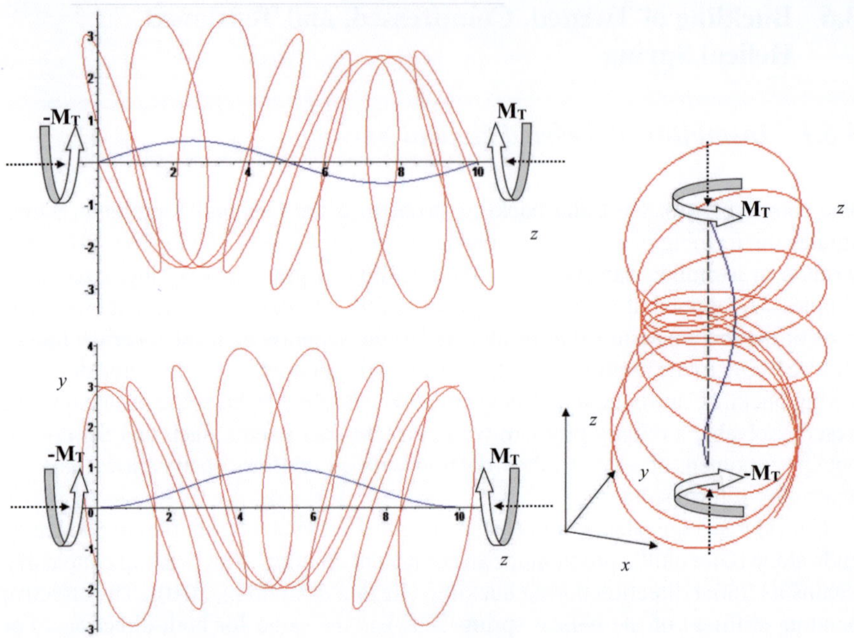

Fig. 3.8 First twist buckling shape (Greenhill) of the simply supported helical spring

$$\langle EI_B \rangle \frac{\mathrm{d}v}{\mathrm{d}z} = -M_T u + c_2.$$

With (3.34) the first buckling shape as follows (Fig. 3.8):

$$u(z) = \sin\frac{2\pi z}{L_0}, \quad v(z) = 1 - \cos\frac{2\pi z}{L_0}, \quad c_1 = c_2 = 0.$$

The critical buckling torque for Greenhill's problem is given by:

$$M_T^* = \frac{2\pi \langle EI_B \rangle}{L_0} = \frac{\pi E \mathrm{d}^4}{16(2+\nu) n_a D}. \tag{3.35}$$

At second, consider the spring with a variable bending stiffness along it axis. For example, the coil diameter or wire diameter vary along the axis of the spring. Introduction of the independent variable ξ:

$$z(\xi) = \int\limits_0^{\xi} \frac{\mathrm{d}t}{F(t)}, \quad F(t) > 0.$$

In the new axes the problem (3.34)–(3.35) transforms to:

$$F(\xi)\langle EI_B\rangle\frac{\mathrm{d}U}{\mathrm{d}\xi} = M_T V + c_1,$$
$$F(\xi)\langle EI_B\rangle\frac{\mathrm{d}V}{\mathrm{d}\xi} = -M_T U + c_2, \tag{3.36}$$

$$U(0) = V(0) = U(L) = V(L) = 0, \tag{3.37}$$

with

$$u(z) = u\left(\int_0^{\xi}\frac{\mathrm{d}t}{F(t)}\right) = U(\xi),$$
$$v(z) = v\left(\int_0^{\xi}\frac{\mathrm{d}t}{F(t)}\right) = V(\xi).$$

The functions $U(\xi)$, $V(\xi)$ are the deflections of an equivalent column with a variable bending stiffness along its length in directions of axes x, y respectively. The value L is the solution of the algebraic equation

$$L_0 = \int_0^L \frac{\mathrm{d}\xi}{F(\xi)}. \tag{3.38}$$

Eigenvalue problem (3.36)–(3.37) describes the Greenhill's problem with a variable effective bending stiffness of the "equivalent column" over the length:

$$\langle EI_B(t)\rangle = \langle EI_B\rangle \cdot F(t), \quad 0 \le t \le L.$$

The critical eigenvalues of the problems (3.36)–(3.37) and (3.33)–(3.34) match and are equal to the value (3.35). Substitution of (3.38) in the expression (3.35) delivers the analytical expression for critical torque (Kobelev, 2016):

$$M_T^{**} = 2\pi\left(\int_0^L \frac{\mathrm{d}t}{\langle EI_B\rangle F(t)}\right)^{-1} \equiv 2\pi\left(\int_0^L \frac{\mathrm{d}t}{\langle EI_B(t)\rangle}\right)^{-1}. \tag{3.39}$$

This function delivers the critical load of Greenhill's problem for a twisted helical spring with a variable effective bending stiffness of the length L. Other clamping conditions could be examined (Kobelev, 2017).

3.6.2 Instability of Helical Springs Under Torque and Axial Force

Consider a geometrically perfect helical spring supported on two spherical hinges, loaded by axial compressive force F and torque M_T. Both loads are assumed to keep their direction during buckling. Adding the expressions for moments due to axial force $(-Fu, -Fv)$ to the Greenhill's twist buckling Eq. (3.33), we get the buckling equations for the helical springs under torque and axial force:

$$\langle EI_B\rangle \frac{\mathrm{d}^2 u}{\mathrm{d}z^2} = M_T \frac{\mathrm{d}v}{\mathrm{d}z} - Fu,$$

$$\langle EI_B\rangle \frac{\mathrm{d}^2 v}{\mathrm{d}z^2} = -M_T \frac{\mathrm{d}u}{\mathrm{d}z} - Fv, \tag{3.40}$$

here $u(z)$, $v(z)$ are the deflections of the equivalent column in directions of axes x, y respectively. The tolerance of the solution could be improved, if we consider the value L as the already compressed length of the helical spring under the action of the axial force F, i.e. $L = L_0 - F/c$.

The ends of the compressed spring $z = 0$ and $z = L$ are assumed to be attached to the supports by ideal spherical hinges and are free to rotate in any directions, Eq. (3.34). We seek the general solution of Eq. (3.40) in the form:

$$u(\mathrm{z}) = A\exp(i\lambda z), \quad v(\mathrm{z}) = B\,\exp(i\lambda z). \tag{3.41}$$

The substitution of (3.41) into (3.40) leads to a system of two homogeneous linear algebraic equations:

$$\begin{bmatrix} F - \langle EI_B\rangle\lambda^2 & i\lambda M_T \\ -i\lambda M_T & F - \langle EI_B\rangle\lambda^2 \end{bmatrix} \begin{bmatrix} A \\ B \end{bmatrix} = \begin{bmatrix} 0 \\ 0 \end{bmatrix}. \tag{3.42}$$

Nontrivial solution of (3.42) is possible, only if the determinant vanishes:

$$\begin{vmatrix} F - \langle EI_B\rangle\lambda^2 & i\lambda M_T \\ -i\lambda M_T & F - \langle EI_B\rangle\lambda^2 \end{vmatrix} = 0. \tag{3.43}$$

For positive torque M_T, two possible solutions of the characteristic equation are:

$$\lambda_{1,2} = \frac{-M_T \pm \sqrt{{M_T}^2 + 4\langle EI_B\rangle F}}{2\langle EI_B\rangle}. \tag{3.44}$$

The general solution is the real or imaginary part of

$$u(\mathrm{z}) = A_1 e^{i\lambda_1 z} + A_2 e^{i\lambda_2 z},$$

$$v(z) = B_1 e^{i\lambda_1 z} + B_2 e^{i\lambda_2 z}. \tag{3.45}$$

The boundary conditions Eq. (3.34) reduce with the solution (3.45) to:

$$A_1 + A_2 = B_1 + B_2 = e^{i\lambda_1 L} A_1 + e^{i\lambda_2 L} A_2 = e^{i\lambda_1 L} B_1 + e^{i\lambda_2 L} B_2 = 0. \tag{3.46}$$

Four Eq. (3.46) possess the nontrivial solution only if $e^{i\lambda_1 L} = e^{i\lambda_2 L}$. In this case:

$$\lambda_1 = \lambda_2 + \frac{2k\pi}{L}, \quad k = 1, 2, \ldots \tag{3.47}$$

The smallest possible buckling load Eq. (3.47) corresponds to $k = 1$:

$$\frac{-M_T + \sqrt{M_T{}^2 + 4\langle EI_B\rangle F}}{2\langle EI_B\rangle} = \frac{-M_T - \sqrt{M_T{}^2 + 4\langle EI_B\rangle F}}{2\langle EI_B\rangle} + \frac{2\pi}{L}. \tag{3.48}$$

As defined in Sect. 3.4, the slenderness ratio of the free spring is $\zeta = L_0/D$ and relative dimensionless length is $\mu = L/L_0$. Equation (3.48) results in the buckling condition for simultaneous axial compression and torsion of helical spring:

$$\frac{F}{F^*} + \left(\frac{M}{M_T^*}\right)^2 = 1, \tag{3.49}$$

where for the cylindrical spring with the constant wire diameter are:

$$F^* = \langle EI_B\rangle\left(\frac{\pi}{L}\right)^2 \equiv \langle EI_B\rangle\left(\frac{\pi}{L_0 - \frac{F^*}{c}}\right)^2 = \frac{Ed^4}{32D^2 n_a} \cdot \frac{\xi + \sqrt{\zeta^2 - \zeta^{*2}}}{1+\nu}, \tag{3.50}$$

$$M_T^* = 2\pi \frac{\langle EI_B\rangle}{L} \equiv \frac{\pi E d^4}{16(2+\nu) D n_a}. \tag{3.51}$$

F^* is the Euler's critical load for compression,

M_T^* is the Greenhill's critical twist buckling moment, Eq. (3.35).The Greenhill's critical twist buckling moment is independent upon slenderness ratio. The dimensionless value in the expression for Euler's critical compressive load Eq. (3.50) is:

$$\zeta^* = \pi\sqrt{2\frac{1+\nu}{2+\nu}}. \tag{3.52}$$

The value ζ^* in Eq. (3.52) denotes the minimal slenderness ratio for instability, which is analogous to discussed in Sect. 3.5. If the slenderness is less than ζ^*, i.e. $\zeta < \zeta^*$, no Euler buckling is possible.

The relation between both critical loads reads:

$$\frac{M_T^*}{F^*} = \frac{2D}{\pi}\mu^*\zeta, \tag{3.53}$$

where is the character dimensionless length:

$$2\mu^* = 1 - \sqrt{1 - \frac{\zeta^{*2}}{\zeta^2}}. \tag{3.54}$$

The graphs of the critical buckling loads are shown on the following figures. The following values were used for calculations:

$$E = 200\ \text{GPa},\ \nu = 0.25,\ n_a = 10,\ D = 30\ \text{mm},\ L_0 = 120\ \text{mm},\ \ \text{d} = 2\ldots3\ \text{mm}$$

Figure 3.9 displays the relative dimensionless length μ^* and Euler critical buckling load F^* as the functions of slenderness ratio. On Fig. 3.10 the pure Greenhill's critical twist buckling moment M_T^* Eq. (3.51) and pure compressive Euler's critical load F^* Eq. (3.50) are drawn as the functions of wire diameter d.

Figure 3.11 illustrates the simultaneous buckling of twisted and compressed helical spring as described in Eq. (3.49). The functions of the Greenhill's critical twist buckling moment M_T^* and the Euler's critical load for compression F^* ares shown for three wire diameters d and other equal geometrical parameters

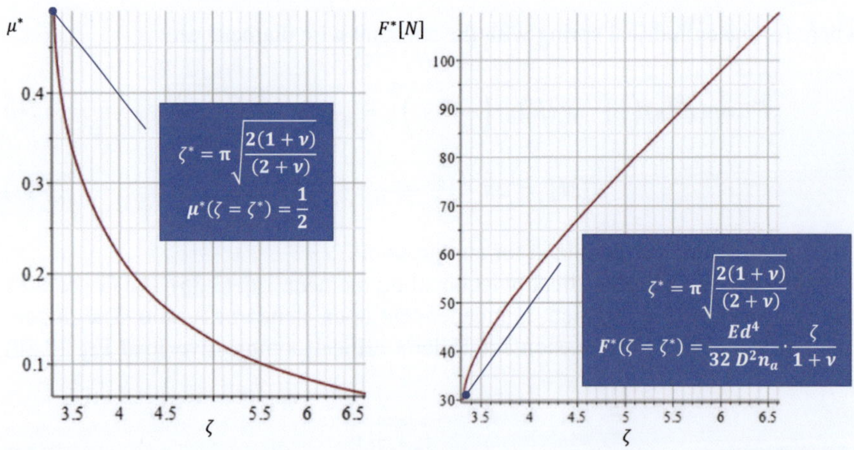

Fig. 3.9 Dependences of the relative dimensionless length μ^* and Euler critical buckling load F^* as functions of slenderness ratio ζ. $\zeta^* = \pi\sqrt{2(1+\nu)/(2+\nu)}$is the minimal slenderness ratio for instability

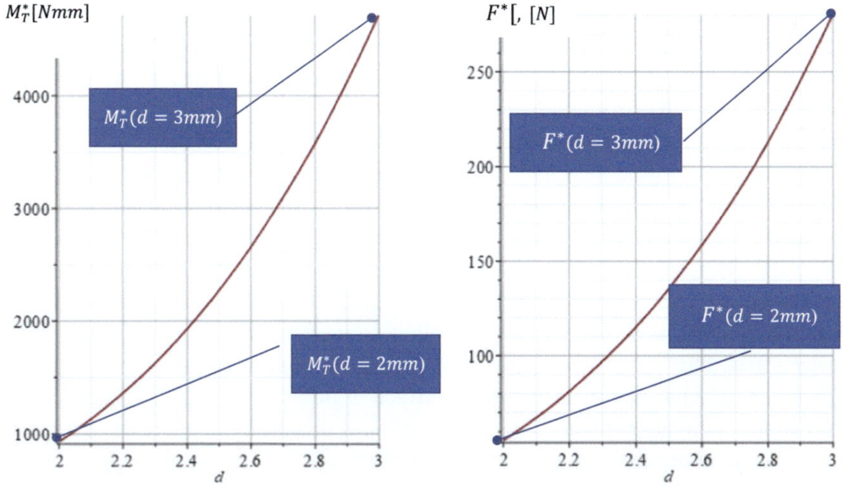

Fig. 3.10 Pure Greenhill's critical twist buckling moment M_T^* and pure Euler's critical load for compression F^* as the function of wire diameter d

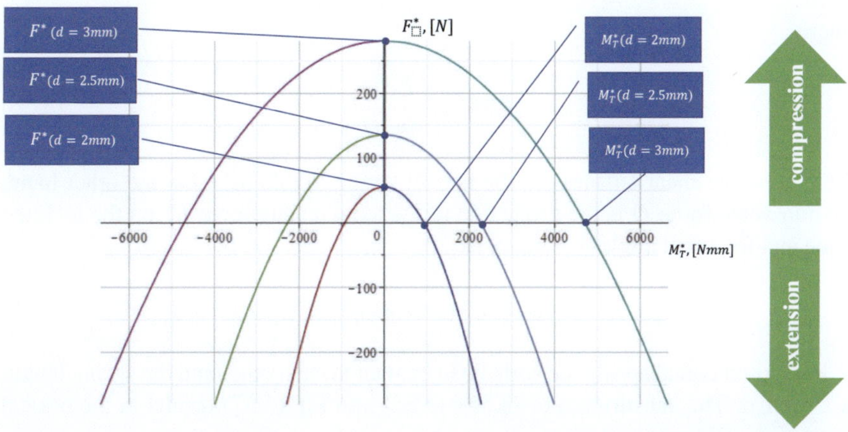

Fig. 3.11 Simultaneous buckling of twisted and compressed helical spring. Greenhill's critical twist buckling moment M_T^* and Euler's critical load for compression F^* ares shown for three wire diameters d and other equal geometrical parameters

3.6.3 Instability of Tension Spring

The other type of instability occurs when a gradually increasing tensile force is applied to the ends of a pre-stressed helical spring (Andreeva, 1962). An elegant explanation of this curios effect was provided by Ponomarev and Andreeva (1980). The instability occurs for the springs with sufficiently large ratios of radius to pitch and twist to bending rigidity). Experimentally observed, that the end-to-end distance suffers a series of discontinuous stretching evolutions. This phenomenon was discovered also somewhat later by Kessler and Rabin (2003). The critical torsional load or angle of rotation that confers instability in a tension spring under combined tension and torsion load was determined in González-Cabrero et al. (2024)

Consider a closed coiled helical spring. The coils of the spring are positioned in planes, normal to the axis of the spring. When the closed coiled helical is tensioned, the sudden inclination of coils happens at certain tensile load. The inclined coils lay no more in the normal plane to the spring axis (Fig. 3.12). The inclination angle between the new coils plane and the normal plane is ϑ. The first instability occurs when the first coil abruptly inclines. The other coils follow to incline with the slightly increasing tension load. Due to pre-stress, each coil remains in contact to its neighboring coils. The inclined coils slide over the upper surface of other coils, such that their centers shift in the inclination plane. The deformation of the shifted coils is the pure shear deformation under the action of the transverse force Q. The shear angle is exactly the inclination angle. The shear angle of an "equivalent column" is given by:

$$\vartheta = \frac{Q}{GS}, \tag{3.55}$$

where

$$\langle GS \rangle = \frac{E\mathrm{d}^4 L}{8n_a D^3} \tag{3.56}$$

is the effective shear stiffness of the helical spring (Table 3.1). On the other hand, the transverse force Q is the projection of the axial tensile force F^* on the inclined plane and for small angle ϑ is equal to:

$$Q = F^* \sin\vartheta \approx F^*\vartheta. \tag{3.57}$$

For closed coiled spring the coils lie in contact to each other and the spring length is $L = n_a\mathrm{d}$. The substitution of (3.55)–(3.56) into Eq. (3.57) results in the critical force of the gliding instability:

$$F^* = GS = \frac{E\mathrm{d}^5}{8D^3}. \tag{3.58}$$

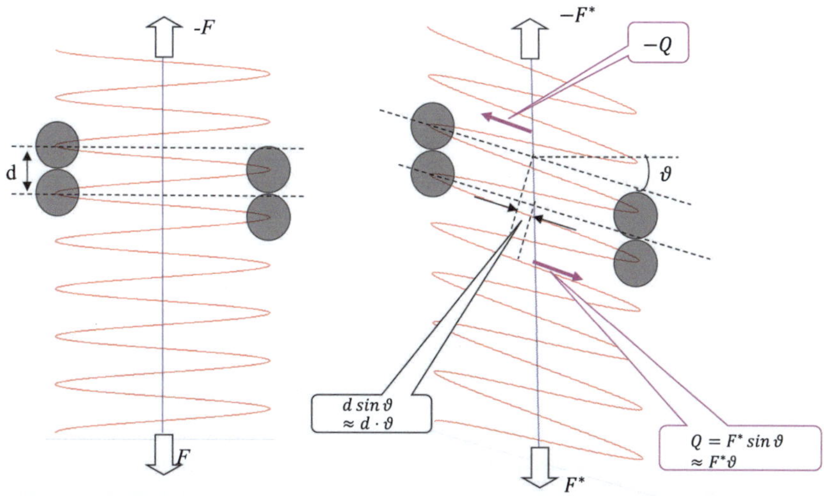

Fig. 3.12 Coil-gliding instability of an extensional helical spring. The coils slide resulting shear force

3.7 Spatial Models for Dynamic Behavior of Helical Springs

In the spatial models do not use the reduction to the hypothetical elastic axis of the spring and equivalent stiffness calculations. The variety of spatial models is explained by the difficulty of the exact equations in the framework of spatial rod theory and by significant complexity with their analytical solution.

Wittrick (1996) derived a set of 12 linear coupled partial differential equations for a uniform helical spring based on the Timoshenko beam theory. Costello (1975) presented a work on the significance of torsional oscillations on the radial expansion of helical springs.

Guido et al. (1978) applied Timoshenko theory for the determination of the transverse frequencies of cylindrical helical springs for different values of slenderness and relative compression. Springs have been considered as having damped ends and lacking end coils. Under these boundary conditions the spring deflection curves result from the superposition of two deflection curves with different wave length and characterized by bending and shear displacements in phase or 180° out of phase. The static behavior of a spatial bar of an elastic and isotropic material under arbitrary distributed loads having a non-circular helical axis and cross-section supported elastically by single and/or continuous supports was studied in Haktanir (1995) by the stiffness matrix method based on the complementary functions approach. By considering the geometrical compatibility conditions together with the constitutive equations and equations of equilibrium, a set of 12 first-order differential equations having variable coefficients was obtained for spatial elements.

Yildirim (1997, 2002) conducted a series of studies to compute the eigenvalues of helical springs of arbitrary shape.

Lee and Thompson (2001) used the dynamic stiffness method to calculate the natural frequencies of helical springs and compared the results with those of the transfer matrix and the finite element method. The equation of free wave motion in a helical spring, derived from Timoshenko beam theory and Frenet formulae, has been used to obtain the dynamic stiffness matrix. The natural frequencies were calculated in the cited Article from this matrix, after applying suitable boundary conditions, by using the Wittrick-Williams method. By computing the axial and transverse transfer stiffness it has been shown how the spring becomes much stiffer at high frequencies, compared to the static stiffness. The effect of the helix angle on three different transition frequencies has been investigated. This effect was adapted by the addition of a static preload.

Becker et al. (2002) investigated the effect of static axial compression upon the natural frequencies of helical springs by the transfer matrix method. The linear disturbance equations governing the resonant frequencies of a helical spring subjected to a static axial compressive load were solved numerically using the transfer matrix method for clamped ends and circular cross-section to produce frequency design charts. The effect of varying the number of turns of the spring was investigated, and in the limit of large numbers of turns, our results validate earlier work on the vibration of helical compression springs in which the helix was modeled as an elastic beam with rigidities corresponding to those of unclosed circular rings.

The pseudospectral method was applied by Lee (2007) to the free vibration analysis of cylindrical helical springs. The displacements and the rotations were approximated by the series expansions of Chebyshev polynomials and the governing equations were collocated. Numerical examples were provided for fixed–fixed, free–free, fixed–free and hinged–hinged boundary conditions.

Taktak et al. (2008) presented a finite element for the dynamic analysis of the cylindrical isotropic helical spring. The hybrid-mixed formulation was used to compute the stiffness matrix. A simple approach was used to calculate the mass matrix. These matrices were used for solving the dynamic equation of the spring to calculate natural frequencies and the dynamic response of a simple or an assembled spring for different types of cross-section.

A numerical solution was presented in Ayadi and Hadj-Taïeb (2008) to describe wave propagations in axially impacted helical springs. The governing equations for such problem were two coupled hyperbolic, partial differential equations of second order. The axial and rotational strains and velocities were considered as principal dependent variables. Since the governing equations were non-linear, the solution of the system of equations can be obtained only by some approximate numerical simulation. The finite element method was applied for the discrete formulation of the mathematical equations leading to a non-linear system of equations solved by an iterative Gauss substitution method.

In the paper of Sorokin (2009), the validity ranges of alternative theories were assessed by comparison of the location of the dispersion curves and a rigorous asymptotic analysis of the exact dispersion equation with two small parameters. It allows for the identification of significant regimes of linear wave motion in a helical spring. In each of these regimes, simple formulae for wave numbers were obtained by the dominant balance method.

An analytical study for free vibration of naturally curved and twisted beams with uniform cross-sectional shapes was carried in Yu et al. (2010) out using spatial curved beam theory based on the Washizu's static model. In the governing equations of motion of the beams, all displacement functions and the generalized warping coordinate were defined at the centroid axis and the effects of rotary inertia, transverse shear deformations and torsion-related warping were included in the proposed model. Explicit analytical expressions were derived for the vibrating mode shapes of a curved, bending-torsional-shearing coupled beam under clamped–clamped boundary condition.

The paper Leamy (2010) presents an efficient intrinsic finite element approach for modeling and analyzing the forced dynamic response of helical springs. The finite element treatment employs intrinsic curvature (and strain) interpolation vice rotation (and displacement) interpolation, and thus can accurately and efficiently represent initially curved and twisted beams with a sparse number of elements. The governing equations of motion contain nonlinearities necessary for large curvatures. In addition, a constitutive model was developed which captures coupling due to non-zero initial curvature and strain. The method was employed to efficiently study dynamically-loaded helical springs.

The governing equations and the associated natural boundary conditions of a pretwisted helical beam with non-circular cross-sections have been derived in Leung (2010) from differential geometry and variational principles. For isotropic materials, the formulation was identical to the existing literature for helical beam with circular cross-sections. Explicit analytical expressions that give the vibrating mode shapes were derived in Yun and Hao (2011) by rigorous application of the Muller root search method was used to determine the natural frequencies. The free vibration analysis of cylindrical helical springs with noncircular cross-sections was carried out by means of an analytical study. In the governing equations of motion of a spring, all displacement functions and a generalized warping coordinate were defined at the centroidal principal axis. The effects of the rotational inertia, axial and shear deformations, including torsion-related warping deformations, were also considered in the formulations.

The paper Frikha et al. (2011) proposes a physical analysis of the effect of axial load on the propagation of elastic wave in helical beams. The model was based on the equations of motion of loaded helical Timoshenko beams. The dimensionless for beams of circular cross-section and the number of parameters governing the problem was reduced to four (helix angle, helix index, Poisson coefficient, and axial strain) were derived and a parametric study was conducted. The outcome of loading was shown to be different in high, medium, and low frequency ranges.

The modeling of non-uniform springs were considered in Renno and Mace (2012). The uniform part of helical springs was modeled using the wave and finite element method since a helical spring can be regarded as a curved waveguide. This model was obtained by post-processing the finite element model of a single straight or curved beam element using periodic structure theory.

The paper Hamza et al. (2013) discusses the vibrations of a coil in helical compression spring which was excited axially. The government equations form the system of four hyperbolic partial differential equations of first order with unknown variables, which were angular and axial deformations and velocities. The numerical resolution was based on the conservative finite difference scheme of Lax-Wendroff. The impedance method was applied to calculate the frequency spectrum. The spring was excited by a sinusoidal axial velocity at its end. The results obtained by using this method were used to analyze the evolution in time of deformations and velocities in different sections. The paper Hamza et al. (2013) studies the resonance of axially excited helical compression springs. The mathematical formulation of the dynamic behavior of the springs was composed of a system of four partial differential equations of first order hyperbolic type, which were the equations of momentum and the laws of constitution. The variables were angular and axial deformations and velocities. To calculate the frequency spectrum and to study the natural frequency response the impedance method was applied.

In the article Yildirim (2012), a set of twelve linearized disturbance dynamic equations in canonical form was derived systematically and in a comprehensive manner based on the first order shear deformation theory to study the buckling and vibration analysis of helical coil springs made of isotropic linear materials. Those complete equations comprise the axial and shear deformation effects together with rotatory inertia effects. The special case of these equations corresponds also to the equations for straight and circular rods.

Principally, all cited models for transverse vibrations of the compressed coil springs could be used for study of buckling behavior of the springs as well.

Namely, the fundamental natural frequency of the transverse vibrations of the column is the function of the conservative axial force, as well as the variable length of the spring. If damping exists, the fundamental frequency was a complex number, which depends on the compressed length of the spring.

The criterion of a static equilibrium state is considered using perturbation method (Godoy, 1999). A perturbation is introduced in the form of a vibration about the static equilibrium. In general, the study of a nonlinear dynamic response is required. The initial linear dynamic responses were indicators of the possible nonlinear dynamic behavior of the perturbed system. This leads to stability in the local sense.

Conclusions

The local dynamic criterion can be stated as follows: An equilibrium state is stable if, for small oscillations around such a state, all oscillation frequencies are real.

Thus, the measurement of the dynamics of the perturbed system is carried out by considering the frequencies of vibration. If at least one vibration frequency is zero, we say that the equilibrium state is critical. In this case, the stability can only be evaluated using nonlinear oscillations. Thus, if the real part of the fundamental natural frequency goes to zero, lateral buckling of the spring occurs. Loss of spring stability is caused by divergence.

If the loading of the spring is not conservative, the loss of stability may be manifested not by the system going into dynamic equilibrium state, but by going into unbounded motion. If at least one frequency is imaginary, we say that the equilibrium state becomes unstable. To cover this possibility, the dynamic behavior of the system must be considered, since stability is essentially a dynamic concept (Bolotin, 1964).

The previous discussion used the linear spring model without coil contacts. The nonlinear, progressive spring behaves differently. The coils of the nonlinear spring make contact and become inactive. The length of the spring must take into account the length of the active coils. The resonant frequency of nonlinear progressive springs increases primarily with the degree of compression. The shorter the spring, the higher the fundamental frequency. However, even for progressive springs, there may be special regions where the eigenfrequency decreases. Thus, even a sufficiently thin progressive spring can suffer from lateral buckling. For accurate study of these phenomena, numerical methods are applicable due to the complexity of the governing equations.

Summary of Principal Results

- The deformation of the central line of the spring characterizes the deformation of the helical wire of the spring.
- This central line is considered as an "equivalent column", which has nearly the same bending, extensional and torsional stiffness as the helical spring itself.
- The static and dynamic equations of the "equivalent column" are derived.
- Closed-form solutions are found for the stability and vibration problems of helical springs.
- The stability of helical springs under compressive and torsional loads is investigated.
- The closed form solutions for critical axial compressive force and torque are derived.

References

Andreeva, L. E. (1962), Elastic elements of instruments, Mashgiz, Moscow, 456 p. Translation: Baruch, A., & Alster, D. Jerusalem, Israel Program for Scientific Translation, Ltd.

Ayadi, S., & Hadj-Taïeb, E. (2008). Finite element solution of dynamic response of helical springs. *International Journal of Simulation Modelling, 7*(1), 17–28. https://doi.org/10.2507/IJSIMM07(1)2.094

Bazant, Z., & Cedolin, L. (2010). *Stability of structures, elastic, inelastic, fracture, and damage theories*. World Scientific.

Becker, L. E., Chassie, G. G., & Cleghorn, W. L. (2002). On the natural frequencies of helical compression springs. *International Journal of Mechanical Sciences, 44*, 825–841.

Biezeno, C. B., & Koch, J. J. (1925). Knickung von Schraubenfedern. *Zeitschrift Für Angewandte Mathematik Und Mechanik, 5*, 279–280.

Bolotin, V. V. (1964). *The dynamic stability of elastic systems*. Holden Day.

Chan, K. T., Wang, X. Q., So, R. M. C., & Reid, S. R. (2002). Superposed standing waves in a Timoshenko beam. *Proceedings of the Royal Society, A, 458*, 83–108.
Chassie, G. G., Becker L. E., Cleghorn, W. L. (1997). On the buckling of helical springs under combined compression and torsion. *International Journal of Mechanical Sciences, 39*(6), 697–704, ISSN 0020-7403. https://doi.org/10.1016/S0020-7403(96)00070-7
Collins, J. A., Busby, H. R., & Staab, G. H. (2010). *Mechanical design of machine elements and machines: A failure prevention perspective.* Wiley.
Costello, G. A. (1975). Radial expansion of impacted helical springs. *Journal of Applied Mechanics, Transaction ASME, 42*, 789–792.
Dick, J. (1942). On transverse vibrations of a helical spring with pinned ends and no axial load. *Philosophical Magazine, 33*(222), 513–519
Encyclopedia of Spring Design. (2013). Spring Manufacturers Institute, 2001 Midwest Road, Suite 106, Oak Brook, Illinois 60523-1335 USA
Frikha, A., Treyssédee, F., & Cartraud, P. (2011). Effect of axial load on the propagation of elastic waves in helical beams. *Wave Motion, 48*(1), 83–92.
Godoy, L. (1999). Theory of elastic stability: Analysis and sensitivity. CRC Press, 450p.
Grammel, R. (1924). Die Knickung von Schraubenfedern. *Zeitschrift für Angewandte Mathematik und Mechanik, 4,* 384-389
Guido, A. R., Della, P. L., & della Valle, S. (1978). Transverse vibrations of cylindrical helical springs. *Meccanica, 13*(2), 90–108.
Haktanir, V. (1995). The complementary functions method for the element stiffness matrix of arbitrary spatial bars of helicoidal axes. *International Journal for Numerical Methods in Engineering, 38*(6), 1031–1056. https://doi.org/10.1002/nme.1620380611
Hamza, A., Ayadi, S., & Hadj-Taieb, E. (2013). Resonance phenomenon of strain waves in helical compression springs. *Mechanics & Industry, 14*, 253–265. https://doi.org/10.1051/meca/2013069
Haringx, J. A. (1948). On highly compressible helical springs and rubber rods, and their application for vibration-free mountings. *Philips Research Reports, 3*, 401–449.
Kessler, D. A., & Rabin, Y. (2003). Stretching instability of helical springs. *Physical Review Letters, 90*, 024301.
Kobelev, V. (2014). Effect of static axial compression on the natural frequencies of helical springs. *Multidiscipline Modeling in Materials and Structures, 10*(3), 379–398.
Kobelev, V. (2016). Isoperimetric inequality in the periodic Greenhill problem of twisted elastic rod. *Structural and Multidisciplinary Optimization, 54*(1), 133–136.
Kobelev, V. (2017). Some exact analytical solutions in structural optimization mechanics based design of structures and machines. *An International Journal, 45*(1). https://doi.org/10.1080/15397733.2016.1143374
Kruzelecki, J., & Zyczkowski, M. (1990). On the concept of an equivalent column in the stability problem of compressed helical springs. *Ingenieur-Archiv, 60*, 367–377.
Leamy, M. J. (2010). Intrinsic finite element modeling of nonlinear dynamic response in helical springs. In *ASME 2010 International Mechanical Engineering Congress and Exposition Volume 8: Dynamic Systems and Control, Parts A and B, Vancouver, British Columbia, Canada*, Vol. 11, pp. 857–867, November 12–18, Paper No. IMECE2010-37434. https://doi.org/10.1115/IMECE2010-37434
Lee, J. (2007). Free vibration analysis of cylindrical helical springs by the pseudospectral method. *Journal of Sound and Vibration, 302*, 185–196.
Lee, J., & Thompson, D. J. (2001). Dynamic stiffness formulation, free vibration and wave motion of helical springs. *Journal of Sound and Vibration, 239*, 297–320.
Lee, C.-Y., Zhuo, H.-C., & Hsu, C.-W. (2009). Lateral vibration of a composite stepped beam consisted of SMA helical spring based on equivalent Euler-Bernoulli beam theory. *Journal of Sound and Vibration, 324*, 179–193.
Leung, A. Y. T. (2010). Vibration of thin pre-twisted helical beams. *International Journal of Solids and Structures, 47*, 177–1195.

Majkut, L. (2009). Free and forced vibrations of Timoshenko beams described by single difference equation. *Journal of Theoretical and Applied Mechanics, 47*(1), 193–210.

Michalczyk, K. (2015). Analysis of lateral vibrations of the axially loaded helical spring. *Journal of Theoretical and Applied Mechanics, 53*(3), 745–755. https://doi.org/10.15632/Jtam-Pl.53.3.745

Ponomarev, S. D., & Andreeva, L. E. (1980). *Calculation of elastic elements of machines and instruments.*

Ponomarev, S. D. (1948). Stability of helical springs under compression and torsion (in Russian). In E. A. Chudakov (Ed.), *Mashinostr.* (Vol. 2, pp. 683–685), Moscow.

Renno, J. M., & Mace, B. R. (2012). Vibration modelling of helical springs with non-uniform ends. *Journal of Sound and Vibration, 331*(12), 2809–2823.

Satoh, T., Kunoh, T., & Mizuno, M. (1988). Buckling of coiled springs by combined torsion and axial compression. *JSME International Journal, 31*, 56–62.

Skoczeń, B., & Skrzypek, J. (1992). Application of the equivalent column concept to the stability of axially compressed bellows. *International Journal of Mechanical Sciences, 34*(11), 901–916. ISSN 0020-7403. https://doi.org/10.1016/0020-7403(92)90020-H

Sorokin, S. V. (2009). Linear dynamics of elastic helical springs: Asymptotic analysis of wave propagation. *Proceedings of the Royal Society A, 465*, 1513–1537. https://doi.org/10.1098/rspa.2008.0468

Stephen, N. G., & Puchegger, S. (2006). On the valid frequency range of Timoshenko beam theory. *Journal of Sound and Vibration, 297*, 1082–1087.

Tabarrok, B., & Xiong, Y. (1992). A spatially curved and twisted rod element for buckling analysis. *International Journal of Solids and Structures, 29*(23), 3011–3023. ISSN 0020-7683. https://doi.org/10.1016/0020-7683(92)90155-M

Taktak, M., Dammak, F., Abid, S., & Haddar, M. (2008). A finite element for dynamic analysis of a cylindrical isotropic helical spring. *Journal of Mechanics of Materials and Structures, 3*(4)

Wittrick, W. H. (1966). On elastic wave propagation in helical springs. *International Journal of Mechanical Sciences, 8*, 25–47.

Yildirim, V. (1997). Free vibration analysis of non-cylindrical coil springs by combined used of the transfer matrix and the complementary functions method. *Communications in Numerical Methods in Engineering, 13*, 487–494.

Yildirim, V. (2002). Expression for predicting fundamental natural frequencies of non-cylindrical helical springs. *Journal of Sound and Vibration, 252*, 479–491.

Yildirim, V. (2012). On the linearized disturbance dynamic equations for buckling and free vibration of cylindrical helical coil springs under combined compression and torsion. *Meccanica, 47*(4), 1015–1033.

Yu, A. M., Yang, C. J., & Nie, G. H. (2010). Analytical formulation and evaluation for free vibration of naturally curved and twisted beams. *Journal of Sound and Vibration, 329*, 1376–1389.

Yun, A. M., & Hao, Y. (2011). Free vibration analysis of cylindrical helical springs with noncircular cross-sections. *Journal of Sound and Vibration, 330*, 2628–2639.

Ziegler, H. (1982). Arguments for and against Engesser's formulas. *Ingenieur Archiv, 52*, 105–113.

Ziegler, H., & Huber, A. (1950). Zur Knickung der gedrückten und tordierten Schraubenfeder. *Zeitschrift Fur Angewandte Mathematik Und Physik, 1*, 183–195.

González-Cabrero J., Font H., Cavas F., Paredes M. (2024) Buckling analysis of a helical extension spring under combined loading, Thin-Walled Structures, 200111914-10.1016/j.tws.2024.111914

Chapter 4
Disk Springs

Abstract In the current chapter, we study the disk spring using the models of thin and moderately thick isotropic shells. The equations developed here are based on common assumptions and are simple enough to be applied to the applied analysis. The analysis of thin-walled disk springs could be performed using basic spreadsheet tools, eliminating the need for cumbersome finite element analysis. The theory of linear and progressive disk wave springs is also presented. This chapter is the fourth section of the first part, which deals with the design and modeling of springs.

Keywords Disk springs with radially free edges · Belleville washers · Shell models · Wave springs

4.1 Thick Shell Model for Disk Springs

4.1.1 Mechanical Models of Elastic Disk Springs

Disk spring (Belleville spring, disk spring washer), is well known for its configuration to nonlinearly withstand a large force with minimum deflection while storing a large amount of energy in relation to the space occupied. Load–displacement formula for a disk spring or a non-slotted disk spring was first developed by Timoshenko and Woinowsky-Krieger (1957) and Almen and Laszlo (1936). The cited theory is based on experimental observations according to which the cross-section of the spring merely rotates about a center point (assumed to be on the middle line of the cross-section) without undergoing an appreciable deflection. The results of Almen and Laszlo agreed with Timoshenko in regarding the radial stresses as negligible and succeeded in calculating tangential stresses and displacements of a disk spring subjected to an axial load uniformly distributed on inner and outer edges. Based on this theory, the dimensions used to define the geometries refer to the middle line of the disk spring cross-section. The equations provided in DIN EN 16984:2017-02 (2017) are sufficiently accurate for evaluating relative flat disk springs of moderate thickness. According to these equations, the deformation behavior of the disk spring

V. Kobelev, *Fundamentals of Springs Mechanics*,
https://doi.org/10.1007/978-3-031-58584-5_4

is treated as a one-dimensional inversion of a circular ring of rectangular cross-section about an inversion center point. The resulting inverted stress condition is overlaid by a bending stress condition caused by the change in the cone angle resulting from the deflection (Hengstenberg, 1983). The cross-section of the disk spring remains rectangular so that force is always applied at the edges I and III.

Alternatively, Hubner (1982, 1984) obtained the equations of static response of disk springs using the shell theory. In the cited articles the static response of truncated disk springs under central load was obtained numerically using shooting method and the Runge–Kutta method. Axisymmetric conical shells under axial forces were investigated numerically in Hübner and Emmerling (1982). While disk springs are rather flat and thick, the shell theory is applicable for steep and thin shells. A geometrically nonlinear approach leads to the Reissner-Meissner equations, which allow the calculation of large deformations. These two nonlinear second order equations have been integrated by a matrix method.

The expressions for determining the stresses, strains and displacements of a truncated or complete thin conical shell with constant thickness and axisymmetric load distributed or concentrated along the meridian are presented in Tavares (1996).

The analysis of non-linear characteristic of a disk spring continues to interest engineers (Niepage, 1983; Niepage et al., 1987). The disk springs with variable thickness were studied in Ferrari (2013) and Rosa et al. (1998). A theoretical analysis on Belleville spring with varying thickness was presented in the study (Saini et al., 2007). The expression for linearly varying thickness used by Rosa et al. (1998) was modified to incorporate a curvature factor and the corresponding load as well as stress equations are derived using the hypothesis of Almen and Laszlo.

In the paper Blom et al., (2009), the field of application of the variable-stiffness concept is extended to three-dimensional conical shells with arbitrary dimensions that can be fabricated using advanced fiber placement machines. Elastic stiffness tailoring of laminated composite panels by allowing the fibers to curve within the plane of the laminae has proven to be beneficial and practical for flat rectangular plate designs.

In this Chapter, we investigate the disk springs using the variation methods. The methods consequently apply the kinematic hypothesis and equations of thin and moderately thick shells. The interesting feature of the presented theory is its possibility to calculate the disk springs of a meridional variable thickness and with free and elastically supported edges.

4.1.2 Geometry of Disk Spring in Undeformed State

Consider a truncated conical shell. Its middle surface in the free state is the frustum ϖ (Fig. 4.1). The inside radius of middle surface of free spring is r_i, the outside radius is r_e. The ratio of outer radius to inside radius reads:

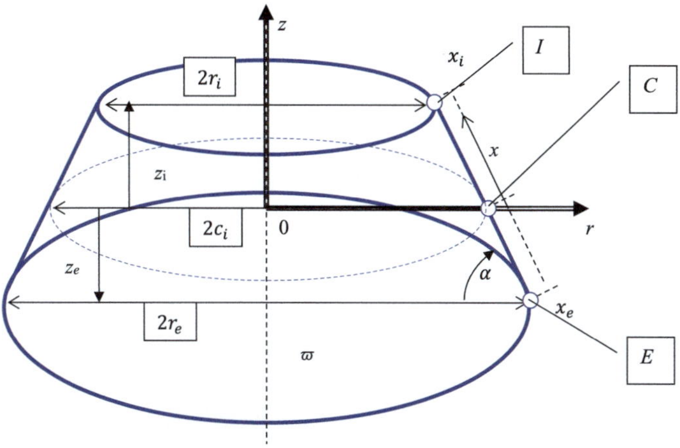

Fig. 4.1 The middle surface of the free disk spring

$$\Delta = \frac{r_e}{r_i} > 0.$$

The slope angle α of the conical shell is constant. The position of the middle surface of the shell relates to a neutral plane C. In cylindrical coordinates (r, θ, z), the middle surface of a free conical shell respectively is described parametrically:

$$r = c_i - x\cos\alpha, \quad z = -x\sin\alpha. \tag{4.1}$$

The arc length x serves as the meridional coordinate:

$$x_e \le x \le x_i.$$

The arc length is measured from a certain inversion point c_i, where

$$x_i = \frac{c_i - r_i}{\cos\alpha} > 0, \quad x_e = \frac{c_i - r_e}{\cos\alpha} < 0. \tag{4.2}$$

The distance from the point with coordinates (x, θ, y) to the axis of the undeformed cone is:

$$\tilde{r} = r + y\sin\alpha = c - x\cos\alpha + y\sin\alpha, \tag{4.3}$$

where y is the normal distance from the point to the middle surface of the shell[1]:

$$-\frac{T}{2} \le y \le \frac{T}{2}.$$

[1] For the material thickness the symbol $$T$$ is used, because symbol $$t$$ is reserved for time.

The thickness of the spring depends on the arc length $T = T(x)$.
From (4.1) and (4.2) follows:

$$r_i = c_i - x_i \cos(\alpha), \quad r_e = c_i - x_e \cos(\alpha), \quad \mathrm{d}r = \cos(\alpha)\mathrm{d}x \tag{4.4}$$

The heights of the inner and outer edges of the middle surface of the free shell are correspondingly (Fig. 4.1):

$$z_i = x_i \sin\alpha > 0, \quad z_e = x_e \sin\alpha < 0. \tag{4.5a}$$

The total height of the middle surface of the unloaded disk spring is:

$$h = z_i - z_e = (\Delta - 1) r_i \tan\alpha. \tag{4.5b}$$

Correspondingly, the principal radii of curvature r_1 and r_2 for the unloaded conical shell are:

$$\frac{1}{r_1} = \frac{\sin\alpha}{r}, \quad \frac{1}{r_2} = 0. \tag{4.6}$$

The width of the middle surface of the disk spring in its free state is:

$$b = r_e - r_i = (\Delta - 1) \cdot r_i.$$

4.1.3 Mass of Disk Spring with Variable Material Thickness

The applied method is valid for an arbitrary thickness distribution over the frustum. We consider the variable thickness along the meridian: $T = T(x)$. Particularly, the thickness function could be represented as follows:

$$T = T_e \cdot \left(\frac{x - x_i}{x_e - x_i}\right)^{\tilde{p}} + T_i \cdot \left(\frac{x_e - x}{x_e - x_i}\right)^{\tilde{p}},$$
$$\tilde{p} > 0, \quad T_e = T(x_e), \quad T_i = T(x_i). \tag{4.7a}$$

If $\tilde{p} \geq 1$, the thickness in the middle of frustum ϖ reduces. Otherwise, if $1 > \tilde{p} > 0$ the thickness in the middle increases. The mass of the disk spring could be reduced using the convex or concave variable thickness (Fig. 4.2) (Kobelev & Hesselmann, 2002). For the optimal design, the parameters $T_e, T_i, \tilde{p}$ serve as the design variables. If $\tilde{p} = 1$, the thickness is the linear function over meridian:

$$T = T_e \cdot \frac{x - x_i}{x_e - x_i} + T_i \cdot \frac{x_e - x}{x_e - x_i}. \tag{4.7b}$$

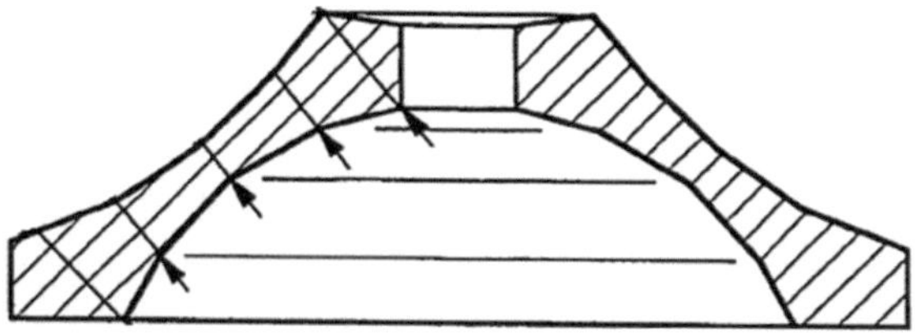

Fig. 4.2 The disk spring with the variable thickness along the meridian (Kobelev & Hesselmann, 2002)

The average thickness of material in this case is $T^* = \frac{T_i + T_e}{2}$.

If $\tilde{p} = 0$, the thickness of material is constant and the cross-section of disk is rectangular. The mass of the disk spring with an axisymmetric variable thickness is:

$$m = \pi \rho_m \int_{x_e}^{x_i} T(x) r \, \mathrm{d}x. \tag{4.8a}$$

For the thickness function (4.7a), the mass of the spring is equal:

$$m = \frac{\pi}{(p+1)(p+2)} [T_e \cdot (r_e p + r_e + r_i) + T_i \cdot (r_i p + r_e + r_i)] \cdot \rho_m \sqrt{h^2 + b^2}. \tag{4.8b}$$

For the linear thickness function (4.7b), the mass of the spring is equal:

$$m = \frac{\pi}{6} [T_e \cdot (2r_e + r_i) + T_i \cdot (r_e + 2r_i)] \, \rho_m \sqrt{h^2 + b^2}. \tag{4.8c}$$

Currently, the mainstream disk springs are manufactured with the constant thickness. Finally, for the constant thickness, Eq. (4.8b) reduces to:

$$m = \pi \rho_m (r_i + r_e) \sqrt{h^2 + b^2} T. \tag{4.8d}$$

4.1.4 Load-Caused Alteration of Strain and Curvature

The main hypothesis of the model is that the slope angle ψ of deformed conical shells changes, but also remains constant over the meridian (Kobelev, 2016). Namely, the point C is the inversion center point for the cross-section of the conical shell. The generatrice rotate about the point C, but remain straight and their lengths do not alter. The points of the middle surface, located on the neutral plane C, do not deflect. This hypothesis is essentially the same, as the hypothesis of Timoshenko and Almen and Laszlo. Analogous to the giant dislocation, which accompanies the distortion of helical springs, the deformation of disk springs is the sum of a giant twist and a giant wedge Volterra disclinations (Nabarro, 1967). The middle surface of the shell in the deformed state is again a frustum Ω, as shown above on (Fig. 4.3).

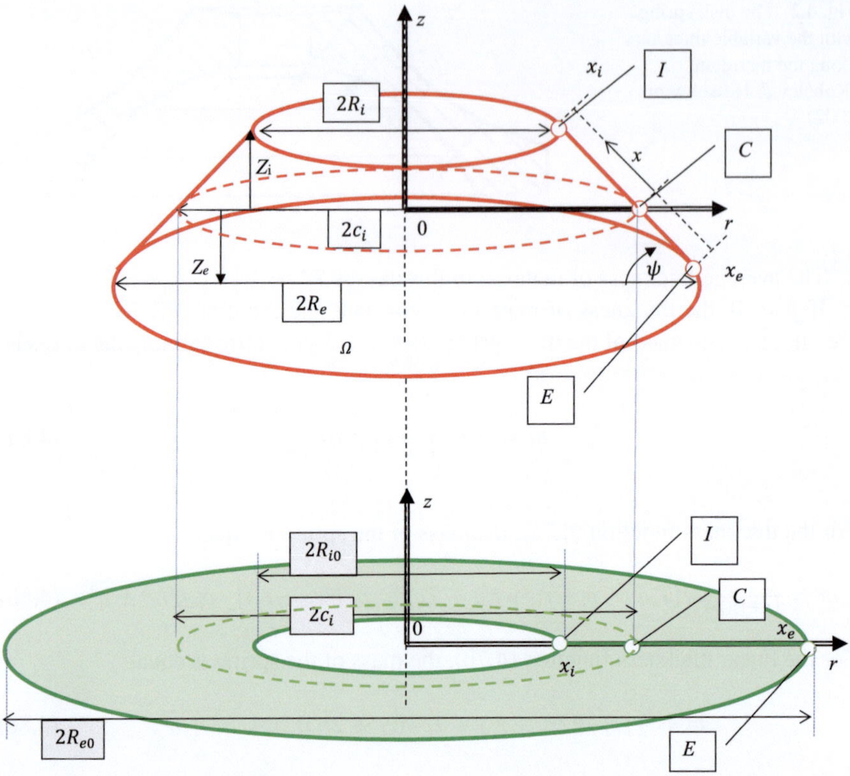

Fig. 4.3 The middle surfaces of the deformed and completely flattened disk spring

In cylindrical coordinates (r, θ, z), the middle surface Ω is given by the parametric equations:

$$R = c_i - x \cos \psi, \quad Z = -x \sin \psi. \tag{4.9}$$

The distance from the point of the thick-walled cone with coordinates (x, θ, y) to the axis of the deformed cone is:

$$\tilde{R} = R + y \sin \psi = c_i - x \cos \psi + y \sin \psi. \tag{4.10}$$

Correspondingly, the principal radii of curvature R_1 and R_2 of the middle surface Ω are:

$$\frac{1}{R_1} = \frac{\sin \psi}{r}, \quad \frac{1}{R_2} = 0. \tag{4.11}$$

The inner and outer radii of Ω are equal to:

$$R_i = c_i - x_i \cos\psi, \quad R_e = c_i - x_e \cos\psi. \tag{4.12}$$

The distances from the plane C to the inner and outer edges of surface Ω read:

$$Z_i = x_i \sin\psi, \quad Z_e = x_e \sin\psi. \tag{4.13}$$

The height and width of middle surface in the deformed state Ω are correspondingly:

$$H = Z_i - Z_e = r_i \cdot \frac{\sin\psi}{\cos\alpha} \cdot (\Delta - 1), \tag{4.14}$$

$$B = R_i - R_e = r_i \cdot \frac{\cos\psi}{\cos\alpha} \cdot (\Delta - 1). \tag{4.15}$$

Upon specializing the strain displacement relations to the case of a conical shell with no transverse shear deformations the following expression for the circumferential mid-surface strain:

$$\varepsilon_\theta = \frac{R - r}{r} = \frac{\cos\alpha - \cos\psi}{c_i - x\cos\alpha} x. \tag{4.16}$$

The circumferential curvature change is:

$$\kappa_\theta = \frac{1}{R_1} - \frac{1}{r_1} = \frac{\sin\psi - \sin\alpha}{r}. \tag{4.17}$$

The slope angle vanishes in the flattened state, as shown below on (Fig. 4.3). With Eq. (4.2), the inner and outer radii of the plane circular strip will be equal nearly to:

$$R_{i0} = c_i - x_i, \quad R_{e0} = c_i - x_e.$$

4.1.5 Disk Springs of Moderate Material Thickness

Considering the thickness of the material, the expressions for height and width must be corrected (Fig. 4.4). The total height and width of the disk spring in its free state from upper inside edge to lower outside are respectively:

$$\tilde{h} = h + T^* \cos\alpha = (\Delta - 1) \cdot r_i \tan\alpha + T^* \cos\alpha, \tag{4.18a}$$

$$\tilde{b} = b - T^* \sin\alpha = (\Delta - 1) \cdot r_i - T^* \sin\alpha. \tag{4.18b}$$

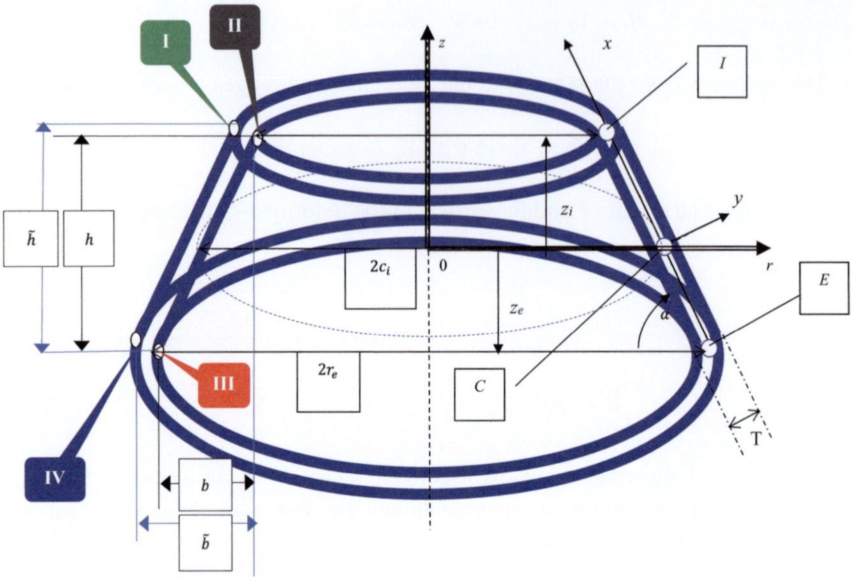

Fig. 4.4 The solid representation the free disk spring

The total height and width of the deformed disk spring from upper inside edge to lower outside are correspondingly:

$$\tilde{H} = H + T^* \cos\psi = r_i \cdot \frac{\sin\psi}{\cos\alpha} \cdot (\Delta - 1) + T^* \cos\psi, \tag{4.18c}$$

$$\tilde{B} = B - T^* \sin\psi = r_i \cdot \frac{\cos\psi}{\cos\alpha} \cdot (\Delta - 1) - T^* \sin\psi. \tag{4.18d}$$

The spring travel on the middle surface is equal to:

$$s = h - H = (\Delta - 1) \cdot r_i \cdot \frac{\sin\alpha - \sin\psi}{\cos\alpha}. \tag{4.19}$$

The spring travel from upper inside edge to lower outside is:

$$\tilde{s} = \tilde{h} - \tilde{H} = (\Delta - 1) \cdot r_i \cdot \frac{\sin\alpha - \sin\psi}{\cos\alpha} + T^* \cdot (\sin\psi - \sin\alpha). \tag{4.20}$$

For the constant thickness $T^* = T$. The influence of rounding on the corners could be also accounted without difficulty.

4.2 Disk Springs of Moderate Thickness

4.2.1 Deformation of Thick Conical Shell

The meridional and circumferential direct stresses relate to strains by means of relations (Marsden & Hughes, 1994):

$$\sigma_\theta = \frac{E}{(1+\nu)(1-2\nu)}((1-\nu)\cdot\mathbf{E}_\theta + \nu\cdot(\mathbf{E}_\mathrm{X} + \mathbf{E}_\mathrm{Y})), \tag{4.21a}$$

$$\sigma_\mathrm{X} = \frac{E}{(1+\nu)(1-2\nu)}((1-\nu)\cdot\mathbf{E}_\mathrm{X} + \nu\cdot(\mathbf{E}_\mathrm{Y} + \mathbf{E}_\theta)), \tag{4.21b}$$

$$\sigma_\mathrm{Y} = \frac{E}{(1+\nu)(1-2\nu)}((1-\nu)\cdot\mathbf{E}_\mathrm{Y} + \nu\cdot(\mathbf{E}_\theta + \mathbf{E}_\mathrm{X})). \tag{4.21c}$$

In Eqs. (4.21a–c) we use the following notations for the principal strains:

$\mathbf{E}_\theta$: the circumferential strain,
$\mathbf{E}_\mathrm{X}$: the meridional strain and
$\mathbf{E}_\mathrm{Y}$: the strain normal to surface of the shell.

All shear strains and shear stresses nullify due to the symmetry conditions. The hypothesis assumes the absence of the stress in the meridional and normal directions respectively:

$$\sigma_\mathrm{X} = 0, \quad \sigma_\mathrm{Y} = 0.$$

From these conditions the meridional strain $\mathbf{E}_\mathrm{X}$ and strain normal to surface of the shell $\mathbf{E}_\mathrm{Y}$ depend solely on the circumferential strain $\mathbf{E}_\theta$:

$$\mathbf{E}_\mathrm{X} = -\nu\mathbf{E}_\theta, \quad \mathbf{E}_\mathrm{Y} = -\nu\mathbf{E}_\theta. \tag{4.22}$$

The strain displacement relations (4.16)–(4.17) lead to the following expressions for the strains in the solid elastic cone with the thickness $T(x)$:

$$\mathbf{E}_\theta = \varepsilon_\theta + y\kappa_\theta = \frac{(\cos\alpha - \cos\psi)\cdot x + (\sin\psi - \sin\alpha)\cdot y}{c_i - x\cos\alpha}, \quad \text{for } -\frac{T}{2} \le y \le \frac{T}{2}. \tag{4.23}$$

4.2.2 Variation Method for Thick Shell Models of Disk Springs

The total potential energy

$$\Pi = U_e + U_f \tag{4.24}$$

is the sum of the elastic strain energy U_e, stored in the deformed body and the potential energy U_f of the applied forces:

$$U_f = \int_{\alpha}^{\psi} M \mathrm{d}\psi, \quad M = F_Z B - F_R H. \tag{4.25}$$

The shell is stressed by the forces in the direction of the rotation axis of the shell or by the forces in radial direction. Normally, the applied load is evenly distributed over the upper inside edge and lower outside edge (Fig. 4.5). The total axial force acting on the inner edge "I" is F_z and the total radial force is F_R.

The strain energy $U_e = U_1$ of the elastically deformed cone is:

$$U_1 = \pi G \int_{x_e}^{x_i} \left[\int_{-T/2}^{T/2} \left(\mathbf{E}_\theta{}^2 + \mathbf{E}_\mathrm{X}{}^2 + \mathbf{E}_\mathrm{Y}{}^2 + \tfrac{\nu}{1-2\nu} \cdot (\mathbf{E}_\theta + \mathbf{E}_\mathrm{X} + \mathbf{E}_\mathrm{Y})^2 \right) r \, \mathrm{d}y \right] \mathrm{d}x. \tag{4.26}$$

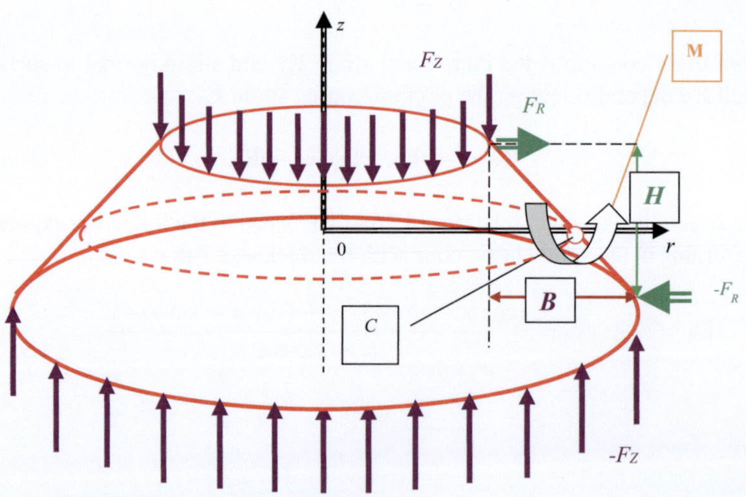

Fig. 4.5 Axial and radial forces on the disk spring

Using Eqs. (4.21a–c) and (4.22) the expression (4.26) for the elastic energy for the nonconstant thickness reads:

$$U_1 = \pi E \int_{x_e}^{x_i} \left[\int_{-T/2}^{T/2} \mathbf{E}_\theta^2 r \, \mathrm{d}y \right] \mathrm{d}x. \tag{4.27}$$

For the constant thickness, the integration (4.27) over the cross-section delivers:

$$U_1 = \frac{\pi E T^3}{12} \cdot \ln\Delta \cdot \frac{(\sin\psi - \sin\alpha)^2}{\cos\alpha} - \frac{\pi E T}{2} \cdot \frac{(\cos\psi - \cos\alpha)^2}{\cos^2\alpha} \cdot \left[\left(\Delta^2 - 1 \right) {r_i}^2 + 4(1-\Delta) r_i c_i + 2{c_i}^2 \ln\Delta \right]. \tag{4.28}$$

The position of the neutral plane is determined from the condition of minimum for elastic energy with respect to c_i:

$$\frac{\partial U_1}{\partial c_i} = -\frac{\pi E T}{2} \cdot \frac{(\cos\psi - \cos\alpha)^2}{\cos^2\alpha} \cdot [4(1-\Delta) r_i + 4 c_i \ln\Delta] = 0. \tag{4.29}$$

Solution of Eq. (4.29) delivers the radius of the inversion center point C:

$$c_1 = \frac{\Delta - 1}{\ln\Delta} r_i \equiv \frac{D_e - D_i}{2\ln(D_e/D_i)}. \tag{4.30}$$

According to the principle of virtual work, the external virtual work is equal to internal virtual work when equilibrated forces and stresses undergo unrelated but consistent displacements and strains. The necessary stationarity condition with respect to ψ delivers the circumferential moment:

$$\begin{aligned} M_1 &= \frac{\partial U_1}{\partial \psi} = E r_i^3 \mu \frac{(\cos\psi - \cos\alpha)\sin\psi}{\cos^3\alpha} \cdot \left[2(\Delta - 1) - \frac{1}{2}\left(\Delta^2 - 1\right) \right] \\ &+ E r_i^3 \mu^3 \cdot \frac{(\sin\psi - \sin\alpha)\cos\psi}{12\cos^3\alpha} \ln\Delta, \\ \mu &= T/r_i. \end{aligned} \tag{4.31}$$

In the absence of the radial force, the axial force on the shell is:

$$F_{1Z} = \frac{1}{B} \frac{\partial U_1}{\partial \psi} = \frac{M_1}{B}. \tag{4.32}$$

The substitution of (4.26) into (4.28) delivers the central formula for the axial force of disk spring $F_{1Z}$and the spring rate of disk spring c_{1Z} of the elastic cone:

$$\begin{aligned} F_{1Z} &= F_{1Z}\big|_{c_i = c_1} = \pi E {r_i}^2 \cdot \frac{\left(F_e \mu + F_f \mu^3\right)}{\cos\psi}, \\ c_{1Z} &= \frac{\mathrm{d}F_{1Z}/\mathrm{d}\psi}{\mathrm{d}s/\mathrm{d}\psi}, \end{aligned} \tag{4.33}$$

$$F_e = \frac{2(1-\Delta) + (1+\Delta)\ln\Delta}{\ln\Delta} \cdot \frac{(\cos\psi - \cos\alpha)}{\cos^2\alpha} \cdot \sin\psi, \tag{4.34}$$

$$F_f = \frac{\ln\Delta}{6(\Delta-1)} \cdot (\sin\alpha - \sin\psi) \cdot \cos\psi. \tag{4.35}$$

The equivalent radial force F_{1R}, which leads to the same stored elastic energy of the spring, relates to the axial force F_{1Z} on the shell as:

$$F_{1R} = -\frac{1}{H}\frac{\partial U_1}{\partial\psi} = -\frac{M_1}{H} = -F_{1Z}\tan\psi. \tag{4.36}$$

If the applied load is evenly distributed over the upper inside edge and lower outside edge the expressions for force and spring rate must be corrected. Accounting the thickness of the material, the axial force $\tilde{F}_{1Z}$ and spring rate $\tilde{c}_{1Z}$ express:

$$\tilde{F}_{1Z} = \frac{1}{\tilde{B}}\frac{\partial U_1}{\partial\psi} = \frac{M_1}{\tilde{B}} = \frac{B}{\tilde{B}}F_{1Z},$$
$$\tilde{c}_{1Z} = \frac{\mathrm{d}\tilde{F}_{1Z}/\mathrm{d}\psi}{\mathrm{d}\tilde{s}/\mathrm{d}\psi}. \tag{4.37}$$

4.2.3 Comparison of Calculation Techniques

The next task is to compare the results, obtained by the variation methods for thin shell and solid cone with the formulas, derived by common stress method.

Firstly, in the total circumferential force in the stress method is derived using the integration of circumferential stress $\sigma_\theta = E \cdot \mathbf{E}_\theta$ over the cross-sectional surface:

$$F_{AL} = E\int_{x_e}^{x_i}\left[\int_{-T(x)/2}^{T(x)/2} \mathbf{E}_\theta \mathrm{d}y\right]\mathrm{d}x. \tag{4.38}$$

Performing the integration (4.38) over the cross-section with the constant thickness, one gets the total circumferential force:

$$F_{AL} = E\mu r_i \cdot \frac{\cos\psi - \cos\alpha}{\cos^2\alpha} \cdot [c_i\ln\Delta - r_i(\Delta-1)]. \tag{4.39}$$

The inversion center radius is determined from the condition of vanishing for circumferential force:

$$F_{AL} = 0.$$

From this condition results the position of the inversion center:

$$c_{AL} = \frac{\Delta - 1}{\ln\Delta} r_i \equiv \frac{D_e - D_i}{2\ln(D_e/D_i)}. \tag{4.40}$$

The comparison of (4.30) and (4.40) demonstrates the equality both expressions for inversion center radii, obtained by variation and stress methods.

Secondly, the circumferential moment in the cross-section is the integral of circumferential stress $\sigma_\theta = E \cdot \mathbf{E}_\theta$ with the arm

$$\rho = x \sin(\alpha - \varphi) + y \cos(\alpha - \varphi).$$

over the cross-section. That is, the moment is the integral over the cross-sectional surface:

$$M_{AL} = E \int_{x_e}^{x_i} \left[\int_{-T(x)/2}^{T(x)/2} \rho \boldsymbol{E}_\theta \mathrm{d}y \right] \mathrm{d}x. \tag{4.41}$$

For the constant material thickness, the integration in (4.41) yields the moment:

$$M_{AL} = Er_i^3 \mu \cdot \frac{(\cos\psi - \cos\alpha)\sin\psi}{\cos^3\alpha} \cdot \left[2(\Delta - 1) - \frac{1}{2}(\Delta^2 - 1) \right] + Er_i^3 \mu^3 \cdot \frac{(\sin\psi - \sin\alpha)\cos\psi}{12\cos^3\alpha} \cdot \ln\Delta. \tag{4.42}$$

Remarkably, that both methods lead to the same result and the expressions (4.31) and (4.41) match.

4.3 Statics of Thin Disk Springs

4.3.1 Forces and Moments in Disk Springs

In this Section the expression for the elastic energy is derived using the thin conical shell model. The meridional and circumferential direct forces and moments for thin shell relate to strains and curvature changes by means of relations (Ventsel & Krauthammer, 2001, Eq. 14.32):

$$N_1 = \frac{ET}{1 - \nu^2}(\varepsilon_1 + \nu\varepsilon_2), \quad N_2 = \frac{ET}{1 - \nu^2} \cdot (\varepsilon_2 + \nu\varepsilon_1), \tag{4.43}$$

$$M_1 = \frac{ET^3}{12(1 - \nu^2)} \cdot (\kappa_1 + \nu\kappa_2), \quad M_2 = \frac{ET^3}{12(1 - \nu^2)} \cdot (\kappa_2 + \nu\kappa_1). \tag{4.44}$$

The values in Eqs. (4.43)–(4.44) are:

$ET/(1-\nu^2)$ the extensional rigidity of shell,
$ET^3/12(1-\nu^2)$ the flexural rigidity.

The second hypothesis assumes the absence of the direct stress in the meridional direction:

$$N_2 = 0.$$

The third hypothesis assumes the absence of the moment in the meridional direction:

$$M_2 = 0.$$

From these conditions the meridional strain ε_2 and curvature change κ_2 depend solely on the circumferential strain ε_1 and curvature change κ_1:

$$\varepsilon_2 = -\nu\varepsilon_1, \tag{4.45}$$

$$\kappa_2 = -\nu\kappa_1. \tag{4.46}$$

Substitution of (4.45) and (4.46) in (4.43) and (4.44) delivers the expressions for the circumferential stress σ_1 and moment M_1:

$$\sigma_1 = \frac{N_1}{T} = E\varepsilon_1, \tag{4.47}$$

$$M_1 = \frac{1}{12}ET^3\kappa_1. \tag{4.48}$$

4.3.2 Strain Energy of Thin Disk Springs

The strain energy of the shell can be determined by means of Kirchhoff's assumptions (Ventsel & Krauthammer, 2001, Eq. 14.51):

$$U_2 = U_{2e} + U_{2f}, \tag{4.49a}$$

$$U_{2e} = 2\pi \int_{x_e}^{x_i} W_2^e r \, \mathrm{d}x,$$

$$U_{2f} = 2\pi \int_{x_e}^{x_i} W_2^f r \, \mathrm{d}x. \tag{4.49b}$$

The extensional and flexural elastic energies in Eqs. (4.49a, 4.49b) respectively are (Libai & Simmonds, 1988):

$$U_{2e} = \pi E \int_{x_e}^{x_i} T(x)\left[(\varepsilon_1 + \varepsilon_2)^2 - 2(1-\nu)\varepsilon_1\varepsilon_2\right] r \, dx, \tag{4.50}$$

$$U_{2f} = \frac{\pi E}{12} \int_{x_e}^{x_i} T^3(x)\left[(\kappa_1 + \kappa_2)^2 - 2(1-\nu)\kappa_1\kappa_2\right] r \, dx. \tag{4.51}$$

Using the formulas (4.45), (4.46) in (4.50) and (4.51), the extensional and flexural surface densities of elastic energy reduce to:

$$2W_2^e = ET\varepsilon_1^2, \quad 2W_2^f = \frac{1}{12}ET^3\kappa_1^2.$$

Accordingly, the extensional and flexural elastic energies are:

$$U_{2e} = \pi E \int_{x_e}^{x_i} r\varepsilon_1^2 T(x)dx, \quad \varepsilon_1 \equiv \varepsilon_\theta, \tag{4.52}$$

$$U_{2f} = \frac{\pi E}{12} \int_{x_e}^{x_i} r\kappa_1^2 T^3(x)dx, \quad \kappa_1 \equiv \kappa_\theta. \tag{4.53}$$

For the constant thickness, the integration (4.52) and (4.53) reduce the extensional and flexural elastic energy components respectively to:

$$U_{2e} = -\frac{\pi ET}{2} \cdot \frac{(\cos\psi - \cos\alpha)^2}{\cos^2\alpha} \cdot \left[(\Delta^2 - 1){r_i}^2 + 4(1-\Delta)r_i c_i + 2{c_i}^2 \ln\Delta\right], \tag{4.54}$$

$$U_{2f} = \pi \frac{ET^3}{12} \cdot \frac{(\sin\psi - \sin\alpha)^2}{\cos\alpha} \cdot \ln\Delta. \tag{4.55}$$

The position of the neutral plane is determined from the requirement of minimum for elastic energy:

$$\frac{\partial U_2}{\partial c} = -\frac{\pi ET}{2} \cdot \frac{(\cos\psi - \cos\alpha)^2}{\cos^2\alpha} \cdot [4(1-\Delta)r_i + 4c_i \ln\Delta] = 0. \tag{4.56}$$

Solution of Eq. (4.56) delivers the radius of the inversion center point C:

$$c_2 = \frac{\Delta - 1}{\ln\Delta} r_i \equiv \frac{D_e - D_i}{2\ln(D_e/D_i)}. \tag{4.57}$$

This method delivers the same expression for inversion center radius as (4.30) and (4.40).

Finally, the circumferential moment is:

$$M_2 = \frac{\partial U_2}{\partial \psi} = E r_i^3 \mu \cdot \frac{(\cos\psi - \cos\alpha)\sin\psi}{\cos^3\alpha} \cdot \left[2(\Delta - 1) - \frac{1}{2}(\Delta^2 - 1)\right]$$
$$+ E r_i^3 \mu^3 \cdot \frac{(\sin\psi - \sin\alpha)\cos\psi}{12\cos^3\alpha} \ln\Delta. \tag{4.58}$$

Comparison demonstrates the identity of the expressions for moment (4.41) and (4.58). Consequently, it is proved that both methods lead to the same expressions for spring forces and spring rates.

For comparison the finite element simulation using program ANSYS (2020) was performed. For the finite element modeling three models were developed (Kobelev, 2016) The finite element simulation with ABAQUS (2020) leads to the similar results.

4.3.3 Almen and Laszlo Method for Thin Disk Springs

The common calculation for the force of disk spring with the constant material thickness T is based on the Almen and Laszlo formulas (Almen & Laszlo, 1936):

$$F_{zDIN} = \frac{4E}{1-\nu^2} \cdot \frac{T^4}{K_1 D_e^2} \cdot \frac{s}{T} \cdot \left[\left(\frac{h}{T} - \frac{s}{T}\right) \cdot \left(\frac{h}{T} - \frac{s}{2T}\right) + 1\right],$$
$$c_{zDIN} = \frac{\mathrm{d}F_{zDIN}}{\mathrm{d}s}. \tag{4.59}$$

These formulas are used in European standard (EN 16984:2017-02, 2017). For comparison one example of spring calculation was considered. The parameters of the disk spring for the example are:

$$D_e = 110\ [\mathrm{mm}];\ D_i = 90\ [\mathrm{mm}];\ h = 6\ \mathrm{mm};\ T = 2\ [\mathrm{mm}],\ E = 200\ [\mathrm{GPa}];\ \nu = 0.3.$$

The forces and spring rates for the example were calculated according Eq. (4.59) for the DIN standard and are shown on Figs. 4.6 and 4.7 with blue dashed lines. The forces and spring rates for the actual theory were calculated according Eq. (4.33) are shown on Figs. 4.6 and 4.7 with solid red lines.

The corner points according to standard (EN 16984:2017-02, 2017) are depicted on Fig. 4.4. The coordinates of corner points are correspondingly:

$$I : \rho = D_i/2,\ y = T/2; \tag{4.60a}$$

$$II : \rho = D_i/2,\ y = -T/2; \tag{4.60b}$$

$$III : \rho = D_e/2,\ y = -T/2; \tag{4.60c}$$

$$IV : \rho = D_e/2,\ y = T/2. \tag{4.60d}$$

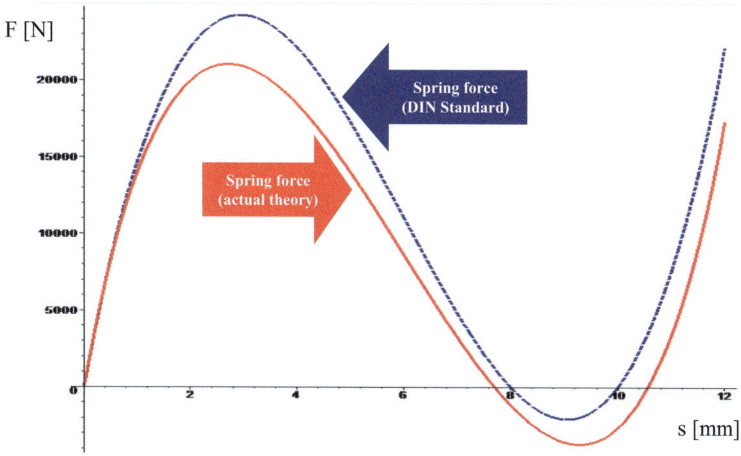

Fig. 4.6 Comparison of axial forces on the disk spring: DIN standard, Eq. (3.19), blue, actual theory, Eq. (4.13), red

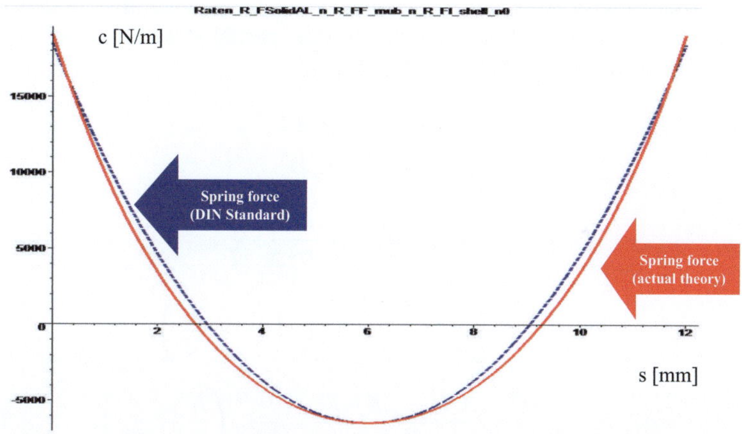

Fig. 4.7 Comparison of spring rates on the disk spring: DIN standard (blue) and actual theory (red)

The stresses in the spring on the corner points I, … IV are pure elastic and are equal to (DIN EN 16984:2017-02, 2017):

$$\sigma_I = -\frac{4E}{1-\nu^2} \cdot \frac{T^2}{K_1 D_e^2} \cdot \frac{s}{T} \cdot \left[K_2 \cdot \left(\frac{h}{T} - \frac{s}{2T} \right) + K_3 \right], \tag{4.61}$$

$$\sigma_{II} = -\frac{4E}{1-\nu^2} \cdot \frac{T^2}{K_1 D_e^2} \cdot \frac{s}{T} \left[K_2 \cdot \left(\frac{h}{T} - \frac{s}{2T} \right) - K_3 \right], \tag{4.62}$$

$$\sigma_{III} = -\frac{4E}{1-\nu^2} \cdot \frac{T^2}{K_1 D_e^2} \cdot \frac{s}{T\Delta} \cdot \left[(K_2 - 2K_3) \cdot \left(\frac{h}{T} - \frac{s}{2T} \right) - K_3 \right], \tag{4.63}$$

$$\sigma_{VI} = -\frac{4E}{1-\nu^2} \cdot \frac{T^2}{K_1 D_e^2} \cdot \frac{s}{T\Delta} \cdot \left[(K_2 - 2K_3) \cdot \left(\frac{h}{T} - \frac{s}{2T} \right) + K_3 \right]. \tag{4.64}$$

The coefficients in Eqs. (4.59)–(4.63) are (DIN EN 16984:2017-02, 2017):

$$K_1 = \frac{1}{\pi} \left(\frac{\Delta - 1}{\Delta} \right)^2 \cdot \left(\frac{\Delta + 1}{\Delta - 1} - \frac{2}{\ln \Delta} \right)^{-1}, \tag{4.65}$$

$$K_2 = \frac{6}{\pi \ln \Delta} \cdot \left(\frac{\Delta - 1}{\ln \Delta} - 1 \right), \quad K_3 = \frac{3(\Delta - 1)}{\pi \ln \Delta}. \tag{4.66}$$

For the considered, ideal rectangular form of the spring the value of the coefficient is $K_4 = 1$.

The introduction of the "partial stresses" eases the estimation of effects of the circumferential bending and uniaxial strain. The "partial stresses" easies could be evaluated from (DIN EN 16984:2017-02, 2017):

$$\sigma_{Bi} = -\frac{4E}{1-\nu^2} \cdot \frac{K_3 T^2}{K_1 D_e^2} \cdot \frac{s}{T}, \tag{4.67}$$

$$\sigma_{Be} = -\frac{4E}{1-\nu^2} \cdot \frac{K_3 T^2}{K_1 D_e^2 \Delta} \cdot \frac{s}{T}, \tag{4.68}$$

$$\sigma_{Ti} = -\frac{4E}{1-\nu^2} \cdot \frac{K_2 T^2}{K_1 D_e^2} \cdot \frac{s}{T} \cdot \left(\frac{h}{T} - \frac{s}{2T} \right), \tag{4.69}$$

$$\sigma_{Te} = -\frac{4E}{1-\nu^2} \cdot \frac{(K_2 - 2K_3) T^2}{K_1 D_e^2} \cdot \frac{s}{T\Delta} \cdot \left(\frac{h}{T} - \frac{s}{2T} \right). \tag{4.70}$$

The physical meaning of partial stresses is the following:

σ_{Be} is the stress due to bending on the outer diameter D_e;
σ_{Bi} is the stress due to bending on the inner diameter D_i.
σ_{Te} is the stress due to circumferential strain on the outer diameter D_e;
σ_{Ti} is the stress due to circumferential strain on the inner diameter D_i.

The stresses on the corner points from Eqs. (4.61)–(4.64) relate to the "partial stresses" Eqs. (4.27)–(4.30):

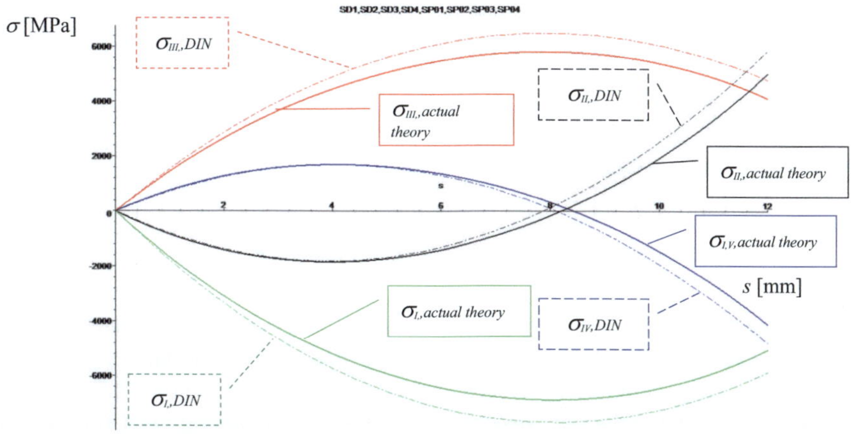

Fig. 4.8 Comparison of corner stresses: DIN standard (blue), actual theory (red)

$$\begin{aligned}\sigma_I &= \sigma_{Bi} + \sigma_{Ti},\\ \sigma_{II} &= -\sigma_{Bi} + \sigma_{Ti},\\ \sigma_{III} &= -\sigma_{Be} + \sigma_{Te},\\ \sigma_{VI} &= \sigma_{Be} + \sigma_{Te}.\end{aligned} \tag{4.71}$$

The stresses on the corner points for the example were calculated according Eqs. (4.61)–(4.64) for the DIN standard and are shown on Fig. 4.8 by dashed lines. The stresses for the actual theory were calculated according Eq. (4.71) are shown on Fig. 4.8 by solid lines.

4.3.4 Stresses in Disk Springs

The stresses in corner points could be evaluated in the shell theory of disk springs. For this purpose, we use Eq. (4.1) for stress–strain dependence and Eq. (4.3) for circumferential strain. The expressions for "partial stresses" follow from these formulas after substitution of coordinates for the corner points (4.60a–d):

$$\begin{aligned}\sigma_{Bi} &= \frac{E}{1-\nu^2}\cdot\frac{\sin\alpha-\sin\psi}{2}\cdot\mu,\\ \sigma_{Be} &= \frac{\sigma_{Bi}}{\Delta},\quad \mu=\frac{T}{r_i},\end{aligned} \tag{4.72}$$

$$\begin{aligned}\sigma_{Ti} &= \frac{E}{1-\nu^2}\cdot\frac{\cos\psi-\cos\alpha}{\cos\alpha}\cdot\frac{1-\Delta+\ln\Delta}{\ln\Delta},\\ \sigma_{Te} &= \frac{1-\Delta+\Delta\ln\Delta}{(1-\Delta+\ln\Delta)\Delta}\sigma_{Ti}.\end{aligned} \tag{4.73}$$

With Eqs. (4.72)–(4.73) for partial stresses, the stresses on the corner points follow from Eq. (4.71). The corresponding travel of the spring provide Eqs. (4.19) and (4.20).

4.4 Disk Wave Springs

4.4.1 Application Fields of Disk Wave Springs

The disk wave springs are the corrugated springs, which comprise an annular spring band. The spring band describes a corrugation line oscillating about a reference surface over the circumference. Such corrugated springs are known as so-called "axial corrugated springs" or "disk wave springs" for absorbing axial forces in a spring fashion wherein the corrugation line oscillates over the circumference relative to a radial base plane. The radial base plane lies normal to the ring axis. In this case, the spring band is generally closed in a ring shape. Such corrugated springs are furthermore known as so-called "radial corrugated springs" in which the spring band oscillates over the circumference relative to a ring cylinder which lies coaxial to the ring axis. Corrugated springs of this type are often slotted over the circumference. Common corrugated springs have an almost linear spring characteristic with a constant spring state (spring stiffness, Fig. 4.9).

The axial corrugated springs or axial radial corrugated springs are utilized as an elastic compensating disk in valve clearance compensating elements in the valve gear mechanism of internal combustion engines. In this case the valve clearance

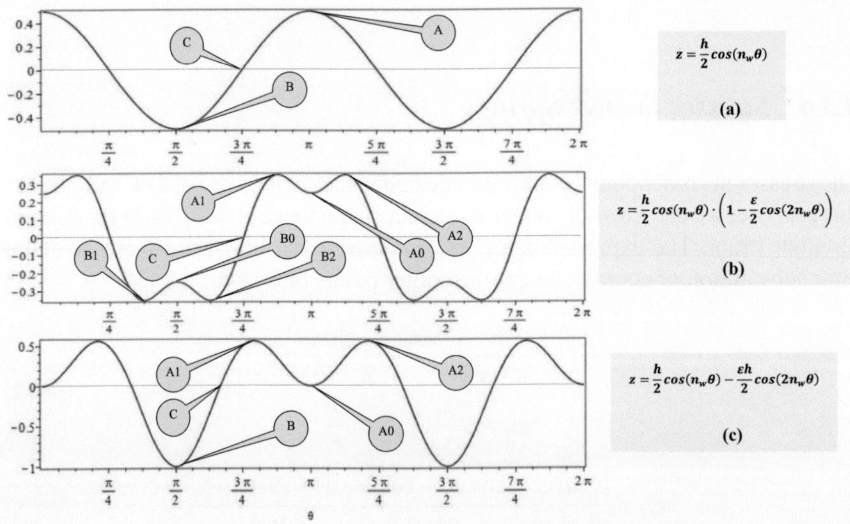

Fig. 4.9 Wave disk springs: **a** linear, **b** nonlinear symmetric, **c** nonlinear asymmetric

compensating elements are constructed as multi-part valve spring plates comprising a cup, a plate and an interposed corrugated spring.

A second preferred utilization for axial corrugated springs consists in their application as a clearance compensating and damping element in multiple-disk clutches of automatic gear mechanisms in motor vehicles. In this case, one or a plurality of corrugated springs are inserted at one end of the disk package of the multiple-disk clutches for coupling the gears, that is on the pressure side or the support side of the disk package. In this case, it is also possible to have a layered arrangement of a plurality of corrugated springs in the same orientation or in the opposite orientations.

For particular applications, certain gradually progressive spring characteristics are desirable. With known corrugated springs this has conventionally only been possible by aggregation of a plurality of corrugated springs have different spring characteristics with interposed flat disks.

The gradually progressive corrugated spring comprises a closed annular spring band which describes a corrugation line oscillating about a radial reference plane over the circumference (Kobelev et al., 2003) (Fig. 4.10). The corrugation line has a plurality of maxima of different height and a plurality of minima of different height, such that the corrugation line comprises a plurality of periods over the circumference. Between two adjacent absolute maxima there is a relative minimum, and wherein between two adjacent absolute minima there is a relative maximum. Every second absolute maximum is immediately adjacent to an absolute minimum and vice versa around the circumference.

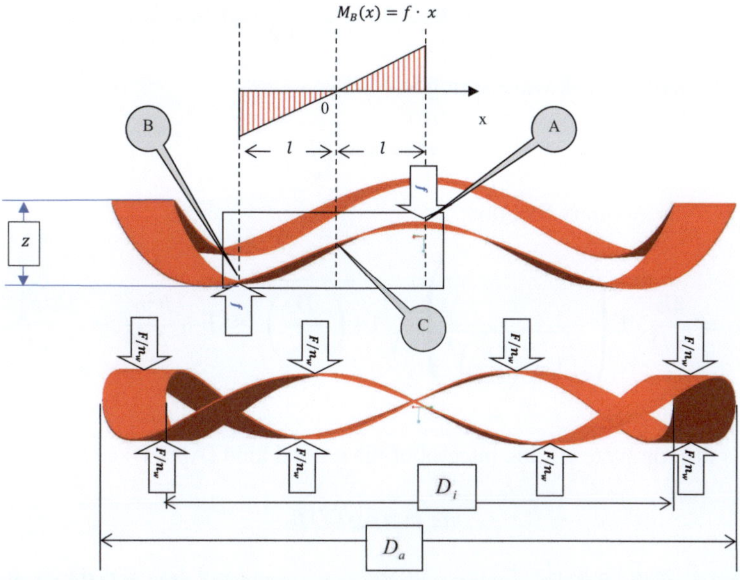

Fig. 4.10 Forces on common linear wave disk spring

4.4.2 Design Formulas for Linear Disk Wave Springs

The central surface of the axial corrugated spring is a curved circular surface with the width:

$$B = \frac{1}{2}(D_e - D_i) \equiv r_e - r_i \tag{4.74}$$

and with the mean diameter

$$D = \frac{1}{2}(D_e + D_i) \equiv r_e + r_i. \tag{4.75}$$

The hollow circular surface is represented for simplification by a flat straight bending strip. The middle surface of the disk wave spring is developable ruled surface. The number of waves is an even integer number n_w, such that the length of a wave is:

$$l_w = \pi \frac{D}{n_w}. \tag{4.76}$$

If the height of the middle surface is h, the directrix line of the middle surface in cylindrical coordinates $\{r, \theta, z\}$ reads (Fig. 4.9a):

$$r = \frac{D}{2}, \quad \theta = 0 \ldots 2\pi, \quad z = \frac{h}{2} \cdot \cos(n_w \theta).$$

The mass of the disk wave spring is:

$$m = m_0 \cdot \varrho,$$

where ϱ is the auxiliary function:

$$\varrho = \frac{4h}{\pi D} \cdot \mathbf{E}\left(\frac{1}{\sqrt{1 + \left(\frac{D}{hn_w}\right)^2}}\right) \cdot \sqrt{1 + \left(\frac{D}{hn_w}\right)^2} \approx 1 + \frac{h^2 n_w^2}{4D^2} - \frac{3h^4 n_w^4}{64D^4}.$$

$\mathbf{E}(m)$ is the complete elliptic integral of the second kind (App. B) and

$$m_0 = \pi \rho_m T D B.$$

is the mass of the flat disk spring with the same projected area πDB and material thickness T.

In the flattened state the projected area of disk wave spring increases in size. The scaling factor is ϱ. Thus, the diameters of flattened strip are: ϱD_e, ϱD_i, ϱD:

For the calculation, the oscillating corrugated shape of disk wave spring is represented by an almost flat bending strip. The arc effects are neglected for stiffness calculation. On the upper and lower tips of waves the disk wave spring contacts to the flat support pistons. Due to the contact reactions, the strip is loaded with pairwise opposite forces. The forces act normally to the strip surface on the upper and lower tips of waves. Overall, the n_w spring forces act perpendicularly downwards and n_w forces perpendicularly upwards. Since the total force at the corrugated spring is F, the partial force acts in one or the other direction at each attachment point reads:

$$f = \frac{F}{n_w}. \tag{4.77}$$

The calculation of the spring rate of the conventional, linear disk wave spring is based on the theory of thin beams. Consider the partial section between two adjacent abutment points (Fig. 4.10). The length of the strip between two adjacent abutment points A and B is:

$$l = \frac{l_w}{2} = \frac{\pi}{2}\frac{D}{n_w}. \tag{4.78}$$

The bending moment in the cross-section of the corrugated spring disappears in the point C lying centrally between points A and B. At this point, we place a local coordinate system with the axis x in the circumferential direction of the bending beam. The bending moment as a function of the distance to the point C is:

$$M_B(x) = f x, \quad 0 \le x \le l. \tag{4.79}$$

The stored elastic energy of the part portion of the elastic plate between points C and A of elastic shaft using (4.79) reads:

$$W = \int_o^{l/2} \frac{M_B^2}{2EI} \mathrm{d}x = \frac{1}{2EI} \int_o^{l/2} (f x)^2 \mathrm{d}x = \frac{1}{2EI} \frac{f^2}{3} \left(\frac{l}{2}\right)^3. \tag{4.80}$$

The second derivative of the stored elastic energy (4.80) yields the spring rate of a single subsection:

$$\frac{1}{c_a} = \frac{\partial^2 W}{\partial f^2} = \frac{1}{EI} \frac{l^3}{24} = \frac{1}{192 EI} \left(\frac{\pi D}{n_w}\right)^3. \tag{4.81}$$

The total spring rate of the linear wave spring with n_w total waves is:

$$c_w = n_w c_a = \frac{192}{\pi^3} \frac{n_w^4 EI}{D^3}. \tag{4.82}$$

The surface moment of inertia of the bending beam with the moderate width B and the material thickness T is

$$I = \frac{BT^3}{12} = \frac{D_e - D_i}{2} \cdot \frac{T^3}{12}. \tag{4.83}$$

For considerably wide springs the formulas for flexural rigidity of thin shells must be applied (Ventsel & Krauthammer, 2001):

$$I = \frac{BT^3}{12(1 - \nu^2)} = \frac{D_e - D_i}{2} \cdot \frac{T^3}{12(1 - \nu^2)}.$$

The spring rate of the linear disk wave spring of a small annular width for small deflections derived with the simplest assumption of an elongated rectangular bending beam. After the substitution (4.83) in (4.82) the spring rate of the conventional, linear disk is:

$$c_w = 16EBn_w^4 \cdot \left(\frac{T}{\pi D}\right)^3 \equiv \frac{64}{\pi^3} \cdot \frac{D_e - D_i}{(D_e + D_i)^3} \cdot En_w^4 T^3. \tag{4.84}$$

The stresses in the disk wave spring oscillate along the circumference of the spring. Maximum bending moments in the cross-section of the corrugated spring are located at the extreme points below the force application points:

$$M_B = \frac{\pi}{n_w}(D_e + D_i)F. \tag{4.85}$$

The moment attains its maximum M_B exactly below the force application points. The stress calculation with (4.85) lead to the highest bending stresses at these points:

$$\sigma = \frac{Fl}{8W_B} = \frac{6\pi D}{8BT^2} \cdot \frac{F}{n_w} = \frac{3\pi}{4T^2} \cdot \frac{D_e + D_i}{D_e - D_i} \cdot \frac{F}{n_w}.$$

The bending moments disappear in the reversal regions between the force application points.

4.4.3 Design Formulas for Non-linear Disk Wave Springs

Progressive wave springs have a gradually progressive characteristic. The wave line has a plurality of maxima of different sizes and a plurality of minima of different sizes. A whole wave section consists of four sub-sections, two sub-sections are connected in parallel and two are connected in series (Fig. 4.11). The spring rate of two partial sections that connected in parallel is exactly twice as high as the spring rate of a single partial section. The spring rate in the series connected partial sections is half

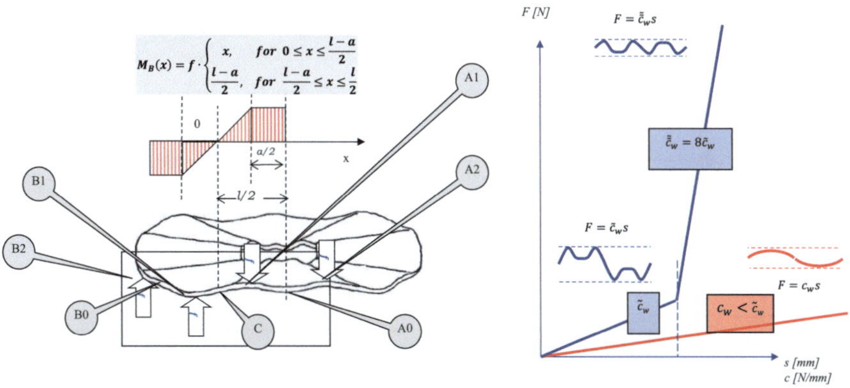

Fig. 4.11 Non-linear wave spring. The spring forces of linear and non-linear wave spring

the spring rate of a section. Consequently, the spring rate of a full shaft is equal to the spring rate of a subsection.

There are two types of the non-linear disk wave springs. For the symmetric spring the directrix line of the middle surface in cylindrical coordinates $\{r, \theta, z\}$ reads (Fig. 4.9b):

$$r = \frac{D}{2}, \quad \theta = 0 \ldots 2\pi,$$
$$z = \frac{h}{2}\cos(n_w\theta) \cdot \left[1 - \frac{\varepsilon}{2}\cos(2n_w\theta)\right].$$

Parameter $\varepsilon>0$ determines the depth of the secondary wave. For the asymmetric spring the directrix line of the middle surface in cylindrical coordinates $\{r, \theta, z\}$ delivers (Fig. 4.9c):

$$r = \frac{D}{2}, \quad \theta = 0 \ldots 2\pi,$$
$$z = \frac{h}{2}\cos(n_w\theta) - \frac{\varepsilon h}{2}\cos(2n_w\theta).$$

The mechanical function of both springs is similar. We explain for briefness only the function of the symmetric disk wave spring. If corrugated springs of this type are installed between two uniform contact surfaces and become increasingly loaded, the tips of waves initially come into contact with the contact surfaces with their absolute maxima and absolute minimums (Fig. 4.12a). The forces act in the vertical direction from the top downwards in the points A1 and A2 and from the bottom upwards in the points B1 and B2. Distance in the circumferential direction between points A1 and A2 is referred to as $2a$. The distance between the points B1 and B2 is the same. We now set the point A0 centrally between the points A1 and A2 and correspondingly the point B0 centrally between the points B1 and B4. To calculate the spring rate, we

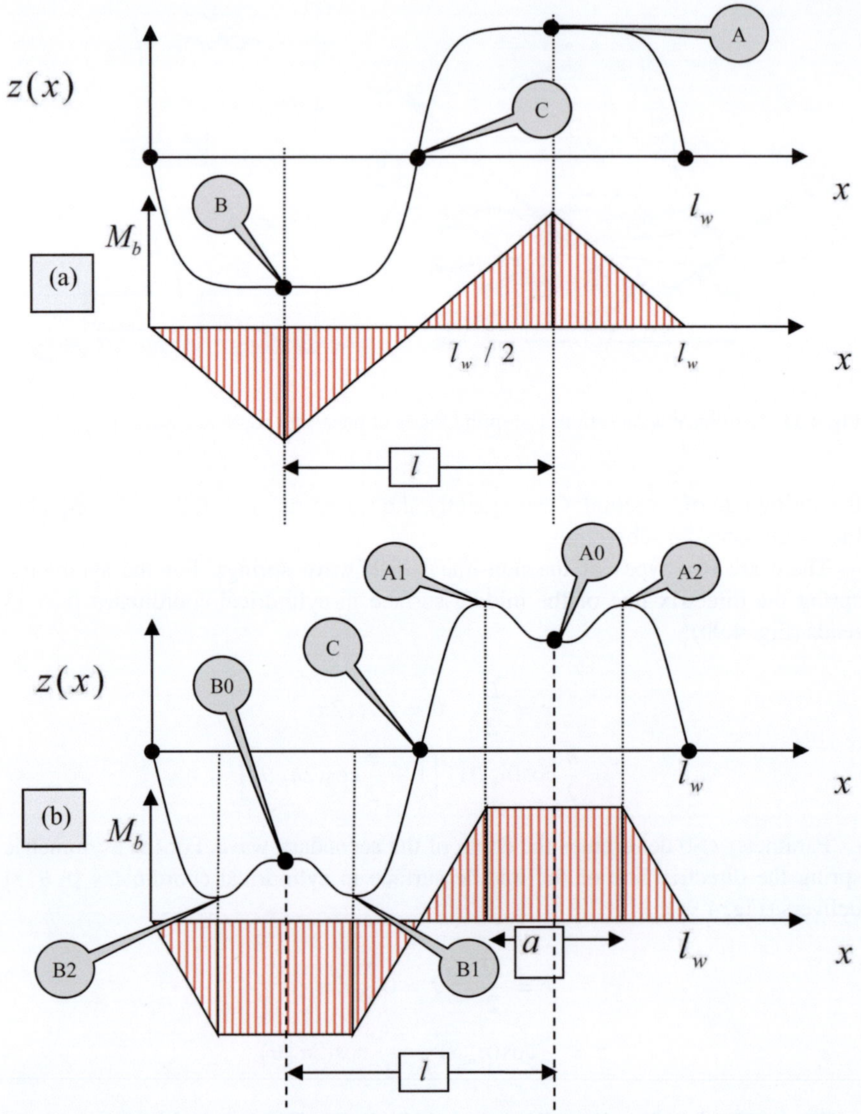

Fig. 4.12 Comparison of bending moments in the linear and progressive wave disk springs

consider the section between two adjacent abutment points A0 and B0 (Fig. 4.12b). The bending moment in the cross-section of the strip vanishes at the point C, because the point C is positioned centrally between points A0 and B0. At the point C there is a local coordinate system with the axis x in the circumferential direction. The bending moment as a function of the distance to the point C is:

$$M_B(x) = \begin{cases} fx, & \text{if } 0 \le x \le \frac{l-a}{2}, \\ f\frac{l-a}{2}, & \text{if } \frac{l-a}{2} \le x \le \frac{l}{2}. \end{cases} \tag{4.86}$$

The stored elastic energy of the beam portion between points C and A is:

$$\begin{aligned} U_e &= \int_o^{\frac{l}{2}} \frac{M_B^2}{2EI}\mathrm{d}x = \frac{f^2}{2EI}\left[\int_o^{\frac{l-a}{2}} x^2\mathrm{d}x + \int_{\frac{l-a}{2}}^{\frac{l}{2}} \left(\frac{l-a}{2}\right)^2 \mathrm{d}x\right] \\ &= \frac{f^2}{16EI}\left[\frac{(l-a)^3}{3} + (l-a)^2 a\right]. \end{aligned} \tag{4.87}$$

The second derivative of the stored elastic energy (4.87) yields the spring rate of a single wave section:

$$\frac{1}{c_a} = \frac{\partial^2 U_e}{\partial f^2} = \frac{1}{8EI}\left[\frac{(l-a)^3}{3} + (l-a)^2 a\right] = \frac{1}{k_w(\lambda_w)}\frac{1}{192EI}\left(\frac{\pi D}{n_w}\right)^3. \tag{4.88}$$

Here $\lambda_w = \frac{a}{l}$ is the length ratio and

$$k_w(\lambda_w) = \frac{1}{(1-\lambda_w)^2(1+2\lambda_w)}. \tag{4.89}$$

is the correction factor. This correction factor accounts the bending moment (4.86) for the double-curved form of half-wave.

The disk wave spring consists exactly of four identical sections. The initial (primary) spring rate of the progressive wave spring that comprises n_w waves is:

$$\tilde{c}_w = n_w c_a = \frac{192}{\pi^3}\frac{n_w^4 EI}{D^3}k_w(\alpha). \tag{4.90}$$

The primary spring rate for small deflections and a small annular width, with the simplest assumption of an elongated rectangular bending beam using (4.83), is therefore:

$$\tilde{c}_w = 16EBn_w\left(\frac{Tn_w}{\pi D}\right)^3 k_w(\lambda_w) \equiv \frac{64}{\pi^3}\frac{D_e - D_i}{(D_e+D_i)^3}En_w^4T^3k_w(\lambda_w). \tag{4.91}$$

The initial (primary) spring rate of the double wave form is greater than the spring rate of the conventional wave spring. For example, for $a = l/2$ and

$$\lambda_w = \frac{a}{l} = \frac{1}{2},$$

if follows from (4.89) that

$$k_w = 2.$$

The primary spring rate of the considered non-linear wave spring is exactly twice the spring rate of the linear wave spring with the same number of full waves. The spring rate increases because the length of elastic bending section of the disk wave spring with the double wave is shorter that the length of elastic bending section of the linear disk wave spring. For the linear spring the bending happens between the tips of sinusoidal oscillating curve. Contrarily, the elastic bending of the disk wave spring with the double wave occurs between the pairs of maxima and minima. That is the conventional, linear disk wave spring with the simple wave form is two times more flexible than the disk wave spring with the double wave.

The disk wave spring with the double wave initially have a linear characteristic with a first lower spring rate, namely primary spring rate $\tilde{c}_w$. With the increasing of force they come into contact with the next larger relative maxima and relative minima at the contact surfaces. At this moment, the spring characteristic of spring changes to a substantially linear spring force. The disk wave spring with the double wave possesses a markedly higher spring rate after the contact event. At moment of the contact, the spring rate jumps to the secondary spring rate $\tilde{\tilde{c}}_w$. The calculation of the secondary spring rate calculation is as follows. At new contact points A0 and B0 suddenly emerge two reaction forces. The number of waves doubles. From initially existing n_w double waves in contact moment develop $2n_w$ simple waves (Fig. 4.11). The new (secondary) spring rate reads:

$$\tilde{\tilde{c}}_w = 16Eb\left(\tfrac{2n_w T}{\pi D}\right)^3 2n_w \equiv \tfrac{64}{\pi^3} \cdot \tfrac{D_e - D_i}{(D_e + D_i)^3} \cdot 16En_w^4 T^3. \tag{4.92}$$

The ratio of primary spring rate (4.91) and secondary spring rate (4.92) at the moment of contact is:

$$\frac{\tilde{\tilde{c}}_w}{\tilde{c}_w} = \frac{16}{k_w(\alpha)}.$$

For a practically relevant corrugated spring the length ratio is:

$$\lambda_w = \frac{1}{2}.$$

The most important case is the disk wave spring with the double wave. For this type spring, the ratio of the spring rate after the contact to the spring rate before the contact is exactly eight (Fig. 4.11). The material utilization of the linear disk spring corresponds approximately to the material utilization of a simple leaf spring with a constant cross-section. The non-linear wave spring possesses a somewhat lower material utilization.

Conclusions

The developed method could be directly extended to slotted disk springs. A slotted disk spring consists of two segments: a disk segment and a number of lever arm segments. The displacements of a disk spring subjected to an axial load are uniformly distributed on the inner and outer edges. Currently, the calculation of a slotted disk spring is based on the SAE formula (SAE, 1996). This formula is limited to a straight slotted disk spring. The dimensions of a disk spring simply refer to the edges of the disk spring section. The same dimensions were also used by Schremmer (1973) in an attempt to propose a new formula for a limited straight slotted disk spring.

In the study Fawazi et al. (2011), a load–displacement formula for the slotted disk spring is newly developed in the form of energy method by considering both rigid and bending deflections of the two segments.

The slotted disk springs are the essential parts of automatic transmissions and clutches of passenger cars. The design of a novel friction clutch utilizes the diaphragm spring to increase the friction area during operation (Shen & Fang, 2007). The load on the spring is reduced for a given transmitted torque. By increasing the friction area, the diaphragm spring can also act as a friction plate, and the number of components required for the clutch can be reduced. The maximum transmittable torque of the clutch is easily adjusted by varying the preload of the cup spring.

As mentioned above, the results of the current study could be immediately extended to the slotted diaphragm springs. Namely, the analytical formulas developed in the current manuscript improve the known formulas for the disk segment of the slotted disk springs for the considerably large cone angles and spring travels, because the current formulas are not based on the usual power series truncation.

Potential engineering applications of the current theory include Bellville and slotted disk springs for automotive and industrial applications.

Summary of Principal Results

- The nonlinear governing equations for the disk spring have been derived.
- The equations describe the deformation and stresses of thin and medium-thickness disk springs with the variable material thickness.
- The variation method is applicable to the springs with free and elastically supported edges.
- The spring rate and stresses for the corrugated spring have been derived.
- The behavior of nonlinear corrugated springs is explained and the expressions for spring forces and spring rate are derived.

References

ABAQUS. (2020). ABAQUS UNIFIED FEA©, Dassault Systèmes Simulia Corp. https://www.3ds.com/products-services/simulia/products/abaqus/

Almen, J. O., & Laszlo, A. (1936). The uniform section disks spring. *Transactions of ASME, 58*(12.4), 305–314.

ANSYS. (2020). ANSYS, Inc. Southpointe, 2600 ANSYS Drive, Canonsburg, PA 15317c. http://www.ansys.com/

Blom, A.W., Tatting, B. F., Hol, J., & Gürdala, Z. (2009). Fiber path definitions for elastically tailored conical shells. *Composites Part B: Engineering, 40*(1), 77–84.

DIN EN 16984:2017-02. (2017). Disc springs—Calculation; German version EN 16984:2016, Beuth Verlag, Berlin. DIN EN 16984-2017-09-Beuth.de.

EN 16984:2017-02. (2017). Disc springs—Calculation; German version DIN EN 16984:2016. Beuth Verlag.
Fawazi, N., Lee, J.-Y., & Oh, J.-E. (2011). A load–displacement prediction for a bended slotted disc using the energy method. *Proceedings of the Institution of Mechanical Engineers, Part C, Journal of Mechanical Engineering Science*, 1–12.
Ferrari, G. (2013). A new calculation method for Belleville disc springs with contact flats and reduced thickness. *International Journal of Manufacturing, Materials, and Mechanical Engineering, 3*(2), 63–73.
Hengstenberg, R. (1983). *Eigenspannungsentstehung in Tellerfedern und Schwingfestigkeit von Tellerfedern großer Scheibendicke*. Diss. RWTH, Fak. Bergbau und Hüttenwesen, Aachen.
Hübner, W. (1982). Deformationen und Spannungen bei Tellerfedern. *Konstruktion, 34*, 387–394.
Hübner, W. (1984). Large deformations of elastic conical shells. *Flexible shells* (pp. 257–270). Springer.
Hübner, W., & Emmerling, F. A. (1982). Axialsymmetrische große Deformationen einer elastischen Kegelschale. *ZAMM—Journal of Applied Mathematics and Mechanics, 62*(8), 404–406.
Kobelev, V., & Hesselmann, B. (2002). Belleville spring, European Patent EP1331416A1, European Patent Office.
Kobelev, V., Hesselmann, B., & Rinsdorf, A. (2003). Corrugated spring with gradual progressive spring characteristic, US 7334784 B2, US Patent Office.
Kobelev, V. (2016). Exact shell solutions for conical springs. *Mechanics Based Design of Structures and Machines, An International Journal, 44*(4). https://doi.org/10.1080/15397734.2015.1066686
Libai, A., & Simmonds, J. G. (1988). *The nonlinear theory of elastic shells*. Academic Press.
Marsden, J., & Hughes, T. (1994). *Mathematical foundations of elasticity*. Dover.
Nabarro, F. R. N. (1967). *Theory of crystal dislocations*. Oxford University Press.
Niepage, P. (1983). Vergleich verschiedener Verfahren zur Berechnung von Tellerfedern – Teil I. Draht, 34, 105–108. *Teil II. Draht, 34*, 251–255.
Niepage, P., Schiffner, K., & Gräb, B. (1987). *Theoretische und experimentelle Untersuchungen an geschlitzten Tellerfedern*. VDI-Verlag.
Rosa, G. L., Messina, M., & Risitano, A. (1998). Tangential and radial stresses of variable thickness Belleville spring. *Journal of Mechanical Design, 123*(2), 294–299. https://doi.org/10.1115/1.1357162
SAE. (1996). *Spring design manual*. Part 5, SAE, HS-158 SAE International.
Saini, P. K., Kumar, P., & Tandon, P. (2007). Design and analysis of radially tapered disc springs with parabolically varying thickness. In *Proceedings of the Institution of Mechanical Engineers, Part C: Journal of Mechanical Engineering Science, 221*(2), 151–158. https://doi.org/10.1243/0954406JMES114
Schremmer, G. (1973). The slotted conical disc spring. *Transactions of the American Society of Mechanical Engineers, 95*, 765–770.
Shen, W., & Fang, W. (2007). Design of a friction clutch using dual Belleville structures. *ASME Journal of Mechanical Design, 129*.
Tavares, S. A. (1996). Thin conical shells with constant thickness and under axisymmetric load. *Computers & Structures, 60*(6), 895–921.
Timoshenko, S., & Woinowsky-Krieger, S. (1957). *Theory of plate and shell*, 2nd ed. McGraw Hill.
Ventsel, E., & Krauthammer, T. (2001). Thin plates and shells, theory, analysis, and applications. Marcel Dekker AG.

Chapter 5
Radially Constrained Disk Springs

Abstract In the current chapter, the calculation of disk springs is studied with free sliding edges and edges with constrained radial motion. The disk springs with constant material thickness are studied using the models of thin and moderately thick isotropic shells. The variation formulations are used to derive load–displacement formulas for the disk springs with different constraints on radial movement on the inner and outer surfaces. The kinematic hypothesis is used for the shell models of conical shells. The motivating feature of the presented theory is its ability to calculate the cup springs with free gliding edges and the edges with constrained radial motion. The equations developed here are based on general assumptions and are suitable for disk springs made of isotropic materials, such as spring steel and light metal alloys. The advantage of the methodology is the derivation of closed-form solutions for several common restrictions on the radial motion of the inner and outer edges. The developed formulas are recommended for industrial calculations of free and restricted disk springs.

Keywords Disk springs with radially constrained edges · Shell models

5.1 Shell Model for Conical Disk Springs

5.1.1 Introduction

In the current Chapter, the disk springs with different degrees of freedom on the inner and outer surfaces are investigated using the variation methods. Such washers are known from the patent (Muhr et al., 2009). The invention relates to a spring plate for a valve spring of an internal combustion engine. The spring plate can be connected directly to the valve stem of a charge changing valve while engaging an annular groove in the valve stem. Spring plates of this type hold and centre the valve springs of charge changing valves which are arranged concentrically relative to the valve stem and which are supported more particularly by a second spring plate at the cylinder head. The valve is actuated by pressure forces being applied to the free end

V. Kobelev, *Fundamentals of Springs Mechanics*,
https://doi.org/10.1007/978-3-031-58584-5_5

of the valve stem. The free end projects beyond the spring plate, by cams, rocker arms or valve levers. Between the actuating elements and the spring plate it is possible to place offsetting devices. The objective is achieved by providing a spring plate for a valve spring of an internal combustion engine. The spring plate can be connected directly to the valve stem of a charge changing valve while engaging an annular groove in the valve stem. An annular disk is arranged on the valve spring and two supporting tongues which are directed radially inwardly from the annular disk and whose inner ends, in a plan view of the Spring plate, comprise recesses for engaging an annular groove. An assembly consisting of a charge changing valve of an internal combustion engine with a valve stem has an annular groove and a spring plate for a valve spring. The spring plate is directly connected to the valve stem while engaging the annular groove. The spring plate is provided with an annular disk for being arranged on the valve spring and with two supporting tongues which are directed radially inwardly from the annular disk and whose inner ends, in a plan view of the annular disk, comprise recesses for engaging the annular groove. By providing only two supporting tongues it is possible to stiffen the spring plate without substantially increasing its mass. Furthermore, the production of the spring plate is simplified and it becomes easier to fit it to the valveshaft. Consequently, the designer needs frequently the calculation formulas for the disk springs, which are fitted upon the bolt shaft or radially restrained movement to outer direction. The latter design case appears, if the washer is placed inside the cylindrical cavity in the rigid metal block. The actual manuscript investigates these restrictions and provides the formulas for the routinely calculations of the assembled disk springs and Belleville washers.[1]

The conical shells of revolution under the axial loads were studied in a couple of recent publications. The purpose of the study (Sobhani & Avcar, 2022a) was to investigate the natural frequencies of perfect and imperfect graphene nanoplatelet reinforced nanocomposite coup structures (conical and cylindrical shells and annular plate structures) resting on elastic bases under general boundary conditions. To create a nanocomposite reinforced polymer matrix with porosity, the material of graphene nanoplates was used. The spring method determines the general boundary conditions for Graphene Nanoplatelet Reinforced Nanocomposite structures. The parameters of the elastic bases were described by two parameters according to Winkler-Pasternak bases. For the determination of the closed solution formulas associated with the Graphene Nanoplatelet Reinforced Nanocomposite coup structures, the Donnell hypothesis and First Order Shear Deformation Theory were used. Similar technique was applied for the evaluation of introduction of nanofillers on the free vibrations of nanocomposite cylindrical shells in (Sobhani and Avcar, 2022b). The common nanofiller materials comprise carbon nanotubes, graphene nanoplates and graphene oxide powders. These materials are usually used to improve the polymer matrix and create nanocomposite materials. To determine the mechanical properties associated with the first type nanocomposite, the modified mixture rule was used. An

[1] V. Kobelev, *Fundamentals of Springs Mechanics*, University of Siegen, Faculty IV: School of Science and Technology. Paul-Bonatz-Str. 9–11, D-57076 Siegen. E-mail: Vladimir.kobelev@uni-siegen.de Homepage: https://www.mb.uni-siegen.de/kobelev/

alternative approach with the Halpin–Tsai scheme, was for evaluation of the mechanical properties of two other types of nanocomposites. The governing equations of motion were derived with Hamilton's principle. The effect of nanofillers on the free vibrations of nanocomposite cylindrical shells was estimated after evaluation of the eigenfrequencies. The parametric examples were presented to explain the effect of the three types of nanofillers on the free vibration responses.

5.1.2 Models of Elastic Disk Springs

The statics of elastic disk springs with the freely gliding in radial direction contacting surfaces is explained in Chap. 4. Evidently, that if both edges can radially glide; the inside edge of the spring moves during the axial compression towards the axis of the disk. The outer edge of the spring moves during the axial compression in the opposite direction away from the axis of the disk. Because both edges can freely glide without friction, both radial reaction forces disappear.

The different load case occurs, if the radial movement of the inside edge is restricted. The cylindrical curb, is assumed to be absolutely rigid. The cylindrical curb is injected into the inner edge of the spring without gap. The spring glides can glide without friction over the outer edge in the radial direction. Another load case happens, if the radial movement of the outer edge is restricted. In this case, the cylindrical curb is inserted outside the outer edge of the spring without gap. Correspondingly, the spring glides can glide without friction over the inside edge in the radial direction.

In this Chapter, the disk springs with different degrees of freedom on the inner and outer surfaces are studied using the variation methods. The methods consequently apply the kinematic hypothesis and equations of thin and moderately thick shells. The motivating feature of the presented theory is its possibility to calculate the disk springs of with free gliding edges and the edges with the constrained radial movement.

The actual case covers the easiest and most common design case and is based on several simplifications and conjectures. The hypotheses are the following. The material deforms linear elastically. The nonlinear elastic and plastic effects of material deformation are ignored. The hereditary effects, which cause the creep and relaxation, are not accounted. The material behaves isotopically. The friction on the inner and outer edges are not present in the actual study. The cross-section of the spring is rectangular, such that the radii on the angels of the rectangular section vanish. Because the small radii in the inner and outer edges are common in the industrial application, the resulted formulas require the corrections for account of radii on the edges. All these influences and enhancements could be evaluated without difficulties using the standard commercial finite element codes, presumed the relevant additional parameters are known form the literature or are properly measured.

5.1.3 Geometry of Disk Spring in Undeformed State

Consider the truncated conical shell with the constant thickness T. Its middle surface in the free state is the surface of revolution Fig. 5.1. Thus, the middle surface is truncated right circular conical surface. The inside radius of middle surface of free spring is $r_i = D_i/2$ the outside radius is $r_e = D_e/2$. The ratios of outer radius to inside radius and the thickness to inside radius correspondingly are:

$$\Delta = \frac{r_e}{r_i} > 1, \quad \mu = \frac{T}{r_i}.$$

Such surface is referred to as the frustum ϖ. The initial slope angle α of the free conical shell is apparently constant. A certain neutral plane is marked by (c).

Two coordinate systems will be used for the description of the geometry of the shell with the thickness T. In the cylindrical coordinate system, there parameters (r, θ, z) determine the location of a certain internal point P of the shell. The origin of cylindrical coordinates is located on the neutral plane (c). The neutral plane (c) with the axial coordinate $z = 0$ overlaps the conical frustum of evolution ϖ. The cross-section of the neutral plane (c) and the frustum ϖ is the circle of radius C. The piercing point of this circle with the coordinate plane (r, z), $\theta = 0$, will be referenced to as the "inversion point". In the cylindrical coordinate system, the middle surface of a free conical shell is given as:

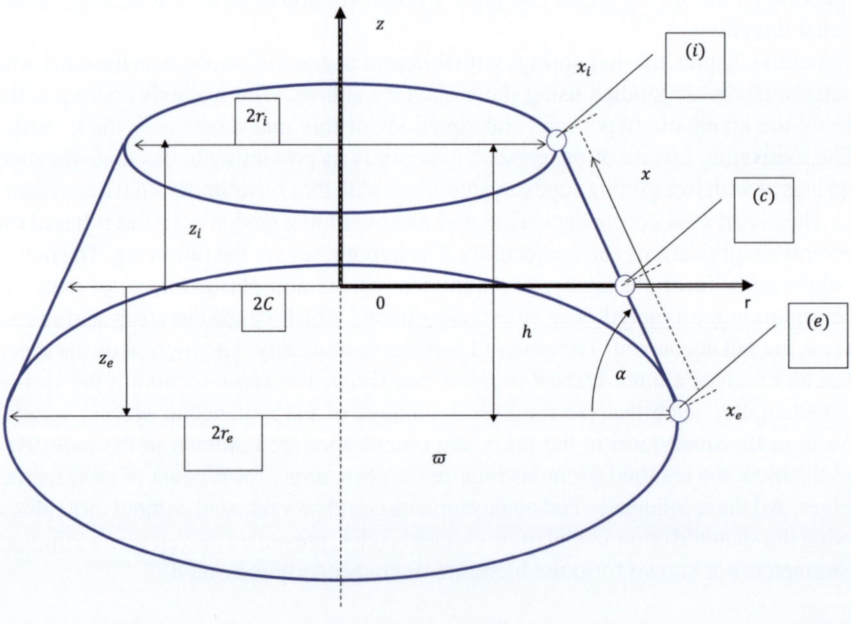

Fig. 5.1 The middle surface ϖ of the free spring

$$x = \frac{C - r}{\cos\alpha}, \quad z = -x\sin\alpha. \tag{5.1}$$

The arc length x serves here as the meridional coordinate. The measure of the arc length starts from the "inversion point". Thus, the neutral plane (c) matches to the arc length $x = 0$. The lengths of the arcs to the inner and outer edges correspondingly are:

$$x_i = \frac{C - r_i}{\cos\alpha} > 0, \quad x_e = \frac{C - r_e}{\cos\alpha} < 0. \tag{5.2}$$

From (5.1) and (5.2) follows:

$$\begin{aligned} &r_i = C - x_i\cos(\alpha),\ r_e = C - x_e\cos(\alpha), \\ &\mathrm{d}r = \cos(\alpha)\,\mathrm{d}x, \quad x_e \le x \le x_i. \end{aligned} \tag{5.3}$$

The heights of the inner and outer edges of the middle surface of the free shell are respectively (Fig. 5.1):

$$z_i = x_i\sin\alpha > 0, \quad z_e = x_e\sin\alpha < 0. \tag{5.4}$$

The total height of the middle surface of the unloaded disk spring is:

$$h = z_i - z_e = (\Delta - 1)r_i\tan\alpha. \tag{5.5}$$

The width of the middle surface of the disk spring in its free state is:

$$b = r_e - r_i = (\Delta - 1)\cdot r_i. \tag{5.6}$$

In the conical coordinate system, the location of the internal point (c) of the conical shell is determined by three parameters (x, θ, y). The normal distance y from the internal point (c) to the middle surface of the shell reads:

$$-\tfrac{1}{2}T \le y \le \tfrac{1}{2}T.$$

The distance from the point (c) with conical coordinates (x, θ, y) to the axis of the cone is:

$$\tilde{r} = r + y\sin\alpha = C - x\cos\alpha + y\sin\alpha. \tag{5.7}$$

Equivalent coordinate system was employed for the description of the vibration of a truncated conical shell in a fluid (Caresta & Kessissoglou, 2008).

The principal radii of curvature r_1 and r_2 for the unloaded conical shell correspondingly are:

$$\frac{1}{r_1} = \frac{\sin\alpha}{r}, \quad \frac{1}{r_2} = 0. \tag{5.8}$$

5.1.4 Variations of Strain and Curvature Due to Axial Contraction of Conical Spring

The main hypothesis of the model is that the slope angle ψ of deformed conical shells changes, but also remains constant over the meridian, as shown in Chap. 4. Namely, the point (c) is the inversion center point for the cross-section of the conical shell with the radius C. The generatrice rotate about the point (c), but remain straight and their lengths do not alter. The points of the middle surface, located on the neutral plane (c), do not deflect. The middle surface of the shell in the deformed state is another frustum Ω, as shown above on (Fig. 5.2). This hypothesis is essentially the same, as the hypothesis of the common models of Timoshenko and Almen and Laszlo.

The principal difference between the actual and the common models consists in the resolution of the position of the neutral plane. The radial travel of the edges hangs on the position of the neutral plane. Consequently, the formal approving of the positions of the neutral plane depends upon kind of the radial restriction.

In cylindrical coordinates (r, θ, z), the middle surface Ω is given by the parametric equations:

$$R = C - x\cos\psi, \quad Z = -x\sin\psi. \tag{5.9}$$

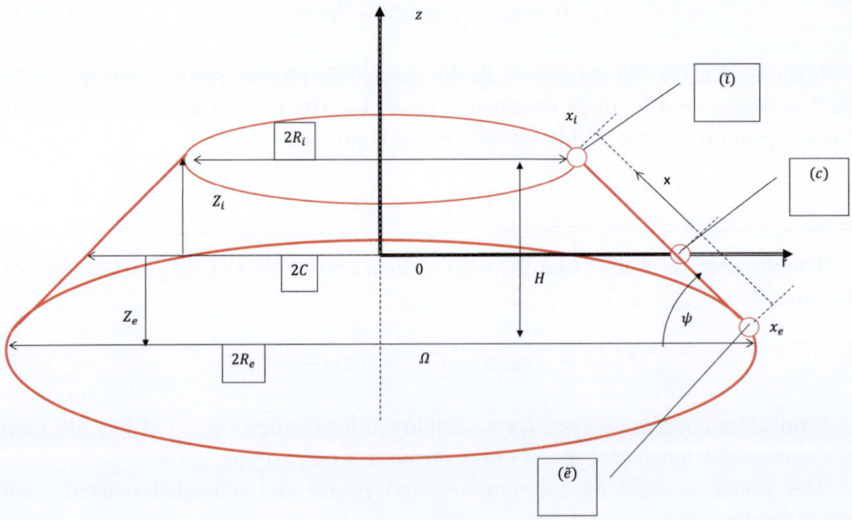

Fig. 5.2 The middle surface **Ω** of the deformed spring

The distance from the point of the thick-walled cone with conical coordinates (x, θ, y) to the axis of the deformed cone is:

$$\tilde{R} = R + y \sin \psi = C - x \cos \psi + y \sin \psi. \tag{5.10}$$

Correspondingly, the principal radii of curvature R_1 and R_2 of the middle surface Ω are:

$$\frac{1}{R_1} = \frac{\sin \psi}{r}, \quad \frac{1}{R_2} = 0. \tag{5.11}$$

The inner and outer radii of Ω are equal to:

$$R_i = C - x_i \cos \psi, \quad R_e = C - x_e \cos \psi. \tag{5.12}$$

The distances from the plane (c) to the inner (i) and outer (e) edges of surface Ω read:

$$Z_i = x_i \sin \psi, \quad Z_e = x_e \sin \psi. \tag{5.13}$$

The height and width of middle surface in the deformed state Ω are respectively:

$$H = Z_i - Z_e = r_i \cdot \frac{\sin \psi}{\cos \alpha} \cdot (\Delta - 1), \tag{5.14}$$

$$B = R_i - R_e = r_i \cdot \frac{\cos \psi}{\cos \alpha} \cdot (\Delta - 1). \tag{5.15}$$

Upon specializing the strain displacement relations to the case of a conical shell with no transverse shear deformations the following expression for the circumferential mid-surface strain:

$$\varepsilon_\theta = \frac{R - r}{r} = \frac{\cos \alpha - \cos \psi}{C - x \cos \alpha} x. \tag{5.16}$$

The circumferential curvature change is:

$$\kappa_\theta = \frac{1}{R_1} - \frac{1}{r_1} = \frac{\sin \psi - \sin \alpha}{r}. \tag{5.17}$$

The slope angle vanishes in the flattened state. With Eq. (5.9), the inner and outer radii of the plane circular strip will be equal nearly to:

$$R_{i0} = C - x_i, \quad R_{e0} = C - x_e.$$

5.1.5 Spring Travel and Heights of Disk Springs

According to Eqs. (5.6) and (5.14), the spring travel on the middle surface is equal to:

$$s = h - H = (\Delta - 1) \cdot r_i \cdot \frac{\sin\alpha - \sin\psi}{\cos\alpha}. \tag{5.18}$$

Considering the constant thickness T of the material, the expressions for height and width must be corrected (Fig. 5.3). The total height and width of the disk spring in its free state from upper inside edge to outer outside are respectively:

$$\begin{cases} \tilde{h} = h + T\cos\alpha = (\Delta - 1) \cdot r_i \tan\alpha + T\cos\alpha, \\ \tilde{b} = b - T\sin\alpha = (\Delta - 1) \cdot r_i - T\sin\alpha. \end{cases} \tag{5.19a}$$

The total height and width of the deformed disk spring from upper inside edge to outer outside are correspondingly:

$$\begin{cases} \tilde{H} = H + T\cos\psi = r_i \cdot \frac{\sin\psi}{\cos\alpha} \cdot (\Delta - 1) + T\cos\psi, \\ \tilde{B} = B - T\sin\psi = r_i \cdot \frac{\cos\psi}{\cos\alpha} \cdot (\Delta - 1) - T\sin\psi. \end{cases} \tag{5.19b}$$

With Eqs. (5.19a) and (5.19b), the spring travel from upper inside edge to lower outside is:

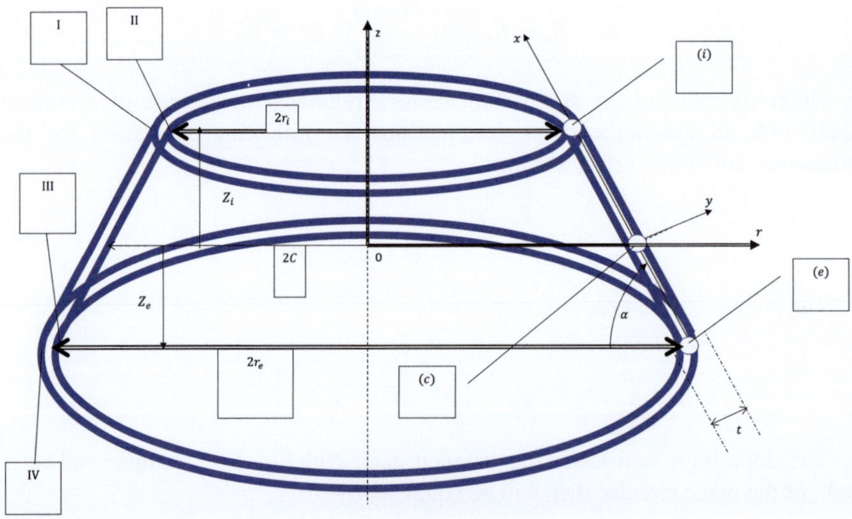

Fig. 5.3 The solid representation the undeformed thick-walled spring

$$\tilde{s} = \tilde{h} - \tilde{H} = (\Delta - 1) \cdot r_i \cdot \frac{\sin\alpha - \sin\psi}{\cos\alpha} + T \cdot (\sin\psi - \sin\alpha). \tag{5.20}$$

The influence of rounding on the corners could be also accounted without difficulty.

5.2 Statics of Disk Springs Using Equations of Axisymmetric Elasticity

5.2.1 Deformation of Conical Shell

In the conical coordinate system, the stresses relate to strains of the isotropic body in the absence of shear stresses by means of relations (Marsden & Hughes, 1994):

$$\begin{cases} \sigma_\theta = \frac{E}{(1+\nu)(1-2\nu)}[(1-\nu)\cdot\mathbf{E}_\theta + \nu\cdot(\mathbf{E}_X + \mathbf{E}_Y)], \\ \sigma_X = \frac{E}{(1+\nu)(1-2\nu)}[(1-\nu)\cdot\mathbf{E}_X + \nu\cdot(\mathbf{E}_Y + \mathbf{E}_\theta)], \\ \sigma_Y = \frac{E}{(1+\nu)(1-2\nu)}[(1-\nu)\cdot\mathbf{E}_Y + \nu\cdot(\mathbf{E}_\theta + \mathbf{E}_X)]. \end{cases} \tag{5.21}$$

In Eq. (5.21) the following notations for the principal strains are used:

$\mathbf{E}_\theta$: the circumferential, or hoop, strain,
$\mathbf{E}_X$: the meridional strain and
$\mathbf{E}_Y$: the strain normal to surface of the shell.

All shear strains and shear stresses nullify due to the symmetry conditions. The hypothesis assumes the absence of the stress in the meridional and normal directions respectively:

$$\sigma_X = 0, \quad \sigma_Y = 0.$$

From these conditions, the meridional strain $\mathbf{E}_X$ and strain normal to surface of the shell $\mathbf{E}_Y$ depend solely on the circumferential strain $\mathbf{E}_\theta$:

$$\mathbf{E}_X = -\nu\mathbf{E}_\theta, \quad \mathbf{E}_Y = -\nu\mathbf{E}_\theta. \tag{5.22}$$

The hoop, or circumferential stress $\boldsymbol{\sigma}_\theta$ is acting circumferential and perpendicular to the axis z and the radius r:

$$\sigma_\theta = E \cdot \mathbf{E}_\theta.$$

The relations (5.21)–(5.22) lead to the following expressions for the strains in the solid elastic cone:

$$\mathbf{E}_\theta = \varepsilon_\theta + y \cdot \kappa_\theta = \frac{(\cos\alpha - \cos\psi)\cdot x + (\sin\psi - \sin\alpha)\cdot y}{C - x\cos\alpha},$$
$$-\tfrac{1}{2}T \le y \le \tfrac{1}{2}T, \quad x_e \le x \le x_i \tag{5.23}$$

The total hoop force in the cross-section F_θ of the conical disk is the integral of the hoop stress over the cross-sectional area:

$$F_\theta = 2\pi \int_{x_e}^{x_i} \left[\int_{-T/2}^{T/2} \sigma_\theta \mathrm{d}y \right] \mathrm{d}x.$$

Equivalently, there is the compression force of the spring to the any curb on one of the constrained edges. Evidently, that the Evidently compression force of the spring is equal to the total hoop force in the cross-section of the truncated conical disk. Particularly, if both edges can glide without resistance, the total hoop force disappears: $F_{G\theta} = 0$.

The MAPLE CAS (MAPLE, 2022) file is provided in (Kobelev, 2023a) for the derivation of the formulas. With the MAPLE code file, the reader can retrieve all subsequent, somewhat bulky calculations of the actual theory.

5.2.2 Variation Method for Shell Models of Disk Springs

The total potential energy

$$\Pi = U_E + U_F \tag{5.24}$$

is the sum of the elastic strain energy U_E, stored in the deformed body and the potential energy U_f of the applied forces:

$$U_F = \int_{\alpha}^{\psi} M \mathrm{d}\psi, \quad M = F_Z B - F_R H. \tag{5.25}$$

The shell is stressed by the forces in the direction of the rotation axis of the shell or by the forces in radial direction. Normally, the applied load is evenly distributed over the upper inside edge and lower outside edge (Fig. 5.4, left). The total axial force acting on the inner edge (i) is F_z and the total radial force is F_R.

The strain energy of the elastically deformed cone in the conical coordinate system is:

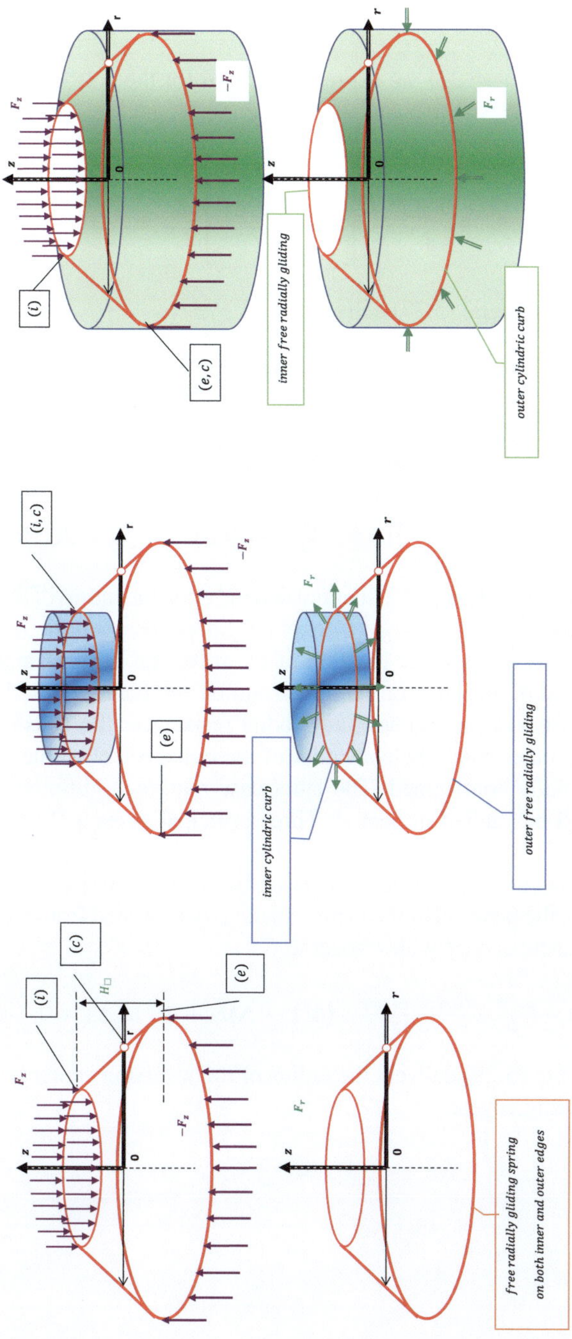

Fig. 5.4 Axial (above) and radial (below) systems of forces on the truncated conical shell

$$U_E = \pi G \int_{x_e}^{x_i} \left[\int_{-T/2}^{T/2} (\mathbf{E}_{\mathbf{X}}^2 + \mathbf{E}_{\mathbf{Y}}^2 + \mathbf{E}_\theta^2 + \tfrac{\nu}{1-2\nu} \cdot (\mathbf{E}_{\mathbf{X}} + \mathbf{E}_{\mathbf{Y}} + \mathbf{E}_\theta)^2) r \, \mathrm{d}y \right] \mathrm{d}x, \quad (5.26)$$

Using the Eqs. (5.21)–(5.22) the expression (5.26) for the elastic energy for the constant thickness reads:

$$U_E = \pi E \int_{x_e}^{x_i} \left[\int_{-T/2}^{T/2} \mathbf{E}_\theta^2 r \, \mathrm{d}y \right] \mathrm{d}x. \quad (5.27)$$

The evaluation of integral (5.27) delivers:

$$U_E = U_b + U_t, \quad (5.28)$$

$$U_b = \tfrac{\pi E T^3}{12} \cdot \ln \Delta \cdot \tfrac{(\sin \psi - \sin \alpha)^2}{\cos \alpha},$$

$$U_t = -\tfrac{\pi E T}{2} \cdot \tfrac{(\cos \psi - \cos \alpha)^2}{\cos^2 \alpha} \cdot \left[(\Delta^2 - 1) r_i^2 + 4(1 - \Delta) r_i C + 2C^2 \ln \Delta \right].$$

Notably, that the elastic energy U_b does not depend upon the radius of the inversion center C. Contrarily, to elastic energy U_t explicitly depends upon the radius of the inversion center C. In its turn, the location of the inversion center hangs on upon the admissible radial movements on the inner and on the outer edges. There exist four possibilities: free gliding without any radial resistance; the constraint of the radial travel on the inner edge; the constraint of the radial travel on the outer edge; both edges are radially constrained. The latter case requires a different kinematic hypothesis and will be studied in Sect. 4. Three principal cases will be considered below.

1°. If both edges of the spring free glides over the inside (i) and outer (e) edges in the radial direction, the position of the neutral plane is determined from the condition of minimum for elastic energy with respect to C:

$$\tfrac{\partial U}{\partial C} = -\tfrac{\pi E T}{2} \cdot \tfrac{(\cos \psi - \cos \alpha)^2}{\cos^2 \alpha} \cdot [4(1 - \Delta) r_i + 4C \ln \Delta] = 0. \quad (5.29)$$

Solution of the Eq. (5.29) delivers the radius of the inversion center point (c) for the spring with gliding edges:

$$C_G = \tfrac{\Delta - 1}{\ln \Delta} r_i \equiv \tfrac{r_e - r_i}{2 \ln(r_e / r_i)}. \quad (5.30)$$

The radius (5.30) of inversion center is given by Almen and Laszlo formula. According to the principle of virtual work, the external virtual work $U_G = U|_{C=C_G}$ is equal to internal virtual work when equilibrated forces and stresses undergo unrelated but consistent displacements and strains. The necessary stationarity condition with respect to ψ delivers the circumferential moment for the spring with gliding edges:

$$M_G = \left.\frac{\partial U}{\partial \psi}\right|_{C=C_G} = M_{t.G} + M_b, \tag{5.31}$$

$$M_{t.G} = Er_i^3 \mu \frac{(\cos\psi - \cos\alpha)\sin\psi}{\cos^3\alpha} \cdot \left[2(\Delta - 1) - \tfrac{1}{2}(\Delta^2 - 1)\right],$$

$$M_b = Er_i^3 \mu^3 \cdot \frac{(\sin\psi - \sin\alpha)\cos\psi}{12\cos^3\alpha} \cdot \ln\Delta .$$

In the absence of the radial force, the axial force on the shell is:

$$F_{G.Z} = F_{t.G} + F_b, \quad F_{t.G} = \frac{M_{t.G}}{B}, \quad F_b = \frac{M_b}{B}. \tag{5.32}$$

The substitution of (5.31) into (5.32) delivers the central formula for the axial force of disk spring $F_{G.Z}$ of the freely gliding disk spring:

$$\begin{cases} F_{G.Z} = \pi E r_i^2 \cdot \frac{\mu F_{t.G} + \mu^3 F_b}{\cos\psi}, \\ F_{t.G} = \frac{2(1-\Delta)+(1+\Delta)\ln\Delta}{\ln\Delta} \cdot \frac{\cos\psi - \cos\alpha}{\cos^2\alpha} \cdot \sin\psi , \\ F_b = \frac{\ln\Delta}{6(\Delta-1)} \cdot (\sin\alpha - \sin\psi) \cdot \cos\psi . \end{cases} \tag{5.33}$$

Evidently, that the inside edge of the spring (i) moves as a result of axial compression towards the axis of the disk (Fig. 5.4, left). The lower, or outer edge of the spring (e) travels in the opposite direction away from the axis of the disk. Because both edges can freely glide without friction, the radial reaction forces on both edges are will not appear.

2°. The different load case occurs, if the radial movement of the inner edge (i) is restricted (Fig. 5.4, middle). One can imagine the absolutely rigid cylindrical curb, inserted inside the inside edge of the spring without gap. The inner edge (i) glides in the vertical direction over the surface of the cylindrical curb, but cannot penetrate inside it. Because the radial movement of the inside edge (i) is constrained, the position inversion point matches the inner radius:

$$C_I = r_i, \tag{5.34}$$

Over the outer edge (e), the spring glides can glide without friction in the radial direction, but its vertical movement vanishes. The substitution of (5.34) into (5.28) gives the elastic energy of the spring on the inner curb:

$$U_1|_{C=C_I} \equiv U_I = U_b + U_{tI}, \tag{5.35}$$

$$U_{tI} = -\frac{\pi ET}{2} \cdot \frac{(\cos\psi - \cos\alpha)^2}{\cos^2\alpha} \cdot \left[(\Delta^2 - 1)r_i^2 + 4(1-\Delta)r_i C_\mathrm{I} + 2C_\mathrm{I}^2 \ln\Delta\right].$$

The necessary stationarity condition with respect to ψ delivers the circumferential moment and the axial force on the radially constrained disk spring on the inside edge:

$$M_I = \frac{\partial U_I}{\partial\psi}, \quad F_{IZ} = \frac{1}{B}\frac{\partial U_I}{\partial\psi} = \frac{M_I}{B}. \tag{5.36}$$

The expression for the axial force F_{IZ} of the disk spring with the absolutely rigid cylindrical curb, inserted inside the inside edge, reads:

$$\begin{cases} F_{IZ} = \pi E r_i^2 \cdot \frac{\mu F_{t.I} + \mu^3 F_b}{\cos\psi}, \quad F_{t.I} = k_I \cdot F_{t.G}, \\ k_I = \frac{\ln\Delta}{\Delta - 1} \cdot \frac{\Delta^2 - 4\Delta 2\ln\Delta + 3}{\Delta\ln\Delta - 2\Delta + \ln\Delta + 2}, \\ F_{I\theta} = 2\pi E T r_i \cdot \frac{\cos\psi - \cos\alpha}{\cos^2\alpha} \cdot (\Delta - \ln\Delta - 1). \end{cases} \tag{5.37}$$

In Eq. (5.37), $F_{I\theta}$ is the compression force of the spring to the inner curb on the inside edge of the spring. The total radial force F_{IR}, acting from the spring to the inner curb, is equal to the total hoop force $F_{I\theta}$.

The Eq. (5.37) demonstrates, that the shift of the inversion point to the inside edge does not influence the bending of the cone. On the contrary, the character of the extensional deformation (circumferential strain) of the middle surface alternates significantly. The extension of the middle surface of free gliding spring occurs outside the inversion point (*e*). The middle surface of the free gliding spring squeezes inside the inversion point. Contrarily, the complete middle surface of the disk spring on the cylindric curb extends. This behavior influences considerably the function of the spring.

3°. The third load case takes place, if the radial movement of the outer edge (*e*) is restricted (Fig. 5.4, right). The outer edge of the spring is inserted without gap into the absolutely rigid hollow cylinder. The inner edge (*i*) of spring can glide without friction in the radial direction. Because the radial movement of the outer edge (*e*) is controlled, the position of the inversion point coincides with outer radius:

$$C_E = r_e. \tag{5.38}$$

The substitution of (5.38) into (5.28) delivers the expression for the elastic energy of the spring inside the outer hollow rigid cylinder:

$$U|_{C=c_E} \equiv U_E = U_b + U_{tE}, \tag{5.39}$$

$$U_{tE} = -\frac{\pi ET}{2} \cdot \frac{(\cos\psi - \cos\alpha)^2}{\cos^2\alpha} \cdot \left[(\Delta^2 - 1)r_i^2 + 4(1-\Delta)r_i C_\mathrm{E} + 2C_\mathrm{E}^2 \ln\Delta\right].$$

Again, the necessary stationarity condition with respect to ψ delivers the circumferential moment and the axial force on the radially constrained disk spring on the outer edge:

$$M_E = \frac{\partial U_E}{\partial \psi}, \quad F_{EZ} = \frac{1}{B}\frac{\partial U_E}{\partial \psi} = \frac{M_E}{B}. \tag{5.40}$$

From Eq. (5.36) follows the expression for the axial force F_{EZ} of the disk spring, inserted on its outer edge into the absolutely rigid cylindrical curb, is the following:

$$\begin{cases} F_{EZ} = \pi E r_i^2 \cdot \frac{\mu F_{t.E} + \mu^3 F_b}{\cos\psi}, \quad F_{t.E} = k_E \cdot F_{t.G}, \\ k_E = \frac{\ln\Delta}{\Delta - 1} \cdot \frac{2\Delta^2 \ln\Delta - 3\Delta^2 + 4\ln\Delta - 1}{\Delta\ln\Delta - 2\Delta + \ln\Delta + 2}, \\ F_{E\theta} = -2\pi E T r_i \cdot \frac{\cos\psi - \cos\alpha}{\cos^2\alpha} \cdot (\Delta\ln\Delta - \Delta + 1). \end{cases} \tag{5.41}$$

In Eq. (5.41), $F_{E\theta}$ is the compression force of the spring to the outerner curb on the outer edge of the spring. The total radial force F_{ER}, acting from the spring to the outer curb, is equal to the total tangential force $F_{E\theta}$.

The extension force of the spring to the outer curb outside the outer edge of the spring follows again form the total tangential force in the cross-section. The only difference is the sign of the total force, acting from inside fixing spring to the outer radial direction. Equation (5.41) reveals, that the move of the inversion point to the outer edge does not influence the bending of the cone. Quite the reverse, the extensional deformation of the middle surface completely alternates. For free gliding spring, the extension of the middle surface of occurs outside the inversion point (*e*). The middle surface of free gliding spring compresses inside the inversion point. Alternatively, the complete middle surface of the disk spring compacts inside the hollow cylindric curb. This behavior influences significantly the force diagram of the spring.

The ratio of the correction coefficients due to circumferential strain of the middle surface reads:

$$\frac{k_E}{k_I} = \frac{2\Delta^2 \ln\Delta - 3\Delta^2 + 4\ln\Delta - 1}{\Delta^2 - 4\Delta + 2\ln\Delta + 3}. \tag{5.42}$$

The ratio of compression force of the spring to the curbs on the inside edge $F_{I\theta}$ to the compression force on the outside edge $F_{E\theta}$ is:

$$\frac{F_{I\theta}}{F_{E\theta}} = \frac{1 - \Delta + \ln\Delta}{1 - \Delta + \Delta\ln\Delta}. \tag{5.43}$$

5.3 Forces and Stresses in Disk Springs

For comparison three examples of spring calculation was considered. The parameters of the disk spring for the example are:

$$D_e = 110\,[\text{mm}],\ D_i = 90\,[\text{mm}],\ T = 2\,[\text{mm}],\ E = 200\,[\text{GPa}],\ \nu = 0.3.$$

Because the height of the neutral surface of the spring significantly influences the behavior of spring forces, three different heights were assumed to the be:

$$h = 2\,[\text{mm}];\ h = 4\,[\text{mm}] \text{ or } h = 6\,[\text{mm}].$$

For the disk springs with restricted radial displacements the forces significantly differ from the standard calculation equations. The axial and radial forces are shown correspondingly on the Figs. 5.5 and 5.6. The forces for the springs with the free gliding edges $F_{AL.Z}$, $F_{AL.R}$ Eq. (5.33) were calculated according the Eq. (5.33) are shown on the Figs. 5.5 and 5.6 with solid red lines. The forces for the springs with the inner cylindric curb $F_{I.Z}$, $F_{I.R}$ were calculated according the Eq. (5.37) are shown on the Figs. 5.5 and 5.6 with dot-dashed blue lines. The forces for the springs with the outer cylindric cavity $F_{E.Z}$, $F_{E.R}$ were calculated according the Eq. (5.41) are shown on the Figs. 5.5 and 5.6 with dot-dashed green lines.

The intermediate radial constraining of the inversion center leads to the considerable variations of the axial spring reactions, as shown on Fig. 5.7. The axial forces of free gliding spring $F_{AL.Z}$ (red surface) and $F_Z(c_i)$ (blue surface) are shown as the functions of spring height H and the radius of the inversion center c_i. The blue surfaces with the variable inversion center coincide with the red surfaces for the position of the Almen and Laszlo inversion center c_{AL}. The inside position of the inversion center corresponds the inner curb. The outer position of the inversion center corresponds the enclosing cylinder.

The graphs of correction coefficients k_E Eq. (5.41), k_I Eq. (5.37) and the ratio of the correction coefficients k_E/k_I Eq. (5.42) are shown as functions of $\Delta = r_e/r_i$ on Fig. 5.8.

The common calculation for the force of disk spring with the constant material thickness T is based on the Almen and Laszlo formulas (Almen & Laszlo, 1936):

$$F_{zDIN} = \frac{4E}{1-\nu^2} \cdot \frac{T^4}{K_1 D_e^2} \cdot \frac{s}{T} \cdot \left[\left(\frac{h}{T} - \frac{s}{T} \right) \cdot \left(\frac{h}{T} - \frac{s}{2T} \right) + 1 \right], \tag{5.44}$$

$$c_{zDIN} = \frac{\mathrm{d}F_{zDIN}}{\mathrm{d}s}. \tag{5.45}$$

These formulas are used in European standard (EN 16984:2017–02, 2017). For the moderate thicknesses and for the moderate heights the force $F_{G.Z}$ correlate with the DIN and SAE standard formulas, as it was shown in (Kobelev, 2016).

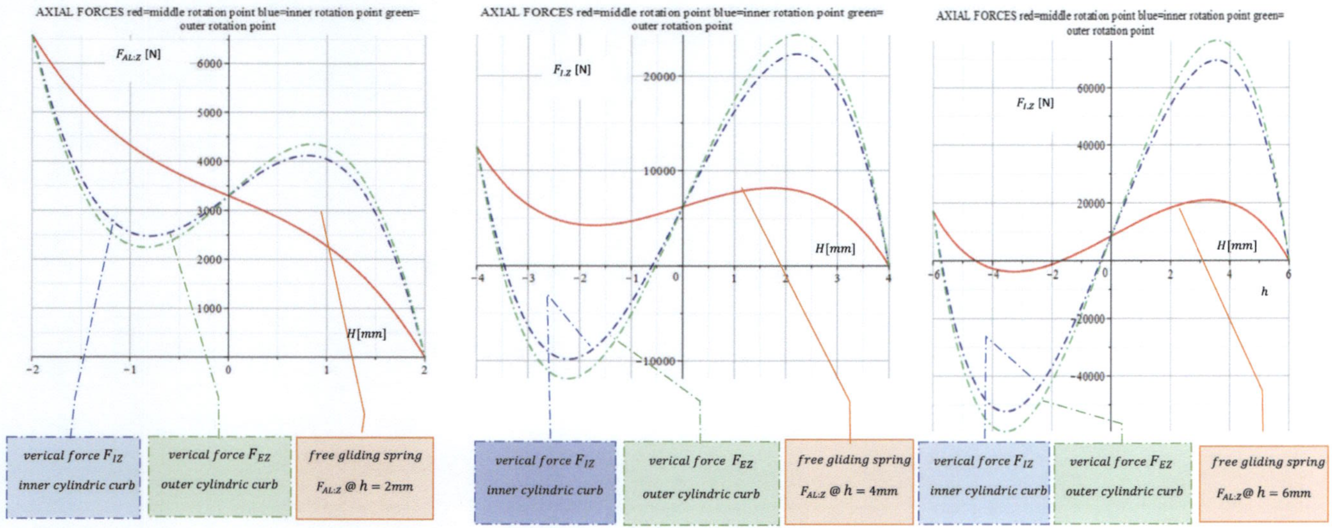

Fig. 5.5 Axial forces: Free gliding spring $F_{AL.Z}$, the spring on the inner curb $F_{I.Z}$ and the spring inside the outer curb as the functions of spring height

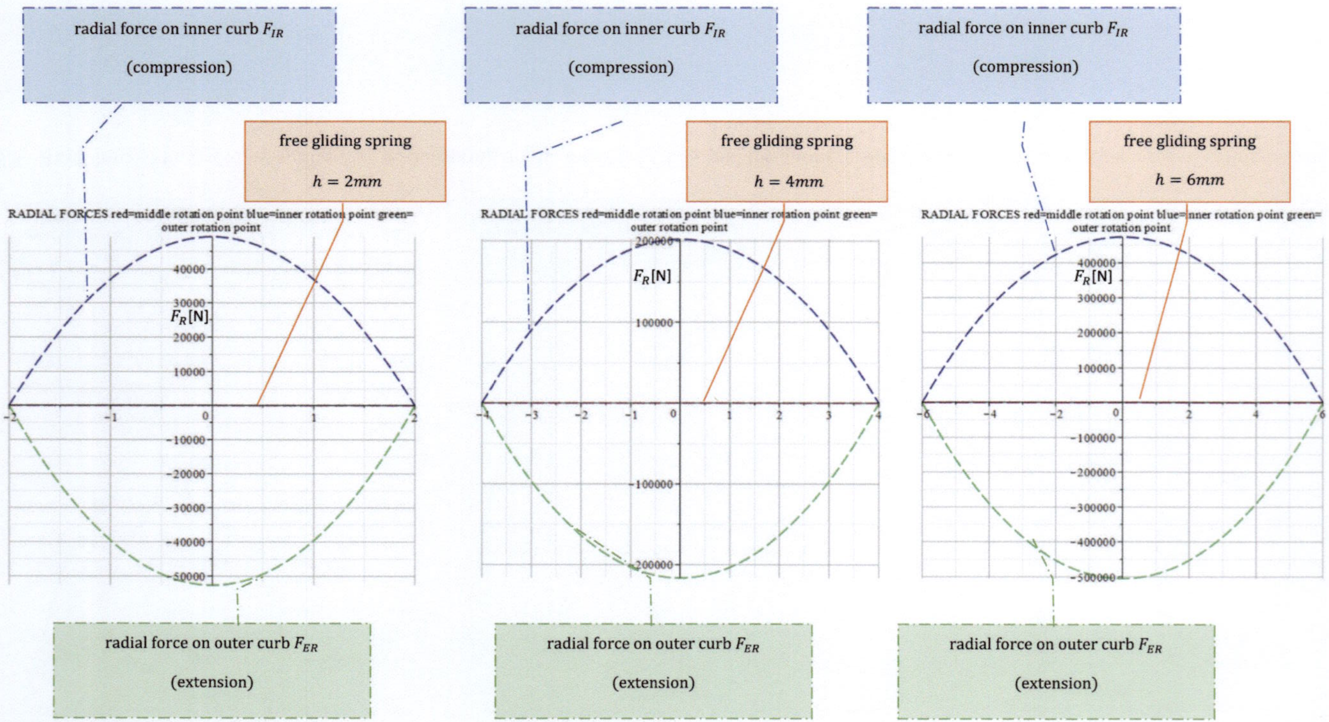

Fig. 5.6 Radial forces: Free gliding spring $F_{AL.R} \approx 0$, the spring on the inner curb $F_{I.R}$ and the spring inside the outer curb $F_{E.R}$ as the functions of spring height

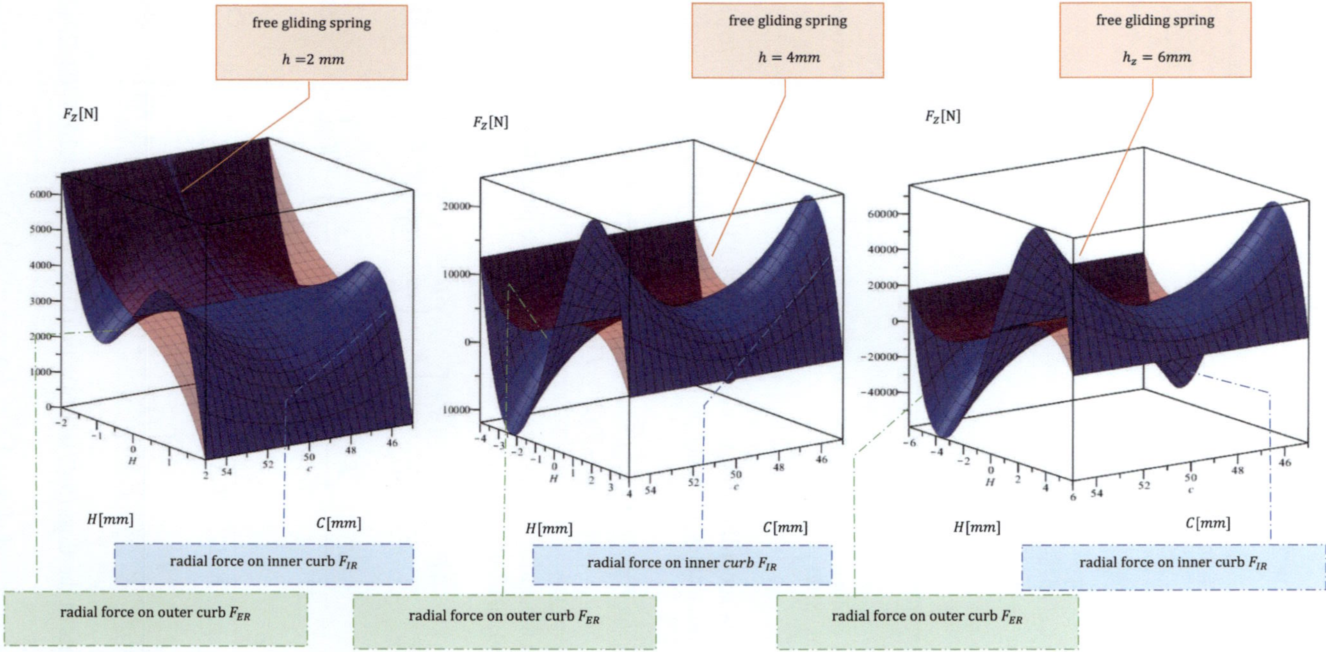

Fig. 5.7 Axial forces: Free gliding spring $F_{AL.Z}$ (red surface) and F_Z (blue surface) as the functions of spring height and the radius of the inversion center C. The blue surfaces with the variable inversion center coincide with the red surfaces for the position of the Almen and Laszlo inversion center C_{AL}. The inside position of the inversion center corresponds the inner curb. The outer position of the inversion center corresponds the enclosing cylinder

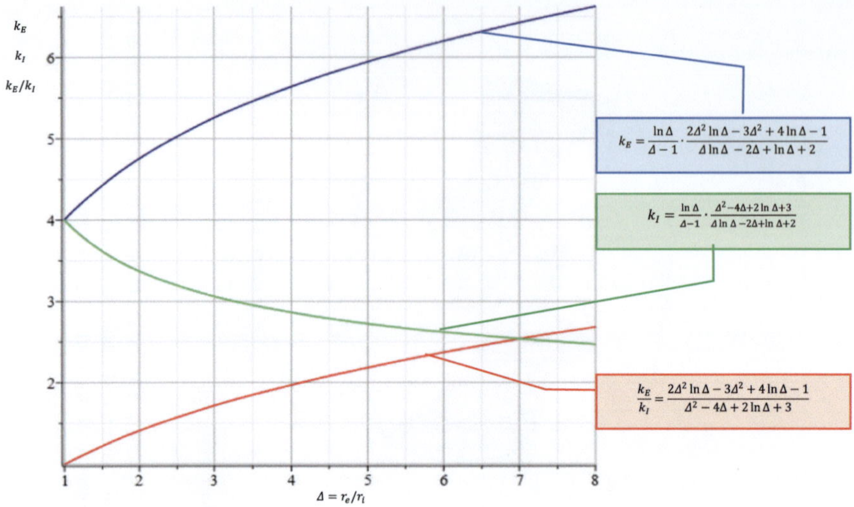

Fig. 5.8 Correction coefficients k_E Eq. (5.41), k_I Eq. (5.37) and the ratio of the correction coefficients k_E/k_I Eq. (5.42)

The corner points according to standard (EN 16984:2017–02, 2017) are depicted on Fig. 5.3. The coordinates of corner points are correspondingly:

$$\begin{aligned} I &: \rho = r_i, y = T/2; \quad II : \rho = r_i, y = -T/2; \\ III &: \rho = r_e, y = -T/2; \quad IV : \rho = r_e, y = T/2. \end{aligned} \tag{5.46}$$

The stresses in the spring on the corner points I,..IV are pure elastic and are equal to [DIN EN 16984:2017–02, 2017]:

$$\sigma_I = -\frac{4E}{1-\nu^2} \cdot \frac{T^2}{K_1 D_e^2} \cdot \frac{s}{T} \cdot \left[K_2 \cdot \left(\frac{h}{T} - \frac{s}{2T} \right) + K_3 \right], \tag{5.47}$$

$$\sigma_{II} = -\frac{4E}{1-\nu^2} \cdot \frac{T^2}{K_1 D_e^2} \cdot \frac{s}{T} \left[K_2 \cdot \left(\frac{h}{T} - \frac{s}{2T} \right) - K_3 \right], \tag{5.48}$$

$$\sigma_{III} = -\frac{4E}{1-\nu^2} \cdot \frac{T^2}{K_1 D_e^2} \cdot \frac{s}{T\Delta} \cdot \left[(K_2 - 2K_3) \cdot \left(\frac{h}{T} - \frac{s}{2T} \right) - K_3 \right], \tag{5.49}$$

$$\sigma_{VI} = -\frac{4E}{1-\nu^2} \cdot \frac{T^2}{K_1 D_e^2} \cdot \frac{s}{T\Delta} \cdot \left[(K_2 - 2K_3) \cdot \left(\frac{h}{T} - \frac{s}{2T} \right) + K_3 \right]. \tag{5.50}$$

The coefficients in the Eqs. (5.47)–(5.50) are (DIN EN 16984:2017–02, 2017]:

$$K_1 = \frac{1}{\pi} \left(\frac{\Delta - 1}{\Delta} \right)^2 \cdot \left(\frac{\Delta + 1}{\Delta - 1} - \frac{2}{\ln \Delta} \right)^{-1}, \tag{5.51}$$

$$K_2 = \frac{6}{\pi \ln \Delta} \cdot \left(\frac{\Delta - 1}{\ln \Delta} - 1 \right), \quad K_3 = \frac{3}{\pi} \cdot \frac{\Delta - 1}{\ln \Delta}. \tag{5.52}$$

For the considered, ideal rectangular form of the spring the value of the coefficient is $K_4 = 1$.

The introduction of the "partial stresses" eases the estimation of effects of the circumferential bending and uniaxial strain. The "partial stresses" could be evaluated from (DIN EN 16984:2017–02, 2017):

$$\sigma_{b.i} = -\frac{4E}{1 - \nu^2} \cdot \frac{K_3 T^2}{K_1 D_e^2} \cdot \frac{s}{T}, \tag{5.53}$$

$$\sigma_{b.e} = -\frac{4E}{1 - \nu^2} \cdot \frac{K_3 T^2}{K_1 D_e^2 \Delta} \cdot \frac{s}{T}, \tag{5.54}$$

$$\sigma_{t.i} = -\frac{4E}{1 - \nu^2} \cdot \frac{K_2 T^2}{K_1 D_e^2} \cdot \frac{s}{T} \cdot \left(\frac{h}{T} - \frac{s}{2T} \right), \tag{5.55}$$

$$\sigma_{t.e} = -\frac{4E}{1 - \nu^2} \cdot \frac{(K_2 - 2K_3) T^2}{K_1 D_e^2} \cdot \frac{s}{T \Delta} \cdot \left(\frac{h}{T} - \frac{s}{2T} \right). \tag{5.56}$$

The descriptions of partial stresses are the following:

$\sigma_{b.e}$ stress due to bending on the outer diameter $D_e = 2r_e$.
$\sigma_{b.i}$ stress due to bending on the inner diameter $D_i = 2r_i$.
$\sigma_{t.e}$ stress due to circumferential strain on the outer diameter D_e;
$\sigma_{t.i}$ stress due to circumferential strain on the inner diameter D_i.

The stresses on the corner points from Eq. (5.46) are the sums of the "partial stresses" (5.53)–(5.56):

$$\begin{aligned} \sigma_I &= \sigma_{b.i} + \sigma_{t.i}, \\ \sigma_{II} &= -\sigma_{b.i} + \sigma_{t.i}, \\ \sigma_{III} &= -\sigma_{b.e} + \sigma_{t.e}, \\ \sigma_{VI} &= \sigma_{b.e} + \sigma_{t.e}. \end{aligned} \tag{5.57}$$

The stresses in corner points could be evaluated in the solid and shell theories of disk springs. For the free gliding spring, the position of inversion center provides Eq. (5.30). The expression (5.30) for the inversion center must be substituted into the stress–strain dependence (5.21) and the circumferential strain (5.23). The formulas for "partial stresses" follow from these formulas after substitution of coordinates for the corner points (5.46):

$$\sigma_{b.i} = \frac{E}{1-\nu^2} \cdot \frac{\sin\alpha - \sin\psi}{2} \cdot \mu, \quad \sigma_{b.e} = \frac{\sigma_{b.i}}{\Delta}, \tag{5.58}$$

$$\sigma_{t.i} = \frac{E}{1-\nu^2} \cdot \frac{\cos\psi - \cos\alpha}{\cos\alpha} \cdot \frac{1-\Delta+\ln\Delta}{\ln\Delta},$$

$$\sigma_{t.e} = \frac{1-\Delta+\Delta \ln \Delta}{(1-\Delta+\ln \Delta)\Delta} \cdot \sigma_{t.i}. \tag{5.59}$$

With the Eqs. (5.58)–(5.59) for partial stresses, the stresses on the corner points follow from Eq. (5.57). Accordingly, the stresses for the springs on the rigid curb enclosed by the hollow cylinder with Eqs. (5.16), (5.17), (5.21), (5.23) after the substitution of the corresponding expressions for the corresponding inversion centers:

$$C_I = r_I, \quad C_E = r_e.$$

The bending partial stresses remain unaffected by the shift of the inversion point to the inside or outside edge. The in-plane stresses due to circumferential strain modify significantly in course of relocation of the inversion points.

5.4 Deformation of Conical Spring with Both Radially Constrained Edges

As it was demonstrated above, if both edges of the disk can freely move in radial direction, total hoop force in the cross-section of the hollow conical disk vanishes, $F_\theta = 0$. Under these conditions, the radius C of the inversion point guarantees the minimum the elastic strain energy U_E among all admissible radii.

If the radial movements of one of the edges are confined, the radius of the inversion point relocates from the preferrable position, which corresponds to the minimum energy of the elastic deformation. On the confined edge appears the reaction as the radial force. If the inner edge (i) freely glides, the total axial force acting on the inner edge (i) is F_z, and the radial reaction force vanishes $F_{I.R} = 0$. Otherwise, if the movement of the inner edge is restricted, as the reaction appears the radial force $F_{I.R}$.

On the outer edge (e), the total axial force acting is $-F_z$ and the total radial force was denoted ad $F_{E.R}$. If the inner edge can freely glide, $F_{E.R} = 0$. If the movement of the outer edge is constrained, the reaction on this constraint is the radial force $F_{E.R}$.

The special situation takes place, if the movements of both inner edges are restricted, the reactions must be opposite $F_{I.R} = -F_{E.R}$. These reactions could not be calculated with the hypotheses of rotating without warping section. The above assumed hypothesis disregards the deformation of the section. The hypothesis of the sectional rotation results the infinite stiffness of the doubly restricted disk spring.

Thus, for the mathematical description of the doubly fixed edges, an alternative hypothesis must be applied. The slope angle α of the conical shell is assumed to be small. The slope of the middle surface of the shell differs slightly from the neutral plane. For the doubly constrained spring, the new hypothesis assumes, that there occurs the membrane deformation of the spring. Because the slope angle is small, the membrane elastic state corresponds to the elastic state of the flat thick cylinder in plane stress state (Ugural & Fenster, 2019). The normal pressure on the inner edge of the disk spring is p_i and on the outer edge is p_e. These normal stresses cause the radial deformation of the disk:

$$u_r(r) = \frac{1-\nu}{E} \cdot \frac{p_i r_i^2 - p_e r_e^2}{r_e^2 - r_i^2} \cdot r + \frac{1+\nu}{E} \cdot \frac{p_i - p_e}{r_e^2 - r_i^2} \cdot \frac{r_i^2 r_e^2}{r}. \tag{5.60}$$

The contraction of the meridional section in radial direction is equal to:

$$\begin{aligned} \delta u_r &= u_r(r_e) - u_r(r_i) \\ &= \frac{(1+\nu) r_e r_i - (1-\nu) r_e^2}{E(r_e+r_i)} p_e - \frac{(1+\nu) r_e r_i - (1-\nu) r_i^2}{E(r_e+r_i)} p_e. \end{aligned} \tag{5.61}$$

As the area of inner surface is $2\pi T r_i$, the total circumferential force on the inner surface of the disk is

$$F_{ir} = 2\pi T r_i p_i.$$

Analogously, the outer surface area is $2\pi T r_a$, and the total circumferential force on the inner edge is

$F_{er} = -2\pi T r_e p_e$.

The axial force is the projection of the circumferential forces on the direction of the spring axis:

$$F_z = F_{ir} \sin\alpha = -F_{er} \sin\alpha. \tag{5.62}$$

Consequently, the normal pressures on the inner and outer surfaces are:

$$p_i = \frac{F_z}{2\pi T r_i \sin\alpha}, \quad p_e = \frac{F_z}{2\pi T r_e \sin\alpha}. \tag{5.63}$$

The substitution (5.63) into (5.61) results the contraction of the meridional section in radial direction:

$$\delta u_r = -\frac{r_e - r_i}{r_e + r_i} \frac{F_z}{\pi E\ T \sin\alpha}. \tag{5.64}$$

The axial contraction of the meridional section is the projection of contraction of the meridional section in radial direction on the direction of the spring axis:

$$\delta u_z = \sin\alpha \cdot \delta u_r = -\frac{r_e - r_i}{r_e + r_i} \cdot \frac{F_z}{\pi E\ T}. \tag{5.64}$$

From (5.64) follows the spring rate of the disk spring with both fixed edges as:

$$c_{z.IE} = -\frac{F_z}{\delta u_r} = \frac{r_e + r_i}{r_e - r_i} \cdot \pi E\ T. \tag{5.65}$$

5.5 Comparison of Spring Constants for Differently Confined Disk Springs

The initial springs rates of unconstrained, constrained on the only inner edge, constrained on the only outer edge and both constrained edges disk springs will be calculated below. These formulas are used for comparison of the load–deflection curves.

At first, the spring rate $c_{z.AL}$ at the begin of deformation $\psi = \alpha$, $z = Z$ of the disk spring with freely gliding edges is estimated. The dependence of force depends upon ψ is given by Eq. (5.33). The expression (5.14) provides the height of the spring. The spring rate at the first moment of the compression results from differentiation of the axial force over height:

$$c_{z.G} = -\left(\frac{\mathrm{d}F_{z.G}}{\mathrm{d}z}\right)_{\psi=\alpha} = -\left(\frac{\mathrm{d}F_{z.G}/\mathrm{d}\psi}{\mathrm{d}z/\mathrm{d}\psi}\right)_{\psi=\alpha} = c_R\phi_G\,, \qquad \left(\frac{\mathrm{d}z}{\mathrm{d}\psi}\right)_{\psi=\alpha} = (\Delta - 1)r_i\,. \tag{5.66}$$

The parameter with the dimension [N/mm] in Eq. (5.66) is:

$$c_R = \frac{\pi\, E\, T}{6(\Delta-1)^2 \cdot \ln\Delta \cdot \cos^3\alpha}. \tag{5.67}$$

The dimensionless function for the gliding spring is:

$$\phi_G = \cos^4\alpha \cdot \mu^2 \cdot \ln^2\Delta - 6\left[\left(1-\Delta^2\right)\ln\Delta + 2(\Delta-1)^2\right] \cdot \sin^2\alpha. \tag{5.68}$$

The spring rates of the springs, which are constrained on the inner and outer edges, are correspondingly:

$$c_{z.I} = -\left[\frac{\mathrm{d}F_{z.I}}{\mathrm{d}z}\right]_{\psi=\alpha} = c_R\phi_I, \quad c_{z.E} = -\left[\frac{\mathrm{d}F_{z.E}}{\mathrm{d}z}\right]_{\psi=\alpha} = c_R\phi_E. \tag{5.69}$$

For the inner and outer restricted disk springs, the matching dimensionless functions are:

$$\phi_I = \cos^4\alpha \cdot \mu^2 \cdot \ln^2\Delta + 6\left(\Delta^2 - 4\Delta + 3 + 2\ln\Delta\right) \cdot \sin^2\alpha, \tag{5.70}$$

$$\phi_E = \cos^4\alpha \cdot \mu^2 \cdot \ln^2\Delta + 6\left(-3\Delta^2 + 4\Delta - 1 + 2\Delta^2\ln\Delta\right) \cdot \sin^2\alpha. \tag{5.71}$$

The set of parameters from Sect. 3 with the fixed height $h_0 = 2$ mm is used for the calculations. With this set of parameters, follow the ratios of the initial spring rates of the springs with constrained edges (5.67) and (5.69) to the spring rate of freely gliding spring (5.66):

$$\frac{c_{z.IE}}{c_{z.G}} = 3838, \quad \frac{c_{z.I}}{c_{z.G}} = 2.42, \quad \frac{c_{z.E}}{c_{z.AL}} = 2.63.$$

The disk spring, which is constrained on both edges possesses a tremendously high spring rate. This conical disk practically incompressible in comparison to the one-side confined or to the unrestricted disk springs.

5.6 Finite Element Simulation of the Disk Springs with the Free Gliding and Radially Constrained Edges

The finite element simulation was performed for verification of the analytical formulas. For comparison the finite element simulation using program ANSYS 2022R2 (ANSYS, 2022) was performed. The APDL code and results files of the finite-element simulation using ANSYS 2022R2 are provided in (Kobelev, 2023b). For the finite element modeling four models were developed.

- Shell Model (0). The middle surface of the spring was meshed with shell elements SHELL181. In the Shell Model (0) the applied load is evenly distributed over the upper edge of middle surface and the vertical displacement on lower edge was restricted. Both inner and outer edges can freely glide in radial directions. No radial reaction forces occur on both edges. All other displacements were assumed to be free.
- Shell Model (1). In the Shell Model (1) the applied load is evenly distributed over the upper edge of middle surface and the vertical displacement on lower edge was restricted. The outer edges can freely glide in radial directions. The radial displacement of inner edge is constrained, such that the radial reaction forces occur on the inner edges. All other displacements were assumed to be free.
- Shell Model (2). In the Shell Model (2) the applied load is evenly distributed over the upper edge of middle surface and the vertical displacement on lower edge was restricted. The inner edges can freely glide in radial directions. The radial displacement of outer edge is constrained, such that the radial reaction forces occur on the outer edges. All other displacements were assumed to be free.
- Shell Model (3). In the Shell Model (3) the applied load is evenly distributed over the upper edge of middle surface and the vertical displacement on lower edge was restricted. The radial displacements of both inner and outer edge are constrained, such that the radial reaction forces occur on inner and outer edges. All other displacements were assumed to be free.

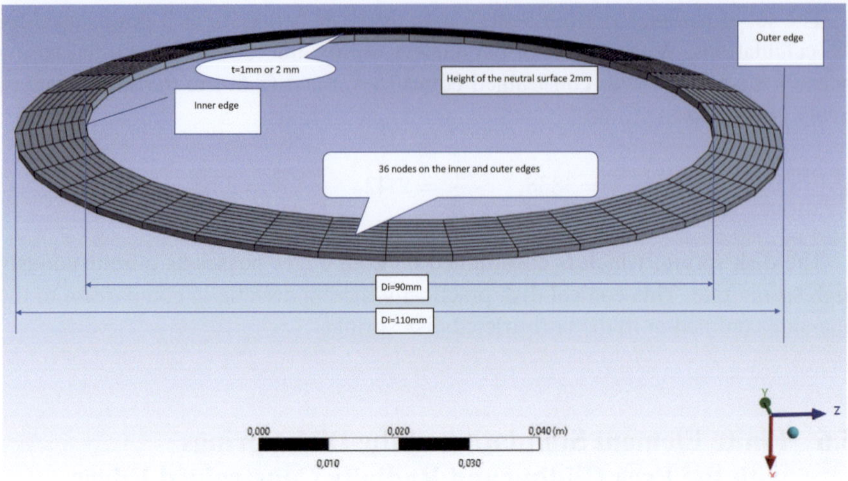

Fig. 5.9 Finite element mesh of the disk spring. Structured rectangular mesh with shell elements

All models demonstrate the influence of thickness on the elastic response of the thick-walled conical springs (Fig. 5.9). There were 396 nodes on the middle surface and 4 nodes for the remote points. For creation of the elements the following APDL commands were used:

```
et,1,181
keyo,1,1,0 ! bending and membrane stiffness
keyo,1,3,2
eblock,19,solid,,360
```

Material properties were set with the ANSYS commands:

```
MP,DENS,1,7850,              ! kg m^-3
MP,ALPX,1,1.2e-05,           ! C^-1
MP,C,1,434,                  ! J kg^-1 C^-1
MP,KXX,1,60.5,               ! W m^-1 C^-1
MP,RSVX,1,1.7e-07,           ! kg m^3 A^-2 s^-3
MP,EX,1,200,000,000,000,     ! Pa
Mp,nuxy,1,0.3,
Mp,murx,1,10,000,
```

The thickness of the shell was fixed using the section command:
/com,*********** Send Sheet Properties ***********

sectype,1,shell
secdata,1.e-003
secoff,mid

No considerable difference due to element type replacement between the obtained results could be stated. The thicknesses of the springs were $T = 1mm$ and $T = 2mm$. For the calculation the following values were used:

$$r_e = 55\,\text{mm},\, r_i = 45\,\text{mm},\, h = 2\,\text{mm},\, E = 200\,\text{GPa},\, \nu = 0.3.$$

Firstly, the spring force from the actual analytical model was compared with the spring force from finite-element models Model (0). The spring force from the analytical model with freely gliding edges was calculated using the Eq. (5.33). The vertical reaction was evaluated using the ANSYS calculation of the reaction force. The correspondence between the spring force (5.33) and the spring force, calculated using the finite element models (0) is remarkable (Fig. 5.10). The order of the relative difference between the numerical results and analytical formulas is of 1% to 1.5%. This confidence demonstrates, that the bending of the middle surface in the radial direction is negligible and could be neglected for the most engineering applications.

Secondly, the spring force from the analytical model was compared with the spring force from finite-element models Model (1). The spring force from the analytical model with the constrained radial displacement of inner edge was calculated using the Eq. (5.37). The vertical reaction was evaluated using the ANSYS calculation of the reaction force. The correspondence between the spring force (5.37) and the

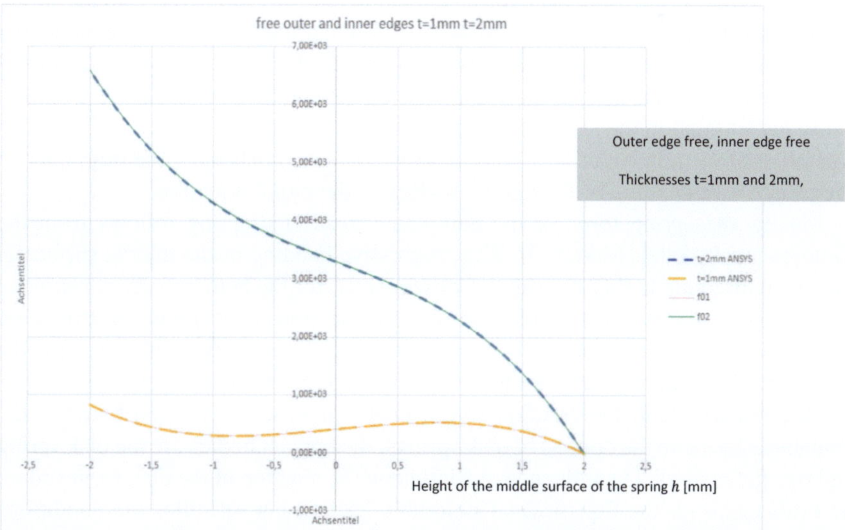

Fig. 5.10 Comparison of analytical formulas for the disk spring with freely radially gliding inner and outer edges to the reaction forces from finite element simulation

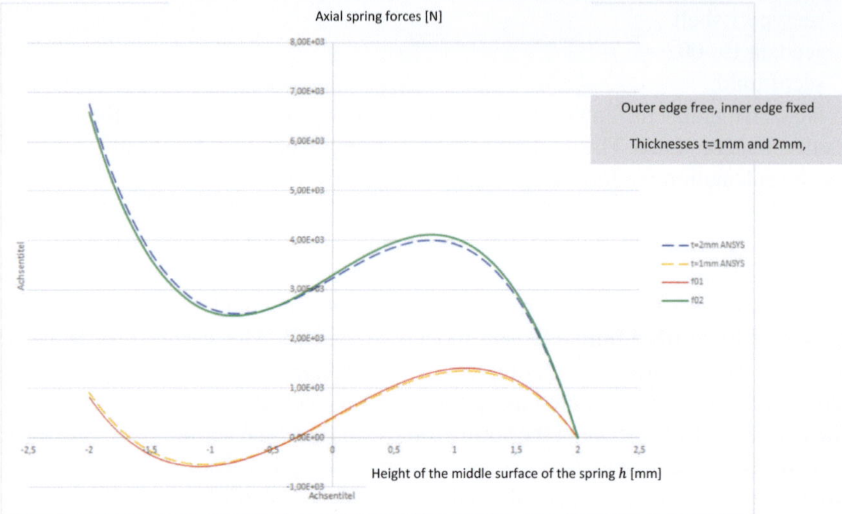

Fig. 5.11 Comparison of analytical formulas for the disk spring with radially gliding outer edge to the reaction forces from finite element simulation. The radial displacement of inner edge is constrained

spring force, calculated using the finite element model (1) is also notable (Fig. 5.11). The order of the relative difference is of 3% to 5%. The slightly increasing error could be explained by the progressive bending of the cone in the radial direction.

Thirdly, the spring force from the analytical model was compared with the spring force from finite-element models Model (2). The spring force from the analytical model with the constrained radial displacement of the outer edge was calculated using the Eq. (5.41). The similar precision of the spring force (5.41) is achieved using the simulation the spring force with the finite element model (2), (Fig. 5.12). The order of the relative difference is of 4–6%. The error has the same origin due to the progressive bending of the middle surface in the radial direction.

Finally, the spring force of the completely restricted spring follows from the finite-element models Model (3). The progressive bending of the middle surface in the radial direction highly increases. Consequently, for the both radially constrained edges the hypothesis of the rotation for the cross-section is not applicable any more. The realistic evaluation of the spring characteristics could be achieved with the application of finite-element methods, as shown on Fig. 5.13.

One remark about the reaction forces in the presence of friction between the contact surfaces. In the contest of disk springs, the contact bodies are the disk spring and two flat pistons. The influence of friction on the reaction of the disk spring could be evaluated with the Eq. (5.25). Coulomb's friction law specifies the conditions that the contact force between two solids. Coulomb's friction depends upon the occurrence of a slip between contact bodies. These two pistons compress the disk spring, one from the top and one from the bottom. For example, one piston glides without friction, while friction occurs on the other contact surface. The contact force

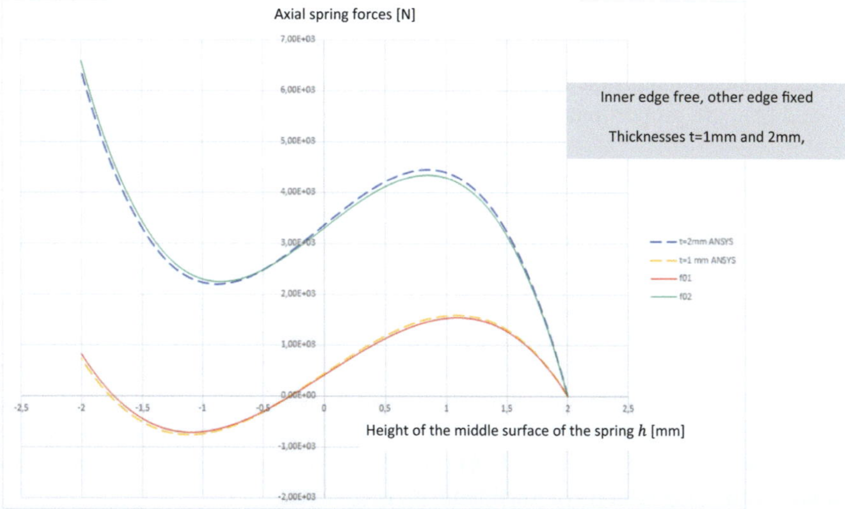

Fig. 5.12 Comparison of analytical formulas for the disk spring with radially gliding inner edge to the reaction forces from finite element simulation. The radial displacement of both springs on the outer edge is constrained

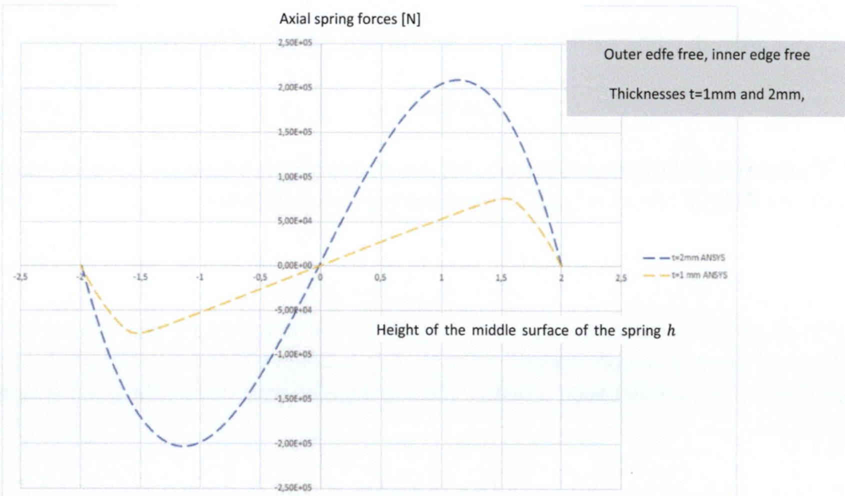

Fig. 5.13 The reaction forces from finite element simulation of two springs with different thickness. Both radial displacements of inner and outer edges are constrained

embraces a normal component F_Z, perpendicular to the contact plane between the disk spring and the contact piston, and a tangential component F_R, parallel to the contact plane on the piston surface. The slip between the spring and piston occurs if F_Z and $|F_R|$ are proportional:

$$|F_R| = \mu F_Z,$$

where μ is the coefficient of friction that characterizes the condition of the two surfaces in contact. In this case, the contact force is said to remain within the Coulomb cone. Provided that the force $|F_R|$ is less than μF_Z, sliding does not occur, and is referred to as adhesion between the two solids. Consequently, friction within the Coulomb cone causes the certain radial forces $F_R = \pm\mu F_Z$. For compression of the spring from the free state, the "plus" sign must be used. Correspondingly, for the extension from the flattened state the "minus" sign is to be applied. The difference between two reaction forces describes the hysteresis due to friction. If the Coulomb's friction occurs, the total meridional moment is equal to:

$$M_F = F_Z B - F_R H = (B \pm \mu H) F_Z = (1 \pm \mu \tan\psi) B\ F_Z. \tag{5.72}$$

The height and width of the cross-section in Eq. (5.72) are given by Eqs. (5.14) and (5.15).

Contrarily, the total meridional moment in the absence of friction reads:

$$M = F_Z B. \tag{5.73}$$

Comparison of the expressions (5.72) and (5.73) for the total meridional moment with and without frictions leads to the friction correction factor:

$$M_F = M \cdot k_Q, \quad k_Q = 1 \pm \mu \tan\psi. \tag{5.74}$$

In all previous calculations, the spring compression forces F_Z were performed without friction. The corresponding forces F_{ZF} account Coulomb's friction. In the presence of friction between a piston and the edge of disk spring, the axial spring forces read:

$$F_{ZF} = F_Z\, k_Q. \tag{5.73}$$

If the middle surface Ω does not yet achieve the flattened state, angle ψ is positive. Thus, the correction factor will be greater than one, $k_Q > 1$. Consequently, friction will increase the reaction forces $F_{ZF} > F_Z$ for the positive height H of the middle surface. When the middle surface Ω flattens, the angle ψ vanishes and the correction factor turns to be one, $F_{ZF} = F_Z$. Thus, friction does not influence the vertical forces of the spring nearby its flattened state.

5.7 Conclusions

The developed method is applicable to disk springs of constant material thickness with radially constrained edges (Tables 5.1, 5.2, 5.3 and 5.4). The vertical displacements of a disk spring result from an axial load uniformly distributed on the inner and outer edges. The verification of the analytical formulas for the free gliding of the spring edges and for the springs with inner or outer constrained edges was carried out using the finite element codes ANSYS (2022) R2. The formulas are precise and simple enough for industrial applications of disk springs made of traditional isotropic materials.

The method could be directly applied to the calculation of slotted disk springs. A slotted disk spring consists of two segments: a disk segment and a number of lever arm segments. At present, the calculation of slotted disk springs is based on the SAE formula (SAE, 1996). This formula is limited to a straight slotted disk spring with free sliding inner and outer edges.

Summary of Principal Results

- The nonlinear governing equations for the cup spring have been derived.
- The equations describe the deformation and stresses of thin and moderately thick disk springs.
- The variation method is applicable to the disk springs with free gliding and rigidly constrained edges.

Table 5.1 Freely gliding disk spring

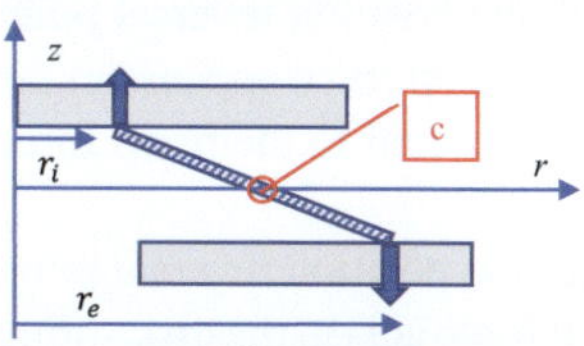 Freely gliding disk spring	Inversion radius:	$C_G = \frac{r_e - r_i}{2\ln(r_e/r_i)}$
	Freely gliding on inside edge	Freely gliding on outside edge
	Spring rate at the first moment of axial compression: $c_{z.G} = c_R \phi_G,\ \Delta = \frac{r_e}{r_i},\ \mu = \frac{T}{r_i},$ $\phi_G =$ $\cos^4\alpha \cdot \mu^2 \cdot \ln^2\Delta - 6\left[\left(1-\Delta^2\right)\ln\Delta + 2(\Delta-1)^2\right]\sin^2\alpha$	

Table 5.2 Disk spring constrained on the inner radius

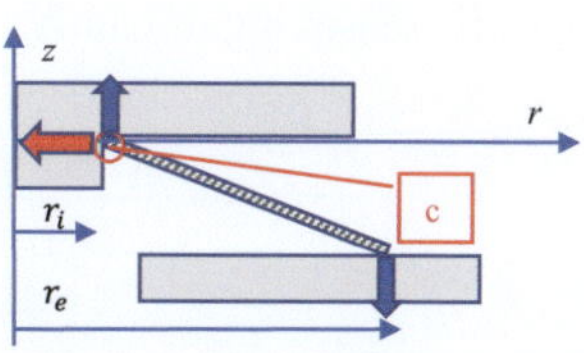 Disk spring constrained on the inner edge	Inversion radius:	$C_I = r_I$
	Reaction on inside edge	Freely gliding outside edge
	Spring rate at the first moment of axial compression: $c_{z.I} = c_R \phi_I,$ $\phi_I =$ $\cos^4\alpha \cdot \mu^2 \cdot \ln^2\Delta + 6\left(\Delta^2 - 4\Delta + 3 + 2\ln\Delta\right)\sin^2\alpha$	

Table 5.3 Disk spring constrained on the outer radius

<table>
<tr><td rowspan="3">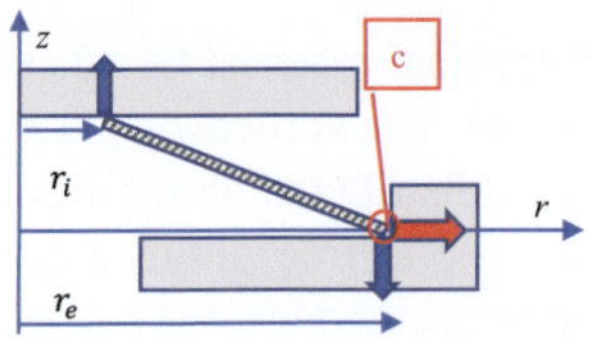Disk spring constrained on the outer edge</td><td>Inversion radius:</td><td>$C_E = r_E$</td></tr>
<tr><td>Freely gliding inside edge</td><td>Reaction on outside edge</td></tr>
<tr><td colspan="2">Spring rate at the first moment of axial compression:
$c_{z.E} = c_R \phi_E,$
$\phi_E = \cos^4 \alpha \cdot \mu^2 \cdot \ln^2 \Delta +$
$6\left(-3\Delta^2 + 4\Delta - 1 + 2\Delta^2 \ln\Delta\right) \sin^2 \alpha$</td></tr>
</table>

Table 5.4 Disk spring constrained on inner and outer radii

<table>
<tr><td rowspan="3">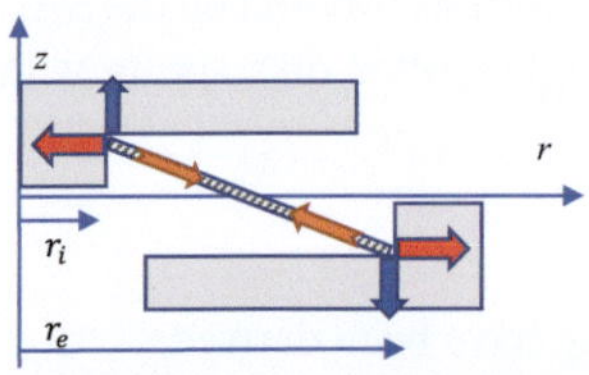Disk spring constrained on both edges</td><td colspan="2">Meridional membrane compression of conical disk</td></tr>
<tr><td>Reaction on inside edge</td><td>Reaction on outside edge</td></tr>
<tr><td colspan="2">Spring rate at the first moment of axial compression:
$c_{z.IE} = \frac{r_e + r_i}{r_e - r_i} \pi E T$</td></tr>
</table>

- If both edges of the spring glide freely over the inner and outer edges in the radial direction, the position of the neutral plane is determined from the condition of the minimum of elastic energy. The spring force in the disk disappears from the condition of minimum elastic energy.
- If the radial movement of the inner edge is restricted, the position reversal point corresponds to the inner radius. In this case, the tensile hoop force appears in the disk spring. The compressive force on the inner edge is equal to the hoop force in the disk.
- When the radial motion of the outer edge is restricted, the position reversal point corresponds to the outer radius. The compressive hoop force acts in the disk spring and causes the extension force on the surface of the outer cylindrical cavity.
- The axial stiffness of the cup spring at the inner edge is higher than the axial stiffness of the free edge spring.
- If the movements of both (inner and outer) edges are restricted, the diaphragm deformation of the spring of the neutral surface prevails. The axial rigidity of the conical disk spring with both radially constrained edges increases significantly.

References

Almen, J. O., & Laszlo, A. (1936). The uniform section disks spring. *Trans ASME, 58*(12.4), 305–314.

ANSYS. (2022). http://www.ansys.com/, ANSYS, Inc. Southpointe, 2600 ANSYS Drive, Canonsburg, PA 15317c

Caresta, M., & Kessissoglou, N. J. (2008). Vibration of fluid loaded conical shells. *The Journal of the Acoustical Society of America.* https://doi.org/10.1121/1.2973237

DIN EN 16984:2017-02 (2017) *Disk Springs - Calculation*; German Version EN 16984:2016, Beuth Verlag, Berlin

Kobelev, V. (2016). Exact shell solutions for conical springs. *Mechanics Based Design of Structures and Machines, An International Journal, 44*(4), 317–339. https://doi.org/10.1080/15397734.2015.1066686

Kobelev, V. (2023a) *MAPLE data for exact shell solutions for conical springs. II. Radial cylindric curb*, Mendeley Data, V1, doi: https://doi.org/10.17632/295pd6trtb.1

Kobelev, V. (2023b). *ANSYS APDL for exact shell solutions for conical springs. II. Radial cylindric curb*, Mendeley Data, V1.https://doi.org/10.17632/f3pscg3pct.1

MAPLE. (2022). https://www.maplesoft.com/. Maplesoft, 615 Kumpf Drive Waterloo, ON Canada N2V 1K8

Marsden, J. E., & Hughes, T. J. R. (1994). *Mathematical foundations of elasticity*. Dover Publications.

Muhr, T., Asbeck, J., Kobelev, V., Westerhoff, K., Brecht, J. D., & Hees, A. (2009). *Valve spring plate with two supporting tongues.* United States Patent No.: US 7566046 B2

SAE. (1996). *Spring design manual.* Part 5, SAE, HS-158 SAE International, Warrendale, PA.

Sobhani E., & Avcar M. (2022a). Natural frequency analysis of imperfect GNPRN conical shell, cylindrical shell, and annular plate structures resting on Winkler-Pasternak Foundations under arbitrary boundary conditions. *Engineering Analysis with Boundary Elements, 144*(145–164), https://doi.org/10.1016/j.enganabound.2022.08.018

Sobhani, E., & Avcar, M. (2022b). The influence of various nanofiller materials (CNTs, GNPs, and GOPs) on the natural frequencies of nanocomposite cylindrical shells: A comparative study. *Materials Today Communications, 33*, 104547. https://doi.org/10.1016/j.mtcomm.2022.104547

Ugural A. C., & Fenster S. K. (2019). *Advanced mechanics of materials and applied elasticity* (International Series in the Physical and Chemical Engineering Sciences); 6. Edition, Pearson Inc. London, ISBN-13:978-0134859286

Chapter 6
Disk Springs with Variable Thickness

Abstract This chapter concludes the treatment of disk springs. In the current chapter, we study the variable thickness cup spring. The thickness is assumed to be variable along the meridional and parallel axes of the conical coordinate system. The calculation of the disk springs includes the cases of free gliding edges and edges on cylindrical curbs, which constrain the radial movement. The equations developed here are based on common assumptions and are simple enough to be applied to industrial calculations.

Keywords Disk springs with variable thickness · Radially variable thickness · Tangential variable thickness · Linear radial dependence of thickness · Quadratic radial dependence of thickness

6.1 Models of Elastic Disk Springs

The Belleville washer, or disk spring, is remarkable for its ability to nonlinearly withstand a large force with minimal deflection while storing a considerable amount of elastic energy relative to the occupied volume. The patent (Hesselmann & Kobelev, 2002) describes a disk spring with non-constant thickness. An inner cross sectional point is located in an inner region of the disk spring near the longitudinal axis, an outer cross sectional point is located in an outer region of the disk spring near the edge, and another cross sectional point is located between the inner cross sectional point and the outer cross sectional point in the region of the neutral point. When the disk spring is installed, the maximum stress is in the area of the inner and/or outer cross section point. In order to ensure a homogeneous stress distribution over the disk spring, the invention provides that the disk thickness of the further cross-sectional point is smaller than the inner cross-sectional point and smaller than the disk thickness of the

V. Kobelev, *Fundamentals of Springs Mechanics*,
https://doi.org/10.1007/978-3-031-58584-5_6

outer cross-sectional point. Chapters 4 and 5 contain the bibliography in the field of disk springs.[1]

In this chapter, we study the disk springs of non-constant thickness under the conditions of different degrees of freedom on the inner and outer surfaces. The variation methods consequently apply the kinematic hypothesis for the elastic deformation of disk springs with meridionally varying thickness. In the former study, the statics of elastic disk springs with constant thickness and contact surfaces that can slide freely were investigated in Chap. 4. It was shown that during axial compression, the inner edge of the spring moves toward the axis of the disk. At the same time, the outer edge of the spring moves in the opposite direction, away from the axis of the disk. Since both edges were free to slide without resistance, both radial reaction forces were absent in the above study.

6.2 Geometry of Disk Spring in Undeformed State

Consider a truncated conical shell. Its middle surface in the free state is created by cutting the cone from the vertex or apex. A plane parallel to the base of the cone cuts the top of the cone or the apex to create a truncated cone ϖ. (Fig. 6.1) The inside radius of middle surface of free spring is r_i, the outside radius is r_e, and $\mathfrak{b} = r_e - r_i$. The ratio of outer radius to inside radius reads:

$$\Delta = \frac{r_e}{r_i} > 1.$$

The slope angle α of the conical shell is constant. The slope angles are shown on Figs. 6.1, 6.2, 6.3 and 6.4. The real disk springs are flatter. Usually, the slope angles of disk springs are not so high and typically vary from 2° to 5°. However, there are other disk springs with significant slope angles. The disk wave springs are the corrugated springs, which comprise an annular spring band. The spring band describes a corrugation line oscillating about a reference surface over the circumference. Such corrugated springs are known as so-called "axial corrugated springs" or "disk wave springs" for absorbing axial forces in a spring fashion wherein the corrugation line oscillates over the circumference relative to a radial base plane. The radial base plane lies normal to the ring axis. In this case, the spring band is generally closed in a ring shape. Such corrugated springs are furthermore known as so-called "radial corrugated springs" in which the spring band oscillates over the circumference relative to a ring cylinder which lies coaxial to the ring axis. The theory of axial and radial corrugated springs was presented in Chap. 5.

---

[1] V. Kobelev, *Fundamentals of Springs Mechanics*, University of Siegen, Faculty IV: School of Science and Technology. Paul-Bonatz-Str. 9–11, D-57076 Siegen. E-mail: Vladimir.kobelev@uni-siegen.de Homepage: https://www.mb.uni-siegen.de/kobelev/

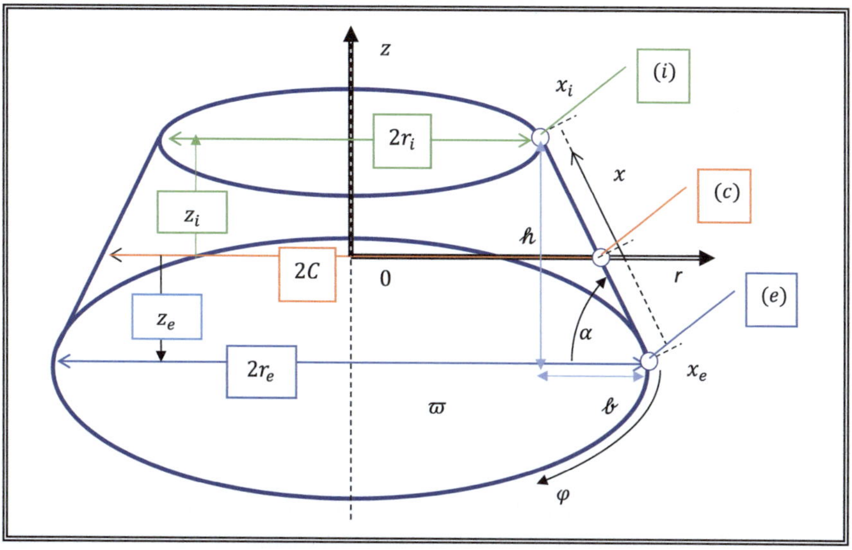

Fig. 6.1 The middle surface ϖ of the free of load Belleville spring

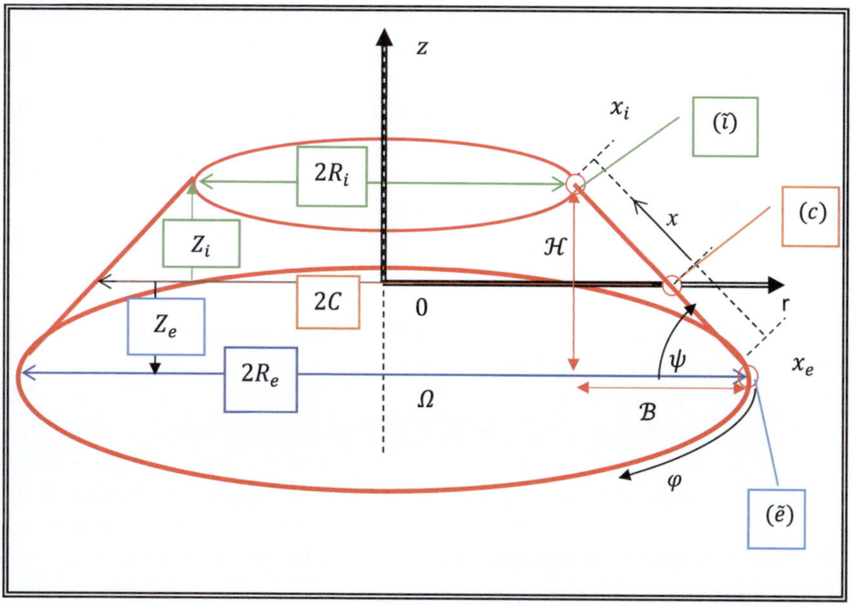

Fig. 6.2 The middle surface $\boldsymbol{\Omega}$ of the axially squeezed Belleville spring

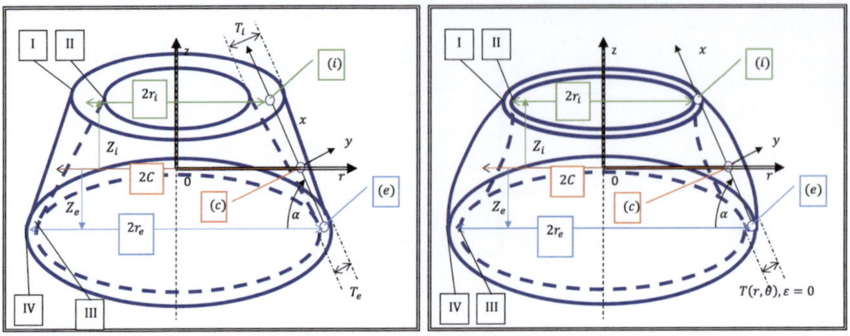

Fig. 6.3 The solid representation the undeformed Belleville spring, Left-linearly meridionally varying thickness, right-quadratically meridionally varying thickness. The special case of constant thickness over the parallels, $\varepsilon = 0$

The slope angle, as shown on Fig. 6.1, is unrealistic for the ordinary disk springs. The chosen illustration with an exaggerated slope angle allows to display the elements of the spring geometry and explain the elastic deformation of the meridional cross-sections. The elastic evolution the shell under load will be described in the cylindrical coordinate system. The origin O of the cylindrical coordinate system is located on the certain neutral plane. As shown later, the position of the neutral plane depends upon the prescribed conditions of the freely gliding or radially restricted edges.

In the cylindrical coordinate system, the radius of inversion centre (c) is C. The radius C follows from the height neutral plane (c), which have to be determined so far. In the cylindrical coordinates (r, φ, z), the middle surface of a free conical shell respectively is described parametrically:

$$r = C - x \cos \alpha, \quad z = -x \sin \alpha. \tag{6.1}$$

The arc length x serves as the meridional coordinate and represents the *latitude* of the point on the middle surface:

$$x_e \le x \le x_i.$$

The angle φ represents the *longitude* of the point. For each cross-section of the spring with a given longitude, there is an inversion point (c). All inversion points with different longitudes stand on the neutral plane. The arc length is measured from the inversion point (c), where:

$$x_i = \frac{C - r_i}{\cos \alpha} > 0.4, \quad x_e = \frac{C - r_e}{\cos \alpha} < 0. \tag{6.2}$$

The distance from the point with coordinates (x, φ, y) to the axis of the undeformed cone $\tilde{r}$:

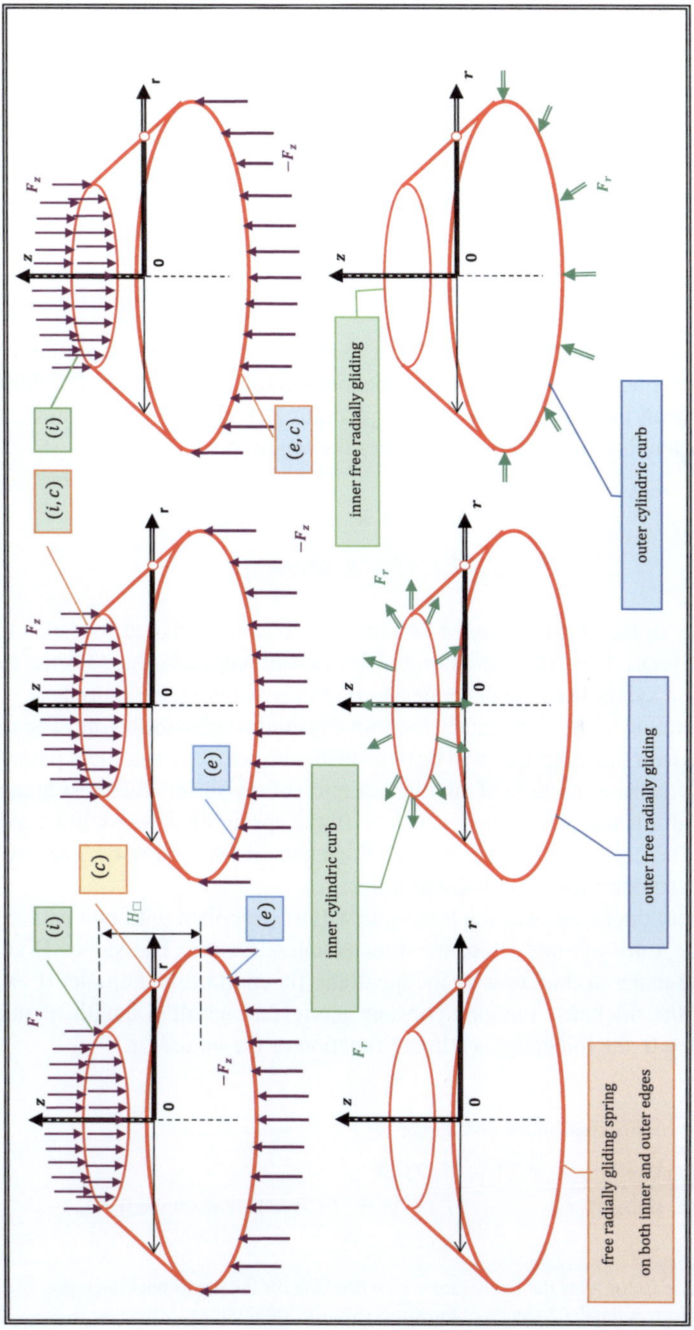

Fig. 6.4 Axial (above) and radial (below) systems of reaction forces on the truncated conical shell

$$\tilde{r} = r + y \sin \alpha = C - x \cos \alpha + y \sin \alpha. \tag{6.3}$$

The value y is the normal distance from the point to the middle surface of the shell:

$$-\tfrac{1}{2}T(x, \varphi) \leq y \leq \tfrac{1}{2}T(x, \varphi).$$

The thickness of the spring washer is variable along the meridian and parallel directions:

$$T(x, \varphi) = \mathscr{M}(x)T_i(\varphi). \tag{6.4}$$

As the reader can verify, the actual theory admits the closed form solution for an arbitrary separable representation of the thickness in form of Eq. (6.4). For definiteness, we perform the calculations for the quadratic dependence of thickness along the latitude x and harmonic dependence over the longitude φ:

$$\begin{aligned} \mathscr{M}(x) &\stackrel{\text{def}}{=} \Theta \cdot X_e + X_i + \mathrm{H} \cdot X_e \cdot X_i, \\ T_i(\varphi) &\stackrel{\text{def}}{=} t \cdot [1 + \varepsilon \cdot \sin(n\varphi)]. \end{aligned} \tag{6.5}$$

The values in Eq. (6.5) are listed in Table 6.1. In the displayed formulas (6.5), the quadratic factor is H. In expression for the meridional thickness, Θ is the linear factor. Factor ε symbolises the dependence of thickness over the longitude φ.

The explanation of the constants in the above formulas is the following. The material thickness over the longitude will be described with one dimensionless parameter ε. The value ε is the amplitude of angular variation of the spring thickness along the longitude, and n is the wave number. If $\varepsilon = 0$, the spring thickness will be axially symmetric. The constant t is the average thickness of the radius r_i. The average thickness on the outer radius r_e is equal to Θt.

The material thickness over the longitude will be described with two parameters Θ and H. The values Θ and H are the dimensionless factors. The values H and Θ determine the material thickness as the quadratic function of the latitude. If $\Theta = 0$ and $\mathrm{H} = 0$, the thickness remains constant along the meridian length of middle surface. If $\mathrm{H} = 0$, the thickness is a linear function of the latitude x.

Table 6.1 Variable thickness of the disk spring

$T(x, \varphi) = T_e(\varphi)X_e + T_i(\varphi)X_i + T_i(\varphi) \cdot \mathrm{H} \cdot X_e \cdot X_i$	
$T_i(\varphi) = [1 + \varepsilon \cdot \sin(n\varphi)] \cdot t$	$T_e(\varphi) = \Theta \cdot T_i(\varphi) \equiv [1 + \varepsilon \cdot \sin(n\varphi)] \cdot \Theta \cdot t$
$X_e(x) = \frac{x_i - x}{x_i - x_e}$	$X_i(x) = \frac{x - x_e}{x_i - x_e}$

H is the quadratic factor, Θ is the linear factor in expression for the meridional thickness
Factor ε characterizes the dependence of thickness over the longitude φ

Table 6.2 Coefficients $a(\mathrm{H}, \Delta), b(\mathrm{H}, \Delta), c(\mathrm{H}, \Delta), d(\mathrm{H}, \Delta)$ of the Eq. (6.49)

$\mathrm{a}(\mathrm{H}, \Delta) = \left[-\Delta^2 + (\mathrm{H}+2)\Delta - 1\right]^3 \cdot \ln\Delta + \left(\mathrm{H}^2 + 15\mathrm{H} + 90\right)\left(1+\Delta^4\right) - \left(9\mathrm{H}^2 + 150\mathrm{H} + 360\right)\Delta\left(1+\Delta^2\right) + \left(46\mathrm{H}^2 + 270\mathrm{H} + 540\right)\Delta^2$
$\mathrm{b}(\mathrm{H}, \Delta) = \frac{\left(-\Delta^2+(\mathrm{H}+2)\Delta-1\right)^3}{(\Delta-1)^6}$
$\mathrm{c}(\mathrm{H}, \Delta) = -6(\Delta+1)\left[-\Delta^2 + (\mathrm{H}+2)\Delta - 1\right] \cdot \ln\Delta + \left[\left(\mathrm{H}^2 - 12\right)\Delta^2 + (10\mathrm{H}+24)\Delta + \mathrm{H} - 12\right](\Delta - 1)$
$\mathrm{d}(\mathrm{H}, \Delta) = \frac{(\Delta-1)(\mathrm{H}+6)}{6[(\Delta^2+1)-\Delta(\mathrm{H}+2)]\ln\Delta+3\mathrm{H}(\Delta^2-1)}$

Finally, if all factors disappear $\varepsilon = 0$, $\Theta = 0$, $\mathrm{H} = 0$, the material thickness will be constant everywhere. In this case, the thickness will be equal to t over the whole surface of the spring.

It is possible to conduct all subsequent computations for the general quadratic function (6.5). For the briefness of formulas, two special cases of the thickness variation will be presented in the manuscript. One case depictures the linearly variable thickness is $\mathrm{H} = 0$. The second case represents the quadratically variable thickness with both equal edge thicknesses $\Theta = 1$.

From (6.1) and (6.2) follows:

$$r_i = C - x_i \cos(\alpha), \quad r_e = C - x_e \cos(\alpha), \quad \mathrm{d}r = \cos(\alpha)\, \mathrm{d}x. \tag{6.6}$$

The heights of the inner and outer edges of the middle surface of the free shell are correspondingly (Fig. 6.1):

$$z_i = x_i \sin\alpha > 0, \quad z_e = x_e \sin\alpha < 0. \tag{6.7}$$

The total height of the middle surface of the free of load disk spring is:

$$h = z_i - z_e = (\Delta - 1) r_i \tan\alpha. \tag{6.8}$$

Correspondingly, the principal radii of curvature r_1 and r_2 for the free of load conical shell are:

$$\frac{1}{r_1} = \frac{\sin\alpha}{r}, \quad \frac{1}{r_2} = 0. \tag{6.9}$$

The width of the middle surface in its free state is:

$$b = r_e - r_i = (\Delta - 1) \cdot r_i. \tag{6.10}$$

All calculations could be verified using the MAPLE CAS (Maple, 2022). The corresponding execution code provided in the cloud-based communal repository Mendeley Data (Kobelev, 2023b).

6.3 Load-Caused Variations of Strain and Curvature

The main hypothesis of the model is that the slope angle ψ of deformed conical shells changes, but still remains equal over the meridional section (Chap. 4). The whole meridional section rotates over the certain point (c). This explains, why the point (c) was referred to as the inversion centre point. Its height in the cylindrical coordinate system was set to zero $z = 0$ and does not vary in course of the elastic evolution of washer. Notable, that the points of the middle surface, located on the neutral plane, do not deflect. In other words, the generatrix rotate about the point (c), but remain straight and its length do not alter. The middle surface of the shell in the deformed state is again a truncated cone Ω, as shown above on (Fig. 6.2).

In cylindrical coordinates (r, φ, z), the middle surface Ω is given by the parametric equations:

$$R = C - x\ \cos\ \psi, \quad Z = -x\ \sin\ \psi. \tag{6.11}$$

The distance from the point of the thick-walled cone with coordinates (x, φ, y) to the axis of the deformed cone is:

$$\tilde{R} = R + y\ \sin\ \psi = C - x\ \cos\ \psi + y\ \sin\ \psi. \tag{6.12}$$

Correspondingly, the principal radii of curvature R_1 and R_2 of the middle surface Ω are:

$$\frac{1}{R_1} = \frac{\sin\ \psi}{r}, \quad \frac{1}{R_2} = 0. \tag{6.13}$$

The inner and outer radii of Ω are equal to:

$$R_i = C - x_i\ \cos\ \psi, \quad R_e = C - x_e\ \cos\ \psi, \quad \mathscr{B} = R_e - R_i. \tag{6.14}$$

The distances from the plane (c) to the inner (i) and outer (e) edges of surface Ω read:

$$Z_i = x_i\ \sin\ \psi, \quad Z_e = x_e\ \sin\ \psi. \tag{6.15}$$

The height and width of middle surface in the deformed state Ω are correspondingly:

$$H = Z_i - Z_e = r_i \cdot \frac{\sin \psi}{\cos \alpha} \cdot (\Delta - 1), \tag{6.16}$$

$$B = R_i - R_e = r_i \cdot \frac{\cos \psi}{\cos \alpha} \cdot (\Delta - 1). \tag{6.17}$$

Upon specializing the strain displacement relations to the case of a conical shell with no transverse shear deformations the following expression for the circumferential mid-surface strain:

$$\varepsilon_1 = \frac{R - r}{r} = \frac{\cos \alpha - \cos \psi}{C - x \cos \alpha} x. \tag{6.18}$$

The circumferential curvature change is:

$$\kappa_1 = \frac{1}{R_1} - \frac{1}{r_1} = \frac{\sin \psi - \sin \alpha}{r}. \tag{6.19}$$

The slope angle vanishes in the flattened state. With Eq. (6.14), the inner and outer radii of the plane circular strip will be equal nearly to:

$$R_{i0} = C - x_i, \quad R_{e0} = C - x_e. \tag{6.20}$$

According to Eq. (6.7), the arc length x which serves as the meridional coordinate and represents the latitude varies from the negative value on the outer edge $x_e < 0$ to the positive value on the inner edge $x_i > 0$. Accordingly, the inequality $R_{i0} < R_{e0}$. holds in Eq. (6.20). The total length of the meridian is $B = R_{e0} - R_{i0} = x_i - x_e > 0$. If at least one edge can glide freely on one of the compression pistons, because according to Eqs. (6.10) and (6.17) always holds $b = B$. This means, that the meridian length does not change during the elastic deformation. No significant reduction of the meridional length occurs, because the stiffness along the wall is much higher than the bending stuffiness of the disk. However, if radial movements of *both* edges are restricted, the membrane compression in the meridional direction occurs. The membrane compression leads to the substantial growth of the axial stiffness of the spring. This effect was evaluated in Chap. 5 (Kobelev, 2023a).

6.4 Statics of Disk Springs Using Equations of Axisymmetric Elasticity

1°. In this Section the expression for the elastic energy is derived using the thin conical shell model. The meridional and circumferential membrane forces and bending moments for thin shell relate to strains and curvature changes by means of relations (Ventsel & Krauthammer, 2001, Eqs. (6.14–6.32)):

$$N_1 = \frac{ET}{1-\nu^2}(\varepsilon_1 + \nu\varepsilon_2), \quad N_2 = \frac{ET}{1-\nu^2} \cdot (\varepsilon_2 + \nu\varepsilon_1), \tag{6.21}$$

$$M_1 = \frac{ET^3}{12(1-\nu^2)} \cdot (\kappa_1 + \nu\kappa_2), \quad M_2 = \frac{ET^3}{12(1-\nu^2)} \cdot (\kappa_2 + \nu\kappa_1). \tag{6.22}$$

The values in Eqs. (6.21–6.22) are:

$T = T(x, \varphi)$	the thickness of the shell,
$ET/(1-\nu^2)$	the extensional rigidity of shell,
$ET^3/12(1-\nu^2)$	the flexural rigidity.

The second hypothesis assumes the absence of the membrane stress in the meridional direction:

$$N_2 = 0.$$

The third hypothesis assumes the absence of the bending moment in the meridional direction:

$$M_2 = 0.$$

From these conditions the meridional strain ε_2 and curvature change κ_2 depend solely on the circumferential strain ε_1 and curvature change κ_1:

$$\varepsilon_2 = -\nu\varepsilon_1, \tag{6.23}$$

$$\kappa_2 = -\nu\kappa_1. \tag{6.24}$$

Substitution of (6.23) and (6.24) in (6.21) and (6.22) delivers the expressions for the circumferential stress σ_1 and moment M_1:

$$\sigma_1 = N_1/T = E\varepsilon_1, \tag{6.25}$$

$$M_1 = ET^3\kappa_1/12. \tag{6.26}$$

The strain energy of the shell can be determined by means of Kirchhoff's assumptions (Ventsel & Krauthammer, 2001, Eqs. (6.14–6.51)):

$$U_E = U_t + U_b, \tag{6.27}$$

The extensional and flexural elastic energies in Eq. (6.27) respectively are

$$U_t = \frac{E}{2} \int_0^{2\pi} \int_{x_e}^{x_i} T(x, \varphi)\left[(\varepsilon_1 + \varepsilon_2)^2 - 2(1 - \nu)\varepsilon_1\varepsilon_2\right] r \, \mathrm{d}x \, \mathrm{d}\varphi, \tag{6.28}$$

$$U_b = \frac{E}{24} \int_0^{2\pi} \int_{x_e}^{x_i} T^3(x, \varphi)\left[(\kappa_1 + \kappa_2)^2 - 2(1 - \nu)\kappa_1\kappa_2\right] r \, \mathrm{d}x \, \mathrm{d}\varphi. \tag{6.29}$$

Using the formulas (6.25), (6.26) in (6.28) and (6.29), the extensional and flexural surface densities of elastic energy reduce to:

$$U_t = \int_0^{2\pi} \int_{x_e}^{x_i} W_t r \, \mathrm{d}x \, \mathrm{d}\varphi, \quad U_b = \int_0^{2\pi} \int_{x_e}^{x_i} W_b r \, \mathrm{d}x \, \mathrm{d}\varphi. \tag{6.30}$$

$$2W_t = ET\varepsilon_1^2, \quad 2W_b = \frac{1}{12} ET^3 \kappa_1^2. \tag{6.31}$$

Accordingly, the extensional and flexural elastic energies are:

$$U_t = \frac{E}{2} \int_0^{2\pi} \int_{x_e}^{x_i} r\varepsilon_1^2 T \, \mathrm{d}x \, \mathrm{d}\varphi, \quad U_b = \frac{E}{24} \int_0^{2\pi} \int_{x_e}^{x_i} r\kappa_1^2 T^3 \, \mathrm{d}x \, \mathrm{d}\varphi. \tag{6.32}$$

The total potential energy

$$\Pi = U_E + U_f \tag{6.33}$$

is the sum of the elastic strain energy U_E, stored in the deformed body and the potential energy U_f of the applied forces:

$$U_f = \int_\alpha^\psi M \, \mathrm{d}\psi, \quad M = F_Z B - F_R H. \tag{6.34}$$

The total moment M in the meridional direction causes the rotation of the cross-section of the spring. Typically, the radial forces act on the outmost outer and on the outmost inside edges, as shown for the disk spring of moderate thickness on Fig. 6.4, left. The axial loads act on the highest and lowest edges. The outmost outer and the lowest edges are near the outer edge of the middle surface. Accordingly, the outmost inner and the highest edges are near the inner edge of the middle surface. For the common disk springs the thickness of the material T is much lower, than the width of the cross-section b. For briefness of actual calculations, the thickness of spring is assumed to be small enough to neglect the differences in load application edges. In this case, both radial and axial loads are assumed to act on the inner and outer edges of the middle surface. For precise calculations, the loads and actual heights of spring could be easily corrected with the obvious geometric formulas. For this purpose, the diagonal from the upper to lower edges must be used for the height H instead of the height of the middle surface. Similarly, the diagonal from the outer to inner edges used for the width B instead of the width of the middle surface.

In the absence of friction, the shell is stressed by the forces F_Z in the direction of the symmetry axis of the disk. Without any friction, the total meridional moment reduces to:

$$M = F_Z B.$$

2°. For calculation of the radial reaction force F_R, acting from the spring to a curb, we remark, that the radial force is equal to the total hoop force F_φ. The hoop force F_φ is the integral of the hoop, or circumferential stress σ_1. The circumferential stress σ_1 is acting circumferential and perpendicular to the axis and the radius of the cone. Finally, the compression force of the spring to the curb on an edge of the spring follows form the total hoop force in the cross-section of the hollow conical disk:

$$F_\varphi = 2\pi \int_{x_e}^{x_i} N_1 \mathrm{d}x. \tag{6.35}$$

If both edges of the disk can freely move in radial direction, total hoop force in the cross-section of the hollow conical disk vanishes, $F_\varphi = 0$. Under these conditions, the radius C of the inversion point guarantees the minimum the elastic strain energy U_E among all admissible radii.

If the radial movements of the edges are confined, the radius of the inversion point relocates from the preferrable position, which corresponds to the minimum energy of the elastic deformation. and the total radial force is $F_{I.R}$. If the inner edge (i) freely glides, the total axial force acting on the inner edge (i) is F_z, and the radial reaction force vanishes $F_{I.R} = 0$. Otherwise, if the movement of the inner edge is restricted, as the reaction appears the radial force $F_{I.R}$.

On the outer edge (e), the total axial force acting is $-F_z$ and the total radial force will be denoted ad $F_{E.R}$. If the inner edge can freely glide, $F_{E.R} = 0$. If the movement of the outer edge is constrained, the reaction on this constraint is the radial force $F_{E.R}$.

If the movements of both inner edges are restricted, the reactions must be opposite $F_{I.R} = -F_{E.R}$. These reactions could not be calculated in the actual theory, because the actual hypotheses disregard the meridional membrane deformation. The double restricted cone will be enormous axially stiff. The washer will store insignificant amount elastic energy and cannot be considered as a spring. This load case is insignificant for the most industrial applications of disk springs.

6.5 Linearly Variable Thickness of Disk Spring

At first, we assume that the amplitude of the angular variation in parallel direction in Eq. (6.5) vanishes$\varepsilon = 0$. In this case, the thickness along parallels will be constant $T_i(\varphi) = t = const$. The thickness of the spring will be variable solely along the meridian.

For the linearly variable thickness is H = 0. The elastic energy U_b is associated with the bending of the conical shell. The elastic energy U_t is linked to the membrane forces in the middle surface of the shell. The integration (6.30) and (6.31) reduce the extensional U_t and flexural U_b components of the elastic energy respectively to:

$$\begin{cases} U_E(\mathrm{H}=0) = U_b + U_t, \\ U_b = c_b \frac{\pi E t^3}{12} \cdot \frac{(\sin\psi - \sin\alpha)^2}{\cos\alpha}, \\ U_t = -c_t \frac{\pi E t}{2} \cdot \frac{(\cos\psi - \cos\alpha)^2}{\cos^2\alpha}. \end{cases} \tag{6.36}$$

The formulas (6.36) contain with two lengthy coefficients $c_t = c_t(C)$, and c_b. The coefficients are provided in closed form in (Kobelev, 2023b). Notably, that the elastic energy U_b does not depend upon. Otherwise, the elastic energy U_t explicitly depends upon the radial location of the inversion centre C.

The radius of the inversion centre C depends upon the possibility of the radial movements on the inside and the outer edges of the spring. The first principal for applications case appears, if the spring glides on the inside (i) and also on outer (e) edges in the radial direction without resistance. If both edges glide radially freely, the condition of minimum for elastic energy with respect to C determine the radius of the inversion centre:

$$\frac{\partial U_E(\mathrm{H}=0)}{\partial C} = \frac{\pi E t}{2} \cdot \frac{(\cos\psi - \cos\alpha)^2}{(\Delta-1)\cos^3\alpha} \cdot \xi_* = 0,$$
$$\xi_* \overset{\mathrm{def}}{=} 4C(\Delta - \Theta)\cdot \ln\Delta + 4(\Delta-1)\cdot\left(-\frac{(\Theta+1)\Delta r_i}{2} + \left(C + \frac{r_i}{2}\right)\cdot\Theta - C + \frac{r_i}{2}\right). \tag{6.37}$$

Solution of the Eq. (6.37) delivers the radius C of the inversion centre (c) for the spring with both freely gliding edges:

$$C_F = \frac{\Theta+1}{2} \cdot \frac{(\Delta-1)^2}{(\Delta-\Theta)\ln\Delta + (\Theta-1)(\Delta-1)} \cdot r_i. \tag{6.38}$$

The index "F" stays for "free gliding edges". Finally, the value for the radius of inversion centre (c), which was unknown up to now, is determined. The arc length x, which serves as the meridional coordinate and represents the *latitude* is fixed. For this purpose, the value $C = C_F$ should be used in Eq. (6.2). The height of the neutral plane is finally determined. Exactly the same solution as (6.38) follows also from the condition $F_\varphi = 0$ with Eq. (6.36). The substitution of $C = C_F$ in the expression for the elastic energy gives the formula of the elastic energy of the free gliding spring.

$$U_F = U_E(\mathrm{H} = 0, C = C_F), \quad U_F = \min_C U_E(\mathrm{H} = 0). \tag{6.39}$$

According to the principle of virtual work, the external virtual work is equal to internal virtual work when equilibrated forces and stresses undergo unrelated but consistent displacements and strains. The necessary stationarity condition of U_{GL} with respect to ψ delivers the circumferential moment. In the absence of the radial force, the total circumferential moment of the free gliding spring correspondingly is:

$$M_F = \frac{\partial U_F}{\partial \psi} = M_{F.t} + M_{F.b}. \tag{6.40}$$

The total axial force $F_{F.Z}$ is equal to the sum of partial forces due to tension of the middle surface $F_{t.GL}$ and due to bending $F_{b.GL}$:

$$\begin{cases} F_{F.Z} = F_{F.t} + F_{F.b}, \\ F_{F.t} = \frac{M_{t.F}}{\mathcal{B}} = c_t \cdot F_{0.t}, \quad F_{0.t} = \frac{\pi E t}{6(\Delta-1) r_i} \cdot \frac{\cos\psi - \cos\alpha}{\cos^2\alpha} \cdot \frac{\sin\psi}{\cos\psi}, \\ F_{F.b} = \frac{M_{b.F}}{\mathcal{B}} = c_b \cdot F_{0.b}, \; F_{0.b} = \frac{\pi E t^3}{6(\Delta-1) r_i} \cdot (\sin\alpha - \sin\psi), \end{cases} \tag{6.41}$$

with the following coefficients:

$$\begin{cases} \mathrm{c}_t = \frac{4(\Delta-\Theta)\left[\left(\Theta+\frac{1}{2}\right)\Delta+\frac{\Theta}{2}+1\right]\cdot\ln\Delta+(\Delta-1)\cdot\left[\left(\Theta^2-8\Theta-5\right)\Delta+5\Theta^2+8\Theta-1\right]}{\Theta-1+\frac{\Delta-\Theta}{\Delta-1}\cdot\ln\Delta}, \\ c_b = -\left(\frac{\Delta-\Theta}{\Delta-1}\right)^3 \ln\Delta - \frac{\Theta-1}{6(\Delta-1)^2}\left[\left(2\Theta^2+5\Theta+11\right)\Delta^2 - \left(7\Theta^2+22\Theta+7\right)\Delta + 11\Theta^2 + 5\Theta + 2\right]. \end{cases} \tag{6.42}$$

The partial forces due to tension of the middle surface $F_{t.GL}$ are equal to the stresses in the membrane theory of thin shell.

In the limit case $\Theta = 1$, the thickness of the washer turns to be constant. The radius C_A of inversion centre for the disk spring with the constant thickness was given by Almen and Laszlo formula (Almen & Laszlo, 1936):

$$C_A = \lim_{\Theta\to 1} C_F = \frac{\Delta-1}{\ln\Delta} r_i. \tag{6.43}$$

The index "A" stays for "Almen and Laszlo", because the Almen and Laszlo describe the special case with $\Theta = 1$. The coefficients $\mathrm{c_t}$, c_b reduce to:

$$\lim_{\Theta\to 1} \mathrm{c_t} = 6(\Delta - 1)\frac{(\Delta+1)\cdot\ln\Delta - 2(\Delta-1)}{\ln\Delta}, \quad \lim_{\Theta\to 1} \mathrm{c_b} = -\ln\Delta. \tag{6.44}$$

The substitution of (6.31) into (6.32) delivers the formula for the axial force $F_{0.Z}$ of the freely gliding disk spring with constant thickness of material:

$$\begin{cases} F_{A.Z} = F_{0.t} + F_{0.b} \\ F_{A.t} = \lim\limits_{\Theta \to 1} F_{t.F} = 6(\Delta - 1) \cdot \frac{(\Delta+1)\cdot \ln \Delta - 2(\Delta-1)}{\ln \Delta} \cdot F_{0.t}, \\ F_{A.b} = \lim\limits_{\Theta \to 1} F_{t.F} = -\ln \Delta \cdot \ F_{0.b}. \end{cases} \tag{6.45}$$

Evidently, that the inside edge of the spring (i) moves as a result of axial compression towards the axis of the disk (Fig. 6.5, left). The lower, or outer edge of the spring (e) travels in the opposite direction away from the axis of the disk. Because both edges can freely glide without friction, the radial reaction forces on both edges are will not appear.

6.6 Quadratically Variable Thickness of Disk Spring

For the quadratically variable thickness with the equal thicknesses on both edges $\Theta = 1$. The amplitude of the angular variation in Eq. (6.5) will be nullified $\varepsilon = 0$. This assumption leads again to $T_i(\varphi) = t$. Because H is not zero, the thickness is the quadratic function of the meridional coordinate x.

The position of the neutral plane is determined from the condition of minimum for elastic energy $U_E(\Theta = 1)$ with respect to C. Solution of the corresponding equation

$$\frac{\mathrm{d}U_E(\Theta = 1)}{\mathrm{d}C} = 0$$

delivers the radius of the inversion centre (c) for the spring with both freely gliding edges:

$$\widetilde{C_F} = \frac{(\mathrm{H}+6)\rangle(\Delta-1)^2}{6\left[(\Delta^2+1)-\Delta(\mathrm{H}+2)\right]\cdot \ln \Delta + 3\mathrm{H}\cdot(\Delta^2-1)} \cdot r_i\,. \tag{6.46}$$

Similarly, to Eq. (6.43), $\widetilde{C_F}$ reduces in the limit case $\mathrm{H} \to 0$ to the Almen-Laszlo expression for the radius of the inversion centre:

$$C_A = \lim_{\mathrm{H} \to 0} \widetilde{C_F} = \frac{\Delta-1}{\ln \Delta} r_i\,.$$

We use again the principle of virtual work. The external virtual work $\widetilde{U_F} = U_E\left(\Theta = 1, C = \widetilde{C_F}\right)$ is equal to internal virtual work when forces and stresses are equilibrated. The necessary stationarity condition with respect to ψ delivers the circumferential moment. In the absence of the radial force, the total bending moment axial force on the free gliding spring correspondingly are:

$$\widetilde{M_F} = \frac{\partial \widetilde{U_F}}{\partial \psi} = \tilde{M}_{F.t} + \tilde{M}_{F.b}. \tag{6.47}$$

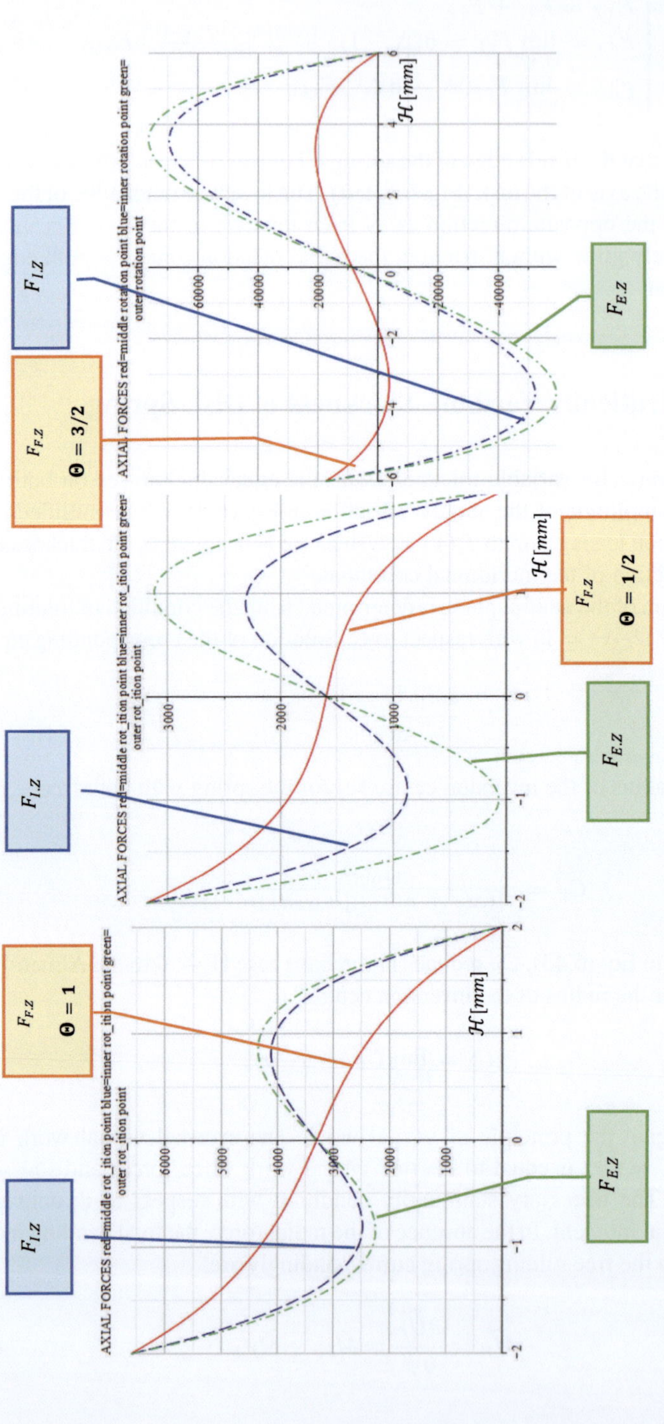

Fig. 6.5 Axial forces of linearly varying springs: Free gliding spring $F_{F.Z}$, fixed inner curb $F_{I.Z}$ and fixed outer curb $F_{E.Z}$ as the functions of Θ. Case $\varepsilon = 0$

In Eq. (6.40) the inversion radius is given by Eq. (6.38), but in Eq. (6.47) the value from Eq. (6.46) must be used. The partial forces due to tension of the middle surface and due to bending correspondingly are:

$$\begin{cases} \tilde{F}_{Z.F} = \tilde{F}_{Z.t} + \tilde{F}_{Z.b}, \\ \tilde{F}_{Z.t} = \frac{\tilde{M}_{F.t}}{\mathcal{B}} = \widetilde{c_t} \cdot F_{0.t}, \\ \tilde{F}_{Z.b} = \frac{\tilde{M}_{F.b}}{\mathcal{B}} = \widetilde{c_b} \cdot F_{0.b}, \end{cases} \tag{6.48}$$

with the coefficients from the Table 6.2:

$$\begin{cases} \widetilde{c_b} = a(H, \Delta) \cdot H \cdot \frac{\Delta^2 - 1}{60(\Delta - 1)^6} + b(H, \Delta) \cdot \ln \Delta, \\ \widetilde{c_t} = c(H, \Delta) \cdot d(H, \Delta). \end{cases} \tag{6.49}$$

6.7 Variable Thickness Along the Parallels

Now the oscillation of thickness along the parallels will be studied. If $\varepsilon \neq 0$, the thickness harmonically oscillates along the parallels:

$$T_i(\varphi) = [1 + \varepsilon \sin(n\varphi)] \cdot t.$$

The angle φ signifies the *longitude* and is measured along the parallels of middle surface. The integrals in Eqs. (6.28) … (6.32) could be easily evaluated for the harmonical oscillation of thickness. For this purpose, the Eq. (6.5) and formulas from Table 6.1 are used:

$$\int_0^{2\pi} T \mathrm{d}\varphi = (\Theta X_e + X_i + \mathrm{H} X_e X_i) \cdot \int_0^{2\pi} T_i(\varphi) \mathrm{d}\varphi = (\Theta X_e + X_i + \mathrm{H} X_e X_i) \cdot 2\pi t, \tag{6.50}$$

$$\int_0^{2\pi} T^3 \mathrm{d}\varphi = (\Theta X_e + X_i + \mathrm{H} X_e X_i) \cdot \int_0^{2\pi} T_i^3(\varphi) \mathrm{d}\varphi = (\Theta X_e + X_i + \mathrm{H} X_e X_i) \cdot 2\pi t^3 \cdot \left(1 + \tfrac{3}{2}\varepsilon^2\right). \tag{6.51}$$

For the derivations of Eqs. (6.50) … (6.51) we used the elementary integrals:

$$\begin{aligned} &\int_0^{2\pi} [1 + \varepsilon \sin(n\varphi)] \mathrm{d}\theta = 2\pi, \\ &\int_0^{2\pi} [1 + \varepsilon \sin(n\varphi)]^3 \mathrm{d}\theta = 2\pi \left(1 + \tfrac{3}{2}\varepsilon^2\right). \end{aligned} \tag{6.52}$$

The above relations show, that the factors in Eqs. (6.41), (6.45) and (6.48), which depend on thickness t linearly, remain unchanged. The factors in Eqs. (6.41), (6.45) and (6.48), which contain the third power of thickness t^3, must be corrected. In Eqs. (6.41), (6.45) and (6.48) the expression t^3 must be replaced by $t^3 \cdot \left(1 + \frac{3}{2}\,\varepsilon^2\right)$. This means, that the the extensional component of the elastic energy U_t is not affected by the variation of thickness along the parallels. The flexural component of the elastic energy U_b increases proportional to angular correction factor $1 + \frac{3}{2}\,\varepsilon^2$. Remarkably, that the correction factor does not depend upon the angular wave number n. The variation of thickness along the parallels could be produced in the modern machinery.

6.8 Radial Forces on the Radially Constrained Disk Springs

1°. n the above studied cases, the radial travel of both edges was not restricted. If the disk spring is squeezed in the axial direction, both edges travel without any resistance in radial direction. The inside edge of the spring (i) moves as a result of axial compression towards the axis of the disk (Fig. 6.5, left). The lower, or outer edge of the spring (e) travels in the opposite direction away from the axis of the disk. Because both edges can freely glide without friction, the radial reaction forces on both edges vanish.

2°. All preceding calculations were performed without friction. The influence of friction on the reaction of the disk spring could be evaluated with the Eq. (6.34). Coulomb's friction law specifies the conditions that the contact force between two solids must verify for there to be a slip between them. Two pistons compress the disk spring, one from the top and one from the bottom. One piston can glide without friction, and friction occurs on the other contact surface. The contact force comprises a normal component F_Z, perpendicular to the contact plane between the exaggerated disk spring and the contact piston, and a tangential component F_R, belonging to the contact plane. The slip between the spring and piston occurs if F_Z and F_R verify the proportionality relationship: $|F_R| = \mu F_Z$, where μ is the coefficient of friction that characterizes the condition of the two surfaces in contact. In this case, the contact force is said to remain within the Coulomb cone. As long as the absolute value of force $|F_R|$ is less than F_Z, sliding does not occur, and is referred to as adhesion between the two solids. During the disk compression, friction causes the radial forces $F_R = -\mu F_Z$. The height H and width H of the cross-section are given by Eqs. (6.16) and (6.17). In the presence of Coulomb's friction, the total meridional moment is equal:

$$M = F_Z B - F_R H = (B + \mu H) \cdot F_Z = \left(1 + \mu \frac{\sin\psi}{\cos\psi}\right) \cdot B \cdot F_Z.$$

During the disk expansion, the direction of frictional resistance alters and the frictional radial forces will be $F_R = \mu F_Z$. Comparison of the above expressions for the total meridional moment with and without frictions leads to the friction correction factor. The correction factor to account friction during the compression reads:

$$k_Q = \begin{cases} 1 + \mu \tan \psi, & \text{for compression movement,} \\ 1 - \mu \tan \psi, & \text{for expansion movement.} \end{cases}$$

To account Coulomb's friction, the resulting expressions for the forces must be multiplied by the correction factor k_Q. If the middle surface Ω does not yet achieve the flattened state, angle ψ is positive. Thus, the correction factor during the compression will be greater than one, $k_Q > 1$. Consequently, friction will increase the reaction forces $k_Q F_Z$ for the positive height $\mathcal{H}$ of the middle surface. If the middle surface Ω flattens, the angle ψ vanishes and the correction factor turns to be one. Friction does not influence the vertical forces of the spring nearby its flattened state. Correspondingly, the correction factor during the expansion will be less than one, $k_Q < 1$.

3°. Further, we investigate the washers with different degrees of freedom on the inner and outer surfaces using the variation methods. The calculation method for the above application applies the kinematic hypothesis and equations of thin and moderately thick shells. The motivating feature of the presented theory is its possibility to calculate the disk springs of with free gliding edges and the edges with the constrained radial movement. In this case, the radial movement of the inner edge (i) is restricted (Fig. 6.5, middle). One can imagine the absolutely rigid cylindrical curb, inserted inside the inside edge of the spring without gap. The inner edge (i) glides in the vertical direction over the surface of the cylindrical curb, but cannot penetrate inside it. Because of the radial support, on the inner cylindric curb appear the radial components of reaction forces. The radial reaction forces on the inner cylindric curb are shown with green vectors. As the radial movement of the inside edge (i) is constrained, the radius of inversion point must be equal to the inner radius:

$$C_I = r_i. \tag{6.53}$$

Over the outer edge (e), the spring glides can glide without friction in the radial direction, but its vertical movement vanishes. Because U_b does not depend upon, the shift of the inversion point to the inside edge does not influence the bending of the cone. However, the character of the extensional deformation (circumferential strain) of the middle surface alternates significantly. The extension of the middle surface of free gliding spring occurs outside the inversion point (e). The middle surface of the free gliding spring squeezes inside the inversion point. Contrarily, the complete middle surface of the washer on the cylindric curb extends. This behaviour influences considerably the function of the spring. For briefness, we will not consider the thickness variations in parallel directions, such that angular amplitude vanishes $\varepsilon = 0$ and $T_i(\theta) = t = const$. If nessesary, the flexural parts in the corresponding formulas could be easily corrected, using the method of Sect. 6.7.

If the radial movement of the inner edge (i) is restricted, the radial reaction on the inner curtailment is equal to:

$$F_{I.R} = k_{I.R} F_{R.0},$$
$$F_{R.0} = \frac{2\pi}{(\Delta-1)^2} \cdot \frac{\cos\psi - \cos\alpha}{\cos^2\alpha} \cdot E \cdot t \cdot r_i, \tag{6.54}$$
$$k_{I.R} = \left[-\Delta^2 + (\mathrm{H}+\Theta+1)\Delta - \Theta\right]\ln\Delta + \tfrac{\Delta-1}{6}\left[(\mathrm{H}+3\Theta+3)\Delta^2 - (5\mathrm{H}+12\Theta)\Delta - 2\mathrm{H}+9\Theta-3\right].$$

4°. The third load case takes place, if the radial movement of the outer edge (e) is restricted (Fig. 6.5, right). We assume, the outer edge of the spring inserted without gap into the absolutely rigid hollow cylinder. The inner edge (i) of spring can glide without friction in the radial direction. Because the radial movement of the outer edge (e) is restrained, the radius of the inversion point has to be equal to outer radius:

$$C_E = r_e. \tag{6.55}$$

The extension force of the spring to the outer curb outside the outer edge of the spring follows again form the total tangential force in the cross-section. The only difference is the sign of the total force, acting from inside fixing spring to the outer radial direction. The translation of the inversion point to the outer edge does not influence the bending of the cone. Quite the reverse, the extensional deformation of the middle surface completely alternates. For free gliding spring, the extension of the middle surface of occurs outside the inversion point (e). The middle surface of free gliding spring compresses inside the inversion point. Alternatively, the complete middle surface of the disk spring compacts inside the hollow cylindric curb. This behaviour influences significantly the force diagram of the spring.

If the radial movement of the outer edge (e) is restricted, the radial reaction on the curbing is equal to:

$$F_{E.R} = k_{E.R} F_{R.0}, \tag{6.56}$$

$$k_{E.R} = \left[-\Delta^2 + (\mathrm{H}+\Theta+1)\Delta - \Theta\right]\ln\Delta$$
$$-\tfrac{\Delta-1}{6}\left[(2\mathrm{H}+3\Theta-9)\Delta^2 + (5\mathrm{H}+12)\Delta - \mathrm{H} - 3\Theta - 3\right].$$

The radial reaction on the outer curbing acts inward, such that the total force in the cylindric coordinate system is negative.

The radial forces of disk springs with the linearly meridionally varying thickness are shown at Fig. 6.6. Free gliding spring $F_{F.R} = 0$, the spring on the inner curb $F_{I.R}$ (6.54) and the spring inside the outer curb $F_{E.R}$ (6.56) are displayed as the functions of the parameter Θ. Three examples of spring calculation will be displayed for comparison. For all simulations, the outer, inner diameter and the height of the middle surface remain the same:

$$D_e = 2r_e = 110\,[mm];\; D_i = 2r_i = 90\,[mm];\; h = 2\,[mm].$$

Young modulus and Poisson coefficient are equal to $E = 200\,[GPa]$; $\nu = 0.3$ in all cases. All examples are calculated for the linearly varying meridional thickness along the meridian, such that:

$$T_i = 2\,[mm],\ \mathrm{H} = 0,\ \varepsilon = 0.$$

Consequently, the behaviour of spring forces will depend solely upon the linear meridional variation of thickness $\Theta = T_e/T_i$. The calculations were provided for the following values of the linear meridional variation of thickness:

$$\Theta = 1, \quad \Theta = \frac{1}{2}; \quad \Theta = \frac{3}{2}.$$

The graphs show, that the forces for the disk springs with restricted radial displacements differ significantly from the forces of springs with freely gliding edges. The axial and radial forces are shown correspondingly on the Figs. 6.5 and 6.6. Because of free gliding, the radial force vanishes $F_{F.R} = 0$. The axial forces for the springs with the free gliding edges $F_{F.Z}$ were calculated according the Eq. (6.41) for different Θ. The forces are shown on the Figs. 6.5 and 6.6 with solid red lines. The forces for the springs with the inner cylindric curb $F_{I.Z}$, $F_{I.R}$ from Eq. (6.54) are shown on the Figs. 6.5 and 6.6 with dashed blue lines. The forces for the springs with the outer cylindric cavity $F_{E.Z}$, $F_{E.R}$ Eq. (6.56) are shown on the Figs. 6.5 and 6.6 with dot-dashed green lines.

With Eqs. (6.54) and (6.56), the ratio of the reaction force on the inner to the program reaction force on the outer edge is equal to:

$$\varrho(\Theta, \mathrm{H}) = \frac{F_{I.R}}{F_{E.R}} \equiv \frac{k_{I.R}}{k_{E.R}} = \frac{f_{11}(\Theta,\mathrm{H})+f_{12}(\Theta,\mathrm{H})}{f_{21}(\Theta,\mathrm{H})-f_{22}(\Theta,\mathrm{H})},$$

where:

$$\begin{aligned} f_{11}(\Theta, \mathrm{H}) &= \left[-\Delta^2 + (\mathrm{H}+\Theta+1)\Delta - \Theta\right] \cdot \ln \Delta, \\ f_{12}(\Theta, \mathrm{H}) &= \tfrac{\Delta-1}{6} \cdot \left[(\mathrm{H}+3\Theta+3)\Delta^2 - (5\mathrm{H}+12\Theta)\Delta - 2\mathrm{H}+9\Theta-3\right], \\ f_{21}(\Theta, \mathrm{H}) &= \left[-\Delta^2 + (\mathrm{H}+\Theta+1)\Delta - \Theta\right] \cdot \ln \Delta, \\ f_{21}(\Theta, \mathrm{H}) &= \tfrac{\Delta-1}{6} \cdot \left[(2\mathrm{H}+3\Theta-9)\Delta^2 + (5\mathrm{H}+12)\Delta - \mathrm{H}-3\Theta-3\right]. \end{aligned} \tag{6.57}$$

The ratio $\varrho = \varrho(\Theta, \mathrm{H})$ from Eq. (6.57) shortens for the *quadratically* varying thicknesses. In this case $\Theta = 1$:

$$\varrho(\Theta = 1, \mathrm{H}) = \frac{f_{11}(1,\mathrm{H})+f_{12}(1,\mathrm{H})}{f_{21}(1,\mathrm{H})-f_{22}(1,\mathrm{H})}. \tag{6.58}$$

The graph of the function $\varrho(\Theta = 1, \mathrm{H})$ from Eq. (6.58) is shown on Fig. 6.7.

The ratio ϱ from Eq. (6.57) simplifies for the *linearly* varying thicknesses. In this case $\mathrm{H} = 0$:

$$\varrho(\Theta, \mathrm{H} = 0) = \frac{f_{11}(\Theta,0)+f_{12}(\Theta,0)}{f_{21}(\Theta,0)-f_{22}(\Theta,0)}. \tag{6.59}$$

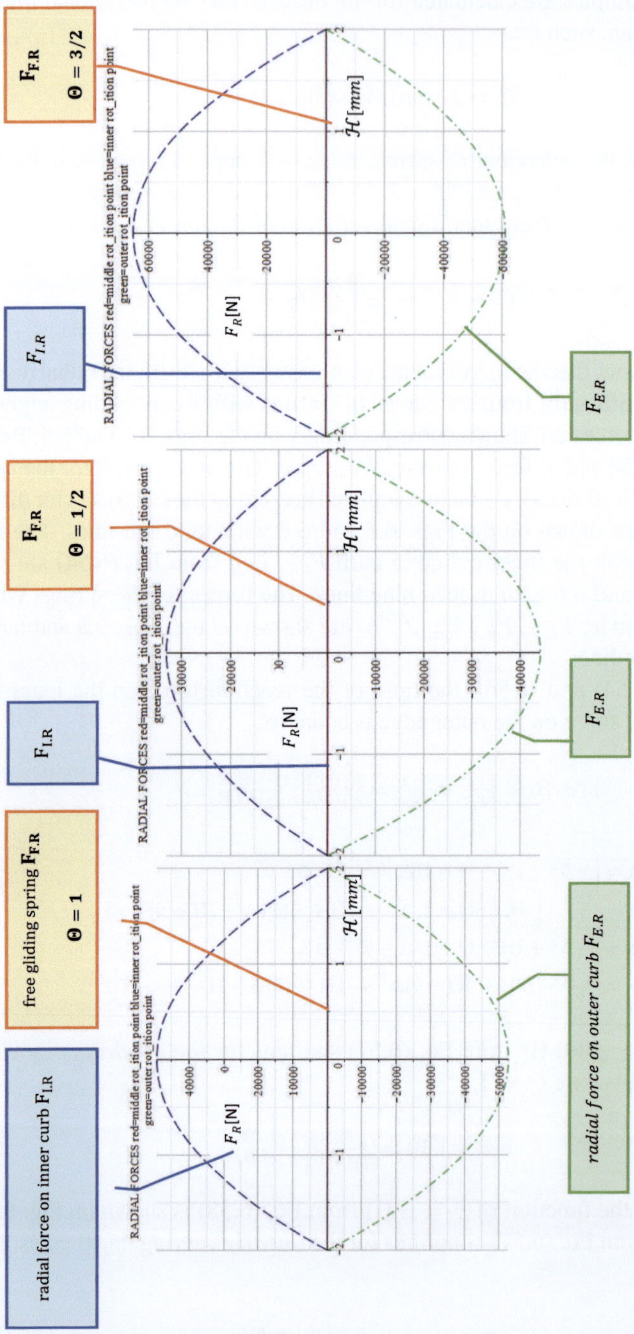

Fig. 6.6 Radial forces of linearly varying springs: Free gliding spring $F_{F.R} \approx 0$, force on the inner curb $F_{I.R}$ and force the outer curb $F_{E.R}$ as the functions of Θ. Case $\varepsilon = 0$

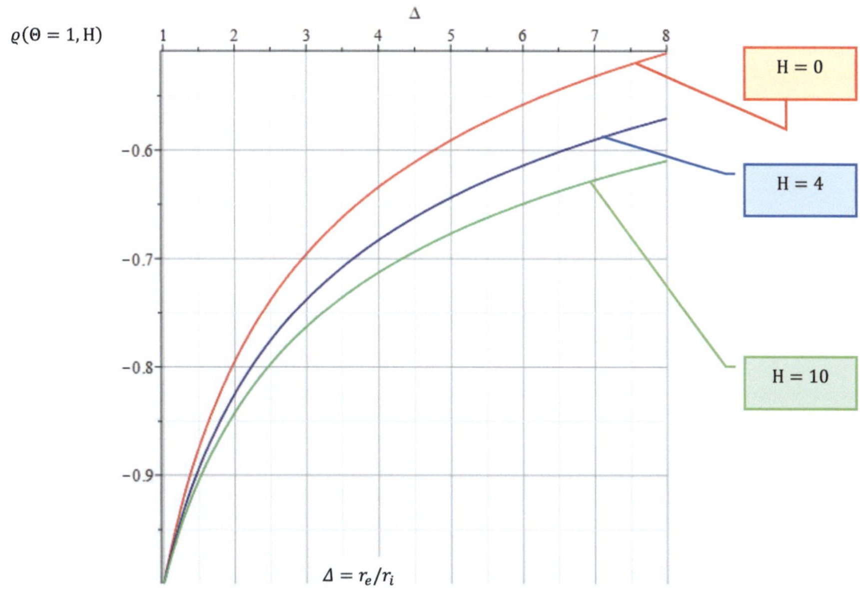

Fig. 6.7 Ratio of the correction coefficients $\varrho(\Theta = 1, \mathrm{H})$, Eq. (6.58) as the function of Δ The angular variation of thickness disappears, $\varepsilon = 0$

The function $\varrho(\Theta, \mathrm{H} = 0)$ from Eq. (6.59) is shown on Fig. 6.8.

In this case $\Theta = 1, \mathrm{H} = 0$ the thickness will be constant and the relation between the reaction forces reduces to:

$$\varrho(\Theta = 1, \mathrm{H} = 0) = \frac{f_{11}(1,0)+f_{12}(1,0)}{f_{21}(1,0)-f_{22}(1,0)} \equiv \frac{1-\Delta+\ln\Delta}{1-\Delta+\Delta\ln\Delta}. \tag{6.60}$$

6.9 Verification of the Analytical Results with the Finite-Element Simulation

For the verification of the above closed-form solutions, the disk springs of different shapes were simulated with the commercial finite-element program. The simulation was performed with the proprietary finite-element system ANSYS, 2022, V2 (2022).

The other geometric parameters are shown in Table 6.3. The thickness of the disk spring in the intermediate diameter $D_m = 100[mm]$ is referred to as T_m[mm]. Totally ten finite element models with the different radial distributions of thickness were build. Each model uses the 3d solid elements. The APDL source code for the simulations is available from the cloud-based communal repository Mendeley Data (Kobelev, 2023b).

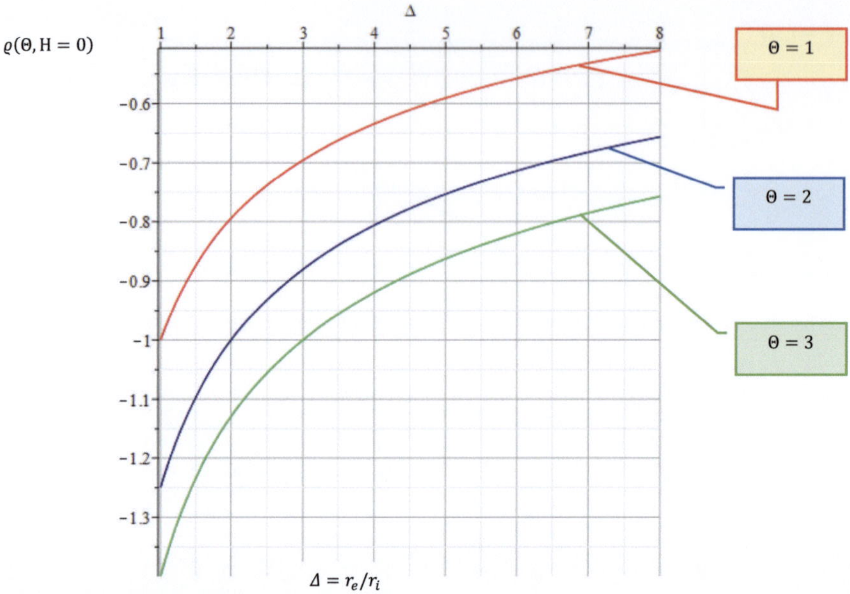

Fig. 6.8 Ratio of the correction coefficients $\varrho(\Theta, \mathrm{H} = 0)$, Eq. (6.59) as the function of Δ The angular variation of thickness disappears,$\varepsilon = 0$

Table 6.3 The geometric parameters for the simulations of the disk springs

Model	Thickness	T_i[mm]	T_e[mm]	T_m[mm]	Θ	$\mathrm{H} = -4 + \frac{8T_m}{T_i+T_e}$
1	Constant	2	2	2	1	0
2	Constant	3	3	3	1	0
3	Linear	1	2	3/2	2	0
4	Linear	2	1	3/2	1/2	0
5	Linear	1	3	2	3	0
6	Linear	3	1	2	1/3	0
7	Quadratic	2	2	1	1	−2
8	Quadratic	2	2	3/2	1	−1
9	Quadratic	2	2	3	1	2
10	Quadratic	2	2	5	1	6

For each model three simulations were performed. The first simulation of each model assumes the freely radially gliding nodes on the inner and on the outer edge of the middle surface.

The second simulation assumes the freely radially gliding nodes on the inner edge of the middle surface, but the radial movement of the nodes on the outer edge was constrained. This condition is equivalent to the moment-free contact of the nodes on the outer edge of the middle surface to the ideal rigid outer, cylindrical cavity.

The third simulation assumes the freely radially gliding nodes on the outer edge of the middle surface, but the radial movement of the nodes on the inner edge was constrained. This condition is equivalent to the moment-free contact of the nodes on the inner edge of the middle surface to the ideal rigid inner cylinder. Thirty simulations were performed in total.

The vertical travel of the outer edge of the middle surface was constrained for all simulations. The vertical travel of the inner edge of the middle surface $h-H$ was the same for all simulated springs. In the intermediate state of the half-deformation, the middle surface of the spring is flattened and its height vanishes $H = 0$. Specifically, the full stroke spring travel of all springs is equal to $2h = 4[mm]$. This means, that the height of the middle surface of the deformed spring in the full stroke state is equal to $h = -H = 2[mm]$. In other words, the middle surface of the spring in the full stroke state is deformed "upside-down". The angular variation of thickness disappears, for all simulated models, such that$\varepsilon = 0$.

The results of the simulation are shown on Figs. 6.9, 6.10, 6.11, 6.12, 6.13, 6.14, 6.15, 6.16, 6.17 and 6.18. The models 1 and 2 possess the constant thickness of the material over the surface of the spring. The vertical compression forces for the springs $F_{F.Z}$, $F_{ANSYS.Z}$ as the function of the height of the middle surface H for the Model 1 are shown on Fig. 6.9. For the Model 2, Fig. 6.10 displays the vertical compression forces for the springs $F_{F.Z}$, $F_{ANSYS.Z}$.

The models 3 to 6 have the linearly varying thickness of the material over radius of the conical middle surface of the spring. The vertical compression forces for the springs $F_{F.Z}$, $F_{ANSYS.Z}$ as the function of the height of the middle surface H for the Model 3 are shown on Fig. 6.11. Figure 6.12 shows the vertical compression forces for the springs $F_{F.Z}$, $F_{ANSYS.Z}$ for the Model 4. The vertical compression forces for the springs $F_{F.Z}$, $F_{ANSYS.Z}$ as the function of the height of the middle surface H for the Model 5 are exposed on Fig. 6.13. Figure 6.14 exhibits the vertical compression forces for the springs $F_{F.Z}$, $F_{ANSYS.Z}$ for the Model 6.

The models 7 to 10 have the parabolically varying thickness of the material over radius of the conical middle surface of the spring. The thickness on the inner and outer radii is equal to $T_i = T_e = 2mm$, but the thickness T_m on the middle radius $r_m = (r_e + r_i)/2$ differs for the considered models. The thickness in the middle of the surface in less than the thickness on the edges for both models 7 and 8. The vertical compression forces for the springs $F_{F.Z}$, $F_{ANSYS.Z}$ as the function of the height of the middle surface H for the Model 7 are shown on Fig. 6.11. The vertical compression forces for the springs $F_{F.Z}$, $F_{ANSYS.Z}$ for the Model 8 are displayed Fig. 6.12.

Alternatively, for both models 9 and 10, the thickness in the middle of the surface in higher than the thickness on the edges. The vertical compression forces for the springs $F_{F.Z}$, $F_{ANSYS.Z}$ as the function of the height of the middle surface H for the Model 9 are exposed on Fig. 6.13. For the Model 10, Fig. 6.14 exhibits the vertical compression forces for the springs $F_{F.Z}$, $F_{ANSYS.Z}$ as well.

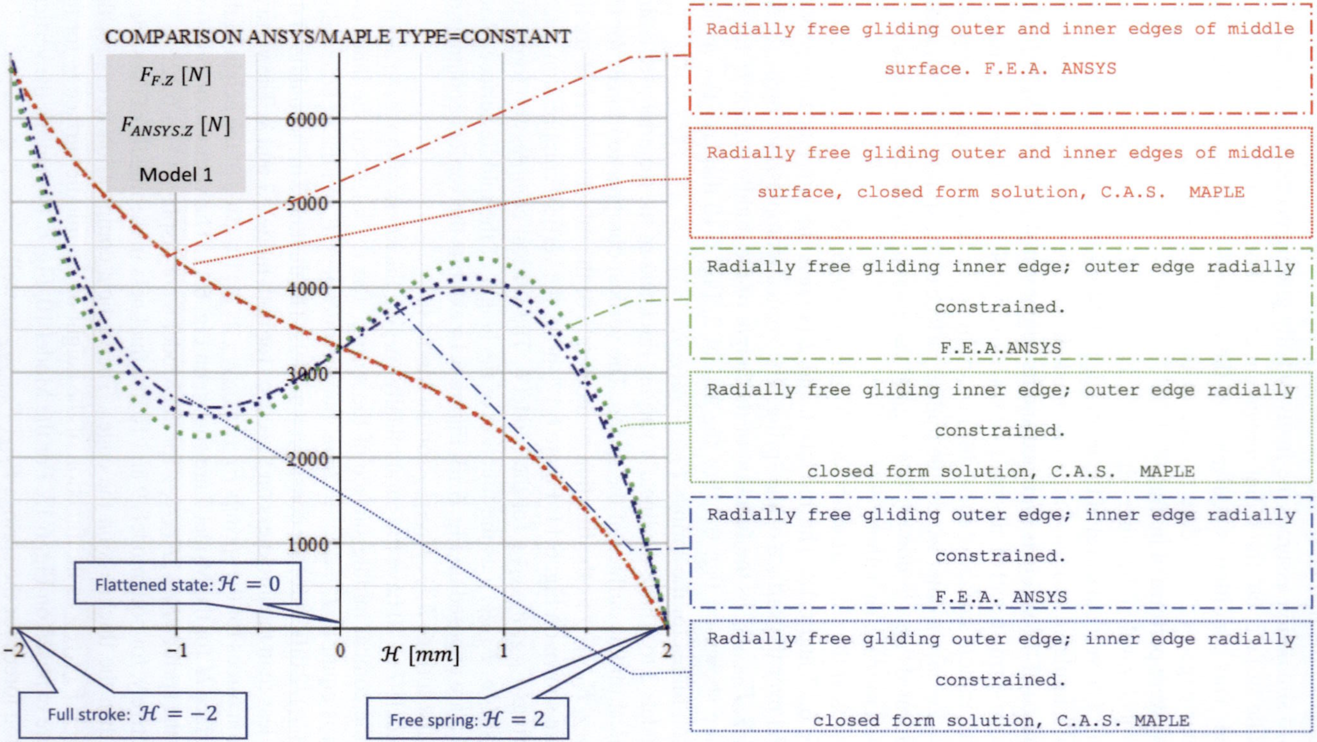

Fig. 6.9 Model 1. Vertical compression forces for the springs $F_{F.Z}$, $F_{ANSYS.Z}$ as the function of the height of the middle surface H

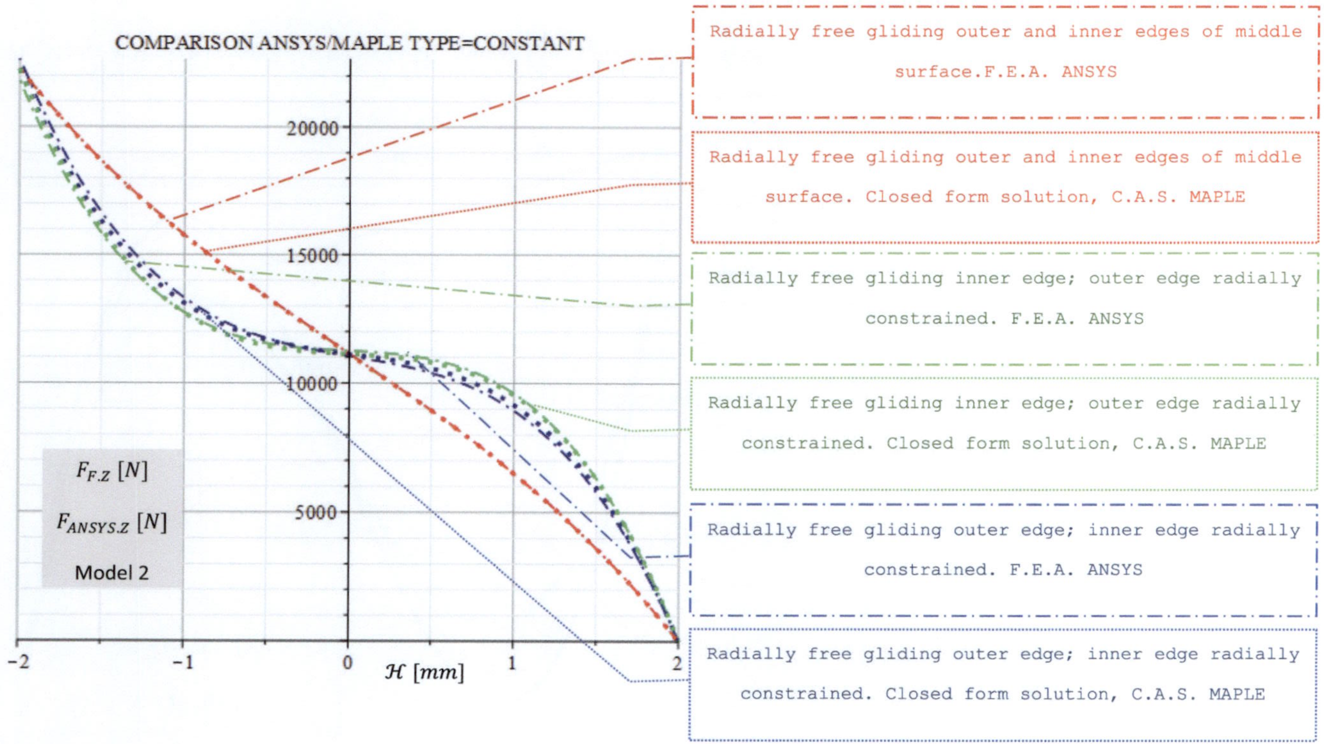

Fig. 6.10 Model 2. Vertical compression forces for the springs $F_{F.Z}$, $F_{ANSYS.Z}$ as the function of the height of the middle surface H

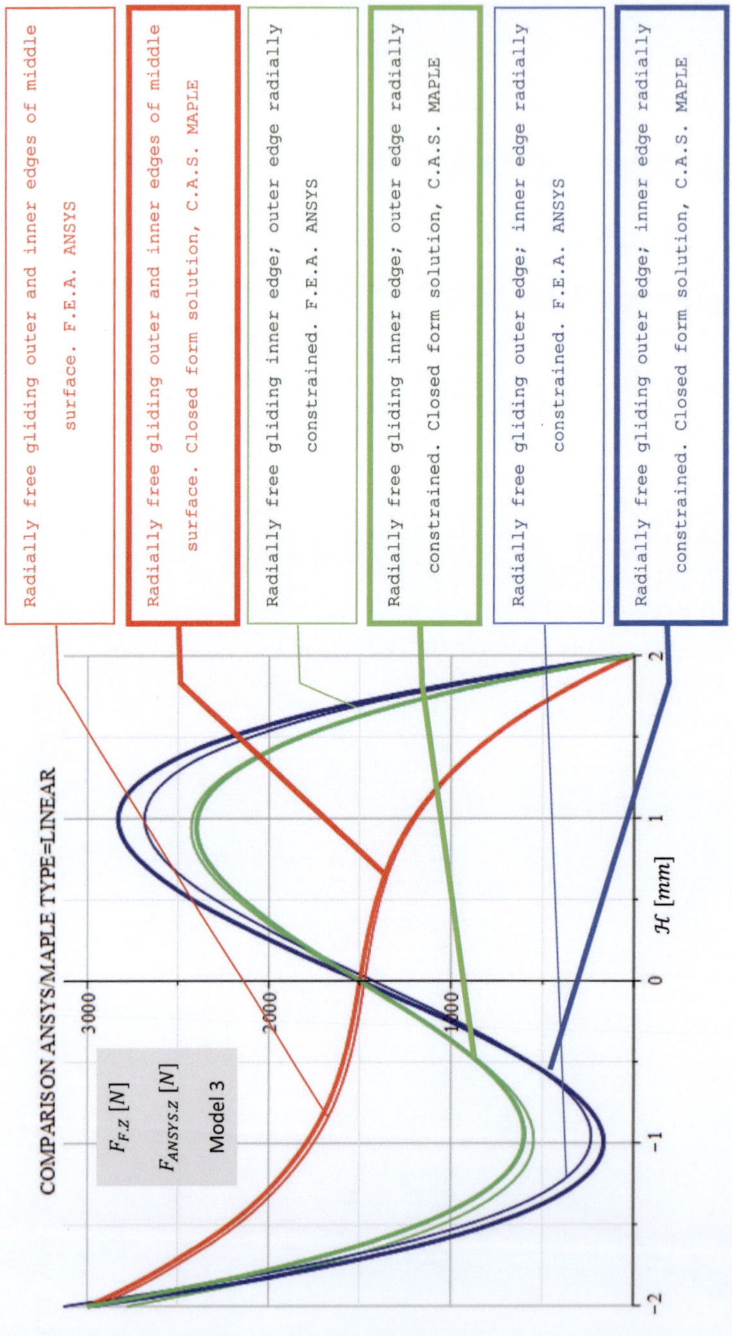

Fig. 6.11 Model 3. Vertical compression forces for the springs $F_{F.Z}$, $F_{ANSYS.Z}$ as the function of the height of the middle surface H

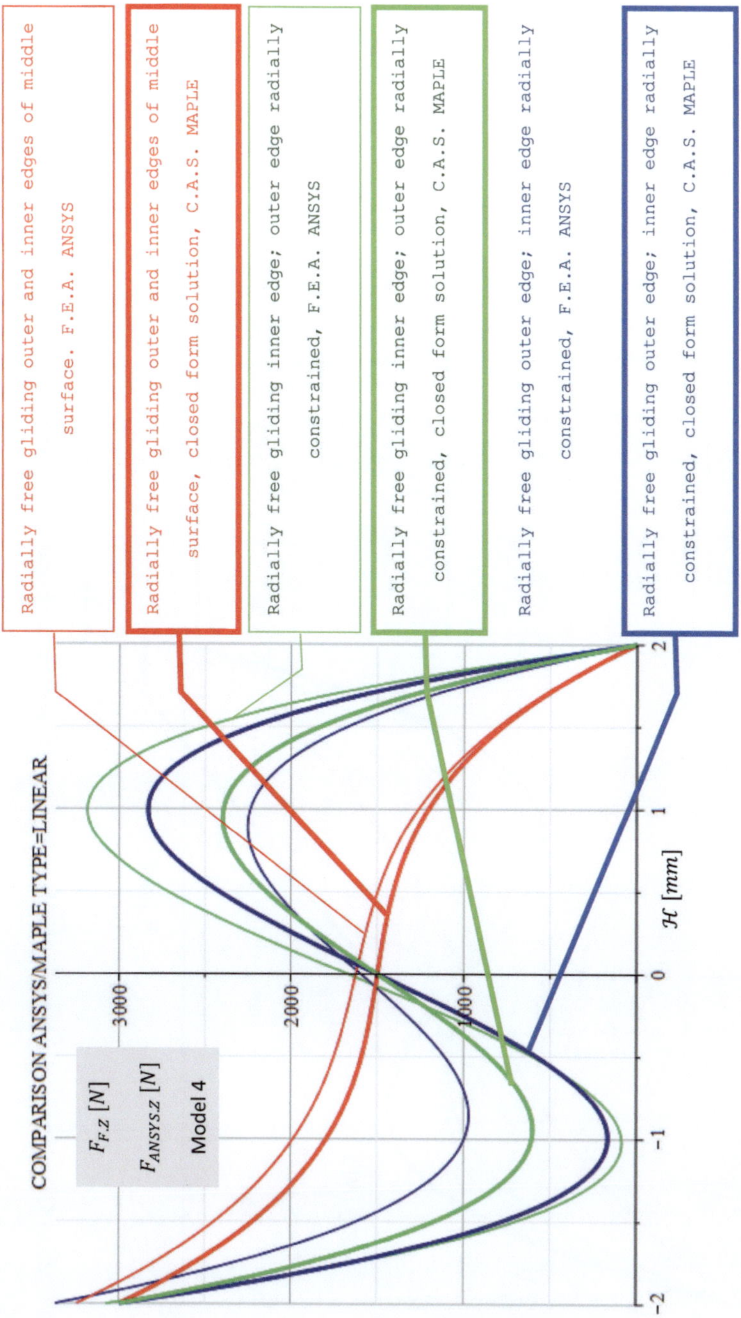

Fig. 6.12 Model 4. Vertical compression forces for the springs $F_{F.Z}$, $F_{ANSYS.Z}$ as the function of the height of the middle surface H

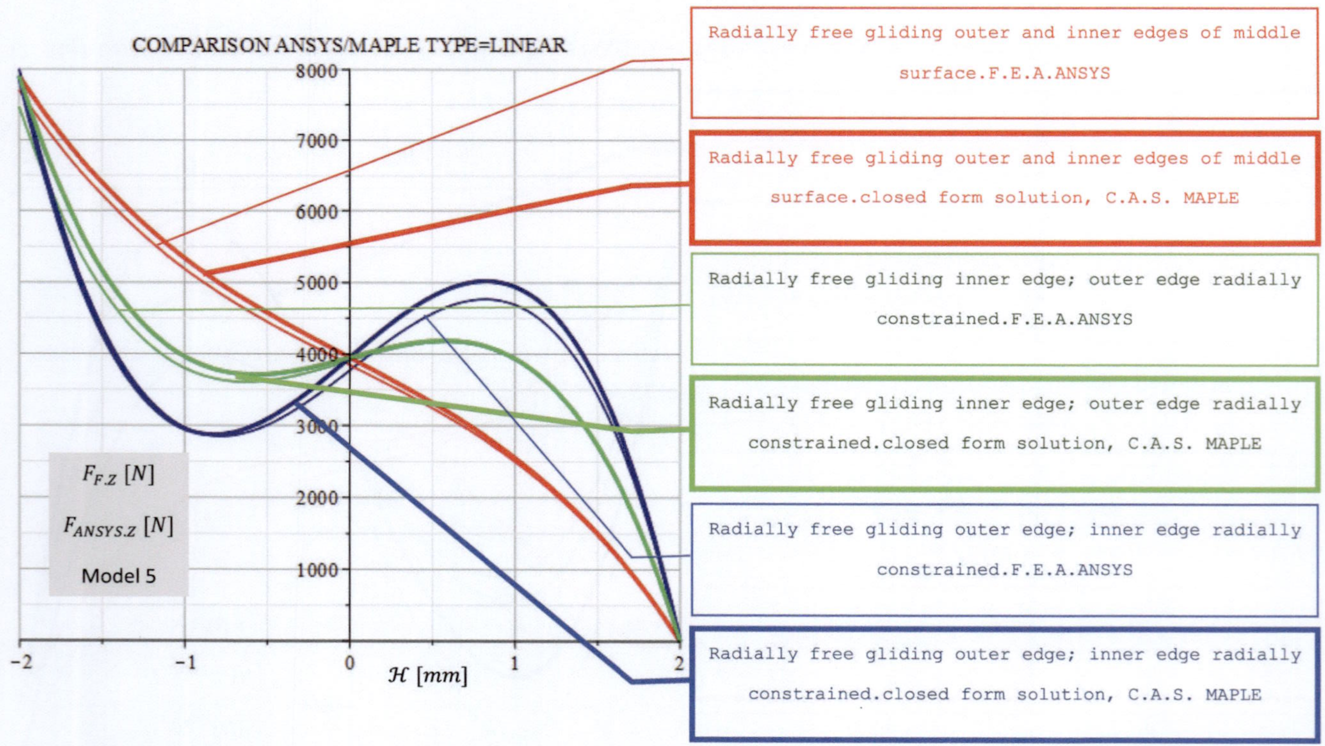

Fig. 6.13 Model 5. Vertical compression forces for the springs $F_{F.Z}$, $F_{ANSYS.Z}$ as the function of the height of the middle surface H

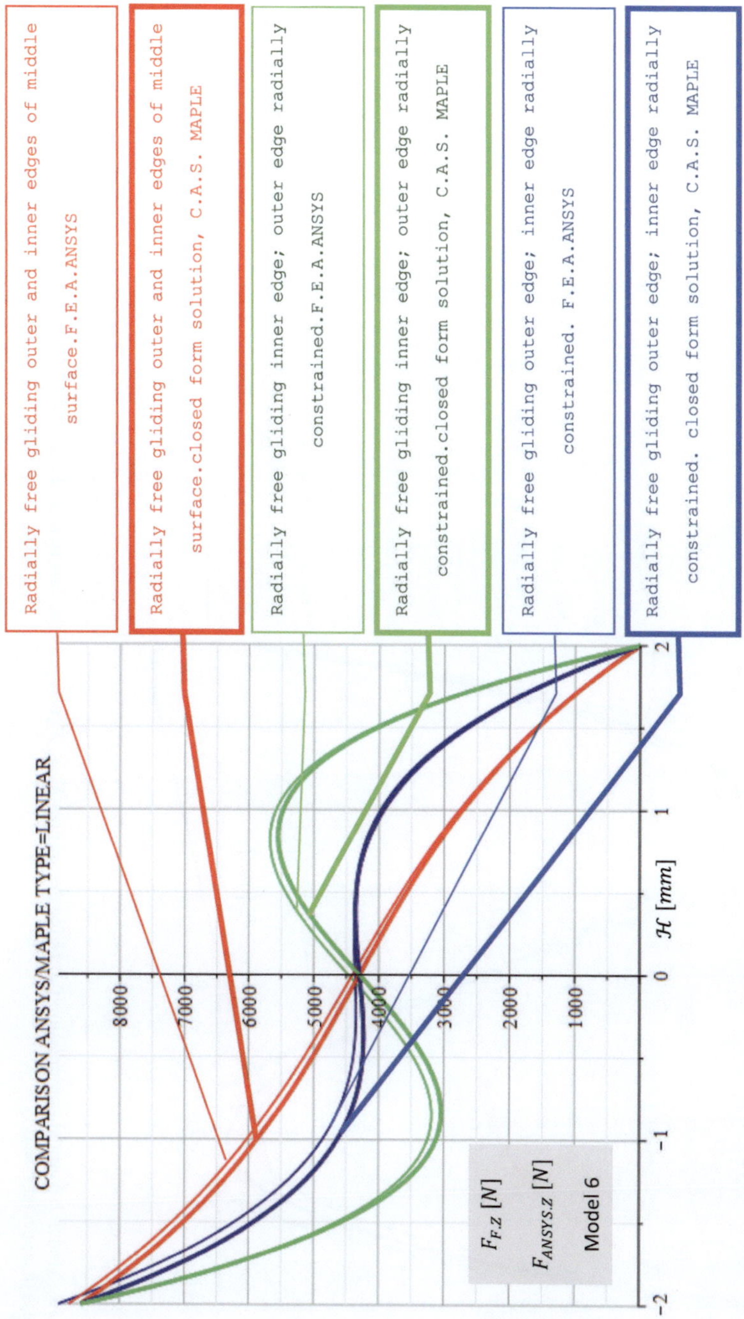

Fig. 6.14 Model 6. Vertical compression forces for the springs $F_{F.Z}$, $F_{ANSYS.Z}$ as the function of the height of the middle surface H

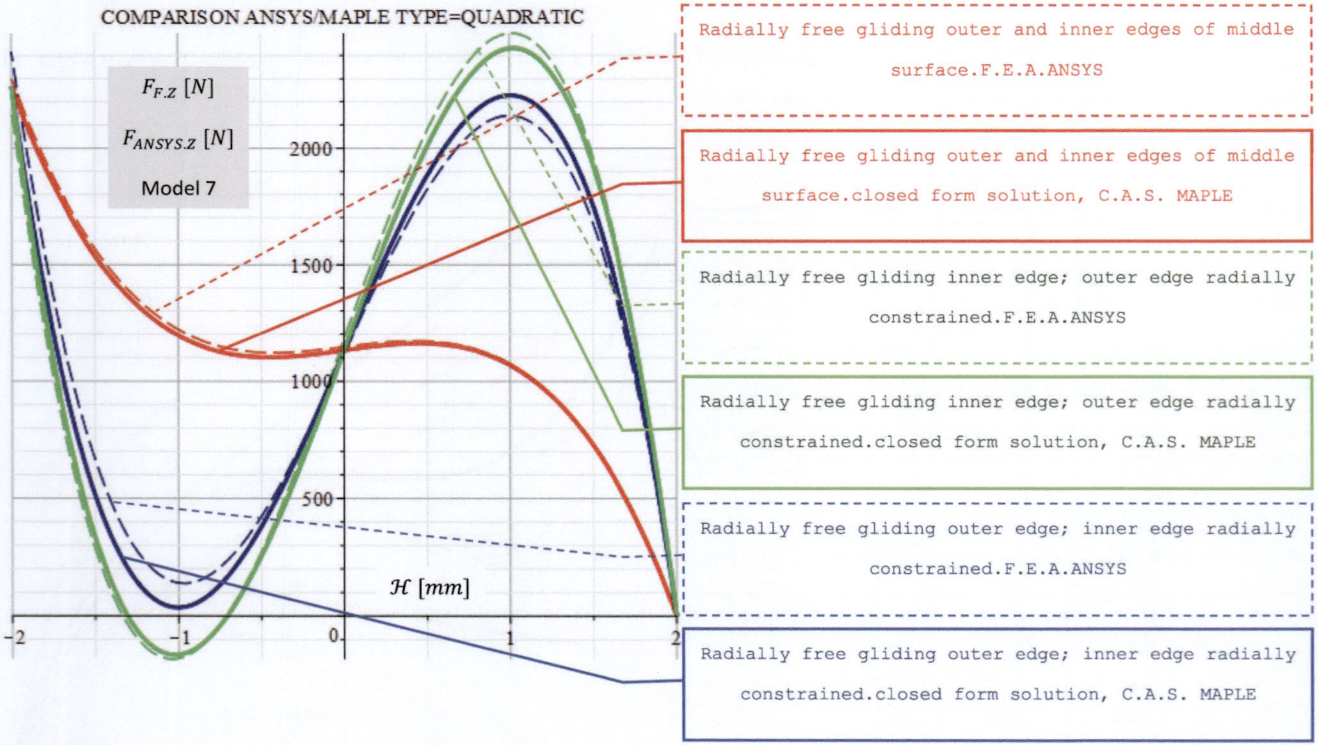

Fig. 6.15 Model 7. Vertical compression forces for the springs $F_{F.Z}$, $F_{ANSYS.Z}$ as the function of the height of the middle surface H The angular variation of thickness disappears, $\varepsilon = 0$

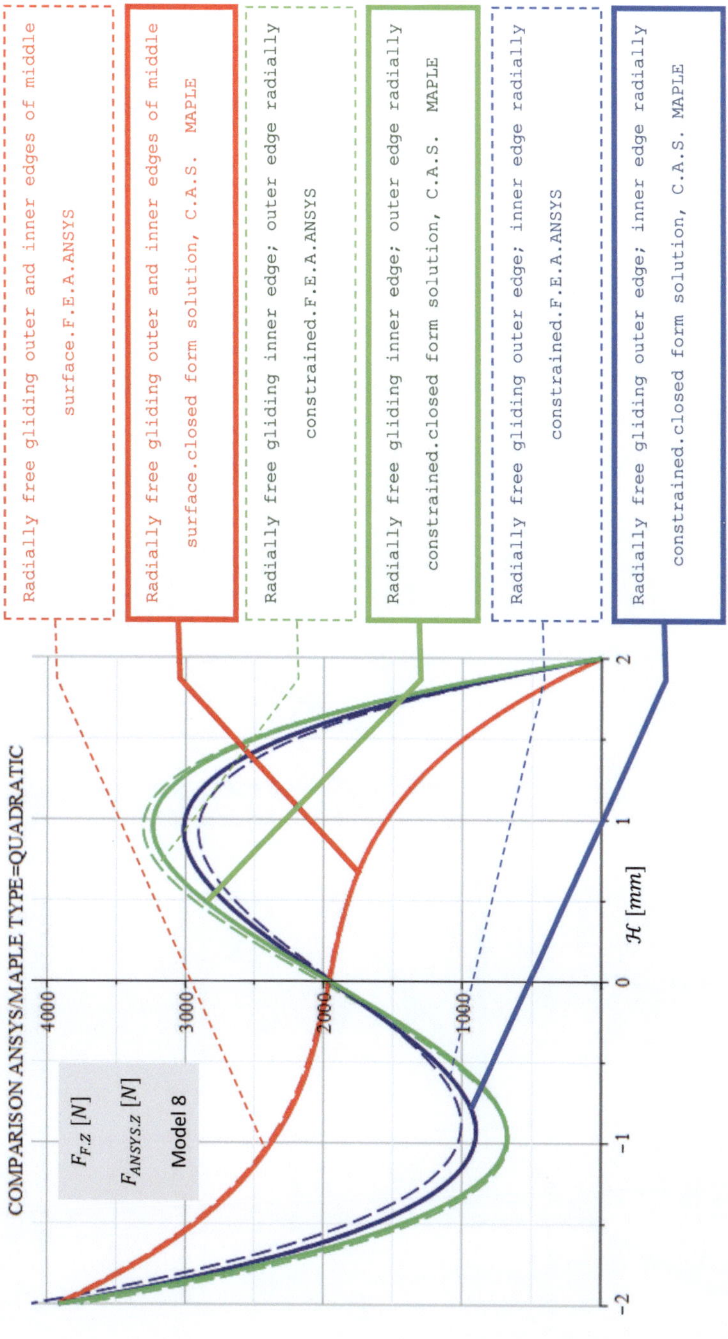

Fig. 6.16 Model 8. Vertical compression forces for the springs $F_{F.Z}$, $F_{ANSYS.Z}$ as the function of the height of the middle surface H

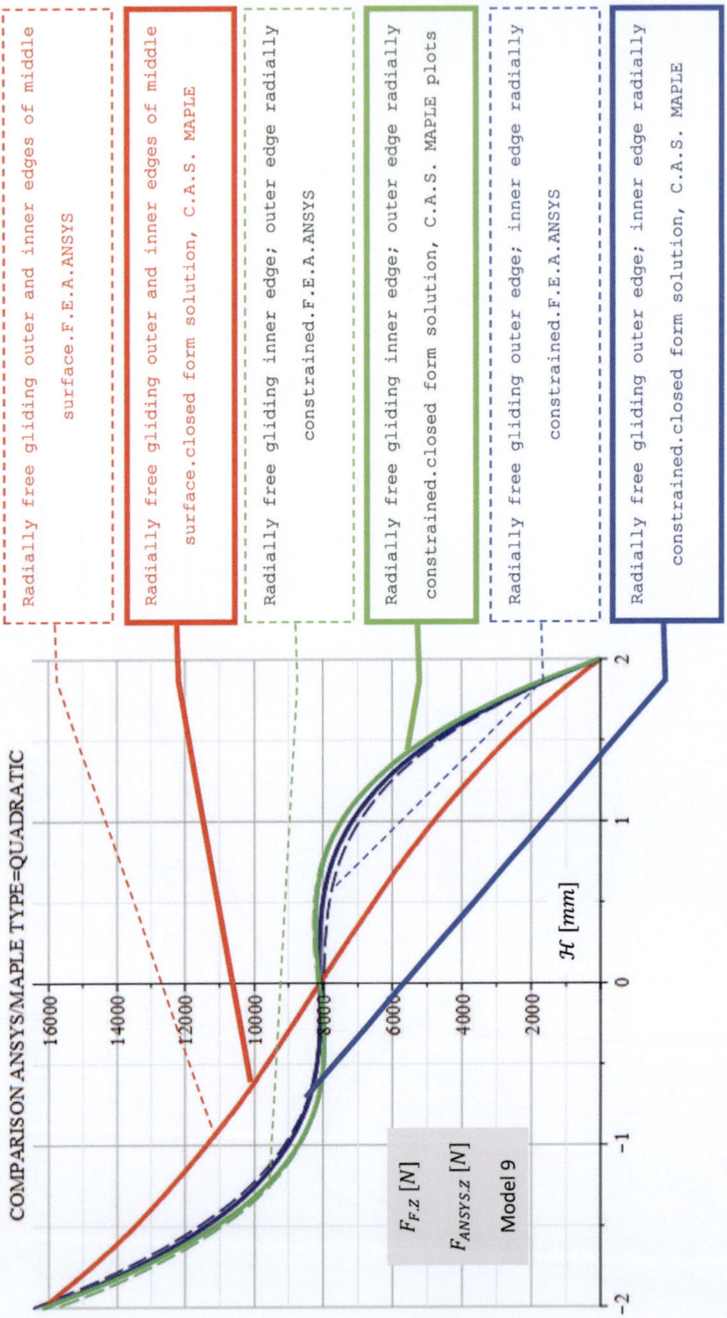

Fig. 6.17 Model 9. Vertical compression forces for the springs $F_{F.Z}$, $F_{ANSYS.Z}$ as the function of the height of the middle surface H

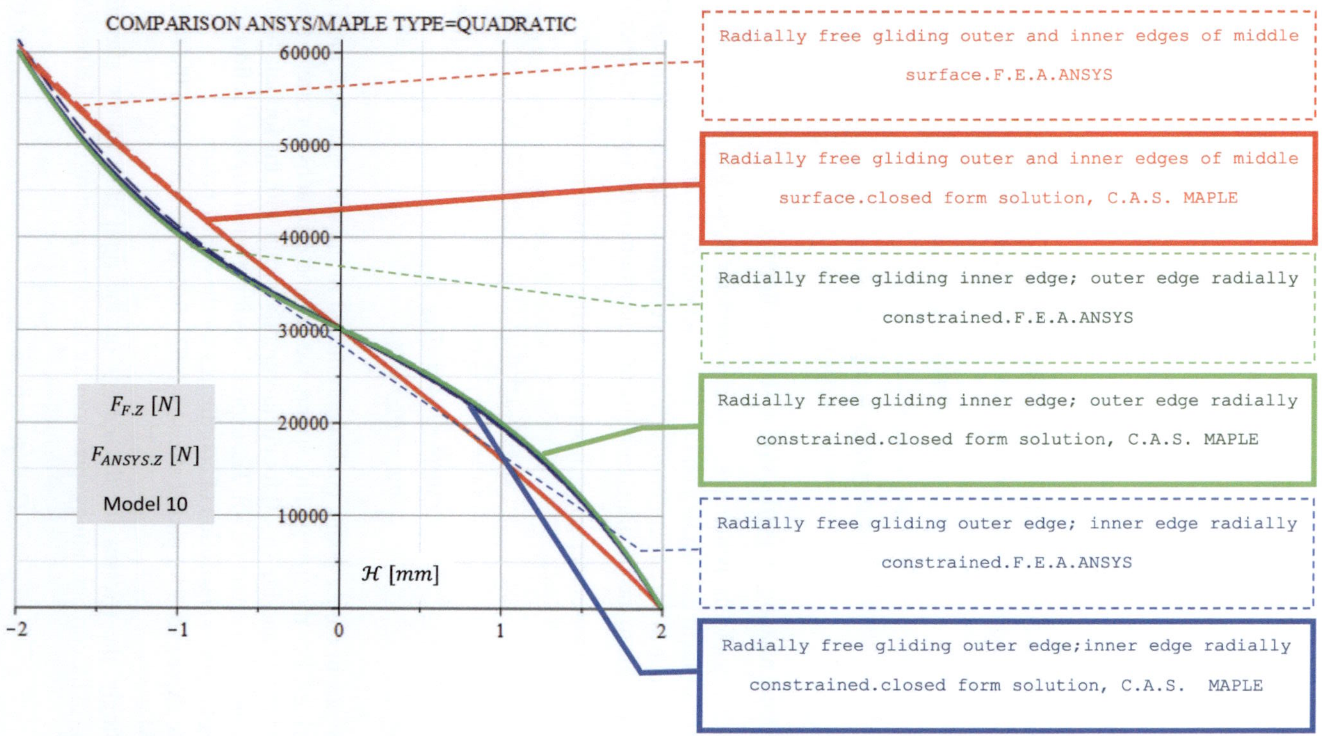

Fig. 6.18 Model 10. Vertical compression forces for the springs $F_{F.Z}$, $F_{ANSYS.Z}$ as the function of the height of the middle surface H

The presented results demonstrate the good correlation between the closed form solutions and the results of FE simulation with program ANSYS. All simulation data were exported as CSV spreadsheets and imported in EXCEL (Microsoft EXCEL, 2023). All output results are stored in the cloud-based communal repository Mendeley Data (Kobelev, 2023b).

6.10 Stresses in Disk Springs

The corner points according to standard (EN 16,984:2017-02, 2017) are depicted on the Fig. 6.3. The coordinates of corner points are correspondingly:

$$\begin{aligned} &I : \rho = r_i, y = \tfrac{1}{2}T(r_i, \varphi); \\ &II : \rho = r_i, y = -\tfrac{1}{2}T(r_i, \varphi), \\ &III : \rho = r_e, y = -\tfrac{1}{2}T(r_e, \varphi); \\ &IV : \rho = r_e, y = \tfrac{1}{2}T(r_e, \varphi). \end{aligned} \tag{6.61}$$

The physical meaning of partial stresses is the following:

$\sigma_{b.e}$ the stress due to the bending on the outer diameter $D_e = 2r_e$;
$\sigma_{b.i}$ the stress due to the bending on the inner diameter $D_i = 2r_i$;
$\sigma_{t.e}$ the stress due to the circumferential membrane strain on the outer diameter D_e;
$\sigma_{t.i}$ the stress due to the circumferential membrane strain on the inner diameter D_i.

The stresses on the corner points from relate to the "partial stresses"

$$\begin{aligned} \sigma_I &= \sigma_{b.i} + \sigma_{t.i}, \\ \sigma_{II} &= -\sigma_{b.i} + \sigma_{t.i}, \\ \sigma_{III} &= -\sigma_{b.e} + \sigma_{t.e}, \\ \sigma_{VI} &= \sigma_{b.e} + \sigma_{t.e}. \end{aligned} \tag{6.62}$$

The stresses in corner points could be evaluated in the solid and shell theories of disk springs. For the free gliding spring, we use the equations for stress–strain dependence (6.21) and Eq. (6.23) for circumferential strain with the formula for the position of inversion centre (6.30). The expressions for "partial stresses" follow from these formulas after substitution of coordinates for the corner points.

With the expressions for partial stresses (6.62), follow the stresses on the corner points (6.61).

Accordingly, the stresses for the springs on the rigid curb follow after the substitution of the corresponding expressions for the corresponding inversion centres:

$$C_I = r_I, \quad C_E = r_e. \tag{6.63}$$

The bending partial stresses remain unaffected by the shift of the inversion point to the inside or outside edge. The in-plane stresses due to circumferential strain modify significantly in course of relocation of the inversion points.

Conclusions

The theory developed is based on experimental observations according to which the cross section of the spring rotates only about a center point (assumed to be on the center line of the cross section) without undergoing any significant deflection. The results of Almen and Laszlo agreed with (Timoshenko, 1957) in that they considered the radial stresses to be negligible and succeeded in calculating the tangential stresses and displacements of a washer subjected to an axial load uniformly distributed on the inner and outer edges. Based on this theory, the dimensions used to define the geometries refer to the center line of the washer cross section. The equations given in (DIN EN 16,984:2017-02, 2017) are sufficiently accurate to evaluate relative flat disk springs of variable thickness along the latitude and longitude of the conical central surface. According to these equations, the deformation behavior of the disk spring is treated as a one-dimensional inversion of a circular ring with a rectangular cross section about certain inversion centers. The derived formulas can be used for design purposes. The formulas do not require CAD modeling of the complex spring geometry or finite element solvers. The formulas can be coded directly into standard spreadsheet programs (e.g. LibreOffice Calc (LibreOffice, 2023), Apache OpenOffice Calc (OpenOffice, 2023), Microsoft EXCEL (2023]). The formulas allow complete evaluation of cup springs with variable material thickness. The developed formulas can be used for design optimization of disk springs and conical spring systems.

Summary of Principal Results

- The closed-form solution derives the formulas for the disks with non-constant thickness along the meridional and angular directions of the disk spring. Both types of radial conditions at the edges are studied, namely free radially sliding edges and edges with restricted radial motion. The influence of friction at the edges on the spring forces is estimated.
- The nonlinear governing equations for the center of the disk spring are derived.
- The equations describe the deformation and stresses of thin and moderately thick washers.
- The variation method is applicable for the disk springs with free gliding and rigidly constrained edges.
- The developed method is applicable to disk springs with radially constrained edges. The vertical displacements of a disk spring result from an axial load uniformly distributed on the inner and outer edges.
- The strain component of the elastic energy is not affected by the thickness variation along the parallels.
- The bending component of the elastic energy increases proportionally to the factor $1 + \frac{3}{2}\varepsilon^2$.

- The correction factor does not depend on the number of waves n of variation in the direction along the parallels.
- The method is also suitable for slotted disk springs with free gliding and rigidly constrained edges.

References

Almen, J. O., & Laszlo, A. (1936). The uniform section disks spring. *Trans ASME, 58*(12.4), 305–314.

ANSYS. (2022). ANSYS, Inc. Southpointe, 2600 ANSYS Drive, Canonsburg, PA 15317c. http://www.ansys.com/

DIN EN 16984:2017–02. (2017). Disc springs—Calculation; German version EN 16984:2016, Beuth Verlag. DIN EN 16984-2017-09—Beuth.de

EN 16984:2017-02 (2017) Disc springs—Calculation; German version DIN EN 16984:2016, Beuth Verlag.

Hesselmann, B., & Kobelev, V. (2002). *Disk spring*, European Patent Application EP1331416. https://worldwide.espacenet.com/patent/search/family/026010978/publication/EP1331416A1?q=pn%3DEP1331416A1

Kobelev, V. (2023b), Exact shell solutions for conical springs. III. disk springs with variable thickness. *Mendeley Data, V1*. https://doi.org/10.17632/gtk67w9kp7.1, https://data.mendeley.com/datasets/gtk67w9kp7

Kobelev, V. (2023a). Exact shell solutions for conical springs. II. Radial cylindric curb. *Mechanics Based Design of Structures and Machines, An International Journal.* https://doi.org/10.1080/15397734.2023.2189942

LibreOffice. (2023). LibreOffice Calc. https://www.libreoffice.org/diskover/calc/

MAPLE. (2022). Maplesoft, a division of Waterloo Maple Inc. 2022, 615 Kumpf Drive Waterloo, ON Canada, N2V 1K8. https://www.maplesoft.com/

Microsoft EXCEL. (2023). Microsoft Corporation, One Microsoft Way, Redmond, WA 98052–6399, USA. https://www.microsoft.com

OpenOffice. (2023). Apache OpenOffice Calc. www.OpenOffice.org/calc

Timoshenko, S., & Woinowsky-Krieger, S. (1957). *Theory of plate and shell* (2nd ed.). McGraw Hill.

Ventsel, E., & Krauthammer, T. (2001). In *Thin plates and shells, theory, analysis, and applications*, Marcel Dekker AG, Basel.

Chapter 7
Thin-Walled Rods with Semi-open Profiles

Abstract The semi-open cross sections are made from standard tubes by flattening them. Due to the possibility of significant variation of mechanical parameters along the length, the flattening could be easily adjusted for optimal design. From the design point of view, the semi-open cross sections fill the gap between the classic closed and open single-wall sections. Due to the high variability of the section geometry, the simple analytical model is essential for primary design purposes and estimation of numerous opposing static effects. The effective torsional stiffness r_t, effective bending stiffness r_z and effective bending spring r_c of the twist beam in terms of section properties of the twist beam with semi-open cross section have been expressed with the analytical formulas. This chapter is the last section of the first part, which studies the elastic elements from the point of view of modeling and design.

Keywords Thin-walled rods · Twist beam axles · Thin-walled rods with open sections

7.1 Thin-Walled Rods with Semi-open Profiles

7.1.1 Open, Closed and Semi-open Wall Sections

A thin-walled beam is an exceptional type of torsion spring. This type of torsion spring serves as a cross member in various industrial and automotive structures. The thin-walled beam is an elongated elastic structural member whose characteristic geometric dimensions are all different orders of magnitude. Thin-walled beams can be classified according to their geometric characteristics. Two classes of thin-walled beams are noteworthy, namely thin-walled beams with open section and thin-walled beams with closed section (Vlasov, 1961; Timoshenko, 1945; Flügge & Marguerre, 1950). First, an intermediate class of thin-walled beams is studied. The cross-section of the beam is closed, but the shape of the cross-section is elongated and curved. The walls forming the section are nearly equidistant. StVenant's free torsional behavior of the beam is the same as the behavior of beams with a closed cross-section. However,

V. Kobelev, *Fundamentals of Springs Mechanics*,
https://doi.org/10.1007/978-3-031-58584-5_7

the total warping function of the semi-open cross section is analogous to the warping function of the open cross section beam.[1]

The difference between the thin-walled beams is the presence of deformation of the section and the corresponding force factor, the bi-moment. It is shown that thin-walled beams with closed sections behave differently depending on the shape of the section. There are two types of thin-walled beams with closed sections. The first type of thin-walled beams has closed profiles with almost equal diameters in all directions, so there is no difference in direction. The torsion of such closed beams is mainly of the StVenant type. The effect of the bi-moment on the torsional behavior of the thin-walled beams of the first type is in most cases negligible compared to the StVenant torsion. The second type of thin-walled beams includes closed sections formed by equidistant walls, elongated sections, and star-shaped sections. The contribution of StVenant's torsional stiffness of beams with this type of profile is of the same order of magnitude, as the contribution of sectoral stiffness and the effect of the bi-moment should be considered.

Also reported are the analytical methods used to design and analyze a rear twist-beam axle in the concept design phase. The twist-beam axle is a specific type of semi-solid suspension. The twist-beam axles are commonly designed with the purpose of improving the vehicle dynamics performance, improving the load carrying capacity, reducing the weight and cost. The main element of the common twist beam axle is the thin-walled twist beam. This beam behaves elastically, providing a certain roll rate for the vehicle, but must be stiff enough to flex and provide the chamber and lateral stiffness of the vehicle. Insufficient camber and lateral stiffness result in oversteering of the vehicle, which is unacceptable from a vehicle dynamics standpoint.

Two classes of thin-walled beam structures are usually distinguished: open and closed (Timoshenko & Gere, 1961; Chilver, 1967; Librescu & Song, 2006). When the locus of points defining the centerline of the wall forms a closed or an open contour (Fig. 7.1), then a thin wall section is said to be closed or open. The thin-walled beam, shown on Fig. 7.1a possesses an open cross-section. Figure 7.1b demonstrates the beam with the closed cross-section. The cross-section of the beam, shown on Fig. 7.1c, will be referred to as semi-opened.

Consider a slender thin-walled structure of cylindrical or prismatic uniform cross-section formed by nearly equidistant walls (Figs. 7.1 and 7.2). Let T be a wall thickness assumed for the beginning to be constant along the beam span but variable along the contour of the cross-section $T = T(s)$, δ is the distance between the walls and L is length of the beam. The middle surface of the wall is defined as the locus of the points equidistant from the upper and the lower bottom surface of the wall. The middle surface is a noncircular cylindrical surface. The straight lines on the middle surface parallel to the beam longitudinal axis are the generators of the middle surface. The intersection of the middle surface with a plane normal to the generators determines the midline of the cross-section contour. The midline of the

[1] V. Kobelev, *Fundamentals of Springs Mechanics*, University of Siegen, Faculty IV: School of Science and Technology. Paul-Bonatz-Str. 9–11, D-57076 Siegen. E-mail: Vladimir.kobelev@uni-siegen.de Homepage: https://www.mb.uni-siegen.de/kobelev/

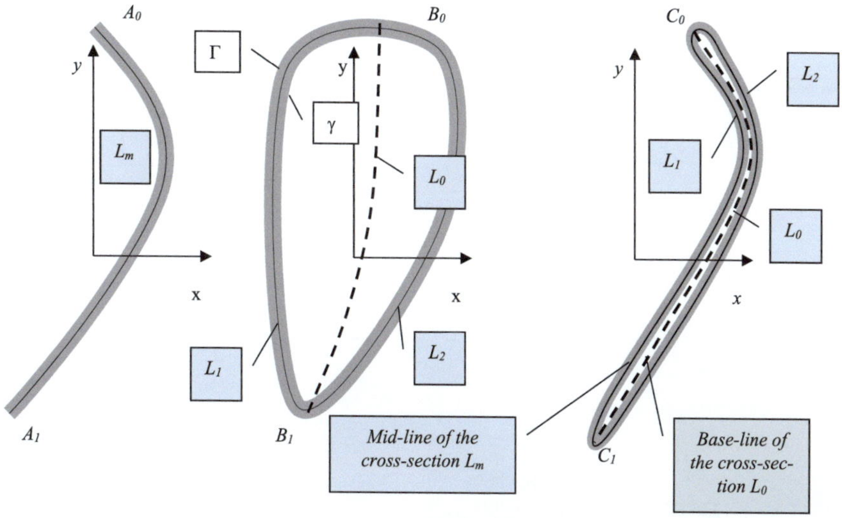

Fig 7.1 Open, closed and semi-opened profiles

open cross-section L_m (Fig. 7.1a) is a plane limited line segment bounded by two end points A_0, A_1. The midline of the closed cross-section is a closed curve (Fig. 7.1b, c). One defines the width of the curve in each direction to be the perpendicular distance between the parallels perpendicular to that direction. The maximal width of the curve is usually defined as its diameter. We denote the most distant points on the curve as B_0, B_1 and refer to them as pole points of the closed mid-line of the cross-section. The points B_0, B_1 divide the closed curve L_m into two open segments L_1 and L_2. We parameterize both segments with the line parameter θ, such that the point B_0 corresponds the value $\theta = 0$, and the point B_1 corresponds the value $\theta = 1$. For each given value of θ there exist two points $B_1(\theta)$ and $B_2(\theta)$ of two segments L_1 and L_2 correspondingly. The position vector of point $B_1(\theta)$ is $\boldsymbol{x}_1(\theta)$ and the position vector of point $B_2(\theta)$ is $\boldsymbol{x}_2(\theta)$. We assume for the beginning the equal thickness of walls at points $B_1(\theta)$ and $B_2(\theta)$. For each $0 \leq \theta \leq 1$ the position vector

$$2\boldsymbol{x}_0(\theta) = \boldsymbol{x}_1(\theta) + \boldsymbol{x}_2(\theta)$$

defines the point on the line segment L_0 with the end points B_0, B_1. This line segment L_0 is referred to as baseline of the closed cross-section. Consequently, for each closed section there exists a line segment L_0 with both endpoints, which correspond to the pole points B_0, B_1 of the mid-line. Position vectors for two corresponding points of the walls L_1 and L_2, are identified by arc coordinates

$$s_1 = s_1(\theta),$$
$$s_2 = s_2(\theta)$$

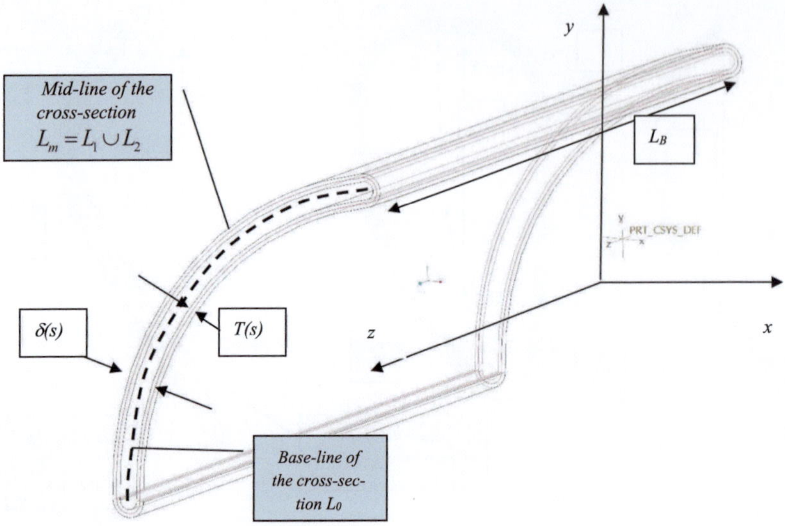

Fig 7.2 Geometry of semi-opened profile and local coordinate system

Generally, for an arbitrary cross-section the introduction of baseline simplifies the modeling only occasionally. For certain practically important case, the mechanical model could be considerably improved with the concept of baseline. Namely, the elongated cross-sections with two nearly equidistant walls, shown on Fig. 7.1b, represent an intermediate case between open and closed cross-sections. The distance δ between points $B_1(\theta)$ and $B_2(\theta)$ is assumed to be much smaller than the length of the curve L_0. The distance between points δ is in the considered case comparative to the material thickness$T$.

The pure torsion behavior of such profiles is in some aspects similar to mechanical behavior of thin-walled beams with closed cross-sections, while the warping function and restricted torsion behavior of these beams are analogous to those of open cross-sections Heilig (1961) and Graße (1965).

7.1.2 Baseline of Semi-open Cross-Section

Each cross-section $z = const$ is a planar figure with the form of a double-walled curved strip. The thickness of each wall is t and the distance between the midlines of walls is δ. The outer boundary of the cross-section is Γ and the inner boundary is γ. Each value of an arc parameter s matches a dimensionless parameter

$$\theta = \theta(s).$$

A point on the baseline L_0 is identified by the arc coordinate s along the baseline

$$\boldsymbol{r}^{(0)}(s) = \boldsymbol{r}^{(0)}(\theta(s)), \ \ s = 0\,..l,$$

where l is the length of the line segment L_0.

7.1.3 Main Hypotheses of Thin-Walled Open-Profile Bars

The proposed technical theory of thin-walled rods with semi-open profile is based on the following set of hypotheses of the kinematic and static nature, analogous to the hypotheses of thin-walled open-profile bars (Slivker, 2007):

- The unchanged-contour hypothesis, according to which the cross-section of the bar does not change its shape in its plane; this means, that the bending of tubular cross-section vanishes. Particularly, the originally circular cross-section remains circular, so that the ovalization could be neglected.
- A no-shear hypothesis: there is no shear in the median surface of the thin-walled bar;
- A no-pressure hypothesis, according to which the longitudinal fibers of the thin-walled bar do not interact in their normal directions;
- A membrane-shell hypothesis: there are no moments in the longitudinal direction, that is, the distribution of normal stresses σ_x over the thickness of the shell is assumed uniform, and tangential stresses are believed negligibly small and therefore approximately equal to zero;
- A uniform-tangential-stress hypothesis, which assumes tangential stress to be uniformly distributed over the thickness of the shell.

The principal hypotheses match the common hypotheses of thin-walled open-profile bars.

The theory of pure torsion of thin-walled bars with closed profiles is based on the following hypothesis: *for the closed cross-sections, tangential stresses are constant over the thickness of the shell.*

On the contrary, the common theory of thin-walled beams with open profiles uses an alternative hypothesis: *tangential stress is supposed to be linear over the shell thickness for thin-walled beams with open profiles* (Laudiero & Savoia, 1990; Bazant & Cedolin, 1991; Wang, 1999).

The present theory could be considered as the application of the theory developed in (Prokic, 2003) for the practically important case of elongated thin-wall profile.

7.2 Deformation Behavior of Cross-Sections

7.2.1 Deformation of Rods with Opened and Closed Profiles

The discussion of the torsion phenomena with comparative behavior of open and closed section is given by (Tamberg & Mikluchin, 1973). Several practical considerations relating to closed sections in torsion are given in (Siev, 1966). This subject is treated in a few textbooks (Boresi & Schmidt, 2003, Chaps. 6–8] and (Salmon & Johnson, 1996), so only a brief treatment follows.

The principal difference of the current model from the common models of open and closed cross-section consists in the assumptions for calculation of the torsion stiffness (Kobelev, 2013). The proposed model could be considered as a special case of the open-profile bar with the double bending stiffness of the beam, but with the state of shear stress typical to the closed cross-sections. Namely, the stress of pure torsion is determined by the Bredt's theorem about shear stress circulation.

The motivation for this conjecture could be explained by the thin, curved, and elongated form of the cross-section. Consider such a cross-section where the distance between the walls vanishes and the walls touch each other along their inner surfaces.

Firstly, due to torsion a relative shift or free gliding between the inner surfaces occurs. In this case, the torque stiffness could be described by the Bredt's theorem about shear stress circulation.

Secondly, the relative shift due to torsion disappears and the gliding of walls is fully constrained. The relative shift vanishes if the walls are welded together over the inner surface. The disappearance of relative shift could be the result of friction between inner surfaces of walls. The effective wall thickness doubles. In this case, the torque stiffness considerably increases.

The present theory could be considered as intermediate model between the closed and opened cross-section theories. The warping of cross-section is essentially the same as the warping of open cross-section profiles, but the torsion stress flow is of Bredt's type.

In other words, two walls that outline the cross-section of semi-open profile bend roughly identically in the direction of the beam axis. This behavior is typical for open cross-sections. Contrarily, the shear stress flow due to torsion is constant over contour L_1 and L_2. With respect to the baseline L_0 the shear stresses are opposite. This stress flow is characteristic for closed cross-sections.

Important for the application of the proposed theory is the lack of bending of the cross-section in its planexy. The distance between the walls does not alter throughout the deformation. Otherwise, if the distance between walls changes, the bending of the cross-section in its plane occurs. This kind of bending and the ovalization leads to the considerable reduction of bending stiffness. If the ovalization or the bending in plane of cross-section xy occurs, the fundamentally different theories, which based on the thin shell theory, must be developed (Ventsel & Krauthammer, 2001).

Another remark concerns the rods with elongated semi-open cross-section, but with curved axis$z$. The hypotheses allow the modelling of the curved rods with semi-open cross-section also. The rigidity of a curved pipe subjected to bending decreases as compared to that of a straight pipe with the same cross-section. It causes significant meridional bending stresses. The classical theory for bending of curved pipes was developed by Karman (1911), where the reasons for significant decrease of their bending rigidity was analyzed. In the cited article and in most subsequent studies it is assumed that actual curvature of longitudinal fibers of a pipe should not be considered. This curvature was considered equal to the centerline curvature. This assumption leads to an error in results if applied to pipes with a small radius of curvature. Clark and Reissner (1951) demonstrated an example in which an actual curvature of longitudinal fibers is given in problem formulation but ignored in problem solution. Without discussing this issue in detail, we refer only to some basic papers on this subject Whatham (1981, 1982) and Cherniy (2001).

7.2.2 Deformation of Rods with Semi-open Profiles

According to the postulation of perfect rigidity of the cross-section in its own plane, the deformation of the cross-section can be described by only three displacement components, namely two translations U_x, U_y and the angle of twist Θ_z of an arbitrary taken pole P (Fig. 7.3). The position vector of the point S_0 of baseline L_0 with coordinates x, y with respect to the point P is $\boldsymbol{R}$. In natural coordinate system $(\boldsymbol{n}, \boldsymbol{t})$ of the curve L_0, the components of vector $\boldsymbol{R}$ are

$$
\begin{aligned}
R_\xi &= (x - x_P)\sin\alpha - (y - y_P)\cos\alpha,\\
R_\eta &= (x - x_P)\cos\alpha + (y - y_P)\sin\alpha.
\end{aligned}
\tag{7.1}
$$

The Cartesian coordinates of the points S_1, S_2 of midlines L_1, L_2 are

$$
\begin{aligned}
x^{(1)} &= x - (\delta/2)\cos\alpha,\\
y^{(1)} &= y - (\delta/2)\sin\alpha
\end{aligned}
$$

and

$$
\begin{aligned}
x^{(2)} &= x + (\delta/2)\cos\alpha,\\
y^{(2)} &= y + (\delta/2)\sin\alpha.
\end{aligned}
$$

The position vectors of these points respectively to the point P are

$$
\begin{aligned}
\boldsymbol{R}^{(1)} &= \boldsymbol{R}_\eta^{(1)} n + \boldsymbol{R}_\xi^{(1)} t,\\
\boldsymbol{R}^{(2)} &= \boldsymbol{R}_\eta^{(2)} n + \boldsymbol{R}_\xi^{(2)} t,
\end{aligned}
$$

where:

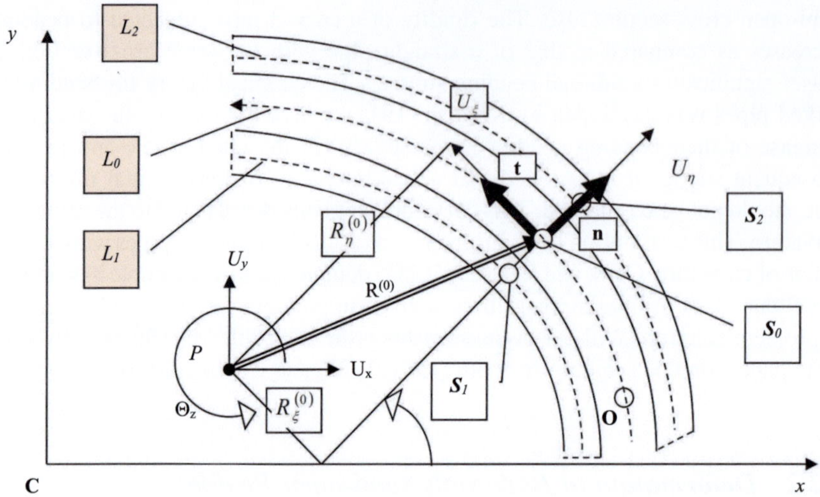

Fig. 7.3 Coordinate system, associated with the base line of the semi-opened profile

$$\begin{aligned}
\boldsymbol{R}_\eta^{(1)} &= \boldsymbol{R}_\eta - \frac{\delta}{2}, \\
\boldsymbol{R}_\eta^{(2)} &= \boldsymbol{R}_\eta + \frac{\delta}{2}, \\
\boldsymbol{R}_\xi^{(1)} &= \boldsymbol{R}_\xi^{(2)} = \boldsymbol{R}_\xi.
\end{aligned}$$

Assuming the planar displacement of the whole cross-section as the solid body, the normal $U_\eta^{(i)}$ and tangential displacements $U_\xi^{(i)}$ of the points $S_i (i = 1, 2)$ read as

$$\begin{aligned}
U_\xi^{(i)} &= U_x \cos\alpha + U_y \sin\alpha + \Theta_z R_\xi^{(i)}, \\
U_\eta^{(i)} &= -U_x \sin\alpha + U_y \cos\alpha + \Theta_z R_\eta^{(i)}.
\end{aligned} \tag{7.2}$$

The counterclockwise rotation, observed from positive z direction, assumed to be positive. In the absence of shear strain, the longitudinal strains at points S_1, S_2 are

$$\begin{aligned}
\varepsilon_z^{(i)} &= \partial_z U_z - x^{(i)} \partial_z^2 U_x - y^{(i)} \partial_z^2 U_y - \omega^{(i)} \partial_z^2 \Theta_z, \\
\partial_z &= \partial/\partial z, i = 1, 2.
\end{aligned} \tag{7.3}$$

The sectorial areas of lines L_i $(i = 1, 2)$ are

$$\omega^{(i)} = \int_0^s R_\eta^{(i)} \mathrm{d}s.$$

Substitution of the normal distance $\delta(s)$ between the lines L_1 and L_2 delivers

$$\begin{aligned}
\omega^{(1)} &= \omega - \frac{1}{2} \int_0^s \delta \, \mathrm{d}s, \\
\omega^{(2)} &= \omega + \frac{1}{2} \int_0^s \delta \, \mathrm{d}s,
\end{aligned}$$

where

$$\omega = \int_0^s R_\eta \, ds$$

is the sectorial area of the baseline L_0.

7.3 Statics of Semi-open Profile Bars

7.3.1 Normal Stresses in Semi-open Profile Bars

The no-pressure hypothesis states, that normal stresses in tangential and normal directions could be neglected in Hooke's law. The expression (7.3) gives the normal stresses in longitudinal direction in both walls

$$\sigma_z^{(i)} = E \cdot \left(\partial_z U_z - x^{(i)} \partial_z^2 U_x - y^{(i)} \partial_z^2 U_y - \omega^{(i)} \partial_z^2 \Theta_z\right), \qquad i = 1, 2. \tag{7.4}$$

The longitudinal force and overall bending moments are

$$N = \int_0^l \left(\sigma_z^{(1)} + \sigma_z^{(2)}\right) T \, ds,$$
$$M_x = \int_0^l \left(\sigma_z^{(1)} + \sigma_z^{(2)}\right) y T \, ds,$$
$$M_y = \int_0^l \left(\sigma_z^{(1)} + \sigma_z^{(2)}\right) x T \, ds,$$

where $T = T(s)$.

After substitution of (7.4) and integration we will have the following expressions

$$N = EA\partial_z U_z - S_{\omega_p} \partial_z^2 \Theta_z,$$
$$M_x = -EI_x \partial_z^2 U_z - I_{x\omega_p} \partial_z^2 \Theta_z,$$
$$M_y = -EI_y \partial_z^2 U_z - I_{x\omega_p} \partial_z^2 \Theta_z, \tag{7.5}$$

with

$$S_{\omega_p} = \int_0^l 2T\omega \, ds, \; A = \int_0^l 2T \, ds,$$
$$I_y = \int_0^l 2Tx^2 \, ds, \; I_x = \int_0^l 2Ty^2 \, ds,$$
$$I_{x\omega_p} = \int_0^l 2Ty\omega \, ds, \; I_{y\omega_p} = \int_0^l 2Tx\omega \, ds.$$

The principal pole P and the principal origin point O could be chosen in such way that the following conditions should hold

$$S_{\omega_p} = 0, \qquad I_{x\omega_p} = 0, \qquad I_{y\omega_p} = 0.$$

The normal stresses in the walls of bar cross-section will be

$$\sigma_z^{(i)} = \frac{N}{A} + \frac{M_x}{I_x} y^{(i)} + \frac{M_y}{I_y} x^{(i)} + \frac{B_m}{I_\omega} \omega^{(i)}, \tag{7.6}$$

where the following expressions for sectorial area and bi-moment are used

$$I_\omega = 2 \int_0^l T\omega^2 \, \mathrm{d}s, \qquad B = \int_0^l \left(\sigma_z^{(1)} + \sigma_z^{(2)}\right) T\omega \, \mathrm{d}s.$$

7.3.2 *Torque and Bi-moment*

According to the theory of torsion for thin-walled closed profiles, the pure-torsion torque M_H is calculated as

$$M_H = GI_T \partial_z \Theta, \tag{7.7}$$

where GI_T is the section's torsion stiffness

$$I_T = \frac{4A_m^2}{\oint\limits_{L_m} T^{-1}(s) \, \mathrm{d}s},$$

where area enclosed by the curve L_m is A_m (Fig. 7.1c).

Bi-moment in cross-section in terms of displacement is

$$B_m = -EI_\omega \partial_z^2 \Theta_z.$$

7.3.3 *Tangential Stresses in bar Cross-Sections*

The tangential stresses in the bar cross-section along the tangent to the profile line is the sum of the average tangential stress τ_S and the stress of pure torsion τ_H (Fig. 7.4). The average tangential stress τ_S is constant over the thickness of the wall. The stress of pure torsion is determined by the Bredt's theorem about shear stress circulation. The magnitude stress of pure torsion is τ_H.

In the closed profiles the tangential stress τ_S and shear stresses of pure torsion τ_H are constant over the thickness of the wall.

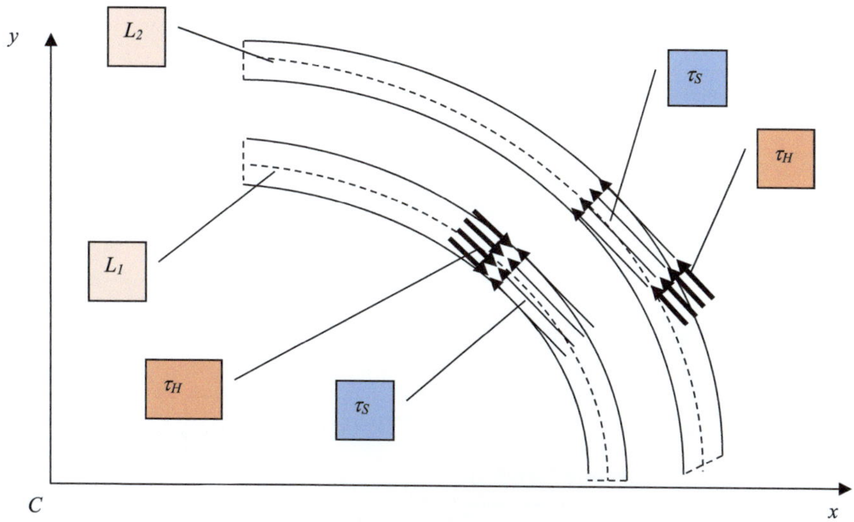

Fig. 7.4 Tangential and shear stresses in the semi-opened profile

Analogously, in semi-open profiles the tangential stress τ_S and shear stress of pure torsion τ_H are also constant over the thickness of the wall.

The tangential stresses in the walls are

$$\tau_{sz}^{(1)} = \tau_S - \tau_H, \quad \tau_{sz}^{(2)} = \tau_S + \tau_H. \tag{7.8}$$

7.3.4 Average Tangential Stress and Equilibrium Conditions

The average tangential stress τ_S could be determined from equilibrium equations. Consider the equilibrium of the two-wall shell-bar element (Fig. 7.5). The element is extracted from the thin-walled bar, so that the element is limited by two longitudinal sections parallel to the generator and located at the distance ds from each other and by two cross-sections at the distance dz from each other.

The equilibrium equation in the absence of external loads for this element for the projection forces on the z-axis reads

$$\partial_s F_\tau + \partial_z F_\sigma = 0, \quad \partial_s = \partial/\partial s, \tag{7.9}$$

where

$$F_\tau = F_\tau(z, s) = \left(\tau^{(1)} + \tau^{(2)}\right)T \tag{7.10}$$

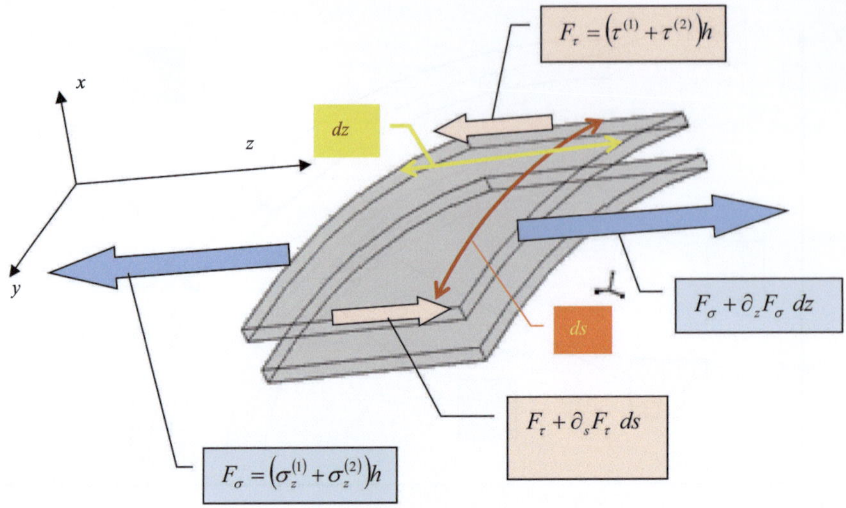

Fig. 7.5 Equilibrium of stresses in the element of semi-opened profile

is the total tangential force per unit length (flow of tangential stresses),

$$F_\sigma = F_\sigma(z, s) = \left(\sigma_z^{(1)} + \sigma_z^{(2)}\right)T \tag{7.11}$$

is the total normal force per unit length at the point of profile.

Integration of the Eq. (7.9) delivers the expression of the tangential force per unit length

$$F_\tau = F_0 - \int_0^s \partial_z F_\sigma \,\mathrm{d}s \equiv F_0 - \int_0^s \partial_z\left[\left(\sigma_z^{(1)} + \sigma_z^{(2)}\right)T\right]\mathrm{d}s.$$

Here $F_0 = F_\tau(z, 0)$ is the total normal force per unit length at zero point of profile. In the case of small distance between walls, the normal stresses in both walls are assumed to be equal

$$\sigma_z^{(1)} = \sigma_z^{(2)}.$$

The substitution of the expression (7.5) for normal stress delivers the expression for the flow of tangential stresses

$$F_\tau = F_o - \frac{A}{A}\partial_z N - \frac{S_x}{I_x}\partial_z M_x - \frac{S_y}{I_y}\partial_z M_y - \frac{S_\omega}{I_\omega}\partial_z B_m. \tag{7.12}$$

The following notations were applied

$$A = 2\int_0^s T\mathrm{d}s, \quad S_x = 2\int_0^s T y\mathrm{d}s,$$

Table 7.1 Geometric and stiffness properties of semi-open twist-beam with V1-shaped and Y-shaped cross-sections

	Static and geometric properties of cross-section	V1-profile, Fig. 7.11 $h = 0, b = 0, \beta = B/H$	Y-profile, Fig. 7.12
I_{xc}	Moment of inertia with respect to x-axis	$\frac{TH^3\beta^2}{12}\frac{1+4\beta^2+4\sqrt{1+4\beta^2}}{1+\sqrt{1+4\beta^2}}$	$\frac{TR^3}{12}\left(3-4\cos\alpha+4\cos^2\alpha\right)$
I_{yc}	Moment of inertia with respect to y-axis	$\frac{TH^3}{16}\frac{1+2\beta^2+\sqrt{1+4\beta^2}}{1+\sqrt{1+4\beta^2}}$	$\frac{2TR^3}{3}\sin^2\alpha$
$S_{\omega x}$	Static moment with respect to x-axis	0	0
$S_{\omega y}$	Static moment with respect to y-axis	0	0
x_c	x-coordinate of mass center	0	0
y_c	y-coordinate of mass center	$\frac{H}{2}\frac{\beta\sqrt{1+4\beta^2}}{1+\sqrt{1+4\beta^2}}$	$\frac{1+2\cos\alpha}{6}R$
I_ω	Sectorial moment of inertia	0	0
α_x	x-coordinate of twist center	0	0
α_y	y-coordinate of twist center	0	0
I_T	Torsion moment of inertia	$\left(1+\sqrt{1+4\beta^2}\right)TH\delta^2$	$3TR\delta^2$
A	Area of material part	$\left(1+\sqrt{1+4\beta^2}\right)TH$	$3TR$
A_m	Area enclosed by curve L_m	$\left(1+\sqrt{1+4\beta^2}\right)\delta H/2$	$\frac{3}{2}\delta R$

$$S_y = 2\int_0^s Tx\mathrm{d}s, \quad S_\omega = 2\int_0^s T\omega\mathrm{d}s.$$

The expressions for geometrical integrals for practically important cross-sections are given in Table 7.1.

7.3.5 *Strain Energy of Semi-open Rod*

The expression of the strain energy as the functional of displacements vector reads (Reissner, 1946):

$$U_e = \frac{1}{2}\int_0^{L_B}\left[EA(\partial_z U_z)^2 + EI_x\left(\partial_z^2 U_x\right)^2 + EI_y\left(\partial_z^2 U_y\right)^2 + GI_T(\partial_z\Theta)^2 + EI_\omega\left(\partial_z^2\Theta\right)^2\right]\mathrm{d}z. \tag{7.13}$$

The strain energy can be represented as a quadratic function of stresses:

$$U_e = \frac{1}{2}\int_0^{L_B}\left[\frac{N^2}{EA} + \frac{M_x^2}{EI_x} + \frac{M_y^2}{EI_y} + \frac{M_H^2}{GI_T} + \frac{B^2}{EI_\omega}\right]\mathrm{d}z. \tag{7.14}$$

Anisotropic generalizations of the developed theory could be developed using the variation methods (Kobelev & Larichev, 1988; Kollar & Pluzsik, 2002).

The differential equations of equilibrium can be obtained from (7.15–7.16) applying the Euler equations of the variation calculus (Kobelev, 2013):

$$\partial_z\left[EA \cdot \partial_z U_z\right] = -p_z, \tag{7.15}$$

$$\partial_z^2\left[EI_x \cdot \partial_z^2 U_x\right] = p_x - \partial_z m_y, \tag{7.16}$$

$$\partial_z^2\left[EI_y \cdot \partial_z^2 U_y\right] = p_y + \partial_z m_x, \tag{7.17}$$

$$\partial_z^2\left[EI_\omega \cdot \partial_z^2 \Theta\right] - \partial_z\left[GI_T \cdot \partial_z \Theta\right] = m_D - \partial_z m_\omega. \tag{7.18}$$

The unknowns in Eqs. (7.15–7.18) are four components of the displacement vector and twist of the beam:

$$U_z(z), U_x(z), U_y(z), \Theta(z).$$

The initially known functions of the coordinate z represent respectively external loads per unit length in the x, y, z directions and the externally applied moments per unit length about x, y, z and external distributed bi-moments:

$$p_x, p_y, p_z, m_x, m_y, m_D, m_\omega. \tag{7.19}$$

The stiffness factors

$$EA, EI_x, EI_y, GI_T, EI_\omega$$

in the Eqs. (7.12–7.15) are also given functions of the coordinate z.

7.4 Applications of Thin-Walled Rods with Semi-open Cross-Sections

7.4.1 Semi-solid Axis with Twist Beam

The principal application of the semi-open profile is the twist-beam for semi-solid automotive suspensions. This axle, also known as torsion-beam axle is frequently used as rear suspension on a wide variety of cars with the front wheel drive (Fig. 7.6). In this suspension, the lateral carrier for a compound steering axle has two longitudinal steering arms and a lateral carrier bonded to them, of a sheet metal in a closed profile (EP2281701, 2009). The center section of the carrier has a smaller cross-section surface than the end sections bonded to the steering arms. The center section of cross-section carrier has at least three double-walled struts closed by outer

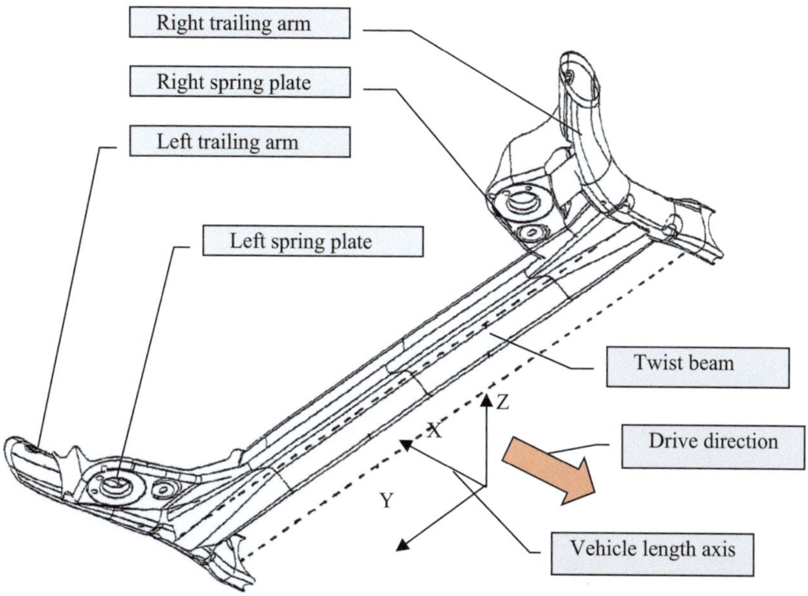

Fig. 7.6 Semi solid axle

bulges, separated from each other by at least partial drawing. The wall thickness of the center section is thinner than at the outer sections.

This suspension is typically described as semi-independent, meaning that the two wheels can move relatively to each other, but their motion is still somewhat interlinked, to a greater degree than in a truly independent rear suspension. The latter can gently compromise the handling of the vehicle. The great advantage of the semi solid axle is its simplicity and maintenance. The known difficulties of the semi-solid axis are the induced effect of suspension stiffness on wheel orientation and oversteer due to deformation of suspension under the action of lateral forces at wheel.

A trail arm of the twist-beam axle is intended to semi-isolate one wheel of a vehicle such as an automobile from the opposite wheel. The elements of the semi-solid axle with twist-beam are shown on Fig. 7.7. The elasticity of twist-beam suspension is based on a large "H" or "C" shaped member, which consists of twist-beam and two trail arms. The front of the "H" is attached to the body via rubber bushings, and the rear of the "H" carries each stub-axle assembly, on each side of the car. The cross beam of the "H" holds the two trailing arms together, and provides the roll stiffness of the suspension by twisting as the two trailing arms move vertically, relatively to each other. Connection points are made to the body mounts, wheel spindles, shocks, and springs, as well as to track bars for lateral stability. Individual component pieces typically making up an axle assembly include the twist beam, trail arms, flanges, spring supports, jounce bumpers, shock mounts, bushing mounts, and torsion bar.

The coil springs usually stand on a spring pad alongside the stub-axle. The shock damper is regularly coaxial with the spring. This location gives them a very high motion ratio compared with most suspensions, which improves their performance and reduces their weight.

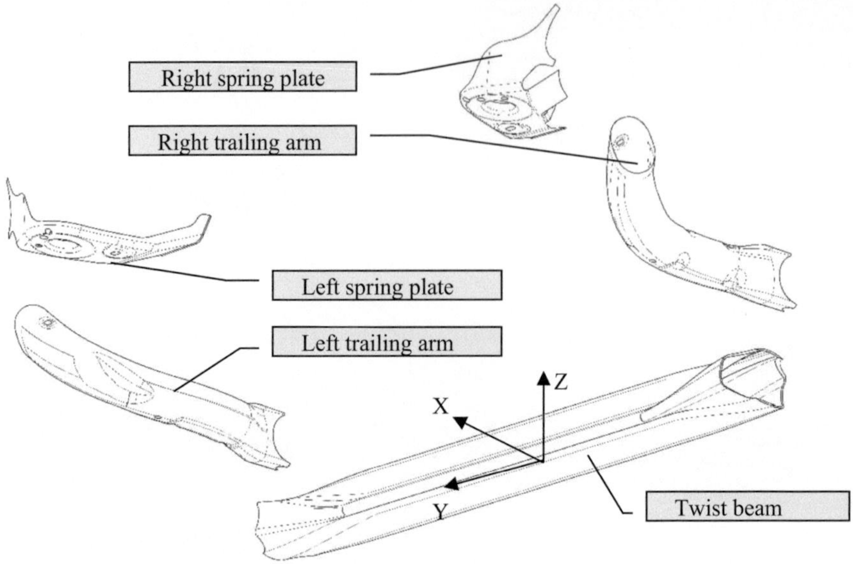

Fig. 7.7 Principal elements of the semi solid axle

7.4.2 Mechanical Models of Twist-Beam Axle

A twist-beam typically comprises a twist-beam for bending loads, coupled with a separate solid torsion rod of different material and characteristics from those of the beam, and extending through the open central portion of the beam for accommodating torsion moment (Reimpell et al., 2001 ; Linnig et al., 2009). The twist-beam is normally a drawn or stamped member, which is often not sufficiently stiff in torsion. Both ends are attached to stubs extending in from the spring seats. The separate rod to control torsion loads is specially fabricated of high strength steel. Its ends are attached separately from, and in a special relationship relative to, the ends of the beam.

The suspension-related kinematic characteristics are of primary importance for vehicle dynamics Gillespie (1992) and Choi et al., (2009). In the present Chapter, we consider only the characteristics of the suspension, relating upon the elastic properties of twist beam. The important characteristics of the axle are the roll stiffness, lateral stiffness, and lateral force camber compliance. The contribution of the stiffness of both trailing arms to the integral stiffness of the axle is usually much higher than the contribution of twist-beam stiffness. For our analysis, we consider the trailing arms stiff and neglect the compliances of the mount bushings. Our aim is to calculate the partial roll stiffness of the axle caused by beam twist. Thus, the contribution of suspension springs on the roll stiffness of the axes is not considered. The total roll stiffness of the axle is just the sum of stiffness due to twist-beam and caused by suspension springs and stabilizer bar.

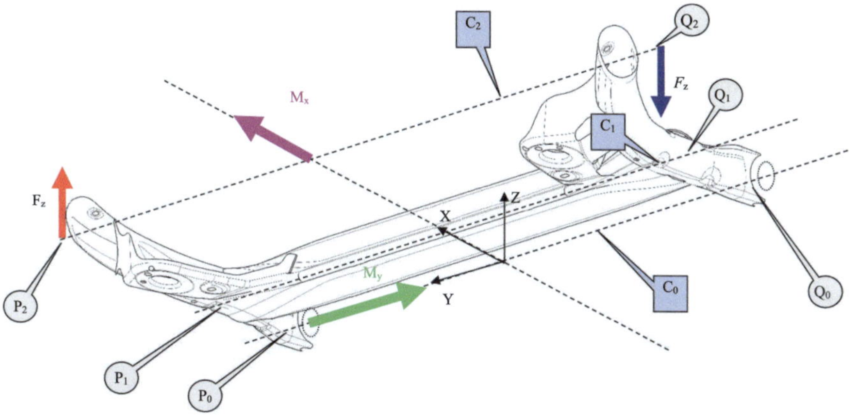

Fig. 7.8 Twist moment of the semi-opened profile M_y and loads on the wheels F_z due to vehicle roll moment M_x

7.5 Elastic Behavior of Twist-Beam Axles Under Load

7.5.1 Loads and Displacements of Twist-Beam Axles

The loads and displacements of the axle relate to the vehicle coordinate system:

- X positive forward; origin at the front axle position (longitudinal direction);
- Z positive up; origin at ground plane (vertical direction);
- Y positive right; centered on symmetry plane (transversal direction).

Consider the twist-beam axle as shown on Fig. 7.8. The twist-beam local axis z is parallel to Y-axis of vehicle coordinate system. The hard points control the static settings and the kinematics of the suspension.

The geometry of the axle is defined by the associated hard points:

The hard points P_0 and Q_0 are the midpoints of rubber bushing mounts and define the mount axis C_0; the points P_1 and Q_1 are located on the shear center axis of the twist beam C_1; the points P_2 and Q_2 are located on the center of the left and right wheels and define the axis C_2. The distance between points P_2 and Q_2 is L_B, the distance between points P_0 and P_2 in the direction of axis x is the effective length of trailing arm L_T.

7.5.2 Roll Stiffness of Twist-Beam Axle

Firstly, we express the roll stiffness of the axle in terms of torsion stiffness of the twist-beam (Fig. 7.8). Let the wheel travel at the points P_2 and Q_2 are u_z and $-u_z$

correspondingly. The associated reaction forces at the points P_2 and Q_2 are F_z and $-F_z$.

The roll moment due to reaction forces is $M_x = 2 \cdot F_z \cdot L_B/2$. The resulting roll angle is:

$$\vartheta_x = \frac{u_z}{L/2}.$$

Consequently, the axle roll stiffness is equal to:

$$r_a = \frac{M_x}{\vartheta_x} = \frac{F_z L^2}{2u_z}.$$

From the other side, for the twist-beam the torsion moment is $M_y \cong F_z L_T$. The twist angle $\vartheta_y = u_z/L_T$. Consequently, the effective torsion stiffness of the twist-beam due to moment and bi-moment is

$$r_t = \frac{M_y}{\vartheta_y} = \frac{F_z L_T^2}{2u_z}. \tag{7.20}$$

The roll stiffness of the axle depends on the effective torsion stiffness of the twist-beam r_t

$$r_a = i_t^2 r_t, \tag{7.21}$$

where

$$i_t = \frac{L_B}{L_T}$$

is the geometrical transmission ratio.

The effective torsion stiffness of the twist-beam r_t of the variable section stiffness could be also easily obtained.

7.5.3 Lateral Stiffness of Twist-Beam Axle

Secondly, we express the lateral stiffness of the axle in terms of bending stiffness of the twist-beam (Fig. 7.9). Let the lateral shift of both wheel spindles under the actions of two equal lateral forces of the magnitude F_y be

$$u_y \cong \theta_z L_T.$$

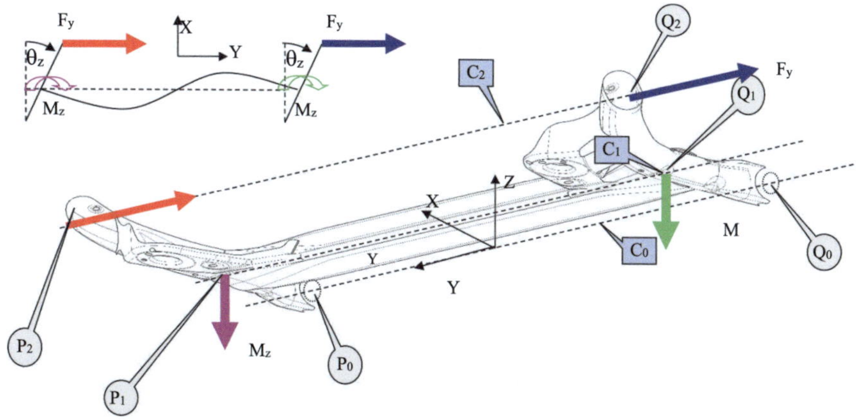

Fig. 7.9 Terminal bending moments of the semi-opened profile M_z due to lateral load F_y in vehicle side direction

The corresponding bending moments on the mount points P_0 and Q_0 are

$$M_z = F_y L_T.$$

The bending moment of the twist-beam can be expressed in terms of its effective bending stiffness r_z:

$$M_z = r_z \theta_z.$$

With this expression the lateral stiffness of the axle is

$$r_l \equiv \frac{F_y}{u_y} = \frac{r_z}{L_T^2}. \tag{7.22}$$

7.5.4 Camber Stiffness of Twist-Beam Axle

Thirdly, the camber stiffness is equal to the effective bending spring rate r_c of twist-beam due to moment in x direction (Fig. 7.10).

$$r_c = \frac{m_x}{\theta_x}. \tag{7.23}$$

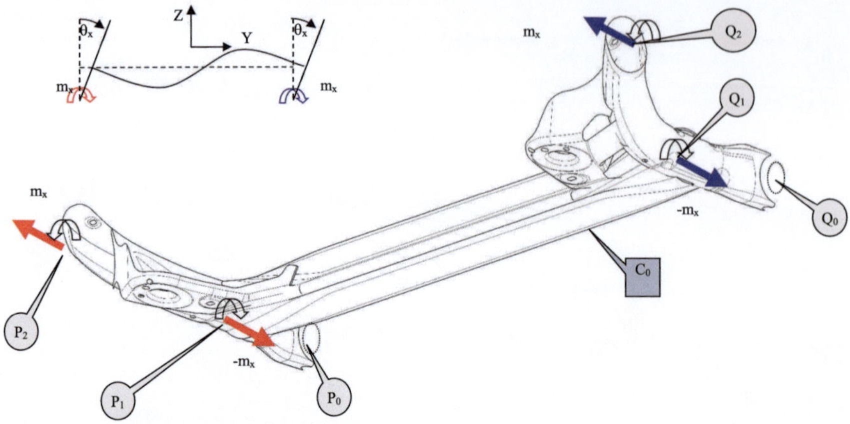

Fig. 7.10 Terminal bending moments of the semi-opened profile m_x due to twist moment on the wheel in vehicle travel direction

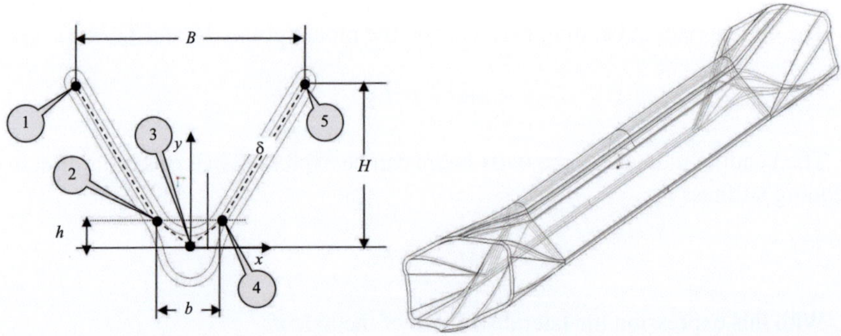

Fig. 7.11 Cross-section of semi-opened twist beam with V1-profile

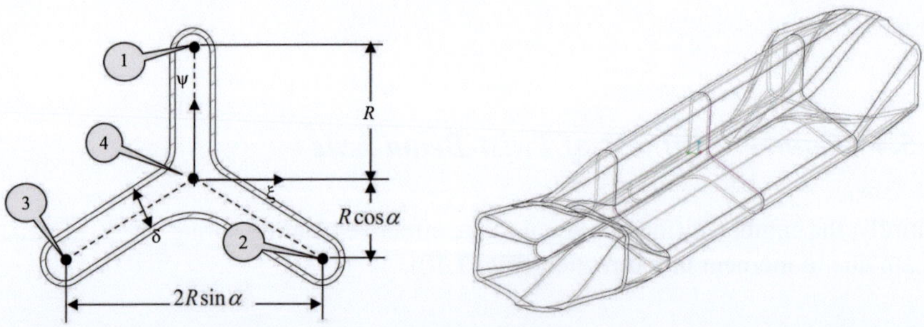

Fig. 7.12 Cross-section of semi-opened twist beam with Y-profile

The stiffness of the rod depends on the sectional shape of the rod and the boundary conditions at the ends.

7.6 Deformation of Semi-open Beam Under Terminal Load

7.6.1 Bending of Semi-open Beam Due to End Moments

Our next aim is to express the effective torsion stiffness r_t, effective bending stiffness r_z and effective bending spring rate r_c of the twist-beam in terms of section properties of the twist-beam with semi-open cross-section.

Consider the semi-open profile beam with the constant material thickness and the identical shape of the cross-section profile along the length of the beam:

$$-L/2 \leq z \leq L/2.$$

Bending of the semi-open profile beam of the length L with simply supported ends $z = -L_B/2$ and $z = L_B/2$, loaded on both ends by the equal terminal moments M_y is given by the formula

$$U_x = \frac{M_y}{12LEI_y}(2z + L_B)(2z - L_B)z.$$

The terminal angles on the supported ends $z = -L_B/2$ and $z = L_B/2$ are

$$\theta_y = \frac{M_y L_B}{6EI_y}.$$

The bending stiffness of the beam, which is defined as angle due to unit terminal moment, reads

$$r_b = \frac{6EI_y}{L_B L}. \tag{7.24}$$

The normal stresses in the direction of z axis in the point with coordinates (x, y) are

$$\sigma_z = \frac{M_y x}{I_y}. \tag{7.25}$$

7.6.2 Torsion Stiffness of Beam with Constant Section Due to Terminal Torques

The equilibrium equation of the rod twisted solely by the torque

$$M_z = M_S + M_H$$

is:

$$\begin{aligned}\partial_z\left[EI_\omega \partial_z^2 \Theta\right] - GI_T \partial_z \Theta &= -M_z, \\ M_z &= -\int_0^z m_D \mathrm{d}z.\end{aligned} \tag{7.26}$$

For the rod with constant stiffness along the span its solution reads:

$$\Theta = a_1 \sinh(\lambda z) + a_2 \cosh(\lambda z) + a_3 - \frac{1}{EI_\omega \lambda} \int_0^z M(z_1) \sinh(\lambda(z - z_1)) \mathrm{d}z_1. \tag{7.27}$$

In Eq. (7.27) a_1, a_2, a_3 are the integration constants. The parameter

$$\lambda_c^2 = \frac{GI_T}{EI_\omega}.$$

defines the characteristic length λ_c^{-1} of the bi-moment influence. If the length λ_c^{-1} much less than the length of the twist beam L_B, such that $\left(\lambda_c^{-1} << L_B\right)$, the influence of the bi-moment on the twist stiffness could be neglected. Otherwise, if the length λ_c^{-1} is comparable to the length of the twist beam L_B, $\left(\lambda_c^{-1} \approx L_B\right)$, the bi-moment stiffens the twist and the influence of bi-moment must be considered for technical purposes.

The unknown constants A_1, A_2, A_3 depend on the boundary conditions on the ends of the rod. For practically important case of the calculation of roll stiffness of the axle, the boundary conditions are:

$$\begin{aligned}\partial_z \Theta|_{z=-L_B/2} &= -\theta, \\ \partial_z \Theta|_{z=L_B/2} &= \theta, \\ \Theta|_{z=0} &= 0.\end{aligned} \tag{7.28}$$

The solution of (7.7) with the conditions (7.9) is

$$\Theta(z) = M_z \frac{-\lambda_c^{-1} \sinh(\lambda_c z) + z \cosh(\lambda_c L_B/2)}{GI_\mathrm{d} \cosh(\lambda_c L_B/2)}. \tag{7.29}$$

The twist angles on the ends of the rod are:

$$\Theta\left(\tfrac{L_B}{2}\right) = -\Theta\left(-\tfrac{L_B}{2}\right) = \left(1 - \frac{\tanh(\lambda_c L_B/2)}{\lambda_c L_B/2}\right) \frac{M_z}{GI_T} \frac{L_B}{2} = \frac{M_z}{GI_T} \frac{L_B}{2\mathrm{K}(\lambda_c L_B)}, \tag{7.30}$$

where

$$\mathrm{K}(\lambda_c L_B) = \left(1 - \frac{\tanh(\lambda_c L_B/2)}{\lambda_c L_B/2}\right)^{-1} > 1$$

is the stiffening factor due to the bi-moment. The stiffness of the twisted rod due to pure torsion stiffness is (Kobelev, 2012):

$$r_t = \frac{M_z}{\Theta(L_B/2)+\Theta(-L_B/2)} = \frac{GI_T}{L_B}\mathrm{K}(\lambda_c L_B). \tag{7.31}$$

The torsion stiffness of the twisted rod without the influence of bi-moment is (Kobelev, 2012):

$$\overline{r}_t = \frac{GI_T}{L_B}. \tag{7.32}$$

The stiffness of the twisted rod is therefore higher that the stiffness without the influence of bi-moment

$$r_t = \mathrm{K}(\lambda_c L_B)\overline{r}_t. \tag{7.33}$$

7.6.3 *Stresses in the Beam with Constant Section Due to Terminal Torques*

The moments due to pure torsion

$$M_H = GI_T \partial_z \Theta,$$

due to constrained torsion

$$M_S = -\partial_z\left[EI_\omega \partial_z^2 \Theta\right]$$

and the bi-moment

$$B_m = EI_\omega \partial_z^2 \Theta$$

correspondingly are

$$M_H(z) = M_z \frac{\cosh(\lambda_c z) - \cosh(\lambda_c L_B/2)}{\cosh(\lambda_c L_B/2)}, \tag{7.34}$$

$$M_S(z) = M_z \frac{\sinh(\lambda_c z)}{\cosh(\lambda_c L_B/2)}, \tag{7.35}$$

$$B_m(z) = M_z \frac{\sinh(\lambda_c z)}{\cosh(\lambda_c L_B/2)}. \tag{7.36}$$

Using the expressions for normal stress, one can obtain the normal stress due to torsion:

$$\sigma_z^{(i)} = \frac{\omega^{(i)}}{I_\omega} B_m = E\omega^{(i)}\partial_z^2\Theta. \tag{7.37}$$

With (7.36) we get from (7.37):

$$\sigma_z^{(i)} = M_z \tfrac{\omega^{(i)}}{I_\omega} \tfrac{\sinh(\lambda_c z)}{\cosh(\lambda_c L_B/2)}. \tag{7.38}$$

Using the equilibrium Eq. (7.11), we get the flow of shear stress

$$F_\tau(z,s) = M_S S_\omega(s)/I_\omega = -\partial_z\left[E S_\omega(s)\partial_z^2\Theta\right].$$

For the wall thickness $T = T(s)$, we calculate from this equation the shear stress τ_S due to bi-moment as

$$\tau_S = \frac{F_\tau}{T} = \frac{S_\omega}{I_\omega T} M_S. \tag{7.39}$$

With (7.39), the shear stress τ_S is

$$\tau_S = M_z \tfrac{S_\omega}{I_\omega T} \tfrac{\cosh(\lambda_c z)}{\cosh(\lambda_c L_B/2)}. \tag{7.40}$$

The shear stress due to pure torsion could be calculated using the BREDT theorem

$$\tau_H = \frac{M_H}{2TA_m} = \frac{GI_\mathrm{d}}{2TA_m}\partial_z\Theta. \tag{7.41}$$

With (7.41) the shear stress due to pure torsion of the twist beam reads

$$\tau_H(z) = \tfrac{M_z}{2TA_m} \tfrac{\cosh(\lambda_c L_B/2) - \cosh(\lambda_c z)}{\cosh(\lambda_c L_B/2)}. \tag{7.42}$$

The total shear stress is (Kobelev, 2012):

$$\tau = \tau_H + \tau_S = \tfrac{M_z}{2TA_m} + \tfrac{M_z}{T}\left(\tfrac{S_\omega}{I_\omega} - \tfrac{1}{2A_m}\right)\tfrac{\cosh(\lambda_c z)}{\cosh(\lambda_c L_B/2)}. \tag{7.43}$$

The area A_m enclosed by the mid-line L_m of the cross-section is

$$A_m = \tfrac{1}{2}\oint_{L_m}\left(-y\tfrac{\mathrm{d}x}{\mathrm{d}s} + x\tfrac{\mathrm{d}y}{\mathrm{d}s}\right)\mathrm{d}s.$$

7.6.4 Equivalent Tensile Stress Due to Simultaneous Bending and Torsion

Normal stress due to bending of the semi-open profile beam, loaded on both ends by the equal terminal moments $\boldsymbol{M} = \{M_x, M_y, M_z\}$ is given by the formula

$$\sigma_z = M_x \frac{y}{I_x} + M_y \frac{x}{I_y} + M_z \frac{\omega^{(i)}}{I_\omega} \frac{\sinh(\lambda_c z)}{\cosh(\lambda_c L_B/2)}. \tag{7.44}$$

Finally, the equivalent tensile stress or von Mises stress for both walls of contour is (Kobelev, 2012):

$$\begin{aligned}
\sigma_{eq}^2 &= \sigma_*^2 + 3\,\tau_*^2, \\
\sigma_* &\overset{\text{def}}{=} M_x \frac{y}{I_x} + M_y \frac{x}{I_y} + M_z \frac{\omega^{(i)}}{I_\omega} \frac{\sinh(\lambda_c z)}{\cosh\left(\frac{\lambda_c L_B}{2}\right)}, \\
\tau_* &\overset{\text{def}}{=} \left[\frac{1}{2TA_m} + \frac{1}{T}\left(\frac{S_\omega}{I_\omega} - \frac{1}{2A_m}\right)\frac{\cosh(\lambda_c z)}{\cosh\left(\frac{\lambda_c L_B}{2}\right)}\right] \cdot M_z\,.
\end{aligned} \tag{7.45}$$

This expression demonstrates that the stress along the span of the beam depends on the span coordinate z.

7.6.5 Stiffness Properties of Semi-open Profiles for Automotive Applications

Various cross-sectional configurations have been suggested for twist-beam and torsion bar elements. The calculation of torsion and bending stiffness of the semi-open requires the calculation of sectorial areas and the corresponding integrals. Several types of semi-open profiles are important for automotive applications (Fig. 7.11 and 7.12). The corresponding formulas are given in Table 7.1.

If the constant material thickness T and constant distance between walls δ over the length of the midline the following relations

$$I_\text{d} = 4A_m^2 T L_m^{-1}, \quad A_m = L_0\delta, \quad A = L_m T, \quad L_m = 2L_0$$

are valid for all types of cross-section.

7.6.6 Semi-open Beams with Variable Cross-Sections

In this chapter, we have initially assumed that the cross-section of the beam remains constant along the span of the beam. However, the majority of the beam sections for the semi-solid axles have a variable cross-section along the beam length. The cross-sections at the ends of the beam have higher local torsional and bending stiffness compared to the cross-sections in the middle regions of the beam. The purpose of increasing the torsional and bending stiffness at the ends is to localize the deformation in the central regions of the beam. Due to the higher torsional and bending stiffness in the end regions, the bending deformation and twisting rate are reduced. The reduction in bending deformation and twist rate results in a reduction in tensile and shear stresses in the end regions. As a result, the weld groove between the twist beam and the trailing beam is subjected to less stress and its fatigue life is significantly increased. To account for the variable cross-section, Eqs. (7.15–7.18) and the corresponding stress formulas must be applied. These equations are valid for any variable cross-section. For design purposes and for estimating spring rates, averaging the reciprocal stiffness over the span gives acceptable results. Precise stress and deformation analysis in the final design phase should be performed using commercial modeling software and finite element codes.

Conclusions

A special type of thin-walled rods with semi-open cross section is proposed and optimized. This class of thin-walled beam-like structures is characterized by a closed but flattened profile. The unusual shape of semi-open thin-walled beams allows efficient optimization due to the large variability of shapes. An intermediate class of thin-walled beam sections is studied. The cross-section of the beam is closed, but the shape of the cross-section is elongated and curved. The walls forming the section are nearly equidistant. The free torsional behavior of the beam according to StVenant is similar to the behavior of beams with a closed cross-section. A technical theory of thin-walled beams with semi-open profile can be based on the following set of hypotheses of kinematic and static nature. The pure torsional behavior of such profiles is in some aspects similar to the mechanical behavior of thin-walled beams with closed cross-sections, while the warping function and the restricted torsional behavior of these beams are analogous to those of open cross-sections.

The application of the semi-open profile is the twist beam for semi-solid automotive suspensions. Twist-beam axles are used not only in compact cars, but more recently also in midsize cars, sport utility vehicles and all-wheel-drive vehicles. The low cost, compact packaging, reduced weight and acceptable axle kinematics often make it a competitive solution. Another advantage of the semi-solid axle is its simplicity and ease of maintenance. The disadvantages of the semi-solid axle are the induced effects of suspension stiffness on wheel alignment and oversteer due to suspension deformation. The longitudinal position of the torque beam controls important parameters of suspension behavior, such as the roll-steer curve and toe and camber compliance. The closer the crossmember is to the axle stubs, the more the camber and toe will change under deflection. A key difference between the camber and toe changes of a twist-beam suspension versus an independent suspension is that the camber and toe changes depend on the position of the other wheel, not the vehicle's chassis. In a traditional independent suspension, camber and toe are based on the position of the wheel relative to the body. If both wheels compress together, their camber and toe will not change. So, if both wheels start off perpendicular to the road and the car compresses together, they will stay perpendicular to the road. The camber and toe changes are the result of one wheel compressing relative to the other. In addition, we analytically express the effective torsional stiffness and effective bending stiffness of the twist beam in terms of the cross-sectional properties of the twist beam with half-open cross-section.

In this chapter, an intermediate class of thin-walled beam sections is studied. The section of the beam is closed, but the shape of the section is elongated and curved in its plane. The walls forming the section are nearly equidistant. StVenant's free torsional behavior of the beam is the same as the behavior of beams with a closed cross-section. A technical theory of thin-walled beams with semi-open section can be based on the following set of hypotheses of kinematic and static nature. The pure torsional behavior of such profiles is in some aspects analogous to the mechanical behavior of thin-walled beams with closed cross-sections, while the warping function and the restricted torsional behavior of these beams are analogous to those of open cross-sections.

The application of the semi-open profile is the twist beam for semi-solid automotive suspensions. The advantage of the semi-solid axle is its simplicity and maintenance. The disadvantages of the semi-solid axle are the induced effects of suspension stiffness on wheel orientation and oversteer due to suspension deformation. The longitudinal position of the torque beam controls important parameters of suspension behavior, such as the roll-steer curve and toe and camber compliance. The closer the crossmember is to the axle stubs, the more camber and toe change under deflection.

The twist beam and conventional independent suspension behave differently under load. In fact, the changes in camber and toe under side load are different for these types of suspensions. The changes in camber and toe depend on the relative position of the wheels.

In a conventional independent suspension, camber and toe are based on the position of the wheel in relation to the body. When both wheels compress together, their camber and toe do not change. So, if both wheels start perpendicular to the road and the car compresses them together, the wheels will remain perpendicular to the road. The camber and toe changes are the result of one wheel compressing relative to the other.

The twist-beam suspension behaves in a more complicated manner. The suspension changes the steering characteristics of the vehicle. The standard terminology used to describe understeer and oversteer is defined by the Society of Automotive Engineers (SAE, 2008) and the International Organization for Standardization (ISO, 2010). In these terms, understeer and oversteer are based on differences in steady-state conditions where the vehicle is following a constant-radius path at a constant speed with a constant steering wheel angle on a flat and level surface. The lateral load on the twist-beam suspension causes oversteer due to its low stiffness. The oversteering effect could be significant if the bending stiffness r_z of the twist-beam is not high enough. However, increasing the bending stiffness leads to high torsional stiffness r_t and roll stiffness of the vehicle. The increasing roll stiffness changes the vehicle dynamics and is generally disadvantageous.

Summary of Principal ResultsSummary of Principal Results

- The semi-open cross sections are flattened tubes.
- The flattening could be easily adjusted for optimal design.
- The analytical model estimates the main static effects of semi-open tubes,
- The main application of semi-open tubes is the twist beam axle,
- For the twist-beam axle the effective torsion stiffness r_t, effective bending stiffness r_z and effective bending spring rate r_c are derived in terms of section properties of tube.

References

Bazant, Z. P., & Cedolin, L. (1991). *Stability of structures.* Oxford University Press.

Boresi, A. P., & Schmidt, R. J. (2003). *Advance mechanics of materials* (6th ed.). Wiley and Sons.

Cherniy, V. P. (2001). Effect of curved bar properties on bending of curved pipes. *Transactions of the ASME, Journal of Applied Mechanics.* 5.1115/1.1357518,68

Chilver, A. H. (1967). *Thin-walled structures.* Wiley & Sons Inc.

Choi, B. L., et al. (2009). Torsion beam axle system design with a multidisciplinary approach. *International Journal of Automotive Technology, 10*, 1.

Clark, R. A., & Reissner, E. (1951). *Bending of curved tubes, advances in applied mechanics* (Vol. II, pp. 93–122). Academic Press.

Flügge, W., & Marguerre, K. (1950). Wölbkräfte in dünnwandigen Profilstäben. *Ing.-Arch., 18*, 23–38.

Gillespie, T. (1992). *Fundamentals of vehicle dynamics*, SAE.

Graße, W. (1965). Wölbkrafttorsion dünnwandiger prismatischer Stäbe beliebigen Querschnitts. *Ing.-Arch., 24*, 330–338.

Heilig, R. (1961). Beitrag zur Theorie der Kastenträger beliebiger Querschnittsform. *Der Stahlbau, 30*, 333–349.

ISO. (2010). International Organization for Standardization, Road vehicles—Vehicle dynamics and road-holding ability—Vocabulary, ISO Standard 8855, Rev. 2010.

Karman, Th. (1911). Über die Formänderung dünnwandiger Rohre, insbesondere federnder Ausgleichrohre. *Z. Ver. Deut. Ing., 55*, 1889–1895.

Kobelev, V. (2012). Thin-walled rods with semi-open profile for semi-solid automotive suspension. *International Journal of Automotive Technology, 13*(2), 231–245.

Kobelev, V., & Larichev, A. D. (1988). Model of thin-walled anisotropic rods. *Mechanics of Composite Materials, 24*, 97–104.

Kobelev, V. (2013). Thin-walled rods with semiopened profiles. *ASME Journal of Applied Mechanics, 80.* 5.1115/1.4006935

Kollar, L. P., & Pluzsik, A. (2002). Analysis of thin-walled composite beams with arbitrary layup. *Journal of Reinforced Plastics and Composites, 21*, 1423.

Laudiero, F., & Savoia, M. (1990). Shear strain effects in flexure and torsion of thin-walled beams with open or closed cross-section. *Journal Thin-Walled Structures, 10*, 87–119.

Librescu, L., & Song, O. (2006). In *Thin-walled composite beams, theory and application*. Springer.
Linnig, W. et al. (2009). The Twist-beam Rear Axle—Design, Materials, Processes and Concepts, ATZ worldwide eMagazines Edition (pp. 2).
Prokic, A. (2003). Stiffness method of thin-walled beams with closed cross-section. *Computers and Structures, 81*, 39–51.
Reimpell, J., Stoll, H., & Betzler, W. (2001). *The automotive chassis: Engineering principles*. SAE.
Reissner, E. (1946). Analysis of shear lag in box beams by the principle of minimum potential energy. *Quarterly of Applied Mathematics, 4*, 268–278.
SAE. (2008). *Vehicle dynamics terminology*, SAE Standard J670, Rev. 2008-01-24.
Salmon, C. G., & Johnson, J. E. (1996). *Steel structures* (4th ed.). HarperCollins College Publishers.
Siev, A. (1966). Torsion in closed sections. *Engineering Journal, AISC, 3*(1), 46–54.
Slivker, V. (2007). In *Mechanics of structural elements, theory and applications*. Springer.
Tamberg, K. G., & Mikluchin, P. T. (1973). *Torsional phenomena analysis and concrete structure design, analysis of structural systems for torsion*, SP-35, (pp. 1–102) American Concrete Institute.
Timoshenko, S., & Gere, J. M. (1961). *Theory of elastic stability*. McGraw-Hill.
Timoshenko, S. (1945). Theory of bending, torsion, and buckling of thin-walled members of open cross-section. *Journal Franklin Institute, 239*, 201–219, 249–268, 343–361.
Ventsel, E., & Krauthammer, T. (2001). *Thin plates and shells, theory, analysis, and applications*, Marcel Dekker AG.
Vlasov, V. Z. (1961). Thin-Walled Elastic Beams, Office of Technical Services, U.S. Department of Commerce, Washington 25, DC, TT-61–11400.
Wang, Q. (1999). Effect of shear lag on buckling of thin-walled members with any cross-section. *Communications in Numerical Methods in Engineering, 15*, 263–272.
Whatham, J. F. (1981) Thin shell equations for circular pipe bends. *Nuclear Engineering and Design, 65*(1), 77–89, ISSN 0029-5493, https://doi.org/10.1016/0029-5493(81)90121-7.
Whatham, J. F. (1982) Thin shell analysis of non-circular pipe bends. *Nuclear Engineering and Design, 67*(2), 287–296, ISSN 0029-5493, https://doi.org/10.1016/0029-5493(82)90147-9.

Part II
Manufacturing of Springs

Chapter 8
Coiling of Helical Springs

Abstract In this chapter, the method for calculation of residual stresses and plastic bending and torsion moments for combined bending and torsion loading is developed. The Bernoulli hypothesis is assumed for the deformation of the beam. The analysis is performed using the deformation theory of plasticity with a nonlinear stress–strain law describing active plastic deformation (SAE AE-22, §8.3.33). The curvature and twist of the beam increase proportionally during plastic loading so that the ratio of curvature to twist remains constant. The complete solutions based on this approximate material law provide a closed analytical solution. The unloading is linear elastic. This chapter is the first section of the second part, which studies mechanical problems arising during the manufacturing of helical springs.

Keywords Coiling of springs · Residual stresses in spring wires · Plastic deformation of spring wire

8.1 Elastic–Plastic Bending and Torsion of Wire

The elastic–plastic problem of combined bending and torsion moments of a straight prismatic bar, made of Levy–Mises material and loaded by a terminal bending couple about the axis of symmetry of the cross-section and a twisting couple, was originally considered in Handelman (1944).

Moment–angle relations are reported in Hill and Siebel (1953) for steel bars of circular section plastically strained by combined bending and couples in constant ratio. The bending and torque approach the theoretical values calculated for the fully plastic state of a plastic-rigid material. Good estimates of the latter values are obtained by bracketing between upper and lower approximate values. A general relation is proposed between the fully plastic values of bending moment, torque, and axial force

V. Kobelev, *Fundamentals of Springs Mechanics*, University of Siegen, Faculty IV: School of Science and Technology. Paul-Bonatz-Str. 9–11, D-57076 Siegen. E-mail: Homepage: https://www.mb.uni-siegen.de/kobelev/

V. Kobelev, *Fundamentals of Springs Mechanics*,
https://doi.org/10.1007/978-3-031-58584-5_8

when all three are applied together. This relation applies for a wide variety of sections and is suitable for plastic limit design.

A long prismatic member is acted on by combinations of bending moments and torques of such a magnitude as to render the member just fully plastic was discussed in Steele (1954). The citing paper takes Handelman's (1944) equation and solves it numerically for a square section member. The moments and torques computed from the numerical solution are compared with a bounded solution due to (Hill & Siebel, 1953). Finally, the stress distributions (shear and bending) are given in a member for two critical combinations of moment and torque.

Combinations of twist and extension of a solid circular cylinder are considered in Gaydon (1952). The Reuss equations are used throughout and these are integrated, for different cases, to give the shear stress and tension in the plastic material. It is shown that the stresses rapidly approach their asymptotic values. A more general case, in which the torsion and extension are such as to make the ratio of axial load to torque constant, is solved numerically. Finally, the residual stresses are evaluated, after partial unloading, for a bar, which has been twisted and extended in constant ratio.

Upper and lower approximations are obtained in Gaydon and Nuttal (1957) to the interaction curve of the bending and twisting couples at yield for the combined bending and twisting of cylinders of ideally plastic-rigid material. Rectangular, I-sections and box-sections are dealt with in detail. For the box section, a comparison is made with the thin tube theory (Hill & Siebel, 1953).

In the paper (Sankaranarayanan & Hodge, 1958), yield criteria expressed in terms of stress resultants are obtained for typical engineering structures, using the first fundamental theorem of limit analysis. These yield criteria, which are often non-linear, are then replaced by inscribed piecewise linear approximations. The complete solutions based on these approximate yield criteria provide lower bounds.

The paper (Imegwu, 1960) deals with the plastic flexure and torsion of prismatic beams loaded by terminal bending and torsion moments. Both moments, acting together, cause full plastic flow. The material is assumed to behave according to the Tresca-Levy–Mises Hypotheses, and in non-hardening and rigid plastic. The results were obtained by numerical solution of the second order non-linear differential equation for a Levy–Mises material. The relationships obtained were found to give points lying virtually on a single interaction curve plotted with non-dimensional coordinates.

The elastic–plastic problem of combined bending and torsion is treated analytically in Ishikawa (1973) for an incompressible isotropic work-hardening material obeying a nonlinear stress strain law. Evolving a theory to satisfy the equilibrium and compatibility condition, the basic nonlinear differential equation in the ordinal Cartesian coordinate system can be linearized, adopting the new parameter in the stress space. The Ramberg–Osgood's law is employed as a nonlinear stress strain relation. The stress components, the bending moment and torque can be evaluated by the numerical calculation.

The paper (Ali, 2005) deals with two aspects of work to examine the elastic–plastic behavior of pre-loaded circular rods subjected to subsequently applied torque within the plastic region. The uniform diameter and reduced section rods of mild steel, fitted with strain gauges, were subjected to initial axial yield loads using a custom-built torque-tension machine. The specimens were gradually twisted, holding the initial axial displacements constant. Then the measurements of the resulting torque and load were recorded using appropriate load cells as well as by the fitted strain gauges. When the axial displacements of the preloaded rods were held constant, the strain gauges readings increase rapidly with the decrease in the initially applied axial loads. The experimental results have been compared with reported in Brooks (1969) theoretical predictions.

An analytical model, which was developed in Baragetti (2006), provides the designer the expressions for the final shape of a wire. The model allows the evaluation with a higher level of accuracy the end shape of wires having different cross-sections after nonlinear bending. The Bernoulli's hypothesis was assumed in the cited paper, such that the model can be used in all the applications where the material behavior of the wire guarantees that plane cross-sections of the wire will remain plane after rotation due to bending.

The residual stresses due to plastic pre-setting the surface of a solid bar were studied in Močilnik et al. (2015).

The results of the elastic–plastic analysis for combined loading were summarized in reference works (Życzkowski, 1967, 1981).

8.2 Modified Ramberg–Osgood's Law

We analyze the plastic loading using deformational theory of plasticity. The deformational theory of plasticity represents active plastic deformation using a nonlinear stress strain law (SAE AE-22, §8.3.33), (Jones, 2009). The curvature and angle of twist pro length unit of the bar during the plastic loading increase proportionally, such that the ratio curvature to twist remains constant.

In this chapter, we use the modified Ramberg–Osgood's law. that is adopted for the analytical calculation of the elastic–plastic problem. For a compressible isotropic work-hardening material, without the distinct yielding point, as observed in the behavior of deformation, the following explicit stress strain expression can be used.:

$$\boldsymbol{s} = 2G_p(\gamma_{oct})\boldsymbol{e}, \tag{8.1}$$

$$3K\varepsilon_{oct} = \sigma_{oct}. \tag{8.2}$$

In (8.1) the tensors $\boldsymbol{s}$ and $\boldsymbol{e}$ are the deviators of stress $\boldsymbol{\sigma}$ and strain tensors $\boldsymbol{\varepsilon}$ respectively. The octahedral normal strain ε_{oct} and octahedral normal stress σ_{oct} are denoted as:

$$3\varepsilon_{oct} \stackrel{\text{def}}{=} \varepsilon_{xx} + \varepsilon_{yy} + \varepsilon_{zz}, \quad 3\sigma_{oct} \stackrel{\text{def}}{=} \sigma_{xx} + \sigma_{yy} + \sigma_{zz}. \tag{8.3}$$

The octahedral shear strain in terms of the second invariant of the deviatoric strain $\mathbf{II}(\boldsymbol{e})$ reads (Appendix A):

$$\gamma_{oct}/2 \stackrel{\text{def}}{=} \sqrt{2\mathbf{II}(\boldsymbol{e})/3}. \tag{8.4}$$

The octahedral shear stress is the double square root of the second invariant of the deviatoric stress $\mathbf{II}(\boldsymbol{s})$:

$$\tau_{oct} \stackrel{\text{def}}{=} \sqrt{2\mathbf{II}(\boldsymbol{s})/3}. \tag{8.5}$$

From the Eq. (8.1) follows the relation between the octahedral shear stress and octahedral shear strain:

$$\tau_{oct} = G_p(\gamma_{oct})\gamma_{oct}. \tag{8.6}$$

The secant modulus G_p is the function of octahedral shear strain γ_{oct}. For linear elastic medium the Eq. (8.6) reduces to

$$\tau_{oct} = G\gamma_{oct}.$$

where G is the elastic shear modulus. The components of deformation tensor increase in the case under consideration proportionally to a single parameter, such that the relation between them remains constant during the deformation history. Under this condition, the deformation theory of plasticity proved to be applicable. In the deformation theory of plasticity, the active plastic deformation the empirical stress–strain relation (8.6) describes fully the plastic deformation law (SAE AE-22, §8.3.33).

In the present Chapter, the following dependence of secant modulus upon the octahedral strain γ_{oct} is used (Kobelev, 2011):

$$G_p(\gamma_{oct}) = \frac{G}{\left(1+(\gamma_{oct}/\gamma_w)^2\right)^k}. \tag{8.7}$$

Particularly, in the case $k = 1/2$ the expression (8.7) for secant modulus reads:

$$G_p(\gamma_{oct}) = \frac{G}{\sqrt{1+(\gamma_{oct}/\gamma_w)^2}}. \tag{8.8}$$

The limit value of measure of stress for $\gamma_{oct} \to \infty$ and for $k = 1/2$ is $\tau_w = G\gamma_w$. For negative values of k, the secant modulus increases with the increasing strain. For positive values of k the secant modulus decreases, as it usually does for common metals and alloys. For $k = 1$, at the limit $\gamma_{oct} \to \infty$ the stress vanishes.

If the stress is reduced, elastic unloading occurs. The elastic unloading of material is characterized by the elastic shear modulus G and Young modulus E.

Essentially the same material model was used in Neuber (1961a, 1961b) for the solution of notch problems.

8.3 Plastic Deformation of Wire During Coiling

A straight cylindrical solid bar with circular cross-section of length L is loaded from a stress-free state by terminal bending moment M_B and the torque M_T. The end sections are $z = 0$ and $z = L$ (Fig. 8.1). The origin of coordinates is chosen at the centroid of area of one cross-section..The polar coordinates in the cross-section of the wire (ρ, ϕ) relate to Cartesian coordinates:

$$x = \rho \sin \phi, \quad y = \rho \cos \phi, \quad 0 < \rho < r, \quad 0 < \phi \leq 2\pi, \quad \rho = \sqrt{x^2 + y^2}.$$

The distribution of stress due to above combined loading is independent of the variable z. Let the curvature of the axis of cylinder in pure bending is κ. In this article the Bernoulli's hypothesis is assumed, such that plane cross-sections of the wire remain plane after rotation due to bending. This means that tensile strain is linearly linked to the distance from the neutral axis:

$$\varepsilon_{zz} = \kappa x.$$

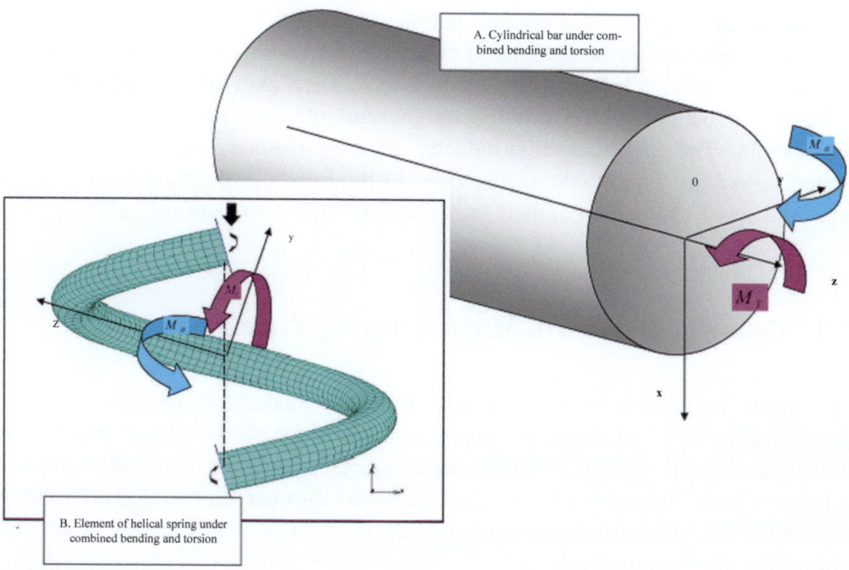

Fig. 8.1 Cylindrical bar under combined bending and torsion

where $R = 1/\kappa$ is radius of curvature of the bar during bending. The shear strain in pure torsion of cylindrical bar with circular cross-section of radius r is:

$$\gamma_{xz} = -\theta y, \ \gamma_{yz} = \theta x.$$

where θ is the angle of twist per unit length of the bar

In polar coordinates the axial and shear deformations in the points of cross-section for the circular rod are respectively:

$$\varepsilon_{zz}(\rho, \phi) = \kappa\rho \sin\phi, \tag{8.9}$$

$$\gamma_{z\varphi}(\rho, \phi) = 2\varepsilon_{z\varphi} = \theta\rho. \tag{8.10}$$

Maximal axial strain

$$\varepsilon_P = \varepsilon_{zz}(r, \pi/2) \equiv r\kappa$$

and the maximal shear strain

$$\gamma_P = \gamma_{z\varphi}(r, \phi) \equiv \theta r$$

attain on the contour of the circular cross-section. Apparently, the axial strain on the contour ε of the cross-section with radius r depends linearly on curvature of the bar during bending κ. The shear on the contour of the cross-section γ is the linear function of the angle of twist per unit length of the bar θ.

From the viewpoint of dislocation character, the coiling process creates during plastic deformation edge and screw dislocations. Their axes are parallel to the axis of helix. The bending of the initially straight wire into a flat torus induces the edge dislocations. The pitch formation induces the screw dislocations. Both dislocation systems overlaps. Finally, the dislocations in the free helical spring have a mixed edge and screw character, but their axes coincide with the axis of the helix.

8.4 Behavior of Wire in Manufacturing Process

The stress distribution in most structural members loaded into the elastic–plastic range is difficult to determine, because the shape of the elastic–plastic interface is itself related to the stress distribution and is, therefore, unknown until the complete solution is found. However, for a solid circular rod subjected to simultaneous bending and torsion, this restriction is removed since the shape of the interface must be annular to preserve axial symmetry.

The non-vanishing stress components for a solid circular rod subjected to combined tension and torsion are $\tau_{z\varphi}$ and σ_{zz}. Substitution of these stress components in the Eqs. (8.1)–(8.3) delivers the following expressions for strains

$$\varepsilon_{rr} = \varepsilon_{\phi\phi} = -\nu_p \varepsilon_{zz}, \qquad \varepsilon_{zz} = \sigma_{zz}/E_p.$$

Both Poisson ratio and Young modulus depend on G_p:

$$\nu_p = \frac{3K - 2G_p}{2(3K + G_p)}, \quad E_p = \frac{3KG_p}{3K + G_p}.$$

The strain and stress tensors for the rod subjected to combined tension and torsion in polar coordinates are

$$\boldsymbol{\varepsilon} = \begin{bmatrix} -\nu_p \varepsilon_{zz} & 0 & 0 \\ 0 & -\nu_p \varepsilon_{zz} & \gamma_{z\varphi}/2 \\ 0 & \gamma_{z\varphi}/2 & \varepsilon_{zz} \end{bmatrix}, \quad \boldsymbol{\sigma} = \begin{bmatrix} 0 & 0 & 0 \\ 0 & 0 & G_p \gamma_{z\varphi} \\ 0 & G_p \gamma_{z\varphi} & E_p \varepsilon_{zz} \end{bmatrix}.$$

The considerable simplification of the solution could be attained with the assumption of incompressibility of the material during the plastic deformation $E_p = 3G_p$, $\nu_p = 1/2$.

The stress tensor reads

$$\boldsymbol{\sigma} = \begin{bmatrix} 0 & 0 & 0 \\ 0 & 0 & G_p(\gamma_{oct})\gamma_{z\varphi} \\ 0 & G_p(\gamma_{oct})\gamma_{z\varphi} & 3G_p(\gamma_{oct})\varepsilon_{zz} \end{bmatrix}. \tag{8.11}$$

The measure of shear strain Γ reduces to

$$\gamma_{oct}^2 = 3\varepsilon_{zz}^2 + \gamma_{z\varphi}^2.$$

Substitution of (8.7) in (8.11) delivers the explicit expressions for stress components in terms of strains:

$$\sigma_{zz} = \frac{E\varepsilon_{zz}}{\left(1 + \frac{3\varepsilon_{zz}^2 + \gamma_{z\varphi}^2}{\gamma_w^2}\right)^k}, \quad \tau_{z\varphi} = \frac{G\gamma_{z\varphi}}{\left(1 + \frac{3\varepsilon_{zz}^2 + \gamma_{z\varphi}^2}{\gamma_w^2}\right)^k}. \tag{8.12}$$

The stresses components in each point (r, ϕ) of the cross-section could be calculated using Eqs. (8.9), (8.10) and (8.12).

The curvature κ and twist θ of the rod increase proportionally to a single parameter, such that their ratio keeps constant during the plastic deformation. The bending and torque moments applied to the end sections of the rod are

$$M_B = \int_0^{2\pi} \left[\int_0^r \sigma_{zz}(\rho, \phi)\rho^2 \sin\phi \mathrm{d}\rho \right] \mathrm{d}\phi,$$

$$M_T = \int_0^{2\pi} \left[\int_0^r \tau_{z\phi}(\rho, \phi)\rho^2 \mathrm{d}\rho \right] \mathrm{d}\phi. \tag{8.13}$$

Using the expression (8.12) for stresses in the cross-section, we can calculate the terminal moments in terms of strains ε_{zz}, $\gamma_{z\varphi}$. Further, with formulas for strains (8.9) and (8.10), the bending and torque moments are derived as the explicit functions of curvature κ and angle of twist per unit length θ. The integrals (8.13) for bending moment M_B and for torque M_T could be explicitly evaluated. The dimensionless functions $P(\lambda, \mu)$ and $Q(\lambda, \mu)$ are listed below for the most interesting case $k = 1/2$:

$$M_B(\kappa, \theta) = \tfrac{4}{3} r^4 G\eta P(\lambda, \mu), \quad M_T(\kappa, \theta) = \tfrac{4}{3} r^4 G\eta Q(\lambda, \mu). \tag{8.14}$$

The dimensionless parameters λ, μ depend on curvature κ and angle of twist per unit length θ:

$$\lambda^2 = \tfrac{3\kappa^2}{3\kappa^2+\theta^2+\eta^2}, \quad \mu^2 = \tfrac{3\kappa^2}{3\kappa^2+\theta^2}.$$

Parameter $\eta = \gamma_w / r$ has the same dimension as θ and κ, namely of an inverse length.

The dimensionless function $P(\lambda, \mu)$ for this case reads

$$\begin{aligned}
P(\lambda, \mu) &= p_K \boldsymbol{K}(\lambda) + p_E \boldsymbol{E}(\lambda) + p_\Pi \boldsymbol{\Pi}\left(\mu^2, \lambda\right) + p_0, \\
p_\Pi &= -\frac{\sqrt{3}\left(\mu^2 - \lambda^2\right)^2}{\lambda^3}, \quad p_E = \frac{\mu^2\sqrt{3}}{\lambda^3}, \\
p_\mathrm{K} &= -\frac{\sqrt{3}\left(\mu^2 - \lambda^4\right)}{\lambda^3}, \quad p_0 = \frac{\pi\mu\sqrt{3}\left(\mu^2 - \lambda^2\right)^{3/2}}{2\lambda^3\sqrt{1-\mu^2}}.
\end{aligned}$$

Here

$$\boldsymbol{K}(k) = \int_0^1 \frac{1}{\sqrt{\left(1 - k^2t^2\right)\left(1 - t^2\right)}} \mathrm{d}t$$

is the complete elliptic integrals of the first kind,(Appendix D)

$$\boldsymbol{E}(k) = \int_0^1 \sqrt{\frac{1 - k^2t^2}{1 - t^2}} \mathrm{d}t$$

is complete elliptic integrals of the second kind,

$$\boldsymbol{\Pi}(\kappa, k) = \int_0^1 \frac{1}{\left(1 - \kappa t^2\right)\sqrt{\left(1 - k^2t^2\right)\left(1 - t^2\right)}} \mathrm{d}t$$

is the complete elliptic integrals of the third kind.

The expression for the dimensionless function $Q(\lambda, \mu)$ is

$$Q(\lambda, \mu) = q_K \boldsymbol{K}(\lambda) + q_E \boldsymbol{E}(\lambda) + q_\Pi \boldsymbol{\Pi}\left(\mu^2, \lambda\right) + q_0,$$
$$q_K = \frac{2\lambda^4 - \lambda^4\mu^2 - 2\lambda^2\mu^2 + \mu^4}{\mu\lambda^3\sqrt{1-\mu^2}}, \quad q_E = -\frac{\mu\left(\mu^2-\lambda^2\right)}{\lambda^3\sqrt{1-\mu^2}},$$
$$q_\Pi = \frac{\left(\mu^2-2\right)\left(\mu^2-\lambda^2\right)^2}{\lambda^3\mu\sqrt{1-\mu^2}}, \quad q_0 = \pi\frac{2\lambda^2-2\mu^2-\lambda^2\mu^2+\mu^4}{2\lambda^3\left(\mu^2-1\right)}\sqrt{\mu^2-\lambda^2}.$$

For the practically significant cases $1/2$, $-1/2$, -1 and -2 of the parameter k the integrals (8.13) are expressed in analytical form Kobelev (2011). The analytical expressions could be analogously found also for integer cases k.

The major advantage of the analytical solutions is that the final formulas could be used for industrial applications and programming of manufacturing machines. These expressions play the fundamental role for the subsequent analysis. With this method we express analytically the bending moment $M_B(\kappa, \vartheta)$ and torque $M_T(\kappa, \vartheta)$ as the functions of curvature of the bar during bending κ and the angle of twist per unit length ϑ. Otherwise, if the moments m_B, m_T are predefined, the κ and ϑ could be determined solving nonlinear algebraic equations

$$M_B(\kappa, \vartheta) = m_B, \quad M_T(\kappa, \vartheta) = m_T.$$

8.5 Elastic Spring-Back and Appearance of Residual Stresses

The simplest definition of residual stresses is as follows: stresses that remain within a part after it has been deformed and all external forces have been removed. More specifically, the deformation must be non-uniform across the material cross-section to give rise to residual stresses. The deformation can result not only from forming operations but also from thermal processes. Phase transformations during heat-treating are known to induce sufficient strain to result in plastic deformation, thereby giving rise to residual stresses. In this chapter, we determine analytically the residual stress, which appear in the cylindrical rod after simultaneous plastic torsion and bending.

One of the principal foundations of mathematical theory of conventional plasticity for rate-independent metals is that there exists a well-defined yield surface in stress space for any material point under deformation. A material point can undergo further plastic deformation if the applied stresses are beyond current yield surface, which is generally referred as plastic loading. If the applied stress state falls within or on the yield surface, the metal will deform elastically only. The unloading of the metal is in this case elastic.

Although it has been always recognized throughout the history of development of plasticity theory that there is indeed inelastic deformation accompanying elastic unloading, which leads to metal's hysteretic behavior, its effects are usually negligible and are ignored in the mathematical treatment.

For a bar, which has been twisted and bended in constant ratio, the residual stresses after unloading are evaluated below. In the state of complete elastic unloading

$$\overline{R} = 1/\overline{k}$$

is the unloaded bending radius and $\overline{\vartheta}$ the angle of twist per unit length of the bar respectively. During the elastic unloading, the decrements of maximal axial strain and shear are

$$\Delta\varepsilon = (\kappa - \overline{\kappa}) \cdot r, \quad \Delta\gamma = \left(\vartheta - \overline{\vartheta}\right) \cdot r.$$

In the final state of complete elastic unloading the bending moment and torque disappear. The condition that bending moment and torque vanish are expressed as follows

$$M_B(\kappa, \theta) - (\kappa - \overline{\kappa}) \cdot EI = 0, \quad M_T(\kappa, \theta) - \left(\vartheta - \overline{\vartheta}\right) \cdot GI_T = 0.$$

The constants I and I_T signify the second and polar moments of inertia of circular wire respectively.

Thus, the curvature and twist per unit length in the final, unloaded state are respectively

$$\overline{\kappa} = \kappa - \frac{M(\kappa, \vartheta)}{EI}, \quad \overline{\vartheta} = \vartheta - \frac{M_T(\kappa, \vartheta)}{GI_T}. \tag{8.15}$$

The residual stresses in the state of complete elastic unloading $\overline{\sigma}_{zz}$, $\overline{\tau}_{\varphi z}$ could be obtain immediately subtracting the elastic stresses from previously calculated stresses σ_{zz}, $\tau_{\varphi z}$ in the state of maximal plasticization:

$$\overline{\sigma}_{zz} = \sigma_{zz} - (\kappa - \overline{\kappa}) \cdot Ex, \quad \overline{\tau}_{\varphi z} = \tau_{\varphi z} - \left(\vartheta - \overline{\vartheta}\right) \cdot G\rho. \tag{8.16}$$

8.6 Post-coiling Shape of Helical Spring

The overwhelming majority of front-wheel-drive automotive suspension systems use helical springs. The process chosen to produce these is determined by quality, performance, price, environmental issues, etc. The industry develops a potentially cost saving cold-coiling process in which less time is spent treating spring metal at elevated temperatures. Generally, there are two ways to coil a spring: hot coiling and cold coiling. Hot coiling implies that the spring is wound from stock at or above the re-crystallization temperature. The strength and fatigue resistance are controlled afterwards by an appropriate heat treatment. Cold-coiling means that the helical winding takes place at a low temperature after the spring has been hardened and tempered. Cold coiling allows the high temperature heat treatments to take place on the bar stock, which is easier to handle than the final coiled spring. The resulting residual stresses can be essentially eliminated by a relatively low temperature tempering treatment following the cold coiling.

The pronounced residual stress pattern within the cold-coiled spring is undesirable for its unpredictable effect on fatigue and corrosion behavior. These stresses are usually by X-ray measurements of the surface stress field along with modeling of the internal stresses. The success of these procedures requires an independent verification of the actual residual stress field over the cross-section of the original wire stock. The only available well-established method for this is neutron diffraction (Allen et al., 1985).

The coiling process itself introduces residual stresses that detrimental to fatigue, creep, and corrosion properties. We apply the obtained formulas for determination of plastic stresses during the manufacturing and residual stresses in helical springs (Kobelev, 2011).

In the moment of coiling, the spring wire undergoes the simultaneous bending and torsion. Curvature and twist of spatially deformed wire could be determined using methods of differential geometry (Eisenhart, 1940). The shape of the wire in the moment of coiling is given by the spatial curve C, which connects the centers cross-sections of wire. The instantaneous curvature κ and torsion χ of the spatial curve C in the moment of coiling are

$$\kappa = \frac{R}{R^2 + p^2}, \quad \chi = \frac{p}{R^2 + p^2},$$

where R is the coiling radius, $p = 2\pi Rtan(\alpha)$ is the pitch of one coil in the moment of coiling.The angle α designates the inclination of the active loaded element of helix with any plane perpendicular to the axis of the coil in moment of coiling.

Assume the flat coil of the spring of radius R, such that there is no pitch of this coil. The length of wire for one coil is $2\pi R$. To produce the pitch of H, the coil must be stretched in the direction normal to the initial plane of the coil. The angle of twist during stretching is H/R, such that angle of twist per unit length is ϑ. Thus, the torsion χ of the spatial curve C relates to angle of twist per unit length as

$$\chi = 2\pi\vartheta.$$

In the moment, which follows coiling, the wire unloads elastically and forms the helical spring. The unloaded shape of the spring is given by the spatial curve $\overline{C}$, which connects the centers cross-sections of final spring. The unloaded curvature $\overline{\kappa}_h$ and residual torsion $\overline{\chi}$ of the spatial curve $\overline{C}$ are calculated with Eq. (8.16). The torsion $\overline{\chi}$ of the spatial curve $\overline{C}$ relates to the residual angle of twist per unit length as

$$\overline{\chi} = 2\pi\overline{\vartheta}.$$

With these values, we calculate finally the unloaded radius and unloaded pitch

$$\overline{R} = \frac{\overline{\kappa}}{\overline{\kappa}^2 + \overline{\chi}^2}, \quad \overline{p} = \frac{\overline{\chi}}{\overline{\kappa}^2 + \overline{\chi}^2}$$

of the manufactured spring, $\overline{p} = 2\pi\overline{R}\tan(\overline{\alpha})$. The angle $\overline{\alpha}$ is the inclination of helix with any plane perpendicular to the axis of the coil after the elastic spring-back of wire.

For comparison, we use the experimental measurements of residual stresses in cold-coiled helical compression springs (Matejicek et al., 2004). The comparison is performed for as-coiled springs in central coils of reported diameter $2\overline{R} = 160\,[\mathrm{mm}]$ and pitch $\overline{H} = 100\,[\mathrm{mm}]$. The diameter of Cr-Si wire, used for the spring manufacturing is of $2r = 14\,[\mathrm{mm}]$. For stress calculation the reported elastic constants are

$$E = 216\,[\mathrm{GPa}], \quad \nu = 0.28.$$

The tensile strength of 1968[MPa], elongation 9.3% and area reduction 32% were applied.

The calculation results for plastic state in the spring material during manufacturing process are shown on Figs. 8.2 and 8.3. On Fig. 8.2 the bending moment and torque moment as the function of load parameter λ are plotted. The bending moment

$$M_B = \lambda M_{B.\max}.$$

and the torsion moment

$$M_T = \lambda M_{T.\max}$$

increase proportionally to load parameter α. During the plastic loading the load parameter increases from 0 to 1, such that for $\alpha = 1$ the maximal plasticization of the material is achieved. The maximal plastic bending moment is 741 [Nm]. The plastic torque moment is 193 [Nm]. Then, during the elastic unloading, the load parameter decreases from 1 to 0, such that for $\lambda = 0$ the free state of the spring with maximal residual stressed of the material is attained. The calculated pitch (Eq. 8.15) at the end of the unload curve ($\lambda = 0$) is $\overline{H} = 106\,[\mathrm{mm}]$ and the unloaded radius is $2\overline{R} = 136\,[\mathrm{mm}]$. For simulation we use the modified Ramberg–Osgood's law.

The stress components and equivalent stress (Eq. 8.16) in the state of maximal plasticization ($\lambda = 1$) are shown on the Fig. 8.3. The maximum of the shear stress is 170 MPa. In the center of the cross-section the stress vanishes. With the increasing

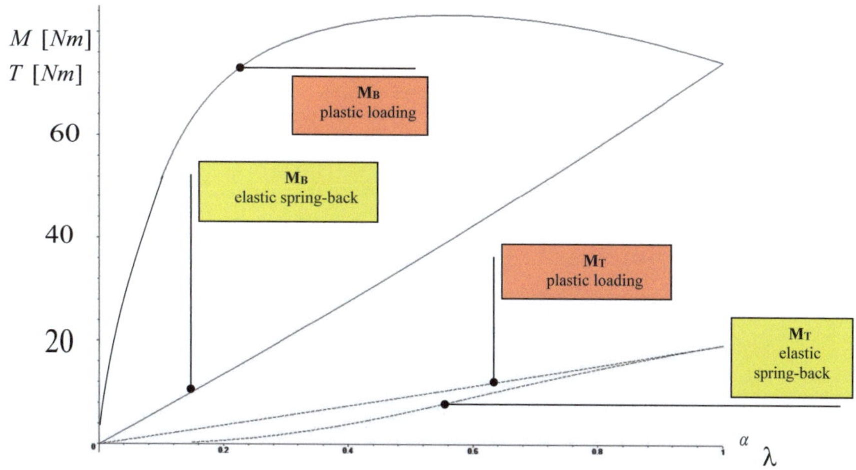

Fig. 8.2 Bending moment and torque as function of loading parameter λ

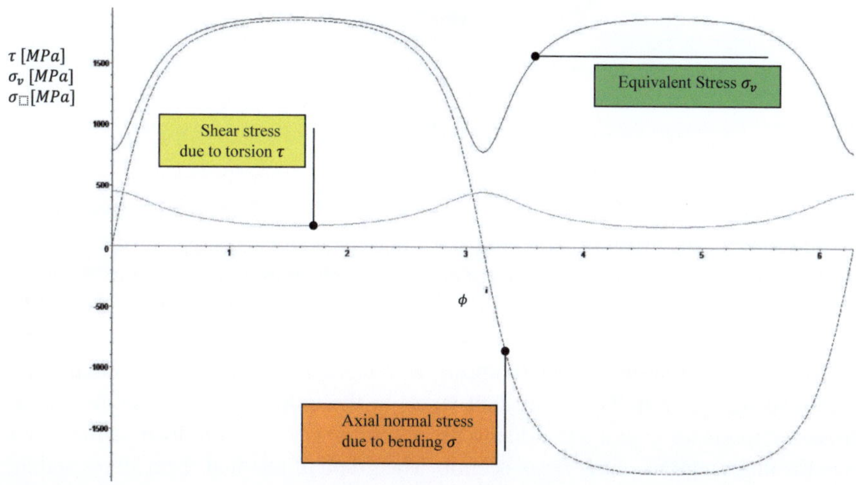

Fig. 8.3 Plastic stresses on the contour of the cross-section in maximal plasticization state

radius the stress increases first linearly with radius of the observation point. On the outer surface of the rod the bending dominates, such that the shear stress stagnates. The normal axial stress at the maximal plasticization point $\lambda = 1$ increases from the value $\sigma_{zz} = -980$ [MPa] on the outmost outer point of the spring body ($x = -r$) to the value $\sigma_{zz} = 980$ [MPa] on the inner point ($x = r$). At $x = 0$ the bending stress disappears. The distribution of bending stress is mirror-symmetric due to the neutral axis $x = 0$. The Mises equivalent stress (Fig. 8.7) in the spring cross-section is symmetric due to the neutral axis $x = 0$. The maximal value of equivalent stress $\sigma_v =$

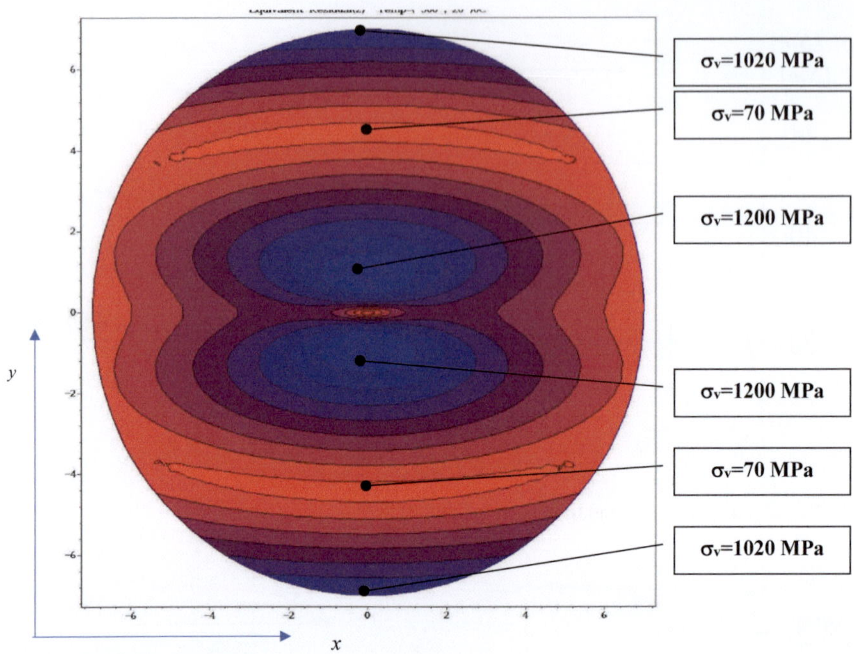

Fig. 8.4 Residual equivalent stress in the unloaded state

1020 [MPa] is attained on the outmost outer and inner points of the cross-section. In the center of the cross-section both shear and axial stress disappear, such that the equivalent stress vanishes. The plots of equivalent, shear and bending stresses on the outer contour of the circular cross-section are given on Fig. 8.3 (Figs. 8.4, 8.5 and 8.6).

Figures 8.6, 8.7 and 8.8 show components and equivalent stress in the final unloaded state of spring ($\lambda = 0$, spring as-coiled).The equivalent residual stress is plotted on Fig. 8.6. The equivalent stress in the unloaded state over the spring cross-section is also symmetric due to the neutral axis $x = 0$, but its maximal value is in the inner regions of the cross-section. The graph of residual shear stress and the axial stress are on Figs. 8.5 and 8.6 correspondingly. The shear stress is symmetric over $x = 0$ axis, but the axial stress due to bending is mirror-symmetric. The plots of equivalent, shear and bending stresses on the outer contour of the circular cross-section are given on Fig. 8.7. The profile of equivalent, shear and bending stresses along the line $y = 0$, also $-r < x < r$ is plotted on Fig. 8.8. On Fig. 8.8 the rectangular points show the measured values of stress reported in the cited paper of Matejicek et al. (2004): The simulated values of axial stress demonstrate an excellent correlation to the measured values. It is evidently exposed, that the equivalent, shear and bending stresses attain their maximal values in the inner region of the cross-section. The maximal values of the residual stress in the inner region are around 10–15% higher, as the corresponding maximal stress on the surface of the wire (Fig. 8.9).

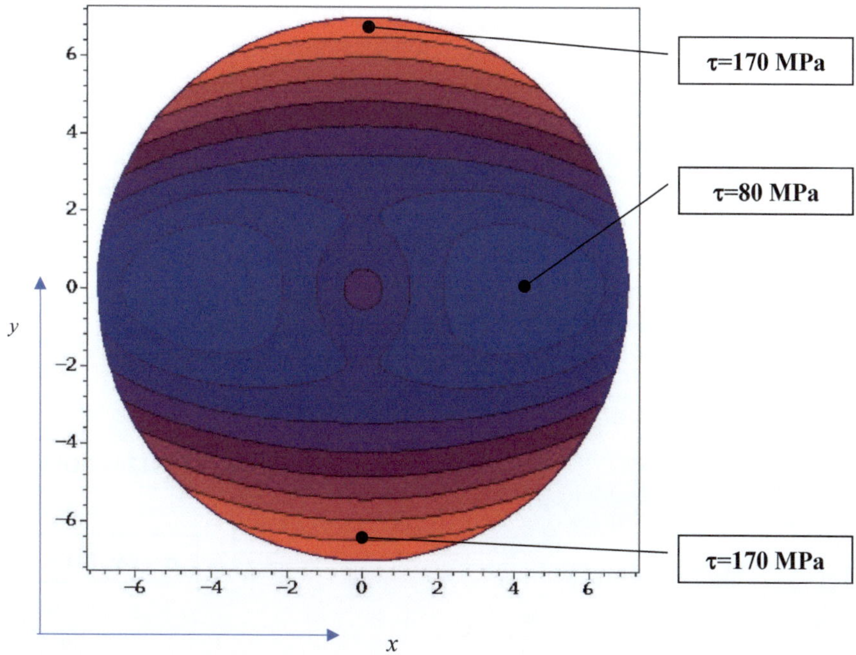

Fig. 8.5 Residual shear stress in the unloaded state

8.7 Conclusions

An important example for manufacturing practice describes the coiling process of a helical spring. The results show the plastic flow and the origin of residual stresses from the viewpoint of the deformational theory of plasticity. The analytical results agree with the experimental values. The proposed method does not require numerical simulation and is well suited for programming of coiling machines. The analytical results show a good correlation with the measured residual stresses. The estimation of loads during the production of cold-wound helical springs and for the dimensioning and wear calculation of coiling tools could be performed with the closed-form solutions.

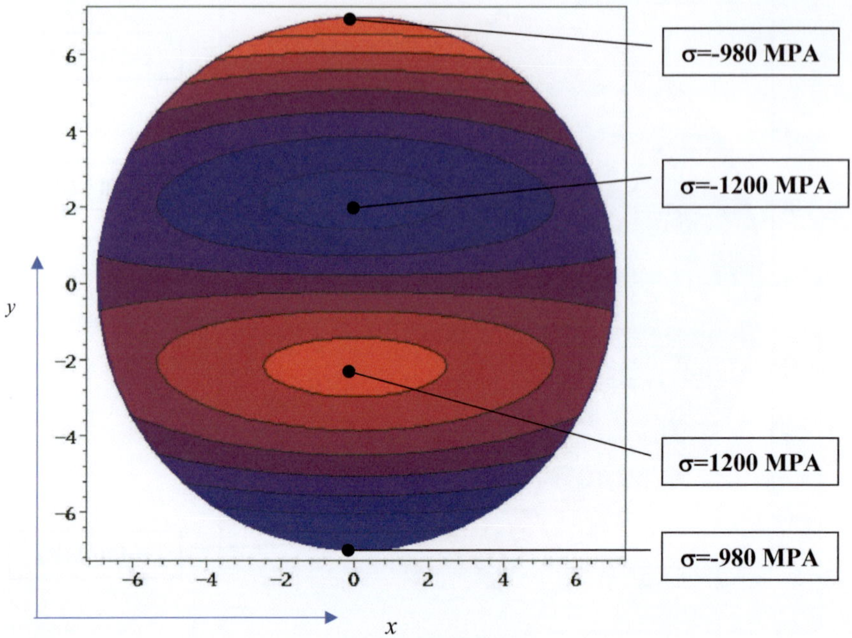

Fig. 8.6 Residual axial normal stress in the unloaded state

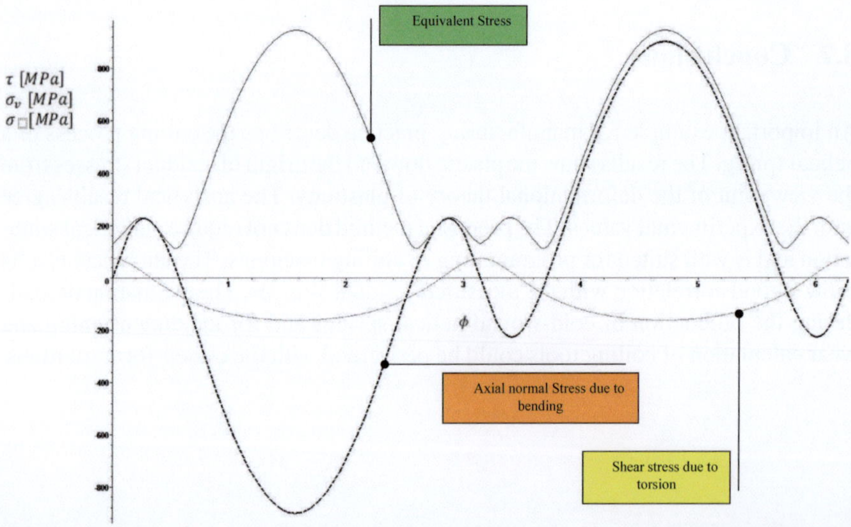

Fig. 8.7 Residual stresses on the contour of the cross-section in the unloaded state

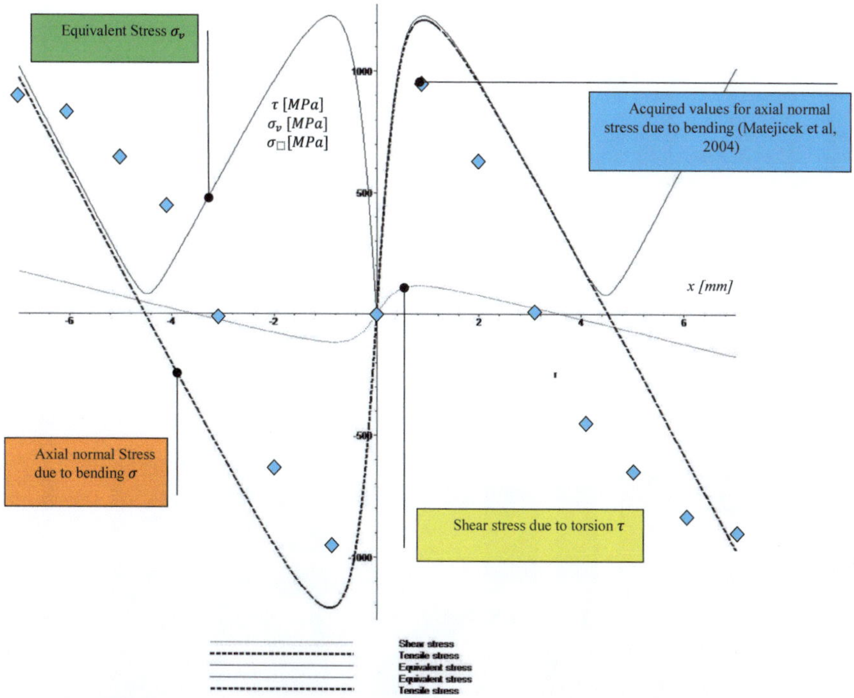

Fig. 8.8 Residual stresses along the positive axis 0 < x < r in the unloaded state

Summary of Principal Results

- The coiling process from the initial straight wire to the final helical spring is analyzed using the deformation theory of plasticity.
- The effects of curvature and pitch during plastic coiling are considered.
- The normal, shear, and octahedral stresses during the coiling process and the resulting residual stresses are presented with formulas.
- The closed form solutions for the moments and forces in the instruments of the wire coiling or winding machine are derived.

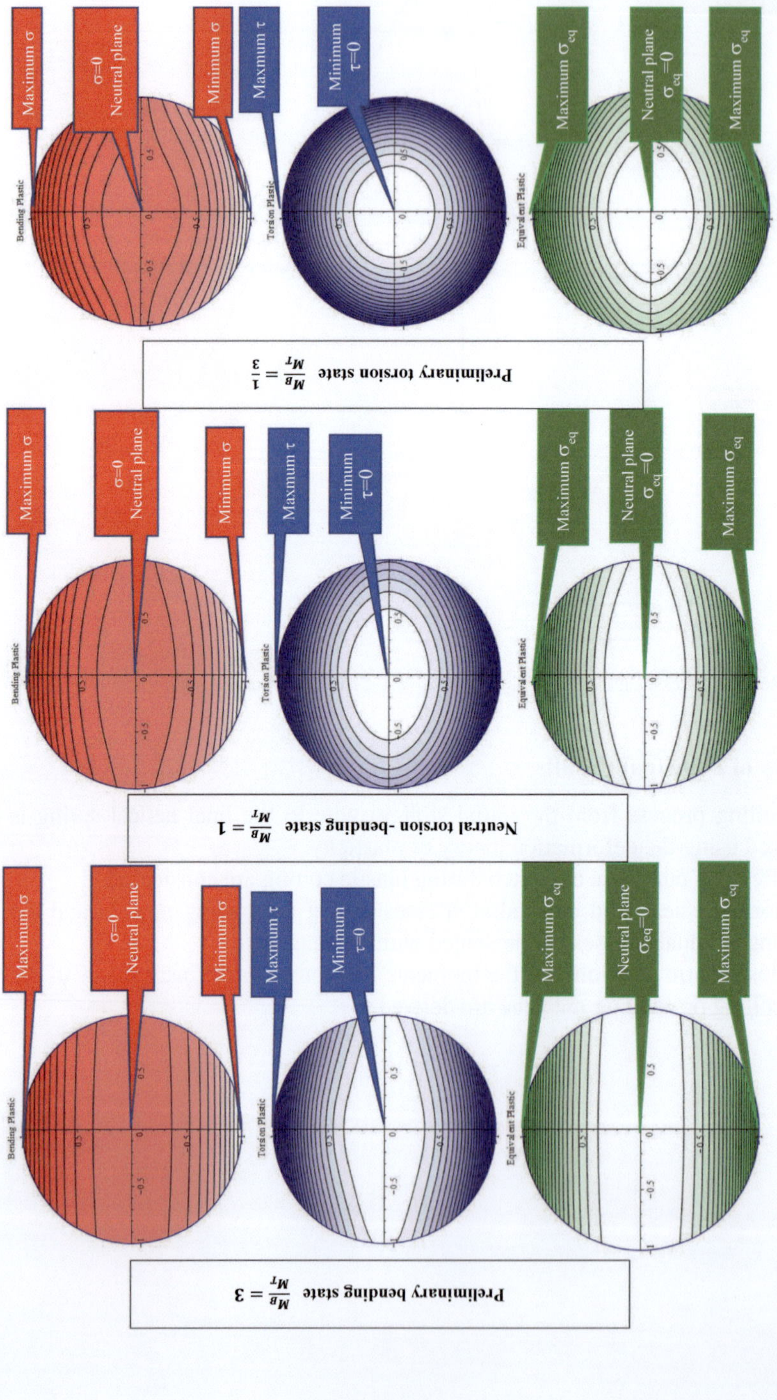

Fig. 8.9 Active stresses along for three different levels of loading Left—bending moment dominates, middle equal torsion and bending moments, right—bending moment dominates

References

Ali, A. R. M. (2005). Plastic deformation and springback of pre-loaded rod under subsequent torsion. *IE (I) Journal MC, 6*, 26–30.

Allen, A. J., Hutchings, M. T., Windsor, C. G., & Andreani, C. (1985). Neutron diffraction methods for the study of residual stress fields. *Advances in Physics, 34*, 445–473.

Baragetti, S. (2006). A theoretical study on nonlinear bending of wires. *Meccanica, 41*, 443–458.

Brooks, D. S. (1969). The elasto-plastic behaviour of a circular bar loaded by axial force and torque in the strain hardening range. *International Journal of Mechanical Science, 11*, 75–85.

Eisenhart, L. P. (1940). *Introduction to differential geometry with use of the tensor calculus*. Princeton.

Gaydon, F. A., & Nuttal, H. (1957). On the combined bending and twisting of beams of various sections. *Journal of the Mechanics and Physics of Solids, 6*, 17–26.

Gaydon, F. A. (1952). On the combined torsion and tension of a partly plastic circular cylinder. *The Quarterly Journal of Mechanics and Applied Mathematics 1*, 29–41

Handelman, G. H. (1944). A variational principle for a state of combined plastic stress. *Quarterly of Applied Mathematics, 1*, 351–353.

Hill, R., & Siebel, M. P. L. (1953). On the plastic distortion of solid bars by combined bending and twisting. *Journal of the Mechanics and Physics of Solids, 1*, 207–214.

Imegwu, E. O. (1960). Plastic flexure and torsion. *Journal of the Mechanics and Physics of Solids*, 8111–146.

Ishikawa, H. (1973). Elasto-plastic stress analysis of prismatic bar under combined bending and torsion. *ZAMM, 68*, 17–30.

Jones, R. M. (2009). *Deformation theory of plasticity*. Bull Ridge Publishing.

Kobelev, V. (2011). Elastoplastic stress analysis and residual stresses in cylindrical bar under combined bending and torsion, ASME. *Journal of Manufacturing Science and Engineering, 133*(4), 044502.

Matejicek, J., Brand, P. C., Drews, A. R., Krause, A., & Lowe-Ma, C. (2004). Residual stresses in cold-coiled helical compression springs for automotive suspensions measured by neutron diffraction. *Materials Science and Engineering, A, 367*, 306–311.

Močilnik, V., Gubeljak, N., & Predan, J. (2015). Surface residual stresses induced by torsional plastic pre-setting of solid spring bar. *International Journal of Mechanical Sciences, 92*. https://doi.org/10.1016/j.ijmecsci.2016.01.004

Neuber, H. (1961). Theory of stress concentration for shear-strained prismatical bodies with arbitrary nonlinear stress–strain law. *Journal of Applied Mechanics, 28*, 544–550.

Neuber, H. (1961b). *Theory of notch stresses: Principles for exact calculation of strength with reference to structural form and material*. United States Atomic Energy Commission, US Office of Technical Services.

Sankaranarayanan, R., & Hodge, P. G. (1958). On the use of linearized yield conditions for combined stresses in beams. *Journal of the Mechanics and Physics of Solids, 7*, 22–36.

Życzkowski, M. (1981). *Combined loadings in the theory of plasticity*. Springer.

Życzkowski, M. (1967). Combined loadings in the theory of plasticity. *International Journal of Non-linear Mechanics*, 173–205

References

Chapter 9
Presetting and Residual Stresses in Springs

Abstract In this chapter, the method for calculating residual stress and permanent deformation of helical springs is developed. The method is based on the deformation formulation of the plasticity theory and general kinematic hypotheses (SAE AE-22, §8.3.33). Two main types of helical springs—compression springs and torsion springs—are studied. For the first type (axial compression or tension springs), the spring wire is twisted. The basic approach neglects the pitch and curvature of the coil and replaces the helical wire with a straight cylindrical rod. The helical spring is in the state of screw dislocation; accordingly, the wire is twisted. The elastic–plastic torsion of the straight rod of circular section is studied. In the second type (torsion helical springs), the helical wire is in the state of edge dislocation, so the wire is in the state of bending. The elastic–plastic deflection of the straight bar with rectangular and circular cross section is studied using Bernoulli's hypothesis. The material for both types of springs is nonlinearly hardening elastic–plastic with elastic unloading. The hyperbolic, Ramberg–Osgood, Ludwik–Hollomon, Swift–Voce, and Johnson–Cook laws for the material are studied. For both problems, the elastic–plastic active deformation and the elastic spring-back allow the closed-form solutions. In addition, this chapter examines the calculation of residual deformation and residual stress for helical springs after a prolonged prestressing process. The article extends the model for the immediate prestressing process by considering the creep deformation of the spring. The method is based on the plasticity theory for the instantaneous flow, which is overexposed by the relaxation over the long-term prestressing. In this article, the following method is used. The plastic deformation of the circular section helical spring occurs instantaneously. As the spring continues to shorten in the tool holder, the stress relaxes and the force of the spring decreases. As a result, after the elastic relaxation of the long-term presetting, the residual stress of the spring also gradually decreases with the compression time. The final length of the springs decreases significantly as the presetting time increases. The advantage of the discovered closed-form solutions is the calculation without the necessity of complex finite element simulation of spring length loss and residual stresses after

V. Kobelev, *Fundamentals of Springs Mechanics*, University of Siegen, Faculty IV: School of Science and Technology. Paul-Bonatz-Str. 9–11, D-57076 Siegen. E-mail: Homepage: https://www.mb.uni-siegen.de/kobelev/.

V. Kobelev, *Fundamentals of Springs Mechanics*,
https://doi.org/10.1007/978-3-031-58584-5_9

prestressing. The analytical expressions are proposed and the exact calibration is applied for evaluation of factors for presetting processes. This chapter is the final section of the second part, which studies the manufacturing processes of helical springs.

Keywords Residual stresses in springs · Time-dependent plastic deformation · Simultaneous creep and plastic deformation

9.1 Elastic–Plastic Deformation During the Presetting Process of Helical Springs

As mentioned above, a helical spring is a spirally wound solid bar, usually with a constant coil diameter and uniform pitch (DIN EN 13906:1, 2002; DIN EN 13906:3, 2002; SAE HS 795, 1989). The process of cold-coiling high tensile steel bars into coil springs for the automotive industry is an advanced technology currently used by leading spring manufacturers. The main manufacturing steps in the cold coil process are: cold coiling, annealing, presetting, shot peening. Residual stress plays an important role in the performance of coil springs. Its effect on the various properties of a material (fatigue, fracture, corrosion, friction, wear, etc.) can be considerable.

In modern spring design, residual stresses must be taken into account. The residual stresses of different origin occur after each manufacturing step. The residual stress fields are usually axisymmetric, that is, the main variation in all specimens occurred along the hoop direction (hoop circumference), at the same time the radial and axial stresses are significantly lower. The simulations and experimental investigations of two-dimensional stress maps of the bar cross section have been collected to elucidate the basic features related to the main cold coiling steps.

The engineering theory of the presetting process was reported by (Graves & O'Malley, 1983). In this article the influence of the chemical composition of the spring material and the achieved shear stress was studied experimentally. The influence of the temperature during the presetting on the spring length reduction was evaluated from the spring tests.

The presetting not only reduced stress concentrations at the internal bore, but contributed to the establishment of advantageous surface residual stress conditions that enhance the fatigue life and resistance to creep and relaxation of the manufactured helical spring (Del Llano-Vizcaya et al., 2006; Wakita et al., 2007). The precise evaluation of presetting parameters is obligatory for the optimal adjustment of the presetting process.

Compression and tension springs are made of an elastic wire material formed into the shape of a helix, which returns after the application of a moderate operating load to its natural length after the release of the load. The helical spring with the constant mean helix diameter, wire diameter $d = 2R$ and active coils number n_a is considered in this article. The principal stresses in helical springs are shear stresses due to twisting of the spring wire. If the spring index $D/2R$ significantly greater than one, the influence of wire curvature on the stress distribution in the cross-section of the helical spring could be neglected. For example, if the spring index is equal to 5, the variation of the shear stress over the circumstance of the cross-section is about 29%. In this case the equations of torsion of the straight wire are applicable for the stress distribution over the cross-section of the helical spring (Chap. 2).

Presetting of the springs is an operation of loading the spring beyond the plastic limits (Atterbury & Diboll, 1960). Before presetting, the initial length of the helical compression spring is equal to L_0 (Fig. 9.1). The spring is rapidly pressed to a fixed preset length L_1 and is hold in this length in the tool for a definite period of time T_P. Usually, the presetting length L_1 is slightly greater the block length to prevent the clash of coils and their slide shift. Afterward, the spring is released, such that the external loads vanish. In this moment the spring unloads elastically and the residual stresses occur. The final length of the spring is L_2, such that $L_1 < L_2 < L_0$. Presetting of the springs induces a favorable residual stress field in the wire. This residual stress field improves the operating load capacity.

The stress profiles during the presetting of the helical spring and the straight torsion bar are similar. In the torsion bar, a relatively large presetting torque is applied to cause an outer region of material under the surface to become plastic while the core of the wire remains elastic. Since part of the wire cross-section has yielded, in the moment the presetting torque is removed, the wire recovers elastically but does not return to its original position. Therefore, the torsion remains enduringly deformed. The corresponding residual shear stresses are "opposite" at the surface. In this context

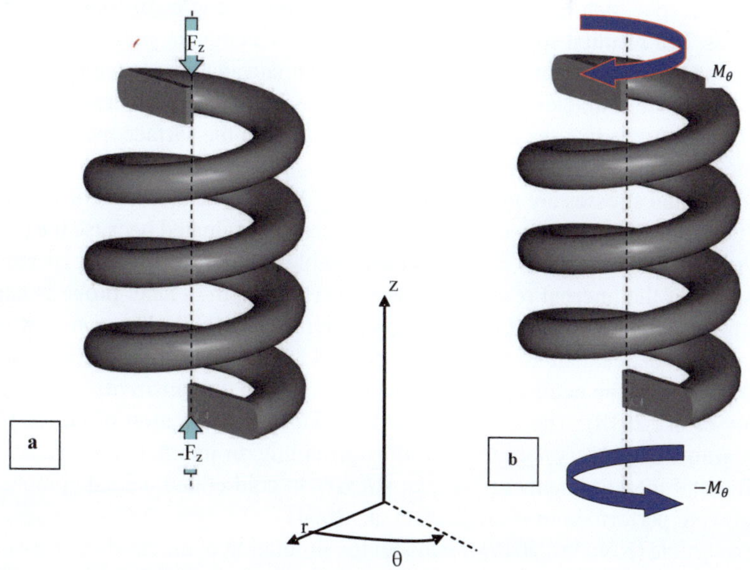

Fig. 9.1 Presetting of the helical compression spring (**a**) and helical torsion spring (**b**)

the word “opposite” means, that the sign of shear stress induced by the applied load, differs from the sign of the residual stress. When the torsion bar is subsequently stressed by the operating load, the residual stresses are subtractive from the stresses attributable to the operating load. Consequently, the surface stresses are less than those which would have occurred in the lack of preset.

Analogously, presetting of the helical compression springs consists of coiling the spring to a longer length than the ultimate desired free length and then compressing it beyond its elastic limit. Presetting allows the use of higher design stresses with the added cost of the secondary operation (Geinitz et al., 2011; Shimoseki et al., 2003). Presetting improves considerably the admissible static loads, decreasing creep and relaxation of springs under operation conditions. Fatigue life is also improved to a certain degree.

The article (Kreuzer, 1984) reports the results of the experimental evaluation of presetting of helical compression springs under the elevated temperatures. The reduction of spring force was evaluated as the function of presetting temperature and test shear stress on the surface of wire. Different spring alloys were studied: Ti318, BS 5216 HD3, BS 5216 M4, Ti314A and CrSi/CrV steel wires.

Yang and Wang (1996) studied experimentally the cyclic creep and fracture behavior of SAE spring steel 5160 for two presetting conditions. In this study was demonstrated, that the cyclic creep rate, cyclic plastic strain range, total plastic strain range and cyclic creep life of the material depend strongly on the preset condition.

The residual stresses due to plastic presetting the surface of a solid bar were studied in (Močilnik et al., 2015). The cited article deals with the residual stresses created at the surface through the plastic presetting. Residual stresses were acquired on the surface of a specimen by x-ray diffraction for different angles of following plastic presetting. The residual stresses were also evaluated using analytical and numerical modelling by finite element methods. The approximate analytical method was based on the torsional characteristic, τ–γ, of the material and tension test results. The cited article reveals that the direction of cold rolling on the surface has a significant influence on residual stresses.

The residual stresses are usually analyzed by X-ray measurements of the surface stress field. The internal stresses could not be directly evaluated because the probing depth of X-rays in metal is low. Another reliable method for direct experimental acquisition of the internal residual stresses over the depth of steel probe is neutron diffraction (Allen et al., 1985; Hutchings & Krawitz, 1992). The non-destructive measurement of residual stress fields is possible. The neutron diffraction uses the probing power of the neutron into steel, combined with the sensitivity of diffraction (Karpov et al., 2018). The measurement evaluates the separation of lattice planes within grains of polycrystalline steel, thus providing an internal strain gauge. The experimental measurements of residual stresses in cold-coiled helical compression springs were performed in (Matejicek et al., 2004).

In the article (Kobelev, 2019) the model for simulation of the residual stresses due to presetting of the helical spring was developed. The analytical formulas permit the assessments of residual stresses and changes of helical spring shape during the manufacturing step. The developed procedures are intended for engineering applications and user interface design of manufacturing equipment and for design of presetting tools. Compression/tension or torsion helical springs are made of an elastic wire material formed into the shape of a helix, which returns to its natural length when unloaded (DIN EN 13906:1, 2002).

The article (Kobelev, 2020) introduced the calculation of remaining deformation and residual stress for helical springs after long-lasting presetting process. The simplified ideal-elastic–plastic model for the immediate presetting took into account the creep deformation of the spring. The method was based on the rate-independent material model without hardening, overexposed by the relaxation over the long-term presetting.

The following method is used in this Chapter. The analysis of active plastic deformation requires a certain nonlinear stress strain law. In this article, the irreversible deformation during the presetting process is analyzed using deformational theory of plasticity. One of the principal foundations of mathematical theory of conventional plasticity for rate-independent metals is that there exists a well-defined yield surface in stress space for any material point under deformation. A material point can undergo further plastic deformation if the applied stresses are beyond current yield surface which is generally referred as plastic loading. If the applied stress state falls within or on the yield surface, the metal will deform elastically only and is undergoing elastic unloading. The time-delayed presetting causes the relaxation of the stresses and also the reduction of the residual stresses. This process requires the analysis methods for creeping solids.

The structure of this Chapter is the following.

In Sect. 9.2 the deformation formulation of plasticity theory with the typical nonlinear flow functions ("hyperbolic" and "Ramberg, Osgood" laws) is briefly introduced. For analysis were used the material properties of the oil hardened and tempered and patented cold drawn spring steel wire of diameter. For experimental assessment, the spring wires have been heat-treated depending on material specifications. The shear stress shear characteristics of common CrSi spring wire were determined using the procedure described in (Reich, 2017).

In Sect. 9.3 the analysis of active plastic torsion and spring-back of the twisted circular wire is performed. This auxiliary solution is used for analysis of presetting for the helical compression springs. The application of the deformation theory of plasticity permits the closed form analytical analysis of the elastic–plastic problem. The analytical derivation makes available the principal formulas for the presetting force over spring travel, active and residual stress on the surface of spring wire and spring length reduction after the completion of the presetting. The major advantage of the analytical solutions is that the final formulas could be used for industrial applications and programming of manufacturing machines. These expressions play the fundamental role for the subsequent analysis. With this method we express analytically the torque as the function the spring travel. The residual stresses in the state of complete elastic unloading on the outer surface of the wire are expressed in closed form.

Section 9.4 studies the helical springs loaded by the axial torque. For analysis the auxiliary problem for the elastic–plastic bending of the wire is solved. The closed-form solutions provide the formulas for the preset moment and elastic spring-back. The analytical expressions for the residual stresses in the state of complete elastic unloading on the outer surface of the wire are provided.

In Sect. 9.5 the alternative method for the description of the material properties. For the simulation of deep plastic deformation, the explicit formulations (Ludwik, Hollomon, Swift and Voce, Johnson and Cook) of the stress–strain curves and strain-hardening of metals (Swift 1952), (Birkert et al, 2013; Kleemola & Nieminen, 1974), were used. The nonlinear stress–strain expression due to Ludwik, Hollomon, Swift and Voce express the stress as the explicit function of strain. Section summarizes the coefficients of the stress–strain equations and the relations between the explicit and implicit formulations. For the closed-form solutions of the plastic deformation problems the relations between Ludwik–Hollomon equation on one side and Ramberg–Osgood equation on the other side are applied. The coefficients of Ludwik–Hollomon and Johnson–Cook equations are expressed through the coefficients of the Ramberg–Osgood relations. The closed form solutions for the problems of torsion and bending of rods are presented. The constitutive laws combine the deformational theory of plasticity with the traditional nonlinear stress–strain laws describing active plastic deformation. The plastic flow during the active load and the origination of the residual stresses are studied.

Section 9.6 studies the relaxation of stresses in the state of time-delayed presetting. After the instantaneous plastic deformation completes, the stage of creep flow occurs. The main hypothesis is the following. The metallic material remains linear-elastic in core region. In the crust region the material experiences plastic flow. During the time-delayed presetting, the strain persists, and in both regions of the elastic and plastic deformation occurs the relaxation of stresses.

The closed form solutions allow the assessments of stresses and geometry changes during manufacturing of helical springs. The formulas could be used for industrial applications and programming of presetting tools.

9.2 Implicit Formulations for the Stress–Strain Curves of Plastic Materials

For the analytical analysis of the elastic–plastic problem, the deformation theory of plasticity with the certain nonlinear flow functions is implemented (SAE AE-22, §8.3.33), (Jones, 2009) Commonly accepted, that the spring steels are isotropic, work-hardening materials with the not distinctly observable yielding points. For such behavior of material, the following nonlinear stress–strain expression can be used:

$$\boldsymbol{e} = f_\tau(\tau_{oct})\frac{\boldsymbol{s}}{\tau_{oct}}. \tag{9.1}$$

The components of deviatoric stress $\boldsymbol{s}$ and strain $\boldsymbol{e}$ tensors respectively are:

$$s_{ij} = \sigma_{ij} - \sigma_{oct}\delta_{ij}, \tag{9.2}$$

$$e_{ij} = \varepsilon_{ij} - \varepsilon_{oct}\delta_{ij}. \tag{9.3}$$

The octahedral normal strain and octahedral normal stress are:

$$\varepsilon_{oct} \equiv \mathbf{I}(\boldsymbol{\varepsilon})/3, \quad \sigma_{oct} \equiv \mathbf{I}(\boldsymbol{\sigma})/3. \tag{9.4}$$

The octahedral shear stress and strain in Eq. (9.1) relate to the second invariants of the corresponding deviatoric tensors $\mathbf{II}(\boldsymbol{e})$, $\mathbf{II}(\boldsymbol{s})$ (Appendix Eq (Appendix A):

$$\tau_{oct} \stackrel{\text{def}}{=} \sqrt{2\mathbf{II}(\boldsymbol{s})/3}, \tag{9.5}$$

$$\gamma_{oct}/2 \stackrel{\text{def}}{=} \sqrt{2\mathbf{II}(\boldsymbol{e})/3}. \tag{9.6}$$

Equation (9.1) expresses the dependence of the strain deviator upon the stress deviator. Because the stress could not be resolved in terms of strain, the stress–strain Eq. (9.1) will be referred to as "implicit stress–strain curve".

From Eq. (9.1) follows the relation between shear stress and shear strain measures for the "implicit stress–strain function":

$$\gamma_{oct} = f_\tau(\tau_{oct}). \tag{9.7}$$

As it is shown later, the accepted relation (9.1) between the deviators leads to the incompressibility of materials during the plastic flow. For this purpose, consider the state of a pure shear and the state of a uniaxial stress. If the sole components are shear stress $\tau_{xz} = \sqrt{3/2}\tau$ and strain $\gamma_{xz} = \sqrt{3/2}\gamma, \gamma = \tau/G$ the intensities of shear stress and strain Eqs. (9.3) and (9.7) read as

$$\tau_{oct} = \tau, \quad \gamma_{oct} = \gamma.$$

In this case the Eq. (9.7) reduces to $\gamma = f_\tau(\tau)$.

Analogously, if the non-zero components are uniaxial stress is $\sigma_{xx} = \sigma$ if follows from (9.1) that the strain $\varepsilon_{xx} = \varepsilon, \varepsilon = \sigma/E$. For the incompressible material $\varepsilon_{yy} = \varepsilon_{zz} = -\varepsilon_{xx}/2$. The octahedral shear stress and strain for the uniaxial state are according to Eq. (9.5) correspondingly

$$\tau_{oct} = \sqrt{2}\sigma/3, \quad \gamma_{oct} = \sqrt{2}\varepsilon.$$

The substitution of these formulas in Eq. (9.7) shows, that the following relation must be fulfilled:

$$\sqrt{3}\varepsilon = f_\tau\left(\sigma/\sqrt{3}\right).$$

Consequently, the formula for the uniaxial stress follows from the equation for the shear stress:

$$\varepsilon = \frac{1}{\sqrt{3}} f_\tau\left(\frac{\sigma}{\sqrt{3}}\right) \stackrel{\text{def}}{=} f_\sigma(\sigma) \tag{9.8}$$

For the subsequent derivations we use the simplest assumptions of Eqs. (9.1) and (9.8). For more precise applications both stress-stain curves should be acquired experimentally. The corresponding constants for each material law could be properly evaluated numerically using the methods of curve fitting. For example, the stress-stain curves were determined experimentally for steel 54CrSi6 (DIN 10089, Material number 1.7102) in Kobelev (2011). The determination was based on backwards calculation the shear stress vector from the vector of torsion moments. For the determination of stress–strain curve the inverse problem has to be solved. For this purpose, the method of piece-wise linear approximation was implemented. The implemented method is suitable for the finite-element simulation of the plastic deformation of steel wire, but is not easily analyzed for the practical application in manufacture.

The consequent derivation of the closed-form expressions is based on the common plastic deformation laws (Kausel, 2017). The first set of expressions for material law will be referenced to as "hyperbolic law":

$$\gamma = \frac{1}{\left[1-(\tau/K_\tau)^n\right]^l} \cdot \frac{\tau}{G}, \quad \varepsilon = \frac{1}{\left[1-(\sigma/K_\sigma)^n\right]^l} \cdot \frac{\sigma}{E}. \tag{9.9}$$

Here G, E are the elasticity moduli, K_τ, K_σ are the solidifying coefficients and n, l are the solidifying exponents. The material constants in formulas (9.9) are habitually assumed to be unrelated to each other. The common setting for the second solidifying exponent is $l = 1/2.$. This material model was used in Neuber (1961a, 1961b) for the solution of notch problems. For the Neuber material, the formula (9.9) reduce to the expressions:

$$\gamma = \frac{\tau}{G\sqrt{1-(\tau/K_\tau)^n}}, \quad \varepsilon = \frac{\sigma}{E\sqrt{1-(\sigma/K_\sigma)^n}}.$$

The second regularly implemented set of expressions reads according to (Ramberg & Osgood, 1943):

$$\gamma = \frac{\tau}{G} + \left(\frac{\tau}{K_\tau}\right)^n, \quad \varepsilon = \frac{\sigma}{E} + \left(\frac{\sigma}{K_\sigma}\right)^n. \tag{9.10}$$

The material constants in formula (9.10) are also not linked in general case.

It should be noted that the independence of plastic deformation Eq. (9.1) upon the hydrostatic stress impose the certain relations on the material constants. From Eq. (9.8) follow the following relationships between the constants of both laws (9.9) and (9.10):

$$E = 3G, \quad K_\sigma = \sqrt{3} K_\tau$$

Thus, the hypothesis (9.8) manifests the material incompressibility of the material in course of the plastic flow. For the incompressible material the bulk modulus is infinite, $K \to \infty$.

Figures 9.2, 9.3 and 9.4 relate to the "hyperbolic law" for shear $\gamma = f_\tau(\tau)$ and for uniaxial state $\varepsilon = f_\sigma(\sigma)$ according to Eq. (9.9). Figures 9.5, 9.6 and 9.7 use the material laws due to Ramberg and Osgood $\gamma = f_\tau(\tau)$ and $\varepsilon = f_\sigma(\sigma)$ as given in Eq. (9.10). We use the same symbols for the constants in material laws (9.9) and (9.10), remembering in mind, that the numerical values of these constants are different for both laws.

For analysis the material properties of the oil hardened and tempered and patented cold drawn spring steel wire of diameter d = 3 mm were used. Wires have been heat-treated depending on material specifications. Actual shear stress shear characteristic was determined using the procedure described in (Reich, 2017). The constants for the Ramberg–Osgood law (9.10) are shown in Table 9.1. The methods of curve fitting deliver the material constants for the hyperbolic law, which are listed in Table 9.2. The plots on Fig. 9.2 display the shear stress-shear strain curves of the implemented materials of the hyperbolic law (9.9). Figure 9.5 demonstrates the shear stress-shear strain curves of the implemented materials for the Ramberg–Osgood law (9.10), which was used for derivation of constants for the hyperbolic law. For the consequent comparison of results, the values in Table 9.2 are derived by the curve interpolation

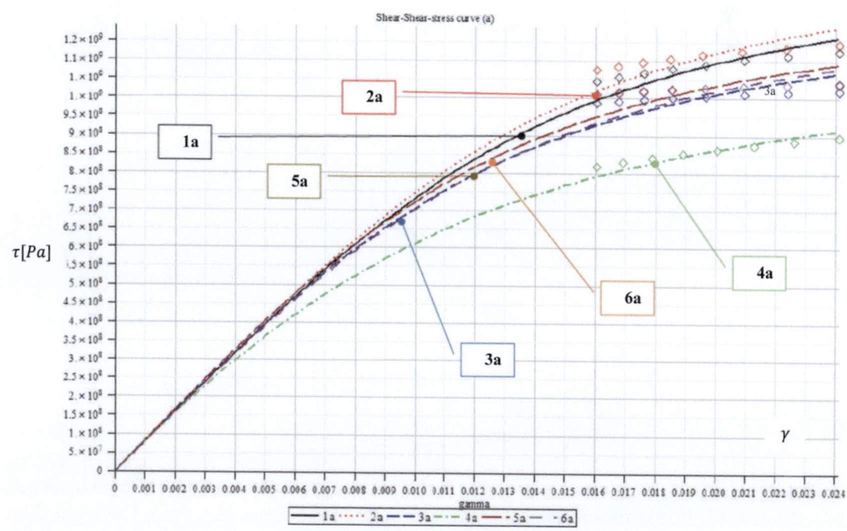

Fig. 9.2 Stress versus shear-stain curve of the oil hardened and tempered spring steel wire (1a, 2a, 3a) and patented cold drawn spring steel wire (4a, 5a, 6a) for the "hyperbolic law"

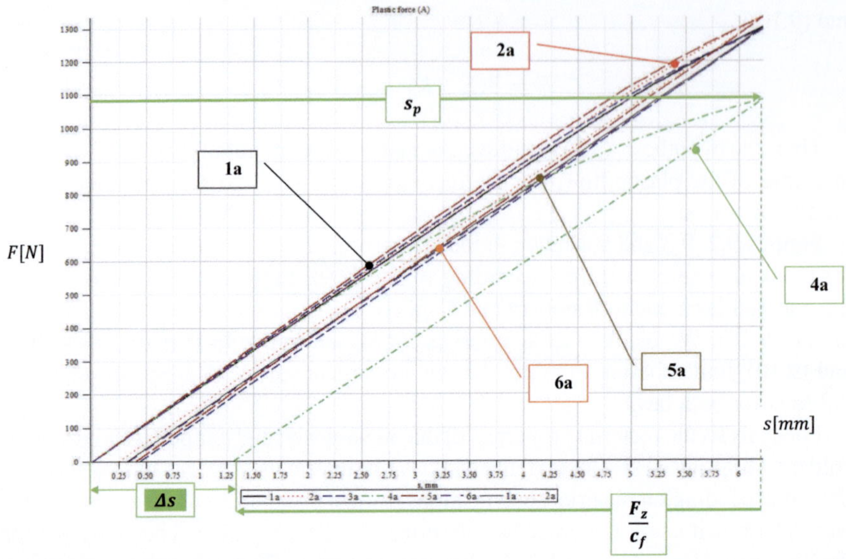

Fig. 9.3 Spring force as functions of the spring stroke for oil hardened and tempered wires (1a, 2a, 3a) and of patented cold drawn spring steel wires (4a, 5a, 6a). For calculation the the "hyperbolic law" is used

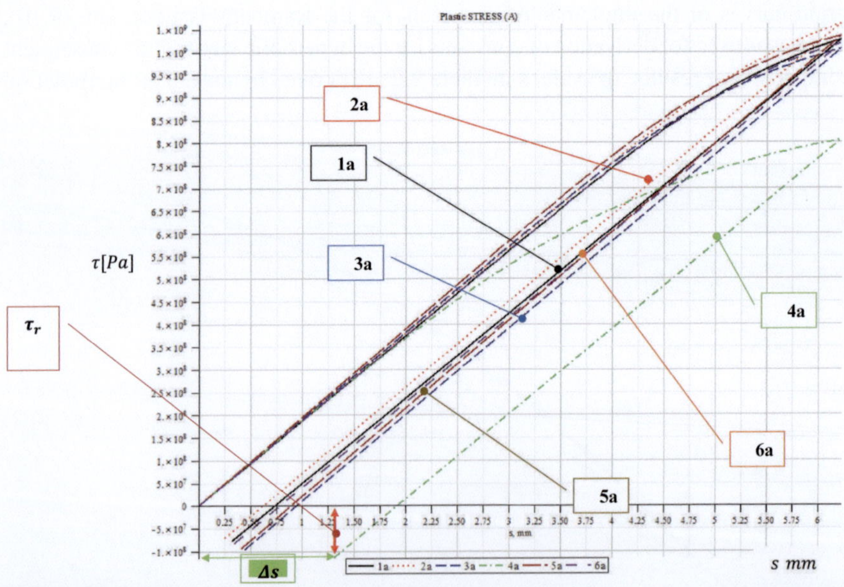

Fig. 9.4 Shear-stress on the surface of spring wire as the function of preset travel for oil hardened and tempered (1a, 2a, 3a) and of patented cold drawn spring steel wires (4a, 5a, 6a). The application of the "hyperbolic law"

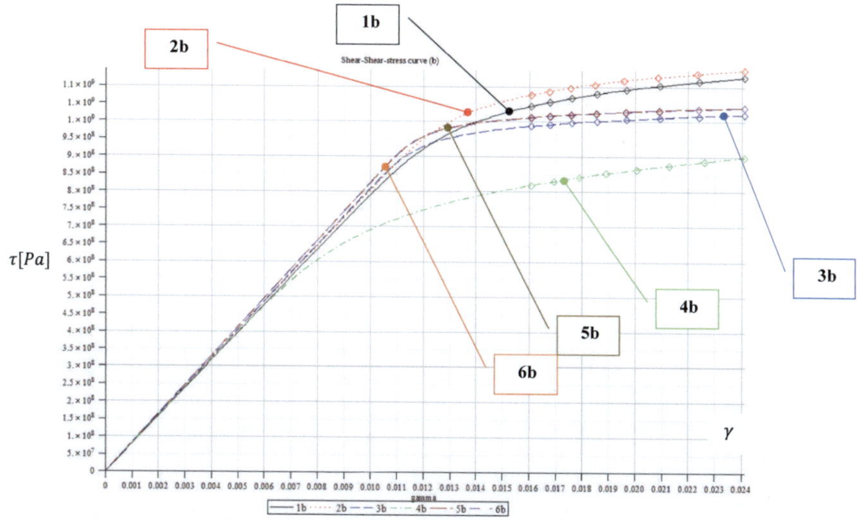

Fig. 9.5 Shear stress as the function of shear for the oil hardened and tempered spring steel wires (1b, 2b, 3b) and patented cold drawn spring steel wires (4b, 5b, 6b) Application of the "Ramberg–Osgood-law"

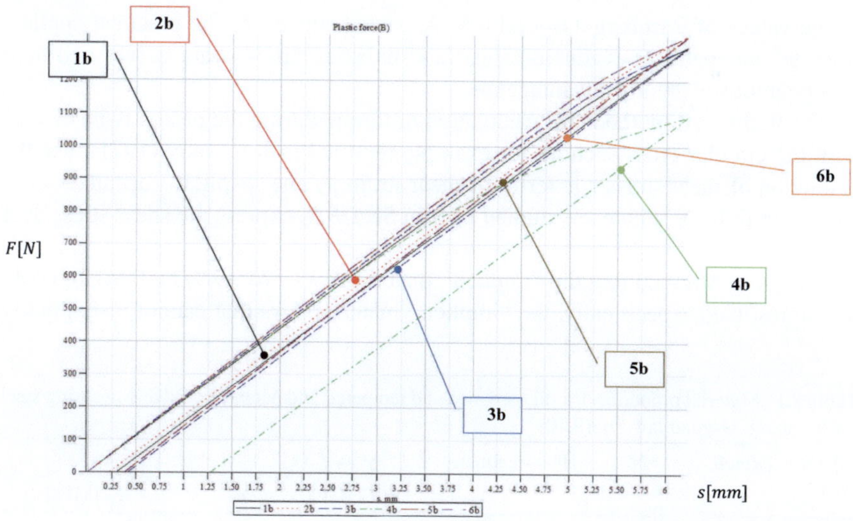

Fig. 9.6 Spring forces as functions of spring travel for oil hardened and tempered (1b, 2b, 3b) and of patented cold drawn spring steel wires (4b, 5b, 6b). For calculations the "Ramberg–Osgood-law"is applied

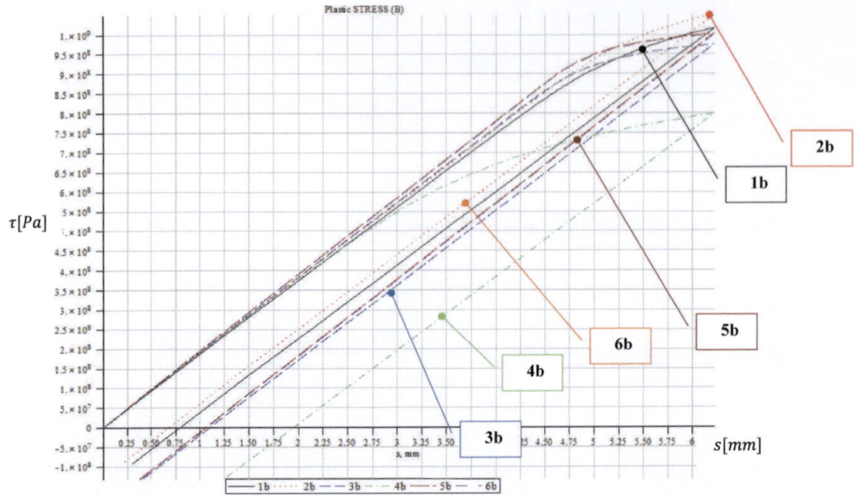

Fig. 9.7 Functional dependence of the shear-stress on the wire surface upon spring travel for oil hardened and tempered (1b, 2b, 3b) and of patented cold drawn spring steel wires (4b, 5b, 6b). The "Ramberg–Osgood-law" is used for simulation

of the values of Ramberg–Osgood law. As stated above, for the practical applications the independent evaluation shear and uniaxial stress–strain curves and their interpolations is greatly advantageous.

The traditional methods for calculation of torque during the plastic twist of wire with the circular cross-section, for example reported in Kachanov (1971), use the resolution of shear stress τ in terms of shear strain γ. The "explicit" formulations of stress–strain laws, which are studied later in Sect. 9.5, express the shear stress over the shear strain.

It must be stressed that the "implicit" material laws, like (9.9) and (9.10), could not be resolved. Specifically, the "implicit" material laws that cannot be explicitly

Table 9.1 Material properties for oil hardened and tempered and patented cold drawn spring steel for Ramberg–Osgood law Eq. (9.10)

Material (Reich, 2017)	N	Heat treatment	$\gamma = f_\tau(\tau)$		
			G [GPa]	n	K_τ [MPa]
Oil hardened and tempered spring steel wire	1	Non-heat-treated	79.396	17.26	1496
	2	350 °C/30'	80.405	20.44	1437
	3	420 °C/30'	80.823	32.25	1172
Patented cold drawn spring steel wire	4	Non-heat-treated	79.422	8.605	1490
	5	200 °C/30'	82.414	39.84	1164
	6	250 °C/30'	82.434	38.91	1166

Table 9.2 Material properties for oil hardened and tempered and patented cold drawn spring steel for hyperbolic law, Eq. (9.9), $l = 1/2$

Material	N	Heat treatment	$\gamma = f_\tau(\tau)$		
			G [GPa]	n	K_τ [MPa]
Oil hardened and tempered spring steel wire	1	Non-heat-treated	79.396	8.650	1219
	2	350 °C/30'	80.405	9.639	1259
	3	420 °C/30'	80.823	8.3672	1169
Patented cold drawn spring steel wire	4	Non-heat-treated	79.422	3.7419	957
	5	200 °C/30'	82.414	8.8377	1201
	6	250 °C/30'	82.434	8.7685	1199

solved for shear stress as the function of shear strain. The offered method allows the closed-form solutions also for "implicit" formulations.

9.3 Analysis of Active Plastic Torsion and Spring-Back of Circular Wire for Presetting Assessment of Helical Compression Springs

9.3.1 Plastic Deformation of Wire During Active Plastic Twisting of a Solid Rod

Consider the helical spring with the mean helix diameter D, wire diameter d and active coils number n_a (Fig. 9.1). The compression an extension springs are subjected to the torsional stresses of their wire. This type of springs is described in (SAE HS 795, 1989, Sections A and B). For brevity, the diameter of wire, the mean diameter of the helix and its pitch are assumed to be constant, such that the helical spring fits into the cylinder with the diameter $D + \mathrm{d}$.

The deflection of the spring is studied on the assumption that it is fixed at the lower end and loaded by the axial load at the upper end. It is assumed that the pitch angle between coils and a plane normal to the axis of the helix is small enough. During the plastic loading, increases the angle of twist pro length unit of the bar proportionally to the vertical component of the deflection. For simplification, the effect of pitch angle and curvature of wire are neglected and the deformation consists only of the twisting of the straight wire (T).

Presetting of the helical compression springs is explained in (SAE HS 795, 1989, Chap. 1, p. 2.10). In this section the mathematical theory of presetting of the helical compression springs is developed. Residual stresses could be induced by the plastic torsion loading and elastic spring-back of a solid bar. A straight cylindrical solid bar with the circular cross-section of the length L is loaded from a stress-free state by the torque M_T. The origin of coordinates is chosen at the centroid of area of one cross-section. The end sections are $z = 0$ and $z = L$.

The shear strain in pure torsion of cylindrical bar with the circular cross-section of the radius $r \equiv \mathrm{d}/2$ reads

$$\gamma_{xz} = -\theta y, \quad \gamma_{yz} = \theta x.$$

Here θ is the angle of twist per unit length of the bar. We use the polar coordinates in cross-section for the circular rod:

$$\begin{cases} x = \rho \sin \phi, \quad & y = \rho \cos \phi, \\ 0 < \rho < r \equiv \frac{\mathrm{d}}{2}, & \rho = \sqrt{x^2 + y^2}. \end{cases}$$

According to the St. Venant's hypothesis, the shear deformation is the linear function of radius (Rubinstein, 1977):

$$\gamma_{z\varphi}(\rho, \phi) = 2\varepsilon_{z\varphi} = \theta \rho. \tag{9.11}$$

The shear on the contour of the cross-section γ_p is the linear function of the angle of twist per unit length of the bar θ. The plastic shear strain for the period of the presetting attains its maximum on the contour of the circular cross-section:

$$\gamma_p = \gamma_{z\varphi}(r, \phi) \equiv \theta r. \tag{9.12}$$

In accordance with Eqs. (9.11) and (9.12), the shear strain is the linear function of radius ρ:

$$\gamma(\rho) = \gamma_p \frac{\rho(\tau)}{r}.$$

9.3.2 *Torque Moment During the Active Plastic Presetting in the Wire Cross Section for Hyperbolic Law*

The sole non-vanishing stress component for a solid circular rod subjected to torsion is τ. The "implicit stress–strain functions" express the strain components in terms of stress. The torque moments applied to the end sections of the rod is:

$$M_T = 2\pi \int_0^r \tau(\rho) \rho^2 \mathrm{d}\rho. \tag{9.13}$$

For application of Eq. (9.13) requires the expressions for the stress components in each point of the circular cross-section. In the state of torsion, the shear stress $\tau(\rho)$ depend on shear strain along the radius of the circular cross-section. The shear stress on the contour of the cross-section τ_p results from the implicit relation: $\gamma_p = f_\tau(\tau_p)$. If the plastic law $\gamma = f_\tau(\tau)$ could be resolved for the shear stress, the solution greatly simplifies, and evaluation of the integral Eq. (9.13) delivers the torque (Kachanov, 1971). Inappropriately, the "implicit stress–strain" laws (9.9) and (9.10) could not be resolved for the shear stress. In this case the change of variables $\rho = \rho(\tau)$ under the integral (9.13) is allows its evaluation:

$$\frac{M_T}{2\pi} = \int_0^r \tau(\rho)\rho^2 \mathrm{d}\rho = \int_0^{\tau_p} \tau\rho^2(\tau)\frac{\mathrm{d}\rho}{\mathrm{d}\tau}\mathrm{d}\tau = \int_0^{\tau_p} \tau\rho^2(\tau)\left(\frac{\mathrm{d}\tau}{\mathrm{d}\rho}\right)^{-1}\mathrm{d}\tau\,. \tag{9.14}$$

For the "hyperbolic law" (9.9) the function $\rho(\tau)$ in Eq. (9.14) reads:

$$\rho(\tau) = r\cdot\frac{\gamma}{\gamma_p} \equiv r\cdot\frac{f_p(\tau)}{f_p(\tau_p)} \equiv r\cdot\frac{\tau}{\tau_p}\cdot\left[\frac{1-(\tau_p/K_\tau)^n}{1-(\tau/K_\tau)^n}\right]^l. \tag{9.15}$$

The new dimensionless variable is introduced for shortness of the formulas:

$$\varphi \stackrel{\text{def}}{=} \frac{\tau_p}{K_\tau}.$$

The integral (9.14) calculates with the variable φ and with the hypergeometric function ${}_2\mathbf{F}_2[]$ as:

$$\begin{aligned} M_T = \frac{\pi\tau_p}{2}\cdot\left(\frac{d}{2}\right)^3\cdot(1-\varphi^n)^{3l}\cdot\Big\{&{}_2\mathbf{F}_2\left(\left[1+3l,\tfrac{4}{n}\right],\left[\tfrac{n+4}{n}\right],\varphi^n\right)\\ &+{}_2\mathbf{F}_2\left(\left[1+3l,\tfrac{n+4}{n}\right],\left[\tfrac{2(2+n)}{n}\right],\varphi^n\right)\cdot\tfrac{4(nl-1)}{n+4}\varphi^n\Big\}. \end{aligned} \tag{9.16}$$

The common setting for the third solidifying exponent is $l = 1/2$. The straightforward way to derive the expression for torque is to substitute the value of the third solidifying exponent into Eq. (9.15) and perform the integration of torque using Eq. (9.14). The expression for torque in the case $l = 1/2$ reads:

$$\begin{aligned} M_T = \frac{\pi\tau_p}{2}\cdot\left(\frac{d}{2}\right)^3\cdot(1-\varphi^n)^{3/2}\cdot\Big\{&{}_2\mathbf{F}_2\left(\left[\tfrac{5}{2},\tfrac{4}{n}\right],\left[\tfrac{2(2+n)}{n}\right],\varphi^n\right)\\ &+{}_2\mathbf{F}_2\left(\left[\tfrac{5}{2},\tfrac{n+4}{n}\right],\left[\tfrac{2(2+n)}{n}\right],\varphi^n\right)\cdot\tfrac{2(n-2)}{n+4}\varphi^n\Big\}. \end{aligned} \tag{9.17}$$

For some values of parameters, the hypergeometric function expresses in terms of the elementary functions. Particularly, the following representations are valid for $n = 2, l = 1/2$:

$$M_T = \frac{2\pi\tau_p}{3\varphi^4}\cdot\left(\frac{d}{2}\right)^3\cdot\left(3\varphi^2 - 2 + 2\sqrt{1-\varphi^2} - 2\varphi^2\sqrt{1-\varphi^2}\right),$$
$$M_T \underset{\varphi\to 0}{=} \frac{\pi\tau_p}{2}\cdot\left(\frac{d}{2}\right)^3\cdot\left(1+\frac{1}{6}\left(\frac{\tau_p}{K_\tau}\right)^2 + \frac{1}{16}\left(\frac{\tau_p}{K_\tau}\right)^4 + \ldots\right); \tag{9.18a}$$

for $n = 4, l = 1/2$:

$$M_T = \frac{\pi\tau_p}{3\varphi^4}\left(\frac{d}{2}\right)^3\left(3\varphi^4 - 1 + \sqrt{1-\varphi^2} - \varphi^4\sqrt{1-\varphi^2}\right),$$
$$M_T \underset{\varphi\to 0}{=} \frac{\pi\tau_p}{2}\left(\frac{d}{2}\right)^3\left(1+\frac{1}{4}\left(\frac{\tau_p}{K_\tau}\right)^4 + \frac{1}{24}\left(\frac{\tau_p}{K_\tau}\right)^8 + \ldots\right); \tag{9.18b}$$

for $n = 8, l = 1/2$:

$$M_T = \frac{\pi\tau_p}{6}\cdot\left(\frac{d}{2}\right)^3\cdot(\varphi^8+3) \equiv \frac{\pi\tau_p}{2}\cdot\left(\frac{d}{2}\right)^3\cdot\left(1+\frac{1}{3}\left(\frac{\tau_p}{K_\tau}\right)^8\right). \tag{9.18c}$$

The limits of the hypergeometric function for the small values of its argument are the following:

$$\lim_{\varphi\to 0}\,{}_2\mathbf{F}_2\left(\left[1+3l, \frac{4}{n}\right], \left[\frac{n+4}{n}\right], \varphi^n\right) = 1,$$

$$\lim_{\varphi\to 0}\,{}_2\mathbf{F}_2\left(\left[1+3l, \frac{n+4}{n}\right], \left[\frac{2(2+n)}{n}\right], \varphi^n\right) = 1.$$

Consequently, for small φ the first term of Taylor series for the twist moment reads:

$$M_T \underset{\varphi\to 0}{=} \frac{1}{2}\pi\tau_p\left(\frac{d}{2}\right)^3 \underset{\theta\to 0}{=} \frac{1}{2}\pi G r^4\theta \quad \text{as} \quad \tau_p \underset{\theta\to 0}{=} G\theta r.$$

This limit case corresponds to the ideal-elastic material with the negligible hardening. The first term of the Taylor series for the moment reveals the moment of the circular rod made of the ideal elastic material. In the case of ideal-elastic deformation, the residual stresses disappear after the elastic spring-back.

9.3.3 Torque Moment During the Active Plastic Presetting in the Wire Cross Section for Ramberg–Osgood Law

For the Ramberg–Osgood law, Eq. (9.10):

$$\rho(\tau) = r\frac{\gamma}{\gamma_p} \equiv r\frac{\tau/G+(\tau/K_\tau)^n}{\tau_p/G+(\tau_p/K_\tau)^n}. \tag{9.19}$$

The torque moment is derived as the explicit function of preset stress:

$$M_T = \pi\tau_0\left(\frac{\mathrm{d}}{2}\right)^3\frac{\Phi(\varphi)}{\Psi(\varphi)}, \tag{9.20}$$

$$\Phi(\varphi) = c_3\varphi^3K_\tau^3 + c_2\varphi^{2+n}K_\tau^2G + c_1\varphi^{1+2n}K_\tau G^2 + c_0\varphi^{3n}G^3,$$
$$\Psi(\varphi) = \left(\varphi K_\tau + \varphi^nG\right)^3\cdot c_4. \tag{9.21}$$

In Eq. (9.21) the coefficients are the polynomials of the solidifying exponent n:

$$\begin{aligned} &c_0 = 4n(n+3)(n+1), && c_1 = 2(2n+1)(3n+1)(n+3),\\ &c_2 = 4(3n+1)(n+2)(n+1), && 2c_3 = c_4 = 2(n+3)(3n+1)(n+1). \end{aligned} \tag{9.22}$$

In the case $G \to \infty$ from (9.19) follows:

$$\rho(\tau) = r\frac{\gamma}{\gamma_p} \equiv r\frac{\tau^n}{\tau_p^n}, \tag{9.23}$$

and the Eq. (9.20) reduces to:

$$M_T \underset{G\to\infty}{=} \pi\tau_p\left(\frac{\mathrm{d}}{2}\right)^3\frac{2n}{3n+1}. \tag{9.24}$$

The limit expression for small φ the twist moment (9.20) is:

$$M_T \underset{\varphi\to 0}{=} \frac{\pi\tau_p}{2}\left(\frac{\mathrm{d}}{2}\right)^3 \underset{\theta\to 0}{=} \frac{\pi G\theta}{2}\left(\frac{\mathrm{d}}{2}\right)^4.$$

The major advantage of the analytical solutions is that the final formulas could be used for industrial applications and programming of manufacturing machines. These expressions play the fundamental role for the subsequent analysis. With this method we express analytically the torque M_T as the function the spring travel.

9.3.4 Elastic Spring-Back and Residual Stresses Appearing

The main stresses generated in compression and tension helical springs are shear stresses due to twisting of the spring wire. In helical springs made of round section wire, the shear stress varies on the outer surface of the wire. The inner surface of the wire is stressed more than the surface of the round straight wire under the action of the same torque. When the spring index is low, the actual stress in the wire is excessive due to the effect of curvature. For the springs with significantly high indices, the stresses are equalized over the outer surface of the wire. The finite element simulations show that the influence of the stress variation on the plastic deformation can be neglected for the spring index higher than six. In this case, the error in torque evaluation is less than 5% for the common spring steels. In the present study, the spring index is assumed to be high enough to neglect the curvature of the wire. The shear stress in the helical wire is assumed to be equal to the shear stress in the straight wire under the action of the same torque.

In terms of the spring wire torque the presetting force reads (Fig. 9.1a):

$$F_z = \frac{2M_T}{D}.$$

With this equation the presetting force of the spring during the plastic deformation results from the expressions (9.16) and (9.20).

The spring travel from the undeformed state to the plastically deformed state is equal to a definite constant value s_p. In the course of the presetting of helical springs the spring force attains the value F_z. When the force disappears, the unloading occurs. The simplest hypothesis for the unloading is the linear elastic spring-back that leads to the increase of the spring length:

$$s_u = \frac{F_z}{c_f}, \quad c_f = \frac{Gd^4}{8D^3 n_a}.$$

Consequently, the reduction of the length of the spring after plastic presetting and elastic unloading reads:

$$\Delta s = s_p - s_u \equiv s_p - \frac{2M_T}{c_f D}. \tag{9.25}$$

The unloading could be evaluated analogously for a more realistic unloading model due to (Masing, 1926), but the Eqs. (9.24) and (9.25) are sufficient for a practical application.

The calculations were performed for the spring DF219 (Reich, 2017) with the mean helix diameter $D = 9$ mm, wire diameter d $= 3$ mm and active coils number $n_a = 7$. Figures 9.2 and 9.5 demonstrate the dependences of active plastic and elastic spring-back forces for the hyperbolic and Ramberg–Osgood laws correspondingly. The final points on the force graphs with the zero force show the spring-back length of the spring, i.e. the reduction of spring length in cause of preset. For the hyperbolic law (9.9) preset length loss Δs is shown for six investigated materials on Fig. 9.4. This figure displays also the shear stress during the plastic deformation and elastic spring-back on the surface of the wire. Figure 9.7 reveals the corresponding results for the Ramberg–Osgood law (9.10) for the same materials.

The simplest definition of residual stress is as follows: stresses that remain in a part after it has been deformed and all external forces have been removed. More specifically, the deformation must be non-uniform across the cross section of the material for residual stresses to occur. Deformation can result not only from forming, but also from thermal processes. It is well known that phase transformations during heat treatment induce sufficient strain to cause plastic deformation, which results in residual stresses. As shown below, the residual stresses that occur in the cylindrical rod after plastic torsion and in the helical spring after prestressing could be determined analytically.

The residual stresses in the state of complete elastic spring-back τ_r is equal to stresses τ_p in the state of maximal plasticization minus the stresses τ_u due to elastic unloading:

$$\tau_r = \tau_p - \tau_u, \quad \tau_u = M_T / W_T, \quad W_T = \pi \mathrm{d}^3/16. \tag{9.26}$$

The residual stresses in the state of complete elastic unloading τ_r on the outer surface of the wire for the hyperbolic law (9.9) is shown for all six investigated materials and the preset stroke of s_p=7.22 mm on Fig. 9.4. Analogously, Fig. 9.7 presents residual stresses in the state of complete elastic unloading τ_r on the outer surface for the Ramberg–Osgood law (9.10) for the corresponding material specifications and an equal stroke s_p=7.22 mm. The values of preset loss and the residual stress on the surface of the wire are displayed in Table 9.3. The corresponding results for the stroke s_p=8.22 mm are shown in Table 9.4.

Figure 9.8 shows the effect of the preset stroke on the loss of spring length and the induced residual stress on the wire surface after the preset step. There is a minimum preset stroke that causes the spring length to decrease after the preset. The greater the preset travel, the greater the length reduction and the greater the negative shear residual stress on the wire surface. The negative residual stress means that the direction of principal stresses is opposite to the direction of shear stresses during plastic deformation of the wire. As shown in Fig. 9.8b, there is a progressive dependence of the length loss on the displacement. Conversely, the dependency of residual stress on preset travel is digressive, as shown in Fig. 9.8a. The published experimental values (Reich, 2019) allow the estimation of the preset reductions (Table 9.3). The correlations with the analytically evaluated values for these parameters are satisfactory. The next step is to fully validate the closed solution with experiments on helical compression springs with different spring indexes and manufacturing regimes.

Table 9.3 Length loss and residual stress on the surface for oil hardened and tempered and patented cold drawn spring steel for hyperbolic and for Ramberg–Osgood law (9.10)

Material	N	Hyperbolic Law (9.9)		Ramberg–Osgood law, (9.10)		Experiment
		Length loss "a", Δs (mm)	Residual stress "a", τ_r (MPa)	Length loss "b", Δs (mm)	Residual stress "b", τ_r (MPa)	Length loss Reich, 2019 (mm)
Oil hardened and tempered spring steel wire	1	0.321	−78	0.302	−91	
	2	0.250	−71	0.228	−86	
	3	0.446	−95	0.426	−129	
Patented cold drawn spring steel wire	4	1.312	−114	1.280	−127	1.07
	5	0.407	−94	0.388	−131	0.5
	6	0.344	−81	0.395	−132	0.68

Preset travel $s_p = 7.22$ mm

Table 9.4 Length loss and residual stress on the surface for oil hardened and tempered and patented cold drawn spring steel for hyperbolic and for Ramberg–Osgood law (9.10)

Material	N	Hyperbolic Law (9.9)		Ramberg–Osgood law, (9.10)	
		Length loss, "a", Δs (mm)	Residual stress „a", τ_r (MPa)	Length loss, "b", Δs (mm)	Residual stress "b", τ_r (MPa)
Oil hardened and tempered spring steel wire	1	1.4216	−154	1.3737	−192
	2	1.3167	−156	1.2797	−201
	3	1.8240	−166	1.7719	−224
Patented cold drawn spring steel wire	4	2.7954	−145	2.7208	−172
	5	1.7777	−171	1.7363	−235
	6	1.6567	−160	1.7458	−232

Preset travel $s_p = 8.22$ mm

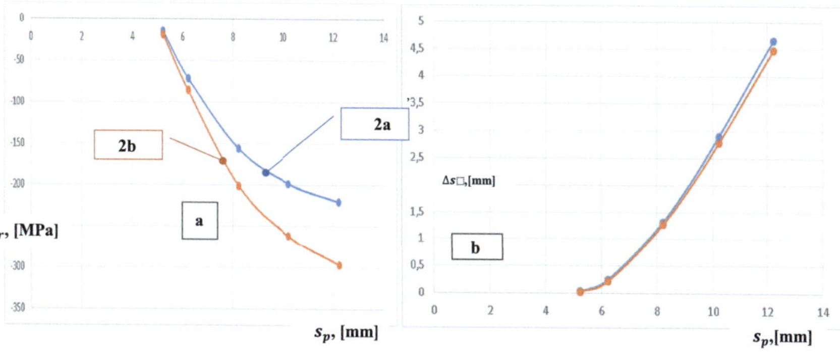

Fig. 9.8 Shear stress on the surface of wire ans loss of the spring length as the functions of preset stroke s_p for oil hardened and tempered wires (2a, 2b). **a** Surface shear-stress τ_r as function of preset stroke s_p, **b** Preset loss Δs as function of preset stroke s_p for two plastic laws for oil hardened and tempered wire

9.4 Evaluation of Helical Torsion Spring Presetting by Analysis of Active Plastic Bending of Rectangular Wire

9.4.1 Plastic Deformation of Wire During Active Plastic Bending Process

Torsion springs accumulate elastic energy when exposed to angular deflection. This type of springs is described in (SAE HS 795, 1989, Section D). For the torsion spring, the torque M_θ is applied on the upper end of the spring as shown of Fig. 9.1b. The angular deflection leads to an increase in the number coils a relative reduction of the spring diameter. The compression an extension springs, considered in the previous section are subjected to the torsional stresses of wires, while torsion springs are subjected to the bending stresses in the spring wire.

The effect of direction of coiling and beneficial residual stress in the helical torsional springs is explained in (SAE HS 795, 1989, Section D2, p. 2.45). In this section the mathematical theory of presetting for helical is developed. The spring wire with rectangular cross-section is loaded from a stress-free state by the bending moment M_B, which is equal to the applied axial torque M_θ. The effect of pitch and curvature of wire for stress calculation will be again neglected for simplicity. This assumption leads to the error in stress estimation less than 5%, if the ratio the mean diameter of spring to the section width is more than 7. The origin of coordinates is chosen at the centroid of area of one cross-section. The uniaxial strain in pure bending of the prismatic bar with the rectangular cross-section of the height T and width B reads:

$$\varepsilon = \kappa x \quad \text{for} \quad -T/2 \leq x \leq T/2. \tag{9.27}$$

In Eq. (9.27), the parameter κ is the curvature of the deformed axis of the bar. The maximal plastic elongation for the duration of presetting:

$$\varepsilon_p = \frac{1}{2}\kappa T \tag{9.28}$$

is attained on the contour of the cross-section. The uniaxial stress on the contour of the cross-section σ_p could be evaluated from the solution of the implicit relation: $\varepsilon_p = f_\sigma(\sigma_p)$.

9.4.2 Bending Moment During the Active Plastic Presetting in the Wire Cross Section for Hyperbolic Law

The sole non-vanishing stress component for a solid prismatic rod subjected to bending is σ. Equation (9.1) expresses the strain components in terms of stress. The bending moment of the wire, applied to the end sections of the rod, is:

$$M_B = B \int_{-T/2}^{T/2} \sigma(x) x \mathrm{d}x. \tag{9.29}$$

For application of Eq. (9.29) uses the stresses components in each point of the rectangular cross-section. In case of bending, the strain depends linearly on coordinate and implicitly leads to the uniaxial strain $\sigma(x)$:

$$\varepsilon(x) = 2\varepsilon_p \tfrac{x(\sigma)}{T} = f_\sigma(\sigma).$$

As already mentioned above, for the "implicit" laws (9.9) and (9.10) could not be resolved for the shear stress. The change of variables $x = x(\sigma)$ under the integral is allows the evaluation of the integral (9.13):

$$\tfrac{M_B}{B} = \int_{-T/2}^{T/2} \sigma(x) x \mathrm{d}x = 2\int_0^{\sigma_p} \sigma x(\sigma) \tfrac{\mathrm{d}x}{\mathrm{d}\sigma} \mathrm{d}\sigma = 2\int_0^{\sigma_p} \sigma x(\sigma) \left(\tfrac{\mathrm{d}\sigma}{\mathrm{d}x}\right)^{-1} \mathrm{d}\sigma. \tag{9.30}$$

For the "hyperbolic law" (9.9) the function $\rho(\tau)$ reads:

$$x(\sigma) = x \cdot \tfrac{\sigma}{\sigma_p} \equiv x \cdot \tfrac{f_\sigma(\sigma)}{f_\sigma(\sigma_p)} \equiv x \cdot \tfrac{\sigma}{\sigma_p} \cdot \left[\tfrac{1-(\sigma_p/K_\sigma)^n}{1-(\sigma/K_\sigma)^n}\right]^l. \tag{9.31}$$

The uniaxial hyperbolic law for the examined materials is shown on Fig. 9.9. We introduce for brevity the new dimensionless variable (dimensionless preset stress):

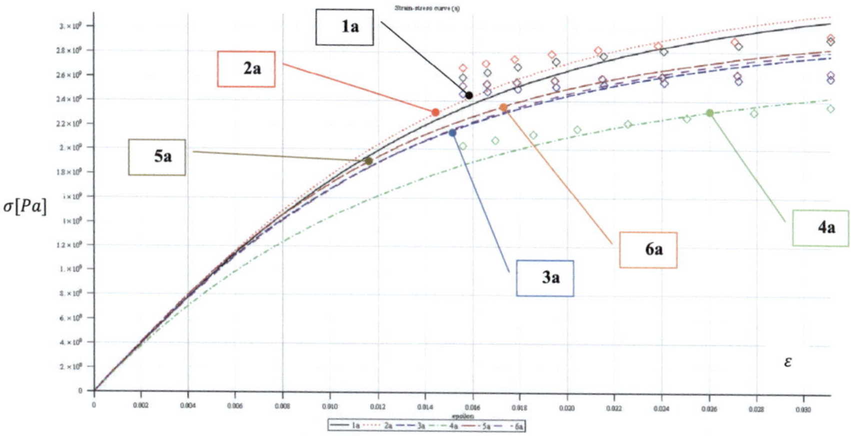

Fig. 9.9 Normal stress as function of tensile stain for the oil hardened and tempered spring steel wires (1a,2a,3a) and patented cold drawn spring steel wires (4a,5a,6a). The "hyperbolic law" is applied

$$\psi \stackrel{\text{def}}{=} \frac{\sigma_p}{K_\sigma}.$$

The integral (9.30) evaluates with Eq. (9.31) to:

$$M_B = \frac{\sigma_p B T^2}{6} \cdot \left(1 - \psi^n\right)^{2l} \cdot \left\{ {}_2\mathbf{F}_2\left(\left[\frac{3}{n}, 1 + 2l\right], \left[\frac{n+3}{n}\right], \psi^n\right) - {}_2\mathbf{F}_2\left(\left[\frac{3+n}{n}, 1 + 2l\right], \left[\frac{2n+3}{n}\right], \psi^n\right) \frac{3(\ln - 1)}{n+3} \psi^n \right\}. \tag{9.32}$$

For the common setting for the third solidifying exponent $l = 1/2$, the bending moment simplifies to:

$$M_B = \frac{\sigma_p B T^2}{12} \cdot \left\{ 3 - \left(1 - \psi^n\right) {}_2\mathbf{F}_2\left(\left[1, \frac{3}{n}\right], \left[\frac{n+3}{n}\right], \psi^n\right) \right\}. \tag{9.33}$$

For $l = 1/2, n = 2$, the hypergeometric function in (9.33) reduces to the elementary functions:

$$M_B = \frac{B T^2 \sigma_p}{4} \cdot \left(\frac{\psi^2 - 1}{\psi^3} \operatorname{arctanh} \psi + \frac{1}{\psi^2} \right). \tag{9.34a}$$

For $l = 1/2, n = 3$, the Eq. (9.33) yields to:

$$M_B = \frac{B T^2 \sigma_p}{4} \cdot \left[1 + \frac{1 - \psi^3}{3\psi^3} \ln\left(1 - \psi^3\right) \right]. \tag{9.34b}$$

In the elastic case the dimensionless variable ψ vanishes, $\psi \to 0$. The formulas for bending moment (9.32) and (9.33) reduce to their limit expression $M_B = \sigma_p BT^2/6$. The active load leads in this case to the pure elastic deformation. Consequently, the residual stresses disappear after the subsequent elastic spring-back.

9.4.3 Bending Moment During the Active Plastic Presetting in the Wire Cross Section for Ramberg–Osgood Law

For the Ramberg–Osgood law, as given by Eq. (9.10), the relation between the tensile stress and the distance to the neutral surface in the center of wire is given by:

$$x(\sigma) = x \cdot \frac{\varepsilon}{\varepsilon_p} \equiv x \cdot \frac{\sigma/E + (\sigma/K_\sigma)^n}{\sigma_p/E + (\sigma_p/K_\sigma)^n}. \tag{9.35}$$

The stress–strain relations of the Ramberg–Osgood law for the examined materials is shown on Fig. 9.12. With this formula, the bending moment in the wire could be expressed in closed form. Application of the Eq. (9.29) provides the expression for the moment:

$$M_B = \frac{BT^2\sigma_p}{6} \frac{\Phi(\psi)}{\Psi(\psi)}, \quad \psi = \frac{\sigma_p}{K_\sigma}, \tag{9.36}$$

$$\begin{aligned} \Phi(\psi) &= c_2\psi^2 K_\sigma^2 + c_1\psi^{1+n} K_\sigma E + c_0\psi^{2n} E^2, \\ \Psi(\psi) &= \left(\psi K_\sigma + \psi^n E\right)^2 \cdot c_3. \end{aligned} \tag{9.37}$$

In Eq. (9.37) the coefficients are the polynomials of the solidifying exponent m:

$$\begin{aligned} c_0 &= 3n(n+2), \quad c_1 = 3(n+1)(2n+1), \\ c_2 &= c_3 = (n+2)(2n+1). \end{aligned} \tag{9.38}$$

In the case $E \to \infty$ from (9.35) follows:

$$x(\sigma) = x \cdot \frac{\varepsilon}{\varepsilon_p} \equiv x \cdot \frac{\sigma^n}{\sigma_p^n} \tag{9.39}$$

and the Eq. (9.36) reduces to:

$$M_B \underset{E\to\infty}{=} \frac{BT^2\sigma_p}{2} \cdot \frac{n}{2n+1}. \tag{9.40}$$

In the elastic case ($\psi \to 0$) the formula for bending moment (9.36) reduces again to

$$M_B = \frac{1}{6}\sigma_p BT^2.$$

9.4.4 Elastic Recovery and Evaluation of Residual Stresses

Consider the helical spring with mean helix diameter D and active coils number n_a (DIN EN 13906:3, 2002). The axially directed torsion moment is applied to the end section the helical spring. The torsion moment is identical to the bending moment in all cross-sections of the spring wire M_B. The expressions (9.32) and (9.36) deliver the presetting moment during the plastic deformation. The torsion moment is statically equivalent to the tangential force F_θ (Fig. 9.1b). In the course of the presetting, the spring force attains the value:

$$F_\theta = 2M_B/D. \tag{9.41}$$

Torsion moment M_B leads to the travel of the end section of the spring in tangential direction. The twist angle in radians of the torsion spring during plastic presetting is θ_p. The tangential travel from the undeformed state to the plastically deformed state is equal to$s_p = D\theta_p/2$.

Residual stresses appear during the elastic spring-back of the wire. The simplest hypothesis for the unloading is the linear elastic spring-back. The increase of the spring twist angle θ_u during the spring-back reads:

$$s_u = \frac{F_\theta}{c_\theta} = \frac{D}{2}\theta_u, \quad c_\theta = \frac{EBT^3}{3\pi D^3 n_a}. \tag{9.42}$$

Consequently, the final reduction of the opening angle of the spring after plastic presetting and elastic unloading is:

$$\Delta s = s_p - s_u \equiv s_p - \frac{2M_B}{c_\theta D}, \tag{9.43}$$

$$\Delta\theta \equiv \frac{2\Delta s}{D} = \theta_p - \theta_u = \theta_p - \frac{4M_B}{c_\theta D^2}.$$

Alternatively, the unloading could be evaluated with the model due (Masing, 1926), but the Eqs. (9.40) and (9.41) are satisfactory for everyday practice.

The calculations were performed for the spring with the mean helix diameter $= 9$ mm, wire dimensions $T = 3$ mm, $B = 3$ mm and active coils number $n_a = 7$. Figures 9.10 and 9.13 demonstrate the dependences of active plastic and elastic spring-back forces for the hyperbolic and Ramberg–Osgood laws correspondingly. The final points on the force graphs with the zero force show the spring-back length of the spring, i.e. the reduction of spring length in cause of preset. For the hyperbolic law (9.9) preset length loss Δs is shown for six investigated materials on Fig. 9.11.

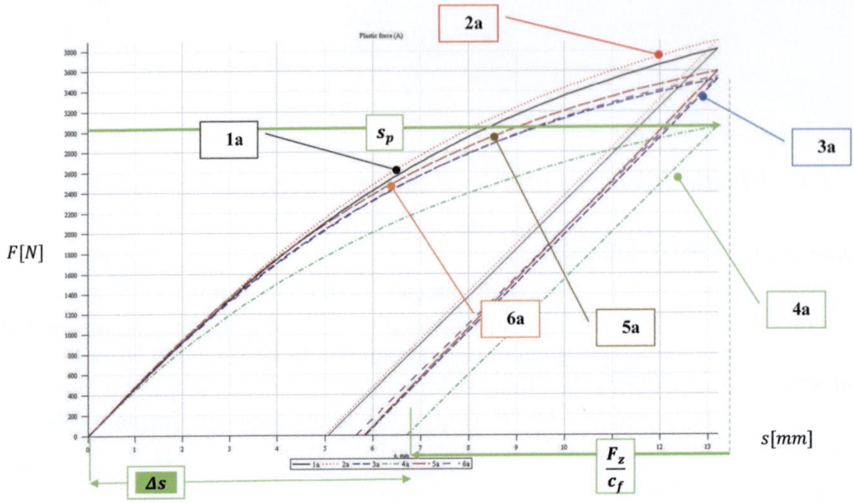

Fig. 9.10 Tangential spring force- as the function of tangential displacement for torsion springs made of oil hardened and tempered (1a, 2a, 3a) and of patented cold drawn spring steel wires (4a, 5a, 6a). The "hyperbolic law" is assumed

This figure displays also the shear stress during the plastic deformation and elastic spring-back on the surface of the wire. Figure 9.14 reveals the corresponding results for the Ramberg–Osgood law (9.10) for the same materials.

The residual stresses in the state of complete elastic unloading σ_r is equal to stresses σ_p in the state of maximal plasticization minus the stresses τ_u due to elastic unloading:

$$\sigma_r = \sigma_p - \sigma_u, \quad \sigma_u = \frac{M_B}{W_B}, \quad W_B = \frac{BT^2}{6}. \tag{9.44}$$

The residual stresses in the state of complete elastic unloading σ_r on the outer surface of the wire for the hyperbolic law (9.9) is shown for all six investigated materials and the preset stroke of s_p=13.22 mm on Fig. 9.11. Analogously, Fig. 9.14 presents residual stresses in the state of complete elastic unloading σ_r on the outer surface for the Ramberg–Osgood law (9.10) for the corresponding material specifications and an equal stroke s_p=13.22 mm. The values of preset loss and the residual stress on the surface of the wire are displayed in Table 9.5.

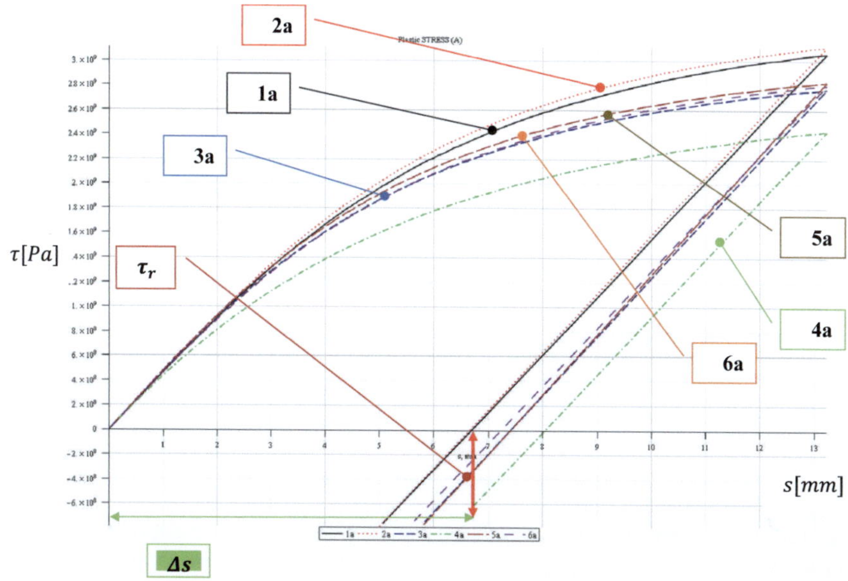

Fig. 9.11 Shear-stress vs.as the function of preset of springs meda of oil hardened and tempered (1a, 2a, 3a) and of patented cold drawn spring steel wires (4a, 5a, 6a). The "hyperbolic law" is used for calculation

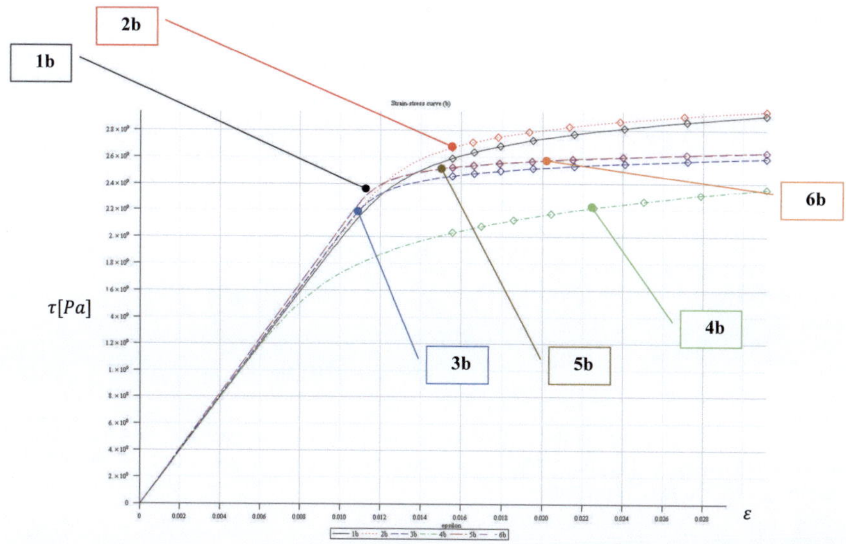

Fig. 9.12 Normal stress as the function of tensile strain curve for the oil hardened and tempered spring steel wires (1b, 2b, 3b) and patented cold drawn spring steel wires (4b, 5b, 6b). Assunption of the "Ramberg–Osgood-law"

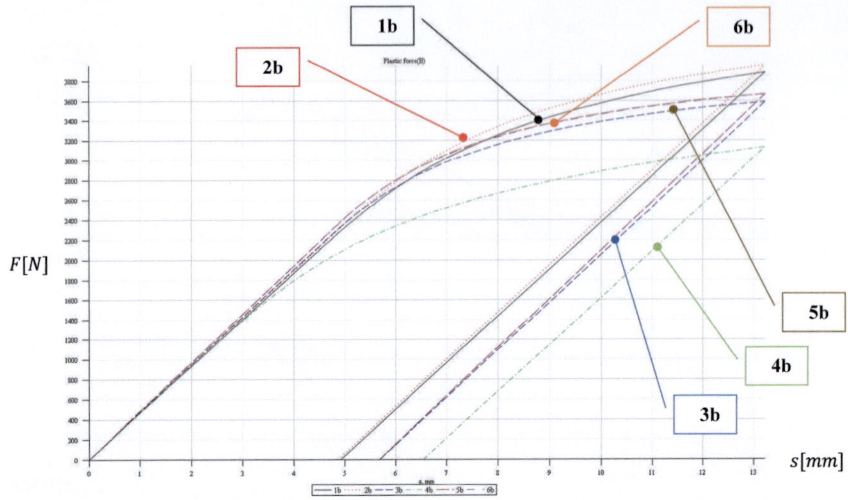

Fig. 9.13 Force- depending on the spring travel for oil hardened and tempered wires (1b, 2b, 3b) and of patented cold drawn spring steel wires (4b, 5b, 6b): Application of the "Ramberg–Osgood-law"

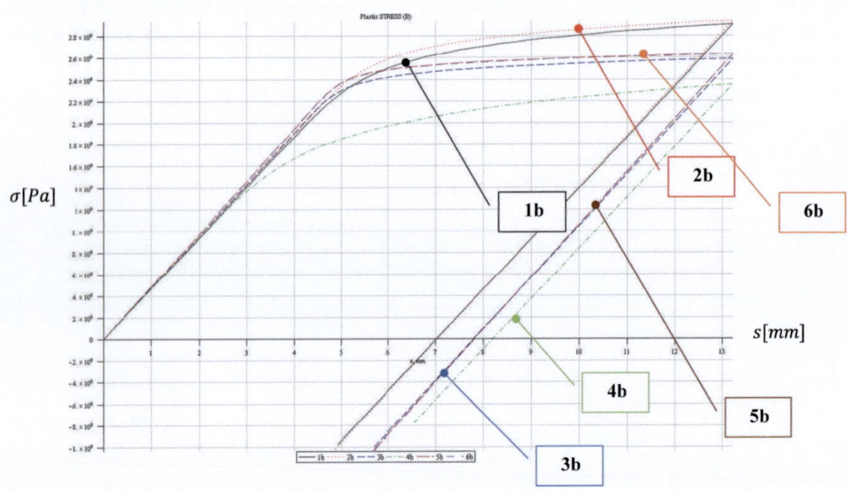

Fig. 9.14 Normal stress vs. on the surface as the function of the tangential spring travel for oil hardened and tempered (1b, 2b, 3b) and of patented cold drawn spring steel wires (4b, 5b, 6b). The "Ramberg–Osgood-law" is presumed

Table 9.5 Length loss and residual stress on the surface for oil hardened and tempered and patented cold drawn spring steel for hyperbolic and for Ramberg–Osgood criterion (9.10) for preset travel $s_p = 13.22$ mm

Material	N	Hyperbolic Law (9.9)		Ramberg–Osgood law, (9.10)	
		Tangential stroke loss "a", Δs [mm]	Residual stress "a", σ_r [MPa]	Tangential stroke "b", Δs [mm]	Residual stress "b", σ_r [MPa]
Oil hardened and tempered spring steel wire	1	7.06	−757	4.92	−976
	2	4.99	−783	4.86	−1018
	3	7.83	−746	7.68	−1000
Patented cold drawn spring steel wire	4	7.71	−618	7.53	−775
	5	7.81	−768	7.67	−1032
	6	7.64	−739	7.68	−1030

9.5 Explicit Formulations for Plastic Stress–Strain Curves

9.5.1 Relationships of Ramberg–Osgood and Johnson–Cook Formulas

For the simulation of deep plastic deformation, the alternative formulations (Ludwik, Hollomon, Swift and Voce, Johnson and Cook) of the stress - strain curves and strain-hardening of metals (Swift 1952), (Birkert et al, 2013; Kleemola & Nieminen, 1974), are commonly used. The nonlinear stress–strain expression due to Ludwik, Hollomon, Swift and Voce express the stress as the explicit function of strain. The stress–strain equations in tensor form read:

$$s = \frac{f_\tau^{-1}(\gamma_{oct})}{\gamma_{oct}} \boldsymbol{e}. \tag{9.45}$$

Equation (9.45) expresses the dependence of the stress deviator on the strain deviator. Since the stress components are functions of the strain components, the stress–strain Eq. (9.45) is referred to as the "explicit stress–strain function". From Eq. (9.45) follows the relationship between shear stress and shear strain intensities for the "explicit stress–strain function" (see Appendix A):

$$\frac{\tau_{oct}}{f_\tau(\tau_{oct})} = \frac{f_\tau^{-1}(\gamma_{oct})}{\gamma_{oct}}.$$

For metallic materials, the relationships between stress and strain are often based on the Johnson–Cook model (Johnson & Cook, 1983). The Johnson–Cook model correctly represents the large deformation, high strain rate, and elevated temperatures. For uniaxial deformation, the Johnson–Cook stress–strain equation is written in the form (A.8):

$$\sigma = F_1(\varepsilon)F_2(T^\circ)F_3(\dot{\varepsilon}). \tag{9.46}$$

The coefficients of the functions in Eq. (9.46) are:

A_σ and B_σ	are the strain coefficients,
T°_m	is the melting temperature,
T°_{ref}	is a reference (ambient) temperature,
m°	is the temperature exponent,
$\dot{\varepsilon}$	is the (plastic) strain rate,
$\dot{\varepsilon}_{ref}$	is a reference (quasi-static) strain rate and.
C	is the strain rate sensitivity parameter.

The term $F_1(\varepsilon)$ is called the Ludwik–Hollomon equation. This term reveals the quasi-static yielding and work hardening behavior:

$$F_1(\varepsilon) = A_\sigma + B_\sigma \varepsilon^{1/n}. \tag{9.47}$$

The term $F_2(T^\circ)$ determines the temperature dependence:

$$F_2(T^\circ) = 1 - \left(\frac{T^\circ - T^\circ_{ref}}{T^\circ_m - T^\circ_{ref}}\right)^m.$$

The third item demonstrates the strain rate sensitivity. The term $F_3(\dot{\varepsilon})$ includes commonly the logarithm of the strain rate, normalized by a reference value. More general approximations are known (Mauerauch & Vöhringer, 1978; El-Magd et al., 1985):

$$F_3(\dot{\varepsilon}) = 1 + C \ln \frac{\dot{\varepsilon}}{\dot{\varepsilon}_{ref}}.$$

For example, the strain rate sensitivity is characterized with two dimensionless exponents:

$$F_3(\dot{\varepsilon}) = \left(1 + C \ln^p \frac{\dot{\varepsilon}}{\dot{\varepsilon}_{ref}}\right)^q.$$

For uniaxial stresses and strains, the Ludwik-Hollomon Eq. (9.47) reads:

$$\sigma = A_\sigma + B_\sigma \varepsilon^{1/n}. \tag{9.48}$$

For pure shear stress, the constants of the Ludwik-Hollomon equation are:

$$\tau = A_\tau + B_\tau \gamma^{1/n}, \tag{9.49}$$

$$A_\tau = \frac{A_\sigma}{\sqrt{3}}, \quad B_\tau = \frac{B_\sigma}{\sqrt{3}} \cdot \left(\frac{1}{\sqrt{3}}\right)^{1/n}. \tag{9.50}$$

Table A.1 (Appendix A) summarizes the coefficients of the stress–strain equations. The formulas in Table A.1 are independent of temperature and strain rate. Dividing the stress in the Ludwik-Hollomon and Ramberg–Osgood equations by the corresponding terms allows these effects to be taken into account. The term $F_2(T^\circ)$ determines the temperature dependence and $F_3(\dot{\varepsilon})$ describes the strain rate sensitivity. Taking both terms into account, we obtain both explicit and implicit stress–strain relations that depend on temperature and strain rate:

$$\frac{\tau_{oct}}{F_2(T^\circ)F_3(\dot{\varepsilon})} = f_\tau^{-1}(\gamma_{oct}), \tag{9.51}$$

$$\gamma_{oct} = f_\tau\left(\frac{\tau_{oct}}{F_2(T^\circ)F_3(\dot{\varepsilon})}\right). \tag{9.52}$$

From the Ludwik–Hollomon Eq. (9.47) one gets once again the Johnson–Cook formula, which expresses the octahedral stress τ_{oct} over the octahedral strain τ_{oct}. From Ramberg–Osgood (9.10) follows the inverse formulation, which expresses the strain τ_{oct} through stress τ_{oct}.

Now study the relations between Ludwik–Hollomon equation on one side and Ramberg–Osgood equation on the other side. It is possible to express the coefficients of Ludwik–Hollomon (9.47) and Johnson–Cook Eq. (9.46) through the coefficients of the Ramberg–Osgood relations. The simplification of the relations could be achieved using the following setting for the constants:

$$B_\sigma = K_\sigma, \quad A_\sigma = \frac{-K_\sigma^2}{K_\sigma + nE} \approx -\frac{K_\sigma^2}{nE}. \tag{9.53}$$

In this case the Ludwik–Hollomon (9.47), Johnson–Cook (9.46) and Ramberg–Osgood (9.10) curves match at$\varepsilon = 1$. Table A.2 recapitulates the simplified stress–strain relations for pure shear and uniaxial elongation states.

9.5.2 *Torsion of the Rod with Circular Cross-Section*

Because the Ludwik–Hollomon Eq. (9.47) expresses stress through the strain, the integrals, which express moments, could be calculated immediately. Using the expressions (9.11) and (9.12) for the radial dependency of shear, the torque for the rod with the circular cross-section results to:

$$M_T = 2\pi \int_0^r \tau(\gamma(\rho))\rho^2 \mathrm{d}\rho \equiv \frac{2\pi}{3}\left(\frac{\mathrm{d}}{2}\right)^3 \cdot \left(A_\tau + \frac{3n}{3n+1}\gamma_p^{1/n} B_\tau\right). \tag{9.54}$$

9.5.3 Bending of the Rod with Circular Cross-Section

Analogously, using the expressions (9.27) and (9.28) for axial strain over normal coordinate, we get the formula for bending moment to the rod with the circular cross-section:

$$M_B = 2\int_0^r \sigma(\varepsilon(x))x\sqrt{\left(\tfrac{d}{2}\right)^2 - x^2}\,dx \equiv \tfrac{2}{3}\cdot\left(\tfrac{d}{2}\right)^3\cdot\left(A_\sigma + \frac{3\sqrt{\pi}\Gamma\left(\frac{1+2n}{2n}\right)}{4\Gamma\left(\frac{1+5n}{2n}\right)}\varepsilon_p^{\frac{1}{n}}B_\sigma\right). \tag{9.55}$$

The function $\Gamma(x)$ in Eq. (9.55) is the Euler gamma function.

9.5.4 Bending of the Rod with Rectangular Cross-Section

The closed-form expression of the bending moment for the prismatic bar with the rectangular cross-section of the height t and width b is:

$$M_B = B\int_{-T/2}^{T/2} \sigma(\varepsilon(x))x\mathrm{d}x \equiv \tfrac{BT^2}{4}\cdot\left(A_\sigma + \frac{2^{(n-1)/2}n}{1+2n}\cdot\varepsilon_p^{1/n}\cdot B_\sigma\right). \tag{9.56}$$

The nonlinear torsion problem for the prismatic rectangular bar is solvable numerically. The method to evaluate the residual stresses and length reduction is developed in Sects. 3.4 and 4.4. The developed method is applicable without difficulties as well for the Eqs. (9.54)–(9.56).

The current industry standard for presenting the acquired mechanical data for the spring wire is the Ramberg–Osgood or similar implicit formulas (Sect. 9.2). The above formulas (9.54)–(9.56) require recalculation to the explicit data representation (such as Johnson–Cook or Ludwik-Hollomon). The modification of the existing material data to a modified representation is necessary for spring wires of different chemical compositions and heat treatment regimes. However, the amount of data currently collected and stored in the spring manufacturing industry is immense. Therefore, the use of Ramberg–Osgood and hyperbolic plastic flow models seems to be appropriate for practice.

9.6 Time-Delayed Presetting

9.6.1 Instantaneous Ideal Elastic-Ideal Plastic Flow

Consider a rod of circular cross section with radius R of wire. The hypothesis of the ideal elastic-perfectly plastic material is applied (Rees, 2006, §7.3). The ideal, elastic-perfectly plastic behavior of the material without hardening is assumed. In the initial moment, the instantaneous deformation occurs. If the stresses are moderate, the purely elastic, reversible deformation occurs. Otherwise, if the stresses are high enough to cause the irreversible deformation, the material flows ideally plastically. For the ideal elastic-ideal plastic material, there is a clear separation between the elastic and plastic zones. The radial distribution of shear stresses after the initial plastic flow is ($r = \mathrm{d}/2$):

$$\tau_p(\rho) = \begin{cases} G\theta\rho, & \text{for } 0 < \rho < r_p, \\ \tau_Y, & \text{for } r \geq \rho \geq r_p. \end{cases} \tag{9.57}$$

In Eq. (9.57) θ is the twist angle per unit length, G is the shear modulus, τ_Y is the yield shear stress.

The radius of boundary between the plastic and elastic zones is:

$$r_p = \begin{cases} \frac{\tau_Y}{\theta G}, & \text{if } \theta \geq \theta_Y, \\ r, & \text{if } \theta < \theta_Y, \end{cases} \qquad \theta_Y \overset{\text{def}}{=} \frac{\tau_Y}{rG}. \tag{9.58}$$

From Eq. (9.58) follows, that if the shear stress on the outer boundary of the wire is less than yield stress, $\tau_p(r = \mathrm{d}/2) \leq \tau_Y$, the material behaves elastically over the complete cross-section, including the skin of the wire. In this case, the twist angle of wire is not greater than θ_Y. Increasing of the twist angle beyond θ_Y leads to the propagation of a plastic zone inwardly from the outer region.

The torque over the circular cross-section M_T^0 is the sum of the partial torque over the elastic region M_E^0 and the partial torque over the plastic region M_P^0:

$$M_T^0 = M_E^0 + M_P^0, \tag{9.59}$$

$$M_E^0 = 2\pi \int_0^{r_p} \rho^2 \tau_p(\rho) \mathrm{d}\rho, \quad M_P^0 = 2\pi \int_{r_p}^{\mathrm{d}/2} \rho^2 \tau_p(\rho) \mathrm{d}\rho.$$

The expressions for the partial torque at the moment of plastic flow $t = 0$ are:

$$M_E^0 = \begin{cases} \frac{\pi \tau_Y^4}{2\theta^3 G^3}, & \text{for } \theta \geq \theta_Y, \\ \frac{\pi \theta G (\mathrm{d}/2)^4}{2}, & \text{for } \theta < \theta_Y, \end{cases} \tag{9.60a}$$

$$M_P^0 = \begin{cases} \frac{2\pi\tau_Y\left((d/2)^3\theta^3G^3-\tau_Y^3\right)}{3\theta^3G^3}, & \text{for } \theta \geq \theta_Y, \\ 0, & \text{for } \theta < \theta_Y. \end{cases} \tag{9.60b}$$

The twist angle pro wire length is

$$\theta = \frac{2s_P}{\pi n_a D^2},$$

where $s_P = L_0 - L_1$.

The twist angle θ depends upon the preset spring travel (height reduction of the spring) s_P during the squeezing the spring in the tool holder.

9.6.2 Equations of Creep During Time-Delayed Presetting

After the instantaneous plastic deformation is completed, the stage of creep flow occurs (Betten, 2008; Krajcinovic, 1983). In this chapter, only the basic equations are presented. A more detailed study of creep is presented in the next Chapter. The main hypothesis is the following. As shown above, the material is in elastic state in the inner region and in plastic state in the outer region. During the time delay, the material is subjected to an infinite strain for a certain period of time. As the strain continues, stress relaxation occurs in both regions of elastic and plastic deformation. The strain rate is defined by the material-specific creep law. Typically for metals and alloys, there are the primary and secondary creep stages. Because the presetting time is too short for the secondary creep flow, there occurs only primary creep.

For the analysis, we assume an isotropic, time-dependent stress function for the strain rate, Eq. (A.9):

$$\dot{e}_{ij} = \frac{s_{ij}}{\tau_{oct}} f_c(\tau_{oct}, t). \tag{9.61}$$

For the common creep laws, the isotropic stress function in (9.61) could be represented as the product of two functions. For details, see Eq. (10.9), Chap. 10. In the case of shear stress, these functions depend on time and octahedral shear stress accordingly:

$$f_c(\tau_{oct}, t) = h_c(t) \cdot g_c(\tau_{oct}). \tag{9.62}$$

The dimensionless time-dependent function $h_c(t)$ in Eq. (9.62) describes the evolution of strain rate:

$$h_c(t) = \left(\frac{t}{\bar{t}}\right)^{\varsigma-1}, \tag{9.63}$$

or

$$h_c(t) = \exp\left[-\left(\frac{t}{\overline{t}}\right)^{\varsigma}\right]. \tag{9.64}$$

The characteristic time constants $\overline{t}, \varsigma$ are introduced in Eqs. (9.63) and (9.64). Usual setting is $\overline{t} = 1[s]$ and $0 < \varsigma < 1$.

The function $g_c(\tau)$ in Eq. (9.62) portrays the stress dependency of the strain rate. The common interpretations for the stress dependency are the Norton-Bailey law (Betten, 2008; Odquist & Hult, 1962):

$$g_c(\tau_{oct}) = \overline{\varepsilon} \cdot \left(\frac{\tau_{oct}}{\overline{\tau}}\right)^{\xi+1}, \tag{9.65}$$

or Garofalo law [Garofalo, 1963]:

$$g_c(\tau_{oct}) = \overline{\varepsilon} \cdot \sinh^{\xi+1}\left(\frac{\tau_{oct}}{\overline{\tau}}\right), \quad \xi \geq 0. \tag{9.66}$$

The dimension of $\dot{e}_{ij}$, of $g_c(\tau)$ and of $\overline{\varepsilon}$ is $\left[s^{-1}\right]$.

9.6.3 Creep Deformation After Instant Plastic Flow

Simultaneously to the plastic flow or instantly after it finishing, happens the relaxation of stresses due to creep. This process starts, if the spring is hold compressed for a definite lengths_P. The strain in the wire during relaxation does not alter, but the stress gradually reduces for the certain period of presetting:$0 \leq t \leq T_P$. To investigate this process analytically, we consider the relaxation of stresses in a wire of the circular cross-section under the persistent twist. For briefness, we neglect the curvature and pitch of the helical spring.

Let $\tau_c(\rho, t)$ is the shear stress in the cross-section of the wire in the state of creep. The total shear strain in any moment of time is $\gamma(\rho, t)$, is the sum of the elastic and the creep components of shear strain:

$$\gamma = \gamma_e + \gamma_c. \tag{9.67}$$

The creep component of shear strain is $\gamma_c(\rho, t)$. The elastic component of shear strain is

$$\gamma_e = \frac{\tau_c}{G}. \tag{9.68}$$

The total deformation remains constant in time during the presetting process. Thus, we consider the total strain $\gamma_0(\rho)$ as function of radius only. However, the elastic and the creep components of strain are the functions as well of radius and of time, such that:

$$\gamma(\rho, t) = \gamma_e(\rho, t) + \gamma_c(\rho, t) \equiv \gamma_0(\rho). \tag{9.69}$$

The shear during presetting process $\gamma_0(\rho)$ remains constant. The time differentiation of (9.65) leads to the differential equation for elastic and creep strain rates:

$$\frac{\mathrm{d}\gamma(\rho, t)}{\mathrm{d}t} = \frac{\mathrm{d}\gamma_e(\rho, t)}{\mathrm{d}t} + \frac{\mathrm{d}\gamma_c(\rho, t)}{\mathrm{d}t} \equiv 0. \tag{9.70}$$

The Eq. (9.68) delivers the expression for the shear strain γ_e. However, in the Eq. (9.70) the strain rate is required. Obviously, the differentiation of the Eq. (9.68) over time delivers the elastic component of strain rate:

$$\frac{\mathrm{d}\gamma_e}{\mathrm{d}t} = \frac{1}{G}\frac{\mathrm{d}\tau_c}{\mathrm{d}t}. \tag{9.71}$$

The substitution of material laws (9.62) and (9.71) in Eq. (9.70) results in the ordinary, nonlinear differential equation of the first order for total shear stress $\tau_c(\rho, t)$:

$$\frac{\mathrm{d}\tau_c}{\mathrm{d}t} + G \cdot f_c(\tau_c, t) = 0, \quad \tau_c(\rho, t = 0) = \tau_p(\rho). \tag{9.72}$$

For the Norton-Bailey creep law (9.65), the variables in the Eq. (9.71) could be separated after the substitution:

$$\frac{\mathrm{d}\tau_c}{\mathrm{d}t} + G \cdot \bar{\varepsilon} \cdot \left(\frac{\tau_c}{\bar{\tau}}\right)^{\xi+1} \frac{\mathrm{d}H}{\mathrm{d}t} = 0,$$

$$h_c(t) = \frac{\mathrm{d}H}{\mathrm{d}t}, \quad H(0) = 0. \tag{9.73}$$

For the compactness of the following expressions an auxiliary function is introduced in Eq. (9.73):

$$H(t) = \int_0^t h_c(t)\mathrm{d}t.$$

The initial condition for the Eq. (9.73) is:

$$\tau_c(\rho, t = 0) = \tau_p(\rho). \tag{9.74}$$

With Eq. (9.74) expresses the creep and plastic stresses match, such that the creep stress originates from the plastic stress.

The solution of the ordinary differential Eq. (9.73) yields the shear stress over the cross-section of the twisted rod as the function of time and radius:

$$\frac{1}{[\tau_c(\rho,t)]^\xi} = \frac{\overline{\varepsilon} G\xi H(t)}{\overline{\tau}^{\xi+1}} + \frac{1}{\tau_p^\xi(\rho)} \equiv \begin{cases} \frac{\overline{\varepsilon} G\xi H(t)}{\overline{\tau}^{\xi+1}} + \frac{1}{(G\theta\rho)^\xi}, & \text{for } \rho < r_p, \\ \frac{\overline{\varepsilon} G\xi H(t)}{\overline{\tau}^{\xi+1}} + \frac{1}{\tau_Y^\xi}, & \text{for } \rho \geq r_p. \end{cases} \tag{9.75}$$

The total torque over the complete cross-section $M_T(t)$ is the function of time. This function is the sum of the partial torque over the elastic region $M_E(t)$ and the partial torque over the plastic region $M_P(t)$:

$$M_T(t) = M_E(t) + M_P(t), \tag{9.76}$$

$$M_E(t) = 2\pi \int_0^{r_p} \rho^2 \tau_c(\rho,t)\mathrm{d}\rho, \quad M_P(t) = 2\pi \int_{r_p}^{\mathrm{d}/2} \rho^2 \tau_c(\rho,t)\mathrm{d}\rho. \tag{9.77}$$

With the expression for total shear stress (9.75) we can evaluate the integrals (9.77) for torque over the elastic and plastic regions correspondingly in terms of the Meijer $\boldsymbol{G}$ -function (Bateman, 1953):

$$M_E(t) = \begin{cases} \frac{2\pi\tau_Y^4}{\xi\Gamma(1/\xi)\theta^3 G^3} \cdot G\left(\left[\left[\frac{\xi-4}{\xi}, \frac{\xi-1}{\xi}\right], []\right], \left[[0], \left[-\frac{4}{\xi}\right]\right], \frac{\xi G \tau_Y^\xi}{\overline{\tau}^{\xi+1}} \overline{\varepsilon} H(t)\right), & \text{for } \theta \geq \theta_Y, \\ \frac{2\pi\theta G(\mathrm{d}/2)^4}{\xi\Gamma(1/\xi)} \cdot G\left(\left[\left[\frac{\xi-4}{\xi}, \frac{\xi-1}{\xi}\right], []\right], \left[[0], \left[-\frac{4}{\xi}\right]\right], \frac{\xi G\cdot(\theta G\mathrm{d}/2)^\xi}{\overline{\tau}^{\xi+1}} \overline{\varepsilon} H(t)\right), & \text{for } \theta < \theta_Y, \end{cases} \tag{9.78}$$

$$M_P(t) = \begin{cases} \frac{2\pi\left((\mathrm{d}/2)^3\theta^3 G^3 - \tau_Y^3\right)}{3\theta^3 G^3} \cdot \left[\frac{G\overline{\varepsilon} H(t)\xi}{\overline{\tau}^{\xi+1}} + \frac{1}{\tau_Y^\xi}\right]^{-1/\xi}, & \text{for } \theta \geq \theta_Y, \\ 0, & \text{for } \theta < \theta_Y. \end{cases} \tag{9.79}$$

For $\xi = 1, 2, 4$ the Meijer $\mathbf{G}$ -function in (9.78) reduces to the elementary functions (Askey, 2010):

$$\frac{\mathbf{G}\left(\left[\left[\frac{\xi-4}{\xi}, \frac{\xi-1}{\xi}\right], []\right], \left[[0], \left[-\frac{4}{\xi}\right]\right], Z\right)}{4n\Gamma(1/\xi)} = \begin{cases} -\frac{2}{3} \cdot \frac{6\ln(1-Z)+6Z+3Z^2+2Z^3}{Z^4} & \text{for } \xi = 1, \\ -\frac{4}{3Z} \cdot \frac{\sqrt{1-Z}-1-Z}{\sqrt{1-Z}+1} & \text{for } \xi = 2, \\ -\frac{4}{3} \cdot \frac{(1-Z)^{3/4}-1}{Z} & \text{for } \xi = 4. \end{cases}$$

The following series expansions are valid for positive ξ:

$$\frac{1}{4\xi\Gamma(1/\xi)}\mathbf{G}\left(\left[\left[\frac{\xi-4}{\xi},\frac{\xi-1}{\xi}\right],[]\right],\left[[0],\left[-\frac{4}{\xi}\right]\right],z\right)=1+\frac{4}{\xi(\xi+4)}z+o\left(z^2\right),$$

$$\frac{1}{4n\Gamma(1/\xi)}\mathbf{G}\left(\left[\left[\frac{\xi-4}{\xi},\frac{\xi-1}{\xi}\right],[]\right],\left[[0],\left[-\frac{4}{\xi}\right]\right],0\right)=1.$$

There is the equivalent representation of the Meijer **G** -function in terms of hypergeometric function (Bateman, 1953):

$$\mathbf{G}\left(\left[\left[\frac{\xi-4}{\xi},\frac{\xi-1}{\xi}\right],[]\right],\left[[0],\left[-\frac{4}{\xi}\right]\right],z\right)=4n\Gamma\left(\frac{1}{\xi}\right)\cdot{}_2\mathbf{F}_2\left(\left[\frac{1}{\xi},\frac{4}{\xi}\right],\left[\frac{\xi+4}{\xi}\right],z\right).$$

According to Eq. (9.73), in the initial moment of creep $H(t=0)=0$. The former formula verifies that the expressions (9.78), (9.79) and (9.80) match at $t=0$:

$$M_P(0)=M_P^0,\quad M_E(0)=M_E^0.$$

The plots of the total torque $M_T^0=M_E^0+M_P^0$ and partial torques over the elastic and plastic regions in the initial moment $t=0$ are shown on Fig. 9.2. The abscissa depictures the twist angle per unit length θ in radians pro meter.

With the expressions for moments (9.76), (9.79) evaluates also the axial force, which is applied on the helical spring. The presetting force reads in terms of the torque and mean diameter of the helical spring:

$$F_z(t)=\frac{2}{D}\begin{cases} M_E^0+M_P^0, & \text{for } t=0,\\ M_E(t)+M_P(t), & \text{for } t>0.\end{cases}\tag{9.80}$$

9.6.4 *Elastic Spring-Back and Occurrence of Residual Stresses*

When the elastic–plastic bar or helical spring is completely unloaded, elastic stresses recover to leave a residual shear stress field. The residual stresses appear in the moment of removal of the torque $M_T(t=t_P)$ in the momentt_P. If the unloading after initial plastic deformation and unloading occurs not immediately after the accomplishing of the plastic process, the relaxation of the stresses appears. The residual stress files is the difference between the shear stress in elastic and plastic zones under the time-dependent torque $M_T(t_P)$, Eq. (9.76) and the elastic shear stress τ_u that recovers in the moment of removal of the moment$M_T(t_P)$.

The spring travel from the undeformed state to the plastically deformed state is equal to a definite constant values_P. In the course of the presetting of helical springs the spring force attains the value F_z. When the force disappears, the unloading occurs. The simplest hypothesis for the unloading is the linear elastic spring-back that leads to the increase of the spring length:

$$s_u^0 = \frac{F_z^0}{c_f}, \quad s_u(T_P) = \frac{F_z(T_P)}{c_f}, \quad c_f = \frac{Gd^4}{8D^3 n_a}, \quad s_u = L_2 - L_1. \tag{9.81}$$

Consequently, the reduction of the length of the spring after plastic presetting and elastic unloading reads:

$$\Delta s^0 = s_P - s_u^0 \equiv s_P - \frac{2M_T^0}{c_f D}, \quad \Delta s = s_P - s_u \equiv s_P - \frac{2M_T(t_P)}{c_f D}. \tag{9.82}$$

Because the torque reduces with time of presetting, the reduction of length increases with time:

$$\Delta s^0 = L_0 - L_2(t_P = 0) \quad \text{or} \quad \Delta s = L_0 - L_2 \quad \text{for } t_P > 0,$$

such that

$$\Delta s^0 < \Delta s.$$

For an infinite creep time the torque vanishes,

$$M_T(t \to \infty) \to 0$$

and the elastic unloading disappears $\Delta s = s_P$. The length of the spring remains equal to its compressed value. In vector form, the corresponding twist angles for the length differences are:

$$\left[\theta_u^0, \theta_u, \Delta\theta^0, \Delta\theta\right] = \frac{2}{\pi n_a D^2}\left[s_u^0, s_u, \Delta s^0, \Delta s\right].$$

The simplest definition of residual stress is as follows: stresses that remain in a part after it has been deformed and all external forces have been removed. More specifically, the deformation must be non-uniform across the cross section of the material for residual stresses to occur. Deformation can result not only from forming, but also from thermal processes. It is well known that phase transformations during heat treatment induce sufficient strain to cause plastic deformation, which results in residual stresses. In this paper, the residual stresses occurring in the cylindrical rod after plastic torsion and in the helical spring after prestressing are determined analytically. The calculation results of the plastic creep deformation and the final instantaneous recovery of the material are shown in Figs. 9.15, 9.16 , 9.17, 9.18, and 9.19.

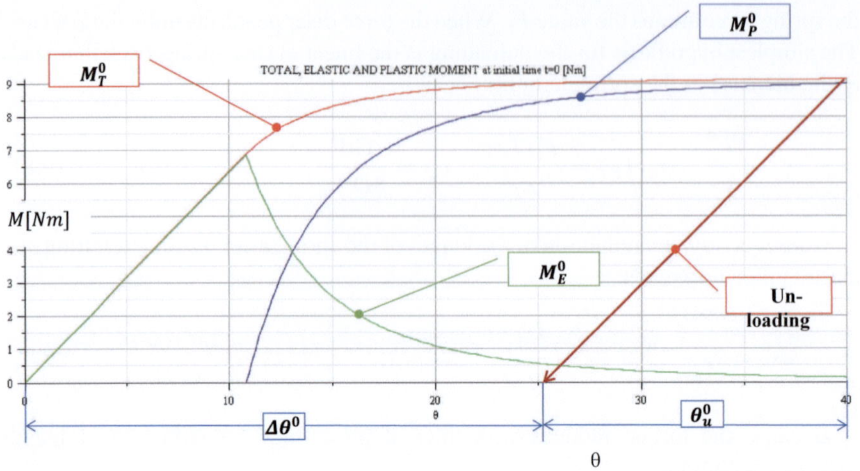

Fig. 9.15 Elastic, plastic and total torque during the plastic deformation as the function of twist angle

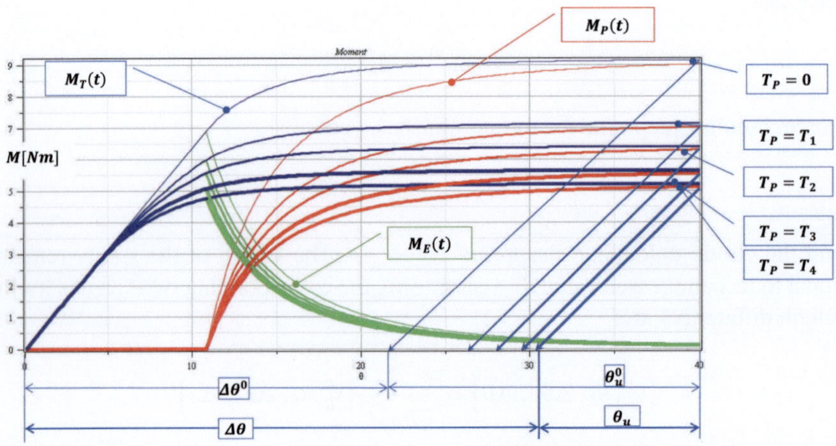

Fig. 9.16 Elastic, plastic and total torque during the creep deformation as the function of twist angle for different preset times $t_1 \equiv 86400s$, $t_2 \equiv 172800s$, $t_3 \equiv 259200s$, $t_4 \equiv 345600s$

If the unloading happens immediately after the plastic deformation and creep does not happen, the residual stress in the state of complete elastic unloading τ_r^0 is equal to stresses τ_p^0 in the state of maximal plasticization minus the stresses τ_u^0 due to elastic unloading:

$$\tau_r^0 = \tau_p^0 - \tau_u^0, \quad \tau_u^0 = \frac{M_T^0}{W_T}, \quad W_T = \frac{\pi \mathrm{d}^3}{16}. \tag{9.83}$$

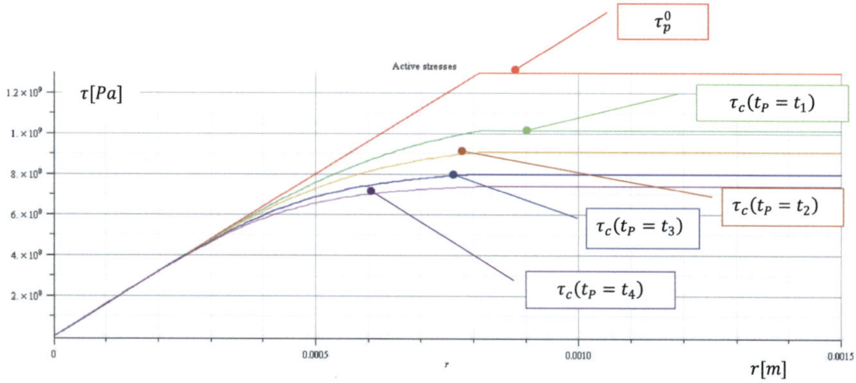

Fig. 9.17 Total shear stress, reduction of stress and residual stress after the plastic deformation (dotted lines) and after preset duration $t_P = 345600s$ (solid lines). Total stress – blue, stress reduction – green, residual stress - red

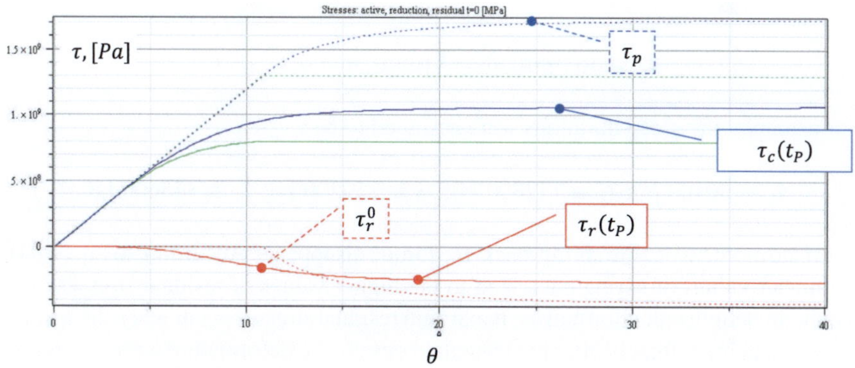

Fig. 9.18 Stress distribution after creep deformation prior to elastic unload for different creep times t

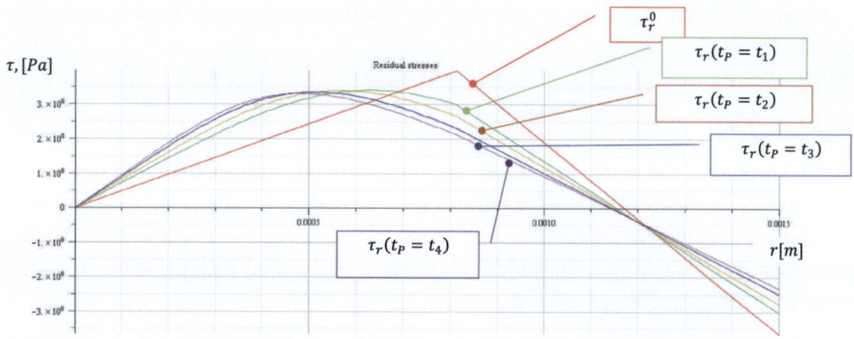

Fig. 9.19 Residual stress distribution over radius after creep deformation for different creep times t

The elastic, plastic and total torque during the plastic deformation as the function of twist angle are shown on Fig. 9.15. If spring is hold in the tool for a certain period of time t, the unloading takes place after the completing of the creep process. The residual stress in the state of complete unloading after creep $\tau_r(T_P)$ is equal to stresses $\tau_c(T_P)$ in the state after creep process minus the stresses $\tau_u(t)$ due to unloading:

$$\tau_r(t_P) = \tau_c(t_P) - \tau_u(t_P), \quad \tau_u(T_P) = \frac{M_T(t_P)}{W_T}. \tag{9.84}$$

The residual stresses in the state of complete elastic unloading $\tau_r(t)$ on the outer surface of the wire depend on the presetting time t_P.

Figure 9.16 demonstrates the effect of creep on the torque for different durations of preset t_P. The ordinate is the value of torques in Nm. For the calculations the following values were used:

$$G = 80 \text{ [GPa]}, \text{h} = 2.5 \cdot 10^{-9} \cdot t^{0.9}, \tau_Y = 1.3 \text{ [GPa]},$$

$$D = 36 \text{ [mm]}, \text{d} = 3 \text{ [mm]}, n = 4, n_a = 7.$$

For example, the creep durations were assumed to be

$$t_1 \equiv 86400 \text{ [s]}; \; t_2 \equiv 172800 \text{ [s]}; \; t_3 \equiv 259200 \text{ [s]}; \; t_4 \equiv 345600 \text{ [s]}.$$

At the moment of wire spring-back, the torque disappears. According to Eq. (9.84), the higher values of spring-back lead to the higher values of residual stresses. The maximum length reduction and the maximum residual stresses occur when the spring-back occurs immediately after the instantaneous plastic deformation without a relaxation period. The longer the relaxation time, the lower the residual stresses and the smaller the elastic recovery of the spring. The final length of the spring after presetting will also be shorter with the higher degree of relaxation.

Figure 9.18 displays the stresses after the instantaneous plastic flow $\tau_p(r = \text{d}/2)$ and the residual stress

$$\tau_r(r = \text{d}/2, t = 0) = \tau_r^0(r = \text{d}/2)$$

on the surface of the wire. For these values the dotted lines of blue and red colors respectively are uses. The stresses after creep period t_P on the surface of the wire are shown with the solid lines: creep stress $\tau_c(r = \text{d}/2, t = t_P)$ and the residual stress $\tau_c(r = \text{d}/2, t = t_P)$. The calculations were performed with the above material parameters and for creep duration:

$$t_P = t_4 \equiv 345600 \text{ [s]}.$$

The corresponding distributions of the shear stresses after the mentioned above presetting times, but before the elastic spring-back over the radius of wire are shown on Fig. 9.17. The resulting radial distributions of the residual stresses are shown on Fig. 9.19.

9.6.5 *Creep Deformation After Instant Plastic Flow for Garofalo Law*

For the Garofalo creep law (Betten, 2008), the variables in the Eq. (9.73) could be separated after the substitution and lead to the initial value problem for shear stress$\tau_c(r, t)$:

$$\frac{\mathrm{d}\tau_c}{\mathrm{d}t} + G \cdot c \cdot \sinh^{\xi+1}\left(\frac{\tau_c}{\overline{\tau}}\right) \cdot \frac{\mathrm{d}H}{\mathrm{d}t} = 0, \quad \tau_c(r, t = 0) = \tau_p(r)\,. \tag{9.85}$$

The initial condition for the Eq. (9.85) presumes the shear stress in the moment $t = 0$ after the instant plastic deformation. For the ideal elastic-ideal plastic material, the shear stresses after the instant plastic flow for the rod with circular cross-section result from Eq. (9.73)

The solution of the ordinary differential Eq. (9.73) for $\xi = 0$ with initial condition (9.74) delivers the shear stress over the cross-section of the twisted rod as the function of time and radius:

$$\tau_c(\rho, t) = \overline{\tau} \cdot \ln\left\{\tanh\left[\frac{cGH(t)}{2\overline{\tau}} + \operatorname{arctanh}\left(\exp\left(\frac{\tau_p(\rho)}{\overline{\tau}}\right)\right)\right]\right\}. \tag{9.86}$$

According to Eq. (9.76), the couple $M_T(t)$ is the function of time is the sum of the couple over the elastic region $M_E(t)$ and the couple over the plastic region $M_P(t)$. With the expression for total shear stress (9.81) the partial couples over the elastic and plastic regions express in terms of polylogarithmic function. The final formulas are too bulky and display the analogous functions of the mechanical values, as result after the assumption of the Norton-Bailey law.

9.7 Conclusions

Spring presetting is the final over-elastic loading of the springs at the end of the manufacturing process at room temperature or hot setting at elevated temperature. When properly applied, presetting has a positive effect on both the relaxation and fatigue behaviour of the springs. The fatigue properties of automotive springs (coil springs, leaf springs or torsion bars), which are stressed in only one working direction (swelling), can be positively influenced by intentional plasticizing (over-elastic deformation) in this direction at the end of the production process. This causes a partial flow in the surface layer, which leads to work hardening and residual compressive stresses in the direction of the operational load when the spring is subsequently relieved. In the case of leaf or parabolic springs, longitudinal compressive residual stresses occur in the edge layer on the bending tension side; in the case of coil springs, corresponding compressive residual stresses occur below 45°, which are related to the tensile stresses from the torsional load stress state of the installed spring.

The change in shape (setting loss) resulting from presetting can be used to set the same overall height for series parts and must be maintained at the original shape of the springs. While the presetting of leaf springs is relatively free, the presetting of helical compression springs is restricted by limiting conditions such as buckling or blocking. To increase the effectiveness of presetting in these cases, it is sometimes carried out at elevated temperatures.

The presetting process of helical springs is essential for manufacturing practice. The analysis was performed using the deformation plasticity theory with several nonlinear stress–strain laws (SAE AE-22, §8.3.33). These laws describe the active plastic deformation of the wire along with the linear elastic recovery. Both twist and curvature of the wire increase continuously during plastic loading and decrease during elastic recovery. The process of plastic flow and the origination of residual stresses are studied. The analytical results are in good agreement with the experimental values. The closed-form solutions allow the estimation of stresses and geometry changes during the manufacturing of helical springs. The final formulas could be used for industrial applications and programming of manufacturing machines, and for dimensioning and wear calculation of presetting tools.

An important aspect of the manufacturing process is the time-dependent presetting of helical springs. The twist of the spring wire increases during plastic loading and decreases during elastic unloading. The combined plasticization-relaxation process causes the formation of residual stresses. At higher temperatures, the relaxation process is accelerated so that at higher temperatures the length reduction is faster and the residual stresses are lower. The proposed method leads to a closed solution. The next step is to validate the formulas with experiments on helical compression springs with different spring indexes and manufacturing regimes. The closed form solutions allow the estimation of relaxation and residual stresses during the manufacturing process of helical springs. The derived formulas facilitate industrial applications.

9.8 Summary of Principal Results

- The instantaneous plastic flow in the initial stage of prestressing is studied in a closed form.
- Fixation of the compressed spring length in a certain period of time leads to relaxation of the stress and reduction of the resulting residual stress.
- The analytical formulas for length loss and stress reduction as functions of prestressing time are derived.

References

Allen, A. J., Hutchings, M. T., Windsor, C. G., & Andreani, C. (1985). Neutron diffraction methods for the study of residual stress fields. *Advances in Physics, 34*, 445–473.

Askey, R. A., & Olde Daalhuis, A. B. (2010) Meijer G-function. In: Olver, F. W. J., Lozier, D. M., Boisvert, R. F., Clark, C. W. (Eds.). *NIST handbook of mathematical functions* (pp. 403–418). Cambridge University Press, Cambridge, MA, ISBN: 978-0521192255

Atterbury, T. J., & Diboll, W. B., Jr. (1960). The effect of presetting helical compression springs. *Journal of Engineering for Industry, 82*(1), 41–44. https://doi.org/10.1115/1.3662990

Bateman, H., & Erdelyi, A. (1953) *Higher transcendental functions* (Vol. I). McGraw–Hill, New York (see § 5.3, "Definition of the G-Function", p. 206)

Betten, J. (2008). *Creep mechanics* (3d ed.). Springer.

Birkert, A., Haage, S., & Straub, M. (2013). *Umformtechnische Herstellung komplexer Karosserieteile.* Springer-Verlag.

Del Llano-Vizcaya, L., Rubio-González, C., Mesmacque, G., & Cervantes-Hernández, T. (2006). Multiaxial fatigue and failure analysis of helical compression springs. *Engineering Failure Analysis, 13*, 1303.

El-Magd, E., Abdel-Ghany, W., & Homayun, M. (1985) Material constants and mechanical behaviour during plastic wave propagation. In *Proceedings of the Numeta '85 Conference/ Swansea*, p. 437/445

Garofalo, F. (1963). An empirical relation defining the stress dependence of minimum creep rate in metals. *Transactions in Metallurgy Society AIME, 227*, 351–356.

Geinitz, V., Weiß, M., Kletzin, U., & Beyer, P. (2011). Relaxation of helical springs and spring steel wires. In: *56th International scientific colloquium*, 12–16 September 2011. Ilmenau University of Technology.

Graves, G. B., & O'Malley, M. (1983). Vorsetzen, insbesondre Warmvorsetzen von Schraubenfedern. *Draht, 34*(4), 171–173.

Hutchings, M. T., & Krawitz, A. D. (1992). *Measurement of residual and applied stress using neutron diffraction.* Springer Science Business Media.

Johnson, G. R., & Cook, W. H. (1983). A constitutive model and data for metals subjected to large strains, high strain rates and high temperatures. In: *Proceedings of 7th International Symposium on Ballistics* (Vol. 21, pp. 541–547).

Jones, R. M. (2009). *Deformation theory of plasticity.* Bull Ridge Publishing.

Kachanov, L. M. (1971). *Foundations of the theory of plasticity.* North-Holland Publishing Company

Karpov, I. D., Em, V. T., Mazalov, P. B., & Sulyanova, E. A. (2018). Characterisation of residual stresses by neutron diffraction at the research reactor IR-8 of NRC "Kurchatov Institute". *IOP Conference Series: Journal of Physics: Conference Series, 1109*, 012046. https://doi.org/10.1088/1742-6596/1109/1/012046

Kausel, E. (2017). *Advanced structural dynamics.* Cambridge University Press.

Kleemola, H. J., & Nieminen, M. A. (1974). On the strain-hardening parameters of metals. *Metallurgical Transactions, 5*, 1863. https://doi.org/10.1007/BF02644152

Kobelev, V. (2020). Delayed presetting of helical springs. *Mechanics Based Design of Structures and Machines, 48*(1), 122–132. https://doi.org/10.1080/15397734.2019.1669457

Kobelev, V. (2011). Elastoplastic stress analysis and residual stresses in cylindrical bar under combined bending and torsion. *Journal of Manufacturing Science and Engineering, 133*(4), 044502.

Kobelev, V. (2019). Elastic–plastic deformation and residual stresses in helical springs. *Multidiscipline Modeling in Materials and Structures,* 16(3), 448–475. https://doi.org/10.1108/MMMS-04-2019-0085

Krajcinovic, D. (1983). Creep of structures—A continuous damage mechanics approach. *Journal of Structural Mechanics, 11*(1), 1–11. https://doi.org/10.1080/03601218308907428

Kreuzer, A. (1984) Warmvorsetzen von Schraubenfeden. *Draht, 35*(7/8), 386–389.

Masing, G. (1926). Eigenspannungen und Verfestigung beim Messing. In: *Proceedings of 2nd International Congress of Applied Mechanics* (pp. 332–325). Zurich.

Matejicek, J., Brand, P. C., Drews, A. R., Krause, A., & Lowe-Ma, C. (2004). Residual stresses in cold-coiled helical compression springs for automotive suspensions measured by neutron diffraction. *Materials Science and Engineering, A, 367*, 306–311.

Mauerauch, E., & Vöhringer, O. (1978). *Z. Werkstofftechnik, 9*(11), 370–391.

Močilnik, V., Gubeljak, N., & Predan, J. (2015). Surface residual stresses induced by Torsional plastic pre-setting of solid spring bar. *International Journal of Mechanical Sciences, 92*. https://doi.org/10.1016/j.ijmecsci.2016.01.004

Neuber, H. (1961a). Theory of stress concentration for shear-strained prismatical bodies with arbitrary nonlinear stress–strain law. *Journal of Applied Mechanics, 28*, 544–550.

Neuber, H. (1961b). *Theory of notch stresses: Principles for exact calculation of strength with reference to structural form and material*. United States Atomic Energy Commission, US Office of Technical Services.

Odquist, F. K. G., & Hult, J. (1962). *Kriechfestigkeit metallischer Werkstoffe*. Springer-Verlag.

Ramberg, W., & Osgood, W. R. (1943). *Description of stress-strain curves by three parameters*. NACA-TN902.

Rees, D. W. A. (2006). *Basic engineering plasticity, an introduction with engineering and manufacturing applications*. Butterworth-Heinemann.

Reich, R. (2017). Prediction of presetting value for helical compression springs. In *9th International congress of Spring industry*, September 29th–October 1th, Taormina/Italy.

Reich, R. (2019). TU Ilmenau, Germany, Private Communication.

Rubinstein, R. J. (1977). On the elastic-plastic torsion problem. *Journal of Engineering and Mathematics, 11*(4), 319–323. https://doi.org/10.1007/BF01537091

Shimoseki, M., Hamano, T., & Imaizumi, T. (Eds.). (2003). *FEM for springs*. Springer-Verlag, 978-3-540-00046-4.

Wakita, M., Kuno, T., Amano, A., Nemoto, A., Saruki, K., & Tanaka, K. (2007). Study on predicting formula for torsional fatigue strength of spring steel, effect of environment, notch and hardness. *Transactions of Japan Society of Spring Engineers, 2007*(52), 1–7.

Yang, Z., & Wang, Z. (1996). Effect of prestrain on cyclic creep behaviour of a high strength spring steel. *Material Science and EngineeriNg, A210*, 83–93.

Part III
Service Life and Durability of Springs

Chapter 10
Creep and Relaxation of Springs

Abstract In this chapter, we study the time-dependent behavior of spring elements under constant and oscillating loading. The common creep laws are implemented to describe the material. The models for stress relaxation under constant strain and for creep under continuous load are established. For simple spring elements, closed-form solutions for the creep and relaxation models are found. The introduction of the "scaled" time variable simplifies the presentation. The experimental procedure for the experimental acquisition of creep models is presented. This chapter concludes the second part, which deals with the manufacturing of springs. The highly undesirable creep can occur during the life cycle of mechanical springs, but proper manufacturing reduces or eliminates such phenomena.

Keywords Creep of springs · Relaxation of springs · Temperature dependent creep laws

10.1 Operational Damage of Spring Elements

During the service time, springs and other machine elements are subjected to various loads. These loads cause different types of stresses in the elements. Safe dimensions for springs are determined by failure assessment methods. Any load-bearing component fails when the induced stress exceeds the elastic limit and permanent deformation of the component occurs. For example, the ultimate strength is the stress limit at which failure and immediate irreversible deformation occur. Depending on the material behavior, a distinction must be made between brittle fracture strength (shear strength) and ductile flow strength (ductile deformation). In the case of ultimate strength, a distinction is made between compressive strength and tensile strength. Compressive strength symbolizes the limit stress at which a material suffers a complete loss of cohesion along a fracture surface. The compressive strength of spring steel alloys is on the order of 3 GPa. Tensile strength refers to the strength of a material under tensile stress and, for steel alloys, is approximately equal to compressive strength. When a certain limit of tensile strength is exceeded in the normal direction, failure

V. Kobelev, *Fundamentals of Springs Mechanics*,
https://doi.org/10.1007/978-3-031-58584-5_10

sometimes occurs by elongation fracture, where the material tears completely along a fracture surface. The common standard tests for determining strength are uniaxial compression and tension tests, triaxial compression tests, and shear tests.

There are two causes of spring damage, depending on the specific application and the nature of the load.

For the first group, static loading causes irreversible deformation as a result of creep. For automotive and industrial springs, the static loads usually do not lead to immediate material failure, but cause the residual deformation of the spring. For example, the clearance of the vehicle gradually decreases over time as a result of creep. On the other hand, the shape does not change if the compression of the spring continues, but the spring load decreases with time and causes the malfunction of the unit. The creep process is characterized by the high ductility of the material. Ductility stabilizes the fracture and compensates the effects of imperfections. The distribution of failures caused by creep is much smaller than the distribution of fatigue failures. Keeping this behavior in mind, we study the creep phenomena in a purely deterministic manner, avoiding statistical analysis of creep. Excessive creep is a common source of spring failure and is discussed in this chapter.

For the second group of springs, the cyclic degradation and fracture of the material is far more relevant. It is accepted that metals fracture under oscillating stresses whose maximum value is less than that required to cause fracture in a static test. However, many ferrous alloys have a "fatigue limit" or stress below which such failure never occurs, even at very high numbers of cycles. Fatigue life under cyclic loading is sensitive to imperfections and manufacturing inaccuracies. The scatter of fatigue life in a large batch of tested springs can be very large. In extreme cases, the size of the scatter can cover about one decimal power. The phenomena of spring fatigue are discussed in Chaps. 11 and 12.

10.2 Common Creep Constitutive Equations

10.2.1 Constitutive Equations for Creep of Spring Elements

When a load is applied to a piece of metal, the metal will initially bend elastically; if the load is high enough, it will also bend plastically. Elastic deflection occurs when the material returns to its previous shape/geometry after relief. Plastic deformation is when the metal undergoes permanent deformation and does not return to its previous shape/geometry. It is the elastic part of the deflection which normally experiences the deflection in durable springs.

The stress/deflection function is represented by Hook's law, which describes the modulus of elasticity E for normal and G for shear stress. A metal of high strength can normally be loaded to a higher stress level without exceeding the limit at which the deflection becomes plastic. The stress value at which the metal begins to deflect plastically is the yield strength σ_Y of the material. The ultimate tensile strength σ_w is the stress at which the metal breaks. In addition to the knowledge of stress–strain relations at low temperatures, the processes that take place at elevated temperatures are particularly interesting. Metals acquire a hardening effect through plastic deformation. If the temperature is increased, recovery and recrystallization processes begin which continuously counteract the hardening. At certain temperatures, the dislocations of recovery processes caused by deformation are continuously compensated. The permanent deformation begins to grow. This process is called "creep", a physical phenomenon that designates a slow plastic deformation, which takes place at stress below the yield strength σ_Y of the material. The basic physical processes are dislocation creep, diffusion creep and diffusion-supported grain boundary sliding. The constitutive models and the solution methods for creep problems were discussed in (Kassner, 2008; Poirier, 1985). Comprehensive reviews of creep laws for common engineering materials were provided in (Betten, 2008; Naumenko & Altenbach, 2007; Yao et al, 2007). The importance of creep effects is even more important for the polymer or composite materials. Such materials exhibit non-linear isochronous stress–strain relationships, the degree of non-linearity increases with stress and time. These materials generally also have a higher deformation resistance under compression than under tensile creep conditions. This behaviour cannot be adequately described by the usual viscoelasticity theory. The corresponding equations enable the prediction of isothermal creep behaviour of non-linear viscoelastic materials under multi-axial time-independent loading (Schlimmer, 1986). The viscoelasticity equations are based on an elastic potential that applies to materials with different behaviour under tensile and compressive loading.

The common material models adequately describe definite creep stages from constant stress uniaxial tests (Naumenko, 2006). These phenomenological models represent an attempt to approximate in the simplest formulas the experimentally acquired data. From the mechanical viewpoint, the phenomenological models must characterize the stress, time and temperature dependences. However, the behavior of the material dramatically changes with temperature and phase transformation (Ashby, 2011). Consequently, the unified approximation with the experimentally accessible number of parameters is possible in a narrow temperature interval. If the character temperature of the structural is fixed, it makes sense to evaluate the material constants for each relevant temperature and for each creep stage.

The parametric methods were developed to compute creep and creep fracture in high temperature applications. The distinctive mathematical properties of the power law allowed the development of analytical methods (Boyle & Spence, 1983, Chap. 3). The corresponding creep equations can be found commercial numerical simulation codes.

Several other models were able to predict the phenomenon of creep focusing on the rupture time and minimum strain rate of a material: "Theta Projection" (Evans et al., 1982), "Wilshire Equations" (Wilshire & Battenbough, 2007; Wilshire & Scharning, 2007) and "Uniaxial Creep Lifting approach" (Wu, 2010). In the "Theta Projection" approach creep, the creep curves under uniaxial constant stress measured over a range of stresses and temperatures. The creep curves can be "projected" to other stress/temperature conditions re-constructing full creep curves.

The "Wilshire Equations" provide the facility to describe the stress-rupture behavior, this together with the proper representation of full creep curves.

The "uniaxial creep-lifting approach" postulates that deformation occurs by grain boundary sliding, dislocation gliding, dislocation climbing, and diffusion mechanisms. The total strain rate for a polycrystalline material is the sum of the contributions from each of these mechanisms.

The newly constructed curves define the necessary creep characteristics. The approach for creep data assessment (ECCC) was elaborated European creep collaborative committee (Holdsworth, 2008).

The obtained results are applied to a practically important problem of engineering, namely the simulation of creep and relaxation of helical and disk springs. A spring is considered to be durable if it does not lose any force, or if it does not shorten or break at a given force, so that the function is disturbed or eliminated. For a spring to be durable, the stress in the material must not exceed the strength of the material. In the spring industry, this phenomenon is called "creep" when a spring loses length under constant load, and "relaxation" when a spring loses load under constant compression. Over time, every spring experiences a loss of stress, which can be referred to as relaxation or creep, depending on the nature of the load on the spring.

Relaxation occurs when the spring is compressed to a constant length and the force decreases over time. Creep occurs when the spring loaded with a constant force experiences an additional loss of length over time, ΔI, so that its overall height decreases.

The significance of creep or relaxation effects depends on temperature, stress in the metal, yield strength of the metal, and time. Increased temperature, stress, and time also increase creep/relaxation. Temperature and stress in particular have a large influence.

In diverse types of springs several kinds of stresses cause creep in material. The helical springs are made up of a wire coiled in the form of a helix and are primarily intended for compressive or tensile loads. The springs store the elastic energy by means of either bending or torsion. Respectively, in material dominates either uniaxial or pure shear stress state. These stresses initiate the creep and loss of length or of spring force. The stress-state in the wire of helical spring is highly inhomogeneous. The stress in the circular wire varies from zero on the axis of the wire to its maximum value on the surface. As was pointed out in (Boardman, 1964), the assumption of the homogeneity of stress is incorrect. The resulting error leads to an erroneous determination of creep law parameters, when with the stress exponent is other than one. However, the frequent values for stress exponents are about four (Gittus, 1971). The cross-section of the wire from which the spring is made may be circular, square, or rectangular. A formal technique was developed in (Chang, 1995) to predict the stress relaxation for compression and torsion springs. The technique uses uniaxial tensile-generated stress-relaxation data for spring wires. Based on the tension-induced stress-relaxation data, the technique was applied to compression springs, where shear stress dominates in predicting the stress relaxation.

The shear stress–strain curve is first constructed from the octahedral stress–strain curve. Stress relaxation is a phenomenon in which part of the elastic strain responsible for the initial stress is replaced by creep strain. The same argument is valid for spring elements in the predominant bending state, such as disk springs or torsion helical springs. The brief study of the creep process during prestressing was already discussed in Chap. 9. In this Chapter we continue to systematically study the creep effects during spring operation. An essential task of creep analysis is to derive the exact closed-form expressions for torsional and bending creep for isotropic materials, which obey the familiar Norton-Bailey, Garofalo and exponential constitutive laws.

10.2.2 Time-Dependent Constitutive Equations

The creep component of strain rate is defined by material specific creep law. The models are based on the time- and strain-hardening constitutive equations for stress that depends on operation time[1] t. Following the common procedure (Betten, 2008), we adopt an isotropic, time-dependent constitutive equation:

$$\frac{\mathrm{d}\boldsymbol{e}_{ij}}{\mathrm{d}t} = \frac{\boldsymbol{s}_{ij}}{\tau_{oct}} f_c(\tau_{oct}, t). \tag{10.1}$$

The constants of the constitutive Eq. (10.1) depend on the type, thermal treatment and grain size spring material, the particular creep mechanism, and temperature. For the spring element, typical trial and operation temperatures are all fixed. Therefore, the parameters of the material law correspond to the definite operation temperature.

Hereafter the incompressible behavior of material in the process of creep deformation ($\dot{\varepsilon}_{kk} = 0$) is presumed. Thus, the octahedral normal strain remains constant during the creep process: $\varepsilon_{oct} = const$.

The constitutive equations Eq. (10.1) comprise the explicit dependence of strain rate upon time. The time dependence expresses the tendency to the strong reduction of creep strain rate in the region of primary creep. Immediately after the application of an external load, the creep rate is maximal, but reduces gradually with time. After a certain period of time, the creep rate stabilizes and remains mainly constant until the final, tertiary creep region achieved and breakage occurs. According to Andrade (1910, 1914), a power or an exponential function of time depicts this reduction of creep strain rate. The reduction does not depend on the stress level. Thus, for all applied stresses the relative strain rate reduction is the same. This feature allows to factorize the function as the product of time function and stress function.

When a plastic material is subjected to a constant load, it deforms with a variable strain rate. The initial strain is roughly predicted by its stress–strain modulus. The experimental results display, that the higher stress in tension accelerates the reduction of strain rate (Poirier, 1985). The interval of time, which leads to the stabilization of

[1] For material thickness, the symbol T is used. The symbol t and t^* are reserved for time variables.

creep rate and transmission from the primary to the secondary creep stage, reduces with the higher stress level.

The most important region for the technical springs is the section of the primary creep. The primary region is the early stage of loading when the creep rate decreases rapidly with time. Further, the material will continue to deform gradually with time, until yielding or rupture causes failure. The total strain rate is the sum of primary, secondary and tertiary partial strain rates:

$$f_c(\tau_{oct}, t) = f_{Ic}(\tau_{oct}, t) + f_{IIc}(\tau_{oct}, t) + f_{IIIc}(\tau_{oct}, t). \tag{10.2}$$

The creep strain rate in these regions is commonly defined the Norton-Bailey law (Betten, 2008). At constant temperature, the creep deformation of metals grows in time following to a power law of stress. According to the Andrade law (Andrade, 1910, 1914), the creep strain rate in the primary creep stage could be described by the relation

$$f_{Ic}(\tau_{oct}, t) = \overline{\varepsilon_I} \cdot \left(\frac{t}{\overline{t}}\right)^{\varsigma-1} \left(\frac{\tau_{oct}}{\overline{\tau}}\right)^{\xi+1}. \tag{10.3a}$$

We use the constants for proper scaling of stress and time. For example, the suitable setting for scaling constants:$\overline{\tau} = 1[\text{MPa}]$, $\overline{\sigma} = 3/\sqrt{2}[\text{MPa}]$, $\overline{t} = 1[s]$. We use this setting. The different setting.

The creep rate in Eq. (10.3) at time moment $t = 0$ is infinite. This context confuses the applicability of the Eq. (10.3). An alternative expression for time function that is free from an initial singularity, uses the exponential time function in primary region:

$$f_{Ic}(\tau_{oct}, t) = \overline{\varepsilon_I} \cdot \exp\left[-\left(\frac{t}{\overline{t}}\right)^{\varsigma}\right]\left(\frac{\tau_{oct}}{\overline{\tau}}\right)^{\xi+1}. \tag{10.3b}$$

The creep strain rate reaches later a steady states which are called the secondary and tertiary creep stages. The recovery and recrystallization processes in this section completely compensate for the strain hardening by dislocations. The creep deformation is proportional to time and the creep rate simultaneously assumes its lowest value. This region therefore frequently covers most of the component's service life. The strain rate functions are:

$$f_{IIc}(\tau_{oct}, t) = \overline{\varepsilon_{II}} \cdot \left(\frac{\tau_{oct}}{\overline{\tau}}\right)^{\xi+1}, \tag{10.4a}$$

$$f_{IIIc}(\tau_{oct}, t) = \overline{\varepsilon_{III}} \cdot \exp\left[\left(\frac{t}{\overline{t}}\right)^{\zeta} - 1\right]\left(\frac{\tau_{oct}}{\overline{\tau}}\right)^{\xi+1}. \tag{10.4b}$$

Here $\overline{\varepsilon_I}$,$\overline{\varepsilon_{II}}$,$\overline{\varepsilon_{III}}$ are the experimental constants for the primary, secondary and tertiary stages of creep. The units of these constants is $[1/s]$.

Function (10.5) could be represented as the product of two functions, which depend correspondingly on time and on stress (Kobelev, 2016):

$$f_c(\tau_{oct}, t) = h_c(t) \cdot g_c(\tau_{oct}). \tag{10.5}$$

The creep law (10.5) pronounces the continuous transition between the primary and secondary creep regimes (Cadek, 1988) Primary creep occurs at the beginning of the creep test and is characterized first by a high strain rate. The creep rate decelerates gradually to a constant value denoting the beginning of secondary creep. This behavior corresponds to the experimentally acquired creep data (Nabarro & de Villers, 1995; Es-Souni, 2000). However, the continuous transition between creep regimes leads to some mathematical difficulties. The regions of primary and secondary creep are frequently being separated for reasons of mathematical simplification (Nezhad & O'Dowd, 2012, 2015).

The derivation of closed form solution is applied to the generalized creep law (10.5). The most general form of creep law with continuous transition between the primary and secondary creep regimes permits the closed form solutions because of simple geometry of considered stress states.

Particularly, the functions in (10.5) are:

$$h_c(t) = \overline{\varepsilon_I} \cdot \left(\frac{t}{\overline{t}}\right)^{\varsigma - 1} + \overline{\varepsilon_{II}}, \text{ or } h_c(t) = \overline{\varepsilon_I} \cdot \exp\left[-\left(\frac{t}{\overline{t}}\right)^{\varsigma}\right] + \overline{\varepsilon_{II}}. \tag{10.6a}$$

The time function (10.6) represents the contribution of primary, secondary and tertiary creep regimes. There are several possibilities to proper exemplifications of the creep strain as the function of time. For the "θ-Projection" (Evans et al., 1982), (Harrison, Evans, 2007) and for the "uniaxial creep lifting" (Wu, 2010) approaches, the secondary strain rate vanishes ($\overline{\varepsilon_{II}} = 0$):

$$h_c(t) = \overline{\varepsilon_I} \cdot \exp\left[-\left(\frac{t}{\overline{t}}\right)^{\varsigma}\right] + \overline{\varepsilon_{III}} \cdot \exp\left[\left(\frac{t}{\overline{t}}\right)^{\zeta} - 1\right]. \tag{10.6b}$$

10.2.3 Experimental Acquisition of Creep Laws

The measurement of the twist of wire under the constant torque provides the straightforward way of the evaluation of creep constant for helical springs (Gooch, 1986). The advantage the torsion creep test over tension test is the relative insensitivity to misalignments and the direct account of the possible radial variations of material properties. The state of twist in a straight wire in the test equipment matches the state in a helical spring. The samples of the real spring wire with the same heat treatment and the same surface handling are used in the experiment. During the torsion creep test the function of the twist angle over time is gained. The determination of the stress exponent requires at least three values of the torque. During the torsion creep test the function of the twist angle over time is gained for each value of applied torque. These at least three functions provide together with corresponding values of torque moments, length and diameter of the wire the basis for determination of creep constants.

For example, assume the Norton-Bailey creep law in a form:

$$\gamma = h_c(t)\tau^{\xi+1}. \tag{10.7}$$

As the shear strain is the linear function of radius: $\gamma = \theta\rho, 0 < \rho < r = \mathrm{d}/2$ the solution of Eq. (10.7) leads to the expression for shear stress:

$$\tau(\rho) = \left(\frac{\theta\rho}{h_c(t)}\right)^{1/(\xi+1)}. \tag{10.8}$$

With Eq. (10.8), the torque reads

$$M_T = 2\pi \int_0^r \tau(\rho)\rho^2 \mathrm{d}\rho = 2\pi \cdot \frac{\xi+1}{4+3\xi} \cdot \rho^{\frac{4+3\xi}{\xi+1}} \cdot \theta^{\frac{1}{\xi+1}} \cdot h_c(t)^{-\frac{1}{\xi+1}}.$$

For a constant value of M_T the twist angle is the function of time:

$$\theta(t) = \frac{M_T^{\xi+1}}{r^{3\xi+4}} \cdot \left(\frac{1}{2\pi}\frac{4+3\xi}{\xi+1}\right)^{\xi+1} \cdot h_c(t).$$

If two functions $\theta_1, \theta_2,$ are gained from experiments for two different moments M_{T1}, M_{T2}, then the stress exponent is:

$$\xi = \frac{\ln(\theta_1/\theta_2)}{\ln(M_{T1}/M_{T2})} - 1.$$

For numerical determination the common curve fitting procedures must be applied. The experimentally determined functions of creep angle over time are shown on Fig. 10.1 (points, right plot, for wire diameter d = 2.7 mm). The lines of different color correspond to different applied constant torque. The experimentally obtained twist angles for three different moments are plotted with solid lines.

The second step is to resolve the time dependence $h_c(t)$. As the stress dependence is already settled, the function that responsible for time dependence reads:

$$h_c(t) = \frac{r^{3m+4}\theta(t)}{M_T^{\xi+1}} \cdot \left(2\pi\frac{\xi+1}{4+3\xi}\right)^{\xi+1}.$$

For commercially available spring materials the time dependence the logarithm of this function is an almost straight line in double logarithmical scale (Fig. 10.1, left):

$$h_c(t) = t^a e^b, \quad \ln(h_c(t)) = at + b.$$

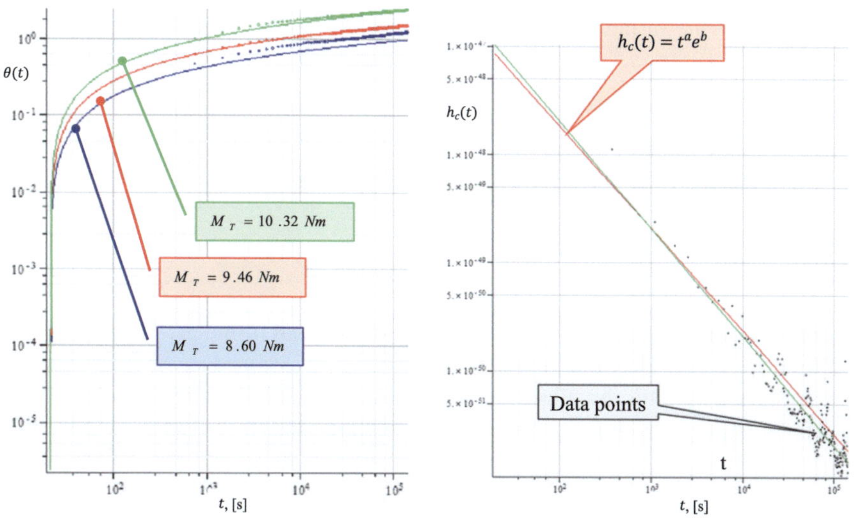

Fig. 10.1 Twist angle as functions of time for different torques (left) and the fitted linear function for time dependency (right)

The values a and b are curve fitting parameters. For example, the coefficients of the Norton creep law for JIS G3561 or VDSiCr at 150 °C are:

$$a = -0.9492, b = -105.5357, e^b = 1.4668 \cdot 10^{-46}, \xi = 3.750.$$

According to Eq. (10.7), the Norton creep law for this material is consequently:

$$\dot{\gamma} = 1.4668 \cdot 10^{-46} t^{-0.9492} \tau^{4.75}.$$

The calculation of the twist angles the acquired material law demonstrates an excellent correspondence to the experimental and calculated values (Fig. 10.1, solid lines, right plot). Typical values for ξ are 4–5. If the fitted value of ξ is much lower or much higher, it could indicate, that the Norton creep law could not satisfactorily fit the stress dependence. The alternative formulations will be proper, as we discuss below.

10.2.4 Time-Invariant Constitutive Equations

The substitution of (10.5) in (10.1) leads to the ordinary differential equation with the separable variables:

$$\frac{de_{ij}}{dt} = \frac{s_{ij}}{\tau_{oct}} h_c(t) \cdot g_c(\tau_{oct}). \tag{10.9}$$

We replace the variables in the Eq. (10.9), using the new variable, "scaled time":

$$t^* = \int_0^t h_c(p)\mathrm{d}p, \quad \frac{\mathrm{d}e_{ij}}{\mathrm{d}t} = \frac{\mathrm{d}t^*}{\mathrm{d}t}\frac{\mathrm{d}e_{ij}}{\mathrm{d}t^*} = \frac{1}{h_c(t)}\frac{\mathrm{d}e_{ij}}{\mathrm{d}t^*}. \tag{10.10}$$

With Eq. (10.10) the creep constitutive Eq. (10.9) reduces to:

$$\frac{\mathrm{d}e_{ij}}{\mathrm{d}t^*} = \frac{s_{ij}}{\tau_{oct}} g_c(\tau_{oct}). \tag{10.11}$$

Introduction of "scaled time" substantially shortens all formulas. The variable t is not used any more. Hereafter, only "scaled time" t^* is assumed as the time variable. For briefness, we omit the subscript "star": $t \equiv t^*$ in the following formulas.

The isotropic, time-invariant constitutive equation for creep reads:

$$\dot{e}_{ij} = \frac{s_{ij}}{\tau_{oct}} \cdot g_c(\tau_{oct}). \tag{10.12}$$

We use the symbol "dot" to denote the time derivative with respect to "scaled time" t^*, such that ($\dot{e}_{ij} = \mathrm{d}e_{ij}/\mathrm{d}t^*$) in Eq. (10.12) and the subsequent equations. Applying Eq. (A.4) and (A.6) to (10.12), we get the scalar differential equation for time dependence of octahedral shear strain:

$$\frac{1}{2}\dot{\gamma}_{oct} = g_c(\tau_{oct}). \tag{10.13}$$

The octahedral shear stress τ_{oct} is the function of "scaled time".

For the beginning, consider the function, which covers the creep process from the beginning to the unlimited flow:

$$g_c(\tau_{oct}) = \bar{\varepsilon} \cdot \left(\frac{\tau_{oct}}{\bar{\tau}}\right)^{\xi+1} \cdot \frac{1}{\left(\frac{\tau_{oct}}{\tau_{th}}\right)^{m_1} - 1} \cdot \frac{1}{1 - \left(\frac{\tau_{oct}}{\tau_Y}\right)^{m_2}}. \tag{10.14}$$

The value τ_{th}[MPa] is the lowest stress, for which creep begins. This stress is referred to as the "creep threshold". Below this threshold, creep cannot be measured. The yield shear stress is for the material τ_Y[MPa].

We evaluate the total creep life from the earliest occurrence of creep until the unlimited flow of the uniformly tensioned specimen. For the uniaxial stress state, we have:

$$\sigma = E\varepsilon, \quad \dot{e}_{11} = \dot{\varepsilon}, \quad \tau_{oct} = \sqrt{2}\sigma/3.$$

There is one differential equation for the axial strain component $\varepsilon(t)$:

$$\dot{\varepsilon} = g_c(E\varepsilon), g_c(E\varepsilon) = \overline{\varepsilon} \cdot \left(\frac{E\varepsilon}{\overline{\sigma}}\right)^{\xi+1} \cdot \frac{1}{\left(\frac{E\varepsilon}{\sigma_{th}}\right)^{m_1} - 1} \cdot \frac{1}{1 - \left(\frac{E\varepsilon}{\sigma_Y}\right)^{m_2}}. \tag{10.15}$$

In the Eq. (10.15), the "creep threshold" and yield stress relate to each other:

$$\sigma_{th} = \frac{3\tau_{th}}{\sqrt{2}}, \qquad \sigma_Y = \frac{3\tau_Y}{\sqrt{2}}.$$

The creep process starts, when the stress achieves the threshold tension value σ_{th}, which is minimum necessary stress for creep:

$$\varepsilon(0) = \varepsilon_{th} \equiv \sigma_{th}/E.$$

The initial condition for the differential Eq. (10.15) expresses this requirement. The solution of Eq. (10.15) with reads:

$$t = t(\varepsilon) = c_1\varepsilon^{m_1+m_2-\xi} + c_2\varepsilon^{m_2-\xi} + c_3\varepsilon^{m_1-\xi} + c_4\varepsilon^{-\xi}. \tag{10.16}$$

The coefficients of Eq. (10.16) are:

$$c_1 = \frac{1}{\varepsilon \cdot (\xi - m_1 - m_2)}\left(\frac{\overline{\sigma}}{E}\right)^{\xi+1}\left(\frac{\sigma_Y}{E}\right)^{-m_2}\left(\frac{\sigma_{th}}{E}\right)^{-m_1}, \tag{10.17}$$

$$c_2 = -\frac{1}{\overline{\varepsilon} \cdot (\xi - m_2)(\sigma_C/E)^{m_2}}\left(\frac{\overline{\sigma}}{E}\right)^{\xi+1}\left(\frac{\sigma_Y}{E}\right)^{-m_2}, \tag{10.18}$$

$$c_3 = -\frac{1}{\overline{\varepsilon} \cdot (\xi - m_1)}\left(\frac{\overline{\sigma}}{E}\right)^{\xi+1}\left(\frac{\sigma_{th}}{E}\right)^{-m_1}, \tag{10.19}$$

$$c_4 = \frac{1}{\overline{\varepsilon} \cdot \xi}\left(\frac{\overline{\sigma}}{E}\right)^{\xi+1}. \tag{10.20}$$

The next task is to calculate the total duration of creep process from the initiation of creep to the final rupture. As pointed above, the creep process starts, when the strain arrives the threshold value $\varepsilon_{th} \equiv \sigma_{th}/E$, i.e. $t(\varepsilon_{th}) = 0$. Creep process finishes with the complete rupture of the specimen, when the continuously increasing strain reaches the value of $\varepsilon_Y \equiv \sigma_Y/E$. Consequently, the total time from the beginning of creep process from the first occurrence of creep to the unabridged flow is:

$$t_{total} = t(\varepsilon_Y) = c_1\varepsilon_Y^{m_1+m_2-\xi} + c_2\varepsilon_Y^{m_2-\xi} + c_1\varepsilon_Y^{m_1-\xi} + c_1\varepsilon_Y^{-\xi}. \tag{10.21}$$

The strain rate as the function of ε reads:

$$\dot{\varepsilon} = \left(\frac{\mathrm{d}t}{\mathrm{d}\varepsilon}\right)^{-1}$$
$$= \frac{\varepsilon^{\xi+1}}{c_1.(m_1 + m_2 - \xi) \cdot \varepsilon^{m_1+m_2} + c_2.(m_2 - \xi) \cdot \varepsilon^{m_2} + c_3 \cdot (m_1 - \xi) \cdot \varepsilon^{m_1} - c_4\xi}. \tag{10.22}$$

For simulation, we used the following material properties:

$$\xi = 3.75, m_1 = m_2 = 1, \sigma_Y = 2\ [\mathrm{GPa}], \overline{\varepsilon} = 10^{-26}s^{-1}, E = 200\ [\mathrm{GPa}],$$
$$\overline{\sigma} = 1\ [\mathrm{MPa}].$$

For calculation we apply different threshold values: $\sigma_{th} = 300{,}350{,}400$ MPa The results for calculation are shown in the Table 10.1 and on Figs. 10.2, 10.3 and 10.4. Figure 10.2 shows the creep strain rate as function of creep strain. Three regions of primary, secondary and tertiary creep are approximately marked for: $\sigma_{th} = 300$ MPa. The regions shorten for the higher stress levels. Figure 10.3 displays the creep strain

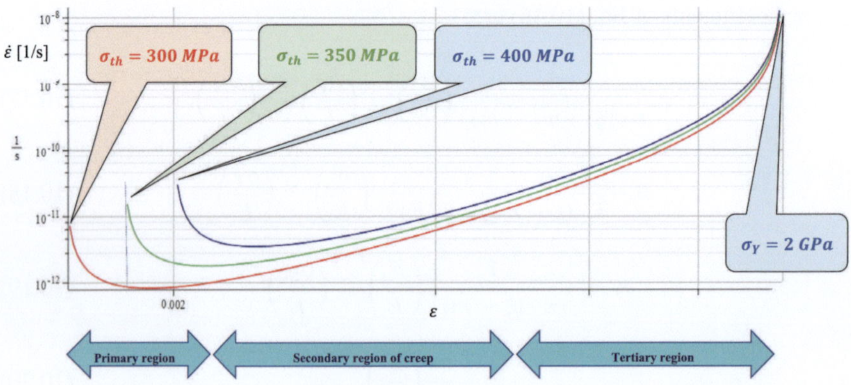

Fig. 10.2 Creep strain rate as function of creep strain. The regions of creep correspond nearly to $\sigma_{th} = 300$ MPa

Table 10.1 Time-independent creep law

	σ_{th} [MPa]	t_{total} [s]	c_1 [s]	c_2 [s]	c_3 [s]	c_4 [s]	$\dot{\varepsilon}[1/s]$
1	300	$1.71 \cdot 10^9 s =$ 54.45 years	2.51×10^4	– 24	– 160	0.176	$\dot{\varepsilon} = -\frac{1.5131\varepsilon^{19/4}}{6.66 \cdot 10^4\varepsilon^2 - 766\varepsilon + 1}$
2	350	$0.896 \cdot 10^9 s =$ 210.42 years	2.15×10^4	– 24	– 137	0.176	$\dot{\varepsilon} = -\frac{1.5131\varepsilon^{19/4}}{5.71 \cdot 10^5\varepsilon^2 - 671\varepsilon + 1}$
3	400	$0.504 \cdot 10^9 s =$ 15.97 years	1.89×10^4	– 24	– 120	0.176	$\dot{\varepsilon} = -\frac{1.5131\varepsilon^{19/4}}{5.00 \cdot 10^5\varepsilon^2 - 600\varepsilon + 1}$

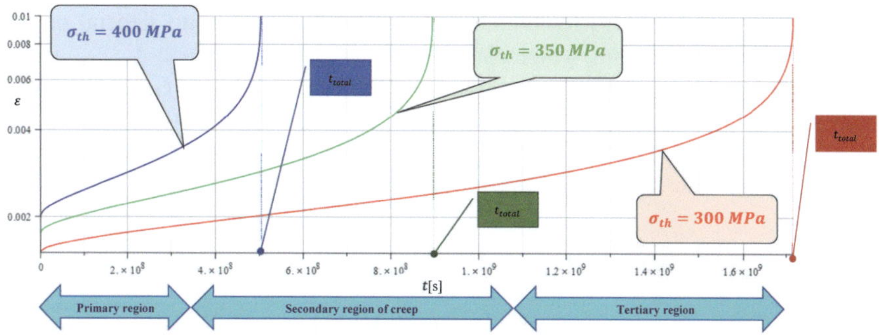

Fig. 10.3 Creep strain as function of time.. The regions of creep correspond nearly to $\sigma_Y = 2$ GPa

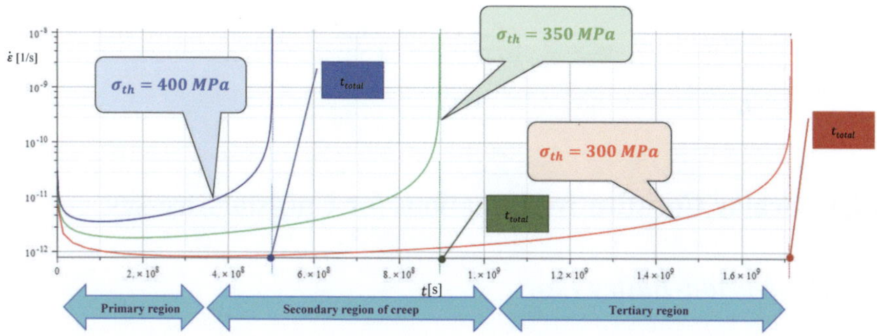

Fig. 10.4 Creep strain rate as function of time.. The regions of creep correspond nearly to $\sigma_Y = 2$ [GPa]

as the function of time over the whole creep region. Figure 10.4 exhibits the creep strain rate as the function of time.

At high stresses, the structural mechanical process determining the exponential growth of temperature is the stress and diffusion-controlled climbing of step dislocations. This is controlled at lower temperatures by diffusion of the lattice atoms along the dislocation lines and at higher temperatures by volume diffusion. The higher the stress, the more pronounced the climbing process is at a given temperature and the more sliding dislocations are available. This is called dislocation or power-law creep. Firstly, consider the isotropic stress function for Norton-Bailey law (Odquist & Hult, 1962):

$$g_c = \overline{\varepsilon} \cdot \left(\frac{\tau_{oct}}{\overline{\tau}} \right)^{\xi+1}. \tag{10.23}$$

Here ξ is the dimensionless experimental constant for the operational temperature of the spring element. The creep constants for several spring materials were acquired in (IGF 18992 BR, 2020). We use the constants $\overline{\sigma}$[MPa], $\overline{t}[s]$ for proper scaling of stress and time. For example, the common setting for scaling constants $\overline{\tau} = 1$[MPa], $\overline{\sigma} = 1$[MPa], $\overline{t} = 1[s]$. The octahedral stress τ_{oct} requires the unit [MPa]. With these settings, the units are $[1/s]$ for both $\overline{\varepsilon}$ and $\dot{e}_{ij}$. All these constants depend on the particular creep mechanism, which changes with the temperature.

Secondly, the isotropic stress function for Garofalo creep law (Garofalo, 1965) reads

$$g_c = \overline{\varepsilon} \cdot \sinh^{\xi+1}\left(\frac{\tau_{oct}}{\overline{\tau}}\right). \tag{10.24}$$

Thirdly, the exponential creep law (Kassner, 2008) is:

$$g_c = \overline{\varepsilon} \cdot \left[\exp\left(\frac{\tau_{oct}}{\overline{\tau}}\right) - 1\right]. \tag{10.25}$$

Finally, Naumenko-Altenbach-Gorash creep law reads (Naumenko et al., 2009):

$$g_c = \overline{\varepsilon} \cdot \left(\frac{\tau_{oct}}{\overline{\tau}}\right) + \overline{\varepsilon} \cdot \left(\frac{\tau_{oct}}{\overline{\tau}}\right)^{\xi+1}.$$

10.3 Scalar Constitutive Equations for Uniaxial Stresses

10.3.1 Norton-Bailey Law

For the uniaxial stress state, there is only non-vanishing component of stress tensor:

$$\sigma_{11} = \sigma, \quad \tau_{oct} = \sqrt{2}\sigma/3.$$

Correspondingly, the non-vanishing components of strain rate are

$$\dot{\varepsilon} \equiv \dot{\varepsilon}_{11} = -2\dot{\varepsilon}_{22} = -2\dot{\varepsilon}_{33}. \tag{10.26}$$

With Eq. (10.23) the uniaxial strain is:

$$\dot{\varepsilon} = \overline{\varepsilon} \cdot \left(\frac{\sigma}{\overline{\sigma}}\right)^{\xi+1}, \quad \overline{\sigma} = \frac{3\overline{\tau}}{\sqrt{2}}. \tag{10.27}$$

For brevity of equations, we introduce the material constant

$$c_\sigma = \frac{\overline{\varepsilon}}{\overline{\sigma}^{\xi+1}}.$$

With this constant the dependence of uniaxial strain rate upon stress reads

$$\dot{\varepsilon} = c_\sigma \sigma^{\xi+1}. \tag{10.28}$$

For pure shear stress state

$$\sigma_{12} = \sigma_{21} = \tau, \quad \tau_{oct} = \sqrt{6}\tau/3.$$

the non-vanishing components of deformation rate are

$$\dot{\varepsilon}_{12} = \dot{\varepsilon}_{21} \equiv \frac{1}{2}\dot{\gamma} = \frac{1}{2}\tilde{\gamma} \cdot \left(\frac{\tau}{\tilde{\tau}}\right)^{\xi+1}.$$

where $\tilde{\gamma} = \sqrt{3}\overline{\varepsilon}$ and $\tilde{\tau} = \overline{\sigma}/\sqrt{3}$.

The creep constant for shear strain reads:

$$c_\tau = \frac{\tilde{\gamma}}{\tilde{\tau}^{\xi+1}}.$$

The Norton-Bailey creep law for pure shear deformation reduces to

$$\dot{\gamma} = c_\tau \tau^{\xi+1}. \tag{10.29}$$

The creep law for spring material Nimonic 80A was experimentally acquired in (Gittus, 1964). In the cited article the similar value of time exponent was reported to the Inconel and EN50 steel.

10.3.2 Garofalo Creep Law

Secondly, the isotropic stress function for Garofalo creep law (10.25) (Garofalo, 1963) for $\xi = 1$ could be represented as

$$g_c(\tau_{oct}) = \overline{\varepsilon} \cdot \sinh \frac{\tau_{oct}}{\overline{\tau}}. \tag{10.30}$$

For the uniaxial stress state, the deformation rate reads

$$\dot{\varepsilon} = \overline{\varepsilon} \cdot \sinh\left(\frac{\sigma}{\overline{\sigma}}\right).$$

or

$$\dot{\varepsilon} \equiv \dot{\varepsilon}_{11} = c_\sigma \sinh\left(\frac{\sigma}{\overline{\sigma}}\right). \tag{10.31}$$

Here $c_\sigma = \overline{\varepsilon}$ is the creep constant for uniaxial strain and possess the dimension $[1/s]$.

For pure shear stress state the deformation rate reads

$$\dot{\varepsilon}_{12} = \dot{\varepsilon}_{21} = \frac{1}{2}\tilde{\gamma} \cdot \sinh\left(\sqrt{3}\frac{\tau}{\overline{\sigma}}\right) \tag{10.32}$$

Finally, the shear strain rate according to Garofalo creep law is

$$\dot{\gamma} \equiv 2\dot{\varepsilon}_{12} = c_\tau \sinh\left(\frac{\tau}{\tilde{\tau}}\right). \tag{10.33}$$

with the constantof dimension $[1/s] : c_\tau = \sqrt{3}\overline{\varepsilon} = \tilde{\gamma} \equiv \sqrt{3}c_\sigma$.

10.3.3 Exponential Law

Equation (10.25) pronounces the isotropic stress function for exponential creep law. For the uniaxial stress state, the strain rate reads:

$$\dot{\varepsilon} \equiv \dot{\varepsilon}_{11} = \overline{\varepsilon} \cdot \left[\exp\left(\frac{\sigma}{\overline{\sigma}}\right) - 1\right] = -2\dot{\varepsilon}_{22} = -2\dot{\varepsilon}_{33}. \tag{10.34}$$

For pure shear stress, the shear deformation rate reduces to:

$$\dot{\gamma} \equiv 2\dot{\varepsilon}_{12} = \tilde{\gamma} \cdot \left[\exp\left(\frac{\tau}{\tilde{\tau}}\right) - 1\right]. \tag{10.35}$$

For the above creep laws the closed form solutions of basic creep problems are derived. Numerical values for the creep constants $\overline{\varepsilon}\left[s^{-1}\right]$, $\tilde{\gamma} = \sqrt{3}\overline{\varepsilon}\left[s^{-1}\right]$ are surely different for diverse creep laws. The constant $\overline{\tau}$[MPa] is used for scaling purposes.

10.4 Creep and Relaxation of Twisted Rods

10.4.1 Constitutive Equations for Relaxation in Torsion

In contrast to the creep test (constant load), the relaxation test keeps the strain constant. This process causes components that have been prestressed to a certain elastic strain at the beginning of their use to lose their prestress over time. The stress decreases (relaxes) as the elastic strain is converted into plastic strain over time. The deformation of body during relaxation does not alter, but the stress gradually reduces. Consider the relaxation problem for a rod with circular cross-section under the constant twist. Let $\tau(r, t)$ is shear stress in the cross-section of rod. The total shear strain in any instant of the time is $\gamma(r, t)$, is the sum of the elastic and the creep components of shear strain:

$$\gamma = \gamma_e + \gamma_c.$$

The creep component of shear strain is $\gamma_c(\rho, t)$. The elastic component of shear strain is

$$\gamma_e = \tau/G. \tag{10.36}$$

Firstly, we investigate the creep for the total deformation that remains constant in time. Thus, we consider the total strain $\gamma_0(\rho)$ as function of radius only. However, the elastic and the creep components of strain are the functions as well of radius and of time, such that:

$$\gamma(\rho, t) = \gamma_e(\rho, t) + \gamma_c(\rho, t) \equiv \gamma_0(\rho). \tag{10.37}$$

The time differentiation of (10.37) leads to the differential equation for elastic and creep strain rates:

$$\dot{\gamma}(\rho, t) = \dot{\gamma}_e(\rho, t) + \dot{\gamma}_c(\rho, t) \equiv 0. \tag{10.38}$$

where dot denotes the time derivative.

The differentiation of the Eq. (10.36) over time delivers the elastic component of strain rate

$$\dot{\gamma}_e = \frac{\dot{\tau}}{G}. \tag{10.39}$$

10.4.2 Torque Relaxation for Norton-Bailey Law

At first, we assume the Norton-Bailey law for the state of shear stress (Boyle, 2012):

$$\dot{\gamma}_c(\rho, t) = c_\tau \tau^{\xi+1}. \tag{10.40}$$

The substitution of material law (10.40) in Eq. (10.38) results in the ordinary nonlinear differential equation of the first order for total shear stress $\tau(r, t)$:

$$\dot{\tau}/G + c_\tau \tau^{\xi+1} = 0. \tag{10.41}$$

The shear stresses in the moment $t = 0$ for the rod with circular cross-section of the diameter $\mathrm{d} = 2r$ are

$$\tau_0(\rho) = G\theta\rho = \frac{\mathrm{T}\varrho}{r}, \quad 0 < \rho < r = \frac{\mathrm{d}}{2},$$

where θ is the twist angle per unit length, $\mathrm{T} = G\theta r$ is the shear stress on the outer surface of the wire in the initial moment of creep. The torque at the initial moment $t = 0$ is pure elastic and is equal to:

$$M_T^0 = \frac{1}{2}\pi \mathrm{T} r^3.$$

The initial condition for the Eq. (10.41) presumes the pure elastic shear stress in the initial moment $t = 0$:

$$\tau(\rho, t = 0) = \tau_0(\rho). \tag{10.42}$$

The average creep time for Norton-Bailey creep is

$$t_c = \frac{1}{\mathrm{T}^{\xi} \cdot G \cdot c_{\tau} \cdot \xi} \equiv \frac{1}{\tilde{\gamma} \cdot \xi} \cdot \frac{\tilde{\tau}^{\xi+1}}{\mathrm{T}^{\xi} \cdot G}.$$

The solution of the ordinary differential Eq. (10.17) with initial condition (10.18) delivers the shear stress over the cross-section of the twisted rod as the function of time and radius:

$$\tau(\rho, t) = \left[\tau_0^{-\xi}(\rho) + c_{\tau} G \xi t\right]^{-1/\xi}. \tag{10.43}$$

The function (10.43) pronounced the relaxation of the shear stress from the initial state (10.42). To calculate the relaxation of the couple, we integrate the shear stress over the cross-section of the bar. The couple as the function of time is

$$M_T(t) = 2\pi \int_0^r \rho^2 \tau(\rho, t) \mathrm{d}\rho.$$

With the expression for total shear stress (10.43) we can calculate the couple

$$M_T(t) = 2\pi \int_0^r \rho^2 \left[\left(\frac{\mathrm{T}\varrho}{r}\right)^{-\xi} + c_{\tau} G \xi t\right]^{-1/\xi} \mathrm{d}\rho. \tag{10.44}$$

For evaluation of the integral (10.44) the formula for $J_p(a, m; X)$ from Appendix B is applied for the case $p = 2$. The integral could be expressed in terms of hypergeometric function (Kobelev, 2014a):

$$\frac{M_T(t)}{M_T^0} = {}_2\mathbf{F}_1\left(\frac{4}{\xi}, \frac{1}{\xi}; \frac{4+\xi}{\xi}; -\frac{t}{t_c}\right). \tag{10.45}$$

10.4.3 *Torque Relaxation for Garofalo Law*

At second, we presume the Garofalo law for pure share state of stress:

$$\dot{\gamma}_c(\rho, t) = c_{\tau} \sinh\left(\frac{\tau}{\tilde{\tau}}\right). \tag{10.46}$$

The solution of the differential Eq. (10.46) with initial condition (10.42) for the Garofalo creep law reads (Kobelev, 2014a):

$$\tau(\rho,t)=\tilde{\tau}\ln\left\{\tanh\left[\frac{Gc_\tau}{2\tilde{\tau}}t+\operatorname{arctanh}\left(\exp\left(\frac{G\rho\theta}{\tilde{\tau}}\right)\right)\right]\right\}. \tag{10.47}$$

Using formula for $I_2(a,b;X)$ from the Appendix, the time dependent torque could be expressed in terms of polylogarithms (Lewin, 1981). The relaxation function for the Garofalo law reads:

$$\frac{M_T(t)}{M_T^0}=\sum_{i=1}^{4}p_i\left(\frac{\tilde{\tau}}{\mathrm{T}}\right)^i, \tag{10.48}$$

$$p_1=\frac{4}{3}\ln(U),\quad U=\frac{\exp(t/t_c)-1}{\exp(t/t_c)+1},\quad V=\exp\left(\frac{\mathrm{T}}{\tilde{\tau}}\right).$$

$$p_2=-4\left[\mathrm{Li}_2\left(-\frac{V}{U}\right)-\mathrm{Li}_2(-V\cdot U)\right],$$

$$p_3=8\left[\mathrm{Li}_3\left(-\frac{V}{U}\right)-\mathrm{Li}_3(-V\cdot U)\right],$$

$$p_4=8\left[\mathrm{Li}_4\left(-\frac{1}{U}\right)-\mathrm{Li}_4(-U)\right]-8\left[\mathrm{Li}_4\left(-\frac{V}{U}\right)-\mathrm{Li}_4(-V\cdot U)\right].$$

In Eq. (10.58), the average time of creep for the Garofalo and exponential law is:

$$t_c=\frac{\tilde{\tau}}{c_\tau \mathrm{G}}. \tag{10.49}$$

10.4.4 Torque Relaxation for Exponential Law

At third, we apply the exponential law for the state of shear stress:

$$\dot{\gamma}_r(\rho,t)=\tilde{\gamma}\cdot\left[\exp\left(\frac{\tau}{\tilde{\tau}}\right)-1\right]. \tag{10.50}$$

The substitution of material law (10.49) in Eq. (10.38) leads to the ordinary differential equation for total shear stress:

$$\frac{\dot{\tau}}{G}+\tilde{\gamma}\cdot\left[\exp\left(\frac{\tau}{\tilde{\tau}}\right)-1\right]=0. \tag{10.51}$$

The solution of the differential Eq. (10.51) with initial condition (10.42) delivers the shear stress over the cross-section of the twisted rod. Relaxation of torque could be expressed again in terms of polylogarithm (Lewin, 1981):

$$\frac{M_T(t)}{M_{T^*}^0} = \sum_{i=0}^{4} p_i \left(\frac{\tilde{\tau}}{\mathrm{T}}\right)^i, \qquad U = \exp\left(\frac{t}{t_c}\right), \tag{10.52}$$

$$p_0 = 1, \quad p_1 = \frac{4t}{3t_c}, \qquad p_2 = 4\,\mathrm{Li}_2(V \cdot (1-U)),$$

$$p_3 = -8\,\mathrm{Li}_3(V \cdot (1-U)), \qquad p_4 = 8\,\mathrm{Li}_4(V \cdot (1-U)) - \mathrm{Li}_4(1-U).$$

10.5 Creep and Relaxation of Helical Coiled Springs

10.5.1 Phenomena of Relaxation and Creep

Compression and tension springs are made of an elastic wire material formed into the shape of a helix, which returns to its natural length when unloaded. Compression springs can be commonly referred to as a coil spring or a helical spring. Coil springs are a mechanical device, which is typically used to store energy and subsequently release it to absorb shock, or to maintain a force between contacting surfaces. The major stresses produced in conical and volute springs are shear stresses due to twisting. The subsequent consideration is applicable for both compression or tension helical springs with the minor alternations. For definiteness, only compression springs will be discussed.

Consider the helical spring with mean helix diameter D, wire diameter d and active coils number n_a. The force $F_z(t)$ applied to a spring that causes a deflection s is:

$$F_z(t) = 2\frac{M_T(t)}{D}.$$

The relaxation of helical springs occurs, when the compressed length of the spring remains constant. During the relaxation process the twist angle per unit length remains constant as well:

$$\theta = \frac{2s}{\pi n_a D^2}.$$

The spring force $F_z(t)$ reduces over time. At initial moment $t = 0$ the spring force due to pure elastic deformation F_z^0 and the shear stress on the outer surface of wire $\tilde{\tau}$ are correspondingly:

$$F_z^0 = \frac{2M_T^0}{D} = \frac{Gd^4 s}{8n_a D^3} \qquad \tilde{\tau} = \frac{G\theta d}{2}. \tag{10.53}$$

2°. Creep of spring. The creep of helical springs occurs, when the spring force remains constant. In this situation the compressed length continuously reduces over time and the deflection increases (Geinitz et al., 2011).

10.5.2 *Relaxation of Helical Springs*

For the solution of relaxation problem we apply the results of relaxation problem for twisted rod with the circular cross-section for helical springs. If the material of helical springs obeys Norton-Bailey law, the spring load as the function of time is (Kobelev, 2014b):

$$F_z(t) = {}_2\boldsymbol{F}_1\left(\frac{4}{\xi}, \frac{1}{\xi}; \frac{4+\xi}{\xi}; -\frac{t}{t_c}\right) F_z^0. \tag{10.54}$$

For certain exponents in Norton-Bailey law, the Eq. (10.54) expresses in terms of elementary functions:

$$F_z(t) = \frac{4}{3}\frac{(1+x)^{3/4} - 1}{x} F_z^0, \quad x = \frac{t}{t_c}, \text{ for } \xi = 4. \tag{10.55}$$

and

$$F_z(t) = \frac{4}{3}\frac{(x-2)\sqrt{1+x} + 2}{x^2} F_z^0, \quad x = \frac{t}{t_c}, \text{ for } \xi = 2. \tag{10.56}$$

The relaxation of spring load over time demonstrates the dimensionless relaxation function:

$$\Phi(t) = \frac{F_z(t)}{F_z^0} = {}_2\mathbf{F}_1\left(\frac{4}{\xi}, \frac{1}{\xi}; \frac{4+\xi}{\xi}; -\frac{t}{t_c}\right). \tag{10.57}$$

On Fig. 10.5 the function Φ is plotted for different twist angles per unit length. For illustration of the procedure, we perform all calculations with Norton-Bailey law and the following set of material constants

$$\begin{gathered} G = 79.74\ [\text{GPa}],\ E = 200\ [\text{GPa}],\ \xi = 4,\ \overline{\varepsilon} = 10^{-24} [\text{sec}]^{-1}, \\ \tilde{\gamma} = 1.73 \cdot 10^{-24} [\text{sec}]^{-1},\ \overline{\sigma} = 2000\ [\text{MPa}],\ \tilde{\tau} = 1154\ [\text{MPa}] \end{gathered}.$$

The three graphs of the function $\Phi(t)$ are shown on the figure for twist angles per unit length ($\theta = 0.1; 0.2; 0.3$). It means that the higher twist angle per unit length,

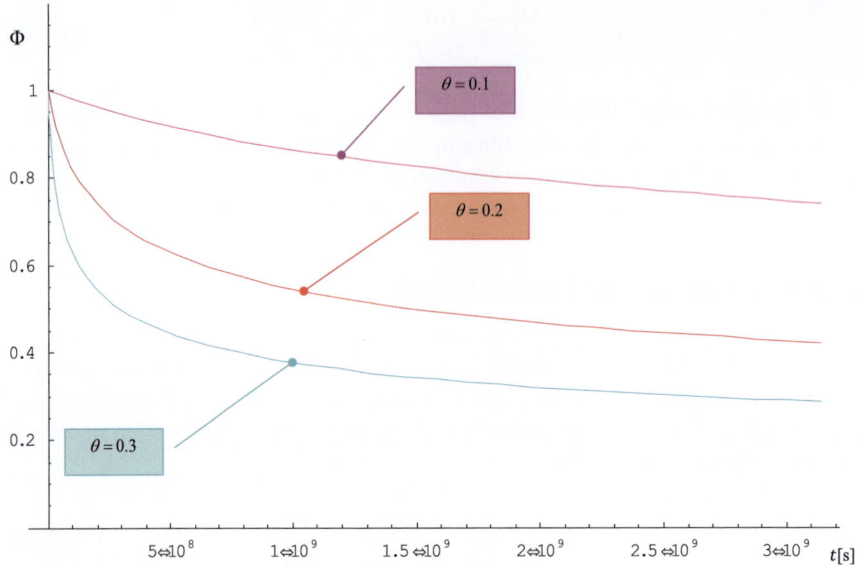

Fig. 10.5 Relaxation function Φ(t) for different twist rates

Fig. 10.5 Relaxation function $\Phi(t)$ for different twist rates

the quickly occurs the relaxation of torque. This apparently happens because the strain rate is higher for higher stresses.

On Fig. 10.6 the function $\Phi(t)$ is plotted for the same twist angle per unit length ($\theta = 0.1$), but for the different exponents

$$\xi = 3.0; \xi = 4.0; \xi = 5.0; \xi = 11.0$$

The rest of material parameters remain the same as in the previous example. Figure 10.6 evidently demonstrates, that the higher the exponent, the higher the relaxation rate.

10.5.3 Creep of Helical Compression Springs

During the creep deformation of springsthe spring force remains constant over time. Correspondingly, the spring length reduces with time. The length decrease rate over time is $\dot{s}$. The shear strain rate varies with time:

$$\dot{\gamma}(t,\rho) = \dot{\theta}\rho = \frac{2\dot{s}\rho}{\pi n_a D^2} \quad \text{for } 0 < \rho < r = \mathrm{d}/2.$$

According to Norton-Bailey law (10.4) the shear stress due to creep is

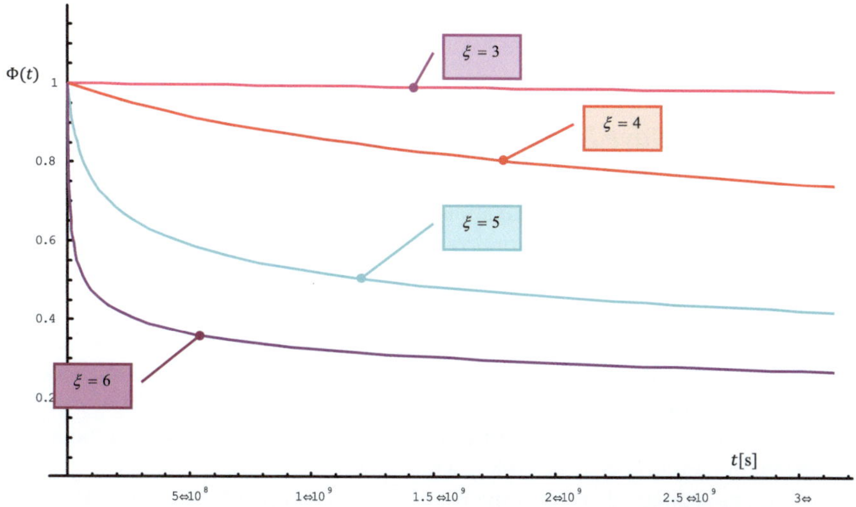

Fig. 10.6 Relaxation function $\Phi(t)$ for twist rate $\theta = 0.1$ and different creep exponents

$$\tau = \left(\frac{\dot{\gamma}}{c_\tau}\right)^{\frac{1}{\xi+1}} \equiv \tilde{\tau} \cdot \left(\frac{2\dot{s}}{\tilde{\gamma}\pi n D^2}\right)^{\frac{1}{\xi+1}} \cdot \rho^{\frac{1}{\xi+1}}.$$

Performing the integration over the area of wire we get the moment due to creep (Kobelev, 2014b):

$$M_T^0 = 2\pi \int_0^r \rho^2 \tau(\rho, t) \mathrm{d}r = 2\pi \cdot \tilde{\tau} \cdot \left(\frac{2\dot{s}}{\tilde{\gamma}\pi n D^2}\right)^{\frac{1}{\xi+1}} \cdot \int_0^r \rho^{2+\frac{1}{\xi+1}} \mathrm{d}r. \tag{10.58}$$

For the solution of Eq. (10.58), we use the following relations:

$$M_T^0 = \frac{D \cdot F_z^0}{2}, \quad \tilde{\gamma} = \tilde{\tau}^{\xi+1} c_\tau,$$
$$\int_0^{r=\mathrm{d}/2} \rho^{2+\frac{1}{\xi+1}} \mathrm{d}r = \frac{\xi+1}{4+3\xi} \cdot \left(\frac{\mathrm{d}}{2}\right)^{\frac{4+3\xi}{1+\xi}}.$$

The solution of Eq. (10.58) provides the reduction rate of spring length. The rate of length reduction for constant spring force F_z^0 is the function of time:

$$\dot{s} = \frac{\pi D^2 n_a c_\tau}{2\mathrm{d}^{4+3\xi}} \cdot \left(\frac{4DF_z^0}{2\pi} \cdot \frac{4+3\xi}{\xi+1}\right)^{\xi+1}. \tag{10.58}$$

10.6 Creep and Relaxation of Beams in State of Pure Bending

10.6.1 Constitutive Equations for Relaxation in Bending

Consider the problem of stress relaxation in the pure bending of a rectangular cross-section ($B \times T$) beam. In the applied Euler–Bernoulli theory of slender beams subjected to a bending moment M_B, a major assumption is that 'plane sections remain plane'. In other words, any deformation due to shear across the section is not accounted for (no shear deformation). During the relaxation, experiment the curvature of the neutral axis of beam κ remains constant over time, such that the bending moment M_B continuously decreases. This case describes the relaxation of bending stress, if the flexure deformation of beam does not alter in time. This problem was considered in (Boyle & Spence, 1983) with the aid of numerical methods.

Let $\sigma(z, t)$ is uniaxial stress in the beam in the direction of beam axis. The total strain in any instant of the time is $\varepsilon(z, t)$; is the sum of the elastic and the creep components of the strain:

$$\varepsilon = \varepsilon_e + \varepsilon_c. \tag{10.59}$$

The elastic component of normal strain is:

$$\varepsilon_e = \frac{\sigma}{E}, \tag{10.60}$$

and $\varepsilon_c(z, t)$. is the creep component of normal strain.

Consider creep under constant in time total strain:

$$\varepsilon(z, t) = \varepsilon_e(z, t) + \varepsilon_c(z, t) \equiv \varepsilon_0(z). \tag{10.61}$$

The normal strain $\varepsilon_0(z) = \varepsilon(z, t = 0)$ is the function of normal coordinate, but remains constant over time during the relaxation process. The time differentiation of (10.61) leads to:

$$\dot{\varepsilon} = \dot{\varepsilon}_e + \dot{\varepsilon}_c \equiv 0. \tag{10.62}$$

We exercise once again the common constitutive models of creep in bending state.

10.6.2 Relaxation of Bending Moment for Norton-Bailey Law

The Norton-Bailey law for a uniaxial state of stress reads:

$$\dot{\varepsilon}_c(z, t) = c_\sigma \sigma^{\xi+1}. \tag{10.63}$$

The substitution of material laws results in the ordinary differential equation for uniaxial stress:

$$\dot{\sigma}/E + c_\sigma \sigma^{\xi+1} = 0. \tag{10.64}$$

The initial condition for the Eq. (10.64) delivers the pure elastic shear stress in the initial moment:

$$\sigma(z, t = 0) = \sigma_0(z). \tag{10.65}$$

For pure elastic bending the following initial distribution of stresses over the cross-section of the beam is valid:

$$\sigma_0(z) = E\kappa z = \frac{2z}{T}\sigma_p, \tag{10.66}$$

where κ is the bending curvature, which presumed to be constant over time and z is the perpendicular distance to the neutral axis. σ_p is the normal stress on the outer surface of the wire in the initial moment of creep.

The solution of the ordinary differential Eq. (10.64) with initial condition (10.65) is

$$\sigma(z, t) = \left[\sigma_0^{-\xi}(z) + c_\sigma E\xi t\right]^{-1/\xi}. \tag{10.67}$$

The function (10.67) pronounced the relaxation of the normal stress from the initial state of pure tension (10.64). To calculate the relaxation of the bending moment, we integrate the normal stress over the cross-section of the beam. The bending moment for the rectangular cross-section of width B and height H is the function of time.

$$M_B(t) = B \int_{-T/2}^{T/2} z\sigma(z, t)\mathrm{d}z. \tag{10.68}$$

With the expression (10.68) we can calculate

$$M_B(t) = 2EB\kappa \int_0^{T/2} z \cdot \left[\frac{1}{z^\xi} + c_\sigma \kappa^\xi E^{\xi+1}\xi t\right]^{-1/\xi} \mathrm{d}z. \tag{10.69}$$

Using the results of Appendix ($J_p(a, m; X)$, case $p = 1$), the integral in (10.69) could be expressed in terms of hypergeometric function (Kobelev, 2014b):

$$\frac{M_B(t)}{M_B^0} = {}_2\mathbf{F}_1\left(\frac{3}{\xi}, \frac{1}{\xi}; \frac{3+\xi}{\xi}; -\frac{t}{t_c}\right), \tag{10.70}$$

where

$$t_c^{-1} = c_\sigma \cdot E \cdot \xi \cdot \Sigma^\xi$$

is the characteristic creep time in bending and

$$M_B^0 = \frac{1}{6}\sigma_p B T^2 \equiv \frac{1}{12}\kappa E B T^3. \tag{10.71}$$

is the elastic bending moment at time $t = 0$. For example, for $\xi = 3$ we have:

$$\frac{M_B(t)}{M_B^0} = \frac{3}{2}\frac{(1+x)^{2/3} - 1}{x},$$
$$x = \frac{t}{t_c}.$$

10.6.3 Relaxation of Bending Moment for Garofalo Law

We consider the now Garofalo law for uniaxial state of stress:

$$\dot{\varepsilon}_c = c_\sigma \sinh\left(\frac{\sigma}{\overline{\sigma}}\right). \tag{10.72}$$

The solution of the ordinary differential Eq. (10.37) with initial condition (10.38) for the Garofalo creep law leads to the expression of normal stress as function of coordinate z and time t (Kobelev, 2014a):

$$\sigma(z,t) = \overline{\sigma}\ln\left\{\tanh\left[\frac{Ec_\sigma}{2\overline{\sigma}}t + \operatorname{arctanh}\left(\exp\left(\frac{2\Sigma}{T\overline{\sigma}}z\right)\right)\right]\right\}. \tag{10.73}$$

The function (10.73) pronounced the relaxation of the normal stress from the initial state of pure tension (10.64). For numerical evaluation, the formula for $I_1(a, b; X)$ from Appendix is to be applied. With this formula, the integral in (10.73) could be expressed in terms of polylogarithm. Relaxation of bending moment could be expressed again in terms of polylogarithm (Lewin, 1981):

$$\frac{M_B(t)}{M_B^0} = \sum_{i=1}^{3} p_i \left(\frac{\overline{\sigma}}{\Sigma}\right)^i,$$
$$U = \coth\left(\frac{t}{2t_c}\right), \quad V = \exp\left(\frac{\Sigma}{\overline{\sigma}}\right),$$
$$p_1 = -\frac{3}{2}\ln(U),$$
$$p_2 = -3\left[\mathrm{Li}_2(-V\cdot U) - \mathrm{Li}_2\left(-\frac{V}{U}\right)\right], \tag{10.74}$$
$$p_3 = 3\mathrm{Li}_3(-V\cdot U) - \mathrm{Li}_3\left(-\frac{V}{U}\right) - 3\mathrm{Li}_3(-U) - \mathrm{Li}_3\left(-\frac{1}{U}\right)$$

10.6.4 Relaxation of Bending Moment for Exponential Law

In this section, we investigate the problem of the pure bending of a rectangular cross-section beam with a modified power law (stress range-dependent constitutive model) subjected to a bending moment. We assume exponential law for the state of uniaxial stress.

The substitution of modified power material law (10.9) results in the ordinary differential equation for uniaxial stress:

$$\frac{\dot{\sigma}}{E} + \overline{\varepsilon} \cdot \left[\exp\left(\frac{\sigma}{\overline{\sigma}}\right) - 1\right] = 0. \tag{10.75}$$

When loaded by a bending moment, the beam bends so that the inner surface is in compression and the outer surface is in tension. The neutral plane is the surface within the beam between these zones, where the material of the beam is not under stress, either compression or tension. The solution of the ordinary differential Eq. (10.75) with initial condition (10.38) delivers the stress over the cross-section of the beam as the function of time and distance z to neutral plane.

For calculation of the bending moment the formula for $J_p(a, m; X)$ from Appendix is applied for the case $p = 1$. With this formula the bending moment in the cross-section could be expressed in terms of hypergeometric function (Kobelev, 2014a):

$$\frac{M_B(t)}{M_B^0} = \sum_{i=0}^{3} p_i \left(\frac{\overline{\sigma}}{\Sigma}\right)^i, \qquad U = 1 - \exp\left(\frac{t}{t_c}\right), \tag{10.76}$$

$$\begin{aligned} p_0 &= 1, \\ p_1 &= \tfrac{3}{2} \cdot \tfrac{t}{t_c}, \\ p_2 &= 3 \cdot \operatorname{Li}_2(V \cdot U), \\ p_3 &= -3 \cdot [\operatorname{Li}_3(V \cdot U) - \operatorname{Li}_3(U)]. \end{aligned}$$

The curvature κ of the beam remains constant in time. In the expressions (10.71), (10.74) and (10.76) the bending moment $M_B(t)$ will be the continuously reducing function of time.

10.6.5 Creep in State of Bending

In contrast, during the bending creep deformation the moment M_B^0 remains constant over time. The curvature $\kappa = \kappa(t)$ will slowly increase with the constant rate of growth. The elongation rate of the strip, which locates on the perpendicular distance z to the neutral axis, is:

$$\dot{\varepsilon}(t, z) = \dot{\kappa} z \quad \text{for} \quad -\frac{T}{2} < z < \frac{T}{2}.$$

According to Norton-Bailey law (10.3) the shear stress due to creep is:

$$\sigma_c = \left(\frac{\dot{\varepsilon}}{c_\sigma}\right)^{\frac{1}{\xi+1}} = \left(\frac{\dot{\kappa} z}{c_\sigma}\right)^{\frac{1}{\xi+1}}.$$

Performing the integration over the area of wire we get the bending moment due to creep (Kobelev, 2014a):

$$M_B^0 = B \int\limits_{-T/2}^{T/2} z\sigma_c(z,t)\mathrm{d}z = \frac{BT^2}{2} \cdot \frac{\xi+1}{3+2\xi} \cdot \left(\frac{T\dot{\kappa}}{2c_\sigma}\right)^{\frac{1}{1+\xi}}. \qquad (10.77)$$

Assuming that bending moment M_B^0 remains constant over time, from this equation the curvature time rate resolves as

$$\dot{\kappa} = \frac{2c_\sigma}{T} \cdot \left(\frac{2M_B^0}{BT^2} \cdot \frac{3+2\xi}{\xi+1}\right)^{\xi+1}. \qquad (10.78)$$

The rate of flexure $\dot{\kappa}$ remains constant, such that the bending radius linearly increases over time. The similar expressions follow for the other studied above creep laws.

The results of this Section are applicable for springs, which stressed typically by bending loads. For example, the leaf springs and torsion springs are in the bending state. Torsion springs may be of helix or spiral type. Accordingly, the results of this Section could be instantly applied to estimate the effects of creep and relaxation of such springs.

10.7 Creep and Relaxation of Disk Springs

10.7.1 Creep of Disk Springs

One important example of the springs with the uniaxial normal stress is the disk springs. In these springs dominates the state of bending of conical part.

For the calculation of disk springs, the theory of the punched flat disks is usually applied. In this theory, the springs are flat circular disks, resting on their external rim and uniformly loaded along its internal rim (Timoshenko, 1948). The shape of the disks is not considered; nevertheless, it has an influence on the shape of the suiting spring diagrams and on stress distribution. The load–deflection curve obtained by calculation is a straight line in this case. The maximum disks flattening load is identical to that obtained by the other calculation methods described hereinafter. The details of the elastic modelling are displayed in Chap. 4. The deformation hypothesis presupposes that the radial stresses are negligible and the cross-section of spring does not distort, but rather that is merely rotates about the inversion point (Sects. 4.1.2 and 4.1.3).

Consider a sector of the disk with the constant thickness T. We use in this chapter the notation T for the thickness of the disk spring, as the notation t is reserved for time. The disk is loaded by a force F_z, which assumed to be constant in time. Consider a sector and in it a strip dx at location x taking the inversion point as the origin. Under the action of force F_z the disk creeps and is deflected through a time dependent angle:

$$\phi = \alpha - \psi,$$

where

$\alpha = h/(r_e - r_i)$ the initial cone angle of disk,

$\psi = H/(r_e - r_i)$ the deformed cone angle of disk,

h the free height of disk, measured as the elevation of the free truncated cone formed by either the upper or lower surface.

H the deformed height of disk, measured as the elevation of the deformed truncated cone formed by either the upper or lower surface,

$r_e = D_e/2$ and $r_i = D_i/2$ are respectively the outer and inner radii of disk.

For small axial deflection of disk s the angle reads

$$\phi = \frac{s}{r_e - r_i},$$

This strip moves slowly into its new deformed position. The resultant tangential strain may be analyzed as the resultant of the radial displacement Δr and the rotation ϕ. The first of these causes a uniform strain throughout the thickness of the disk, if one neglects the small variation in distance to the center of the disk at various points of the section. The second results in a tangential bending strain which is zero in the neutral surface and maximum at the upper and lower surfaces. The tangential stresses produced by these two components of the strain cause a radial moment about point O which resists the moment created by the external and that of the deflected one forces.

For the studied springs, the thickness of spring material T is thin enough. This assumption allows neglecting the tangential stress due to bending. In other words, the influence of tangential bending strain is abandoned.

Calculating the tangential stress due to the radial displacement solely, we can use the expression for tangential strain in terms of rotation ϕ:

$$\varepsilon_\theta = \frac{x}{x - C_i} \cdot \left(\alpha - \frac{\phi}{2}\right) \cdot \phi. \tag{10.79}$$

Here c_i is the distance of inversion point to the disk symmetry axis in the creep state,

Under the creep conditions the rotation $\phi = \phi(t)$ and, consequently, tangential strain $\varepsilon_\theta = \varepsilon_\theta(t)$ are the functions of time. The tangential strain rate results form

(10.79):

$$\dot{\varepsilon}_\theta \equiv \frac{\mathrm{d}\varepsilon_\theta}{\mathrm{d}t} = \frac{x}{x-c_i}\cdot\left[\left(\alpha-\frac{\phi}{2}\right)\cdot\dot{\phi}+\left(-\frac{\dot{\phi}}{2}\right)\cdot\phi\right] \equiv \frac{x}{x-c_i}\cdot(\alpha-\phi)\cdot\dot{\phi}. \tag{10.80}$$

Because it is assumed that the radial stresses are negligible, the tangential stress relates to tangential strain by means the uniaxial Norton- Bailey law:

$$\dot{\varepsilon}_\theta(z,t) = c_\sigma \cdot \sigma_\theta^{\xi+1} \tag{10.81}$$

The tangential stress is the function of rotation angle ϕ and its time derivative $\dot{\phi}$:

$$\sigma_\theta = \left[\frac{(\alpha-\phi)\dot{\phi}}{c_\sigma}\cdot\frac{x}{x-C_i}\right]^{\frac{1}{\xi+1}}. \tag{10.82}$$

The radial moment of the tangential forces in the section about point O is:

$$\mathrm{d}M_\theta = \sigma_\theta T\cdot x\cdot\sin(\alpha-\phi)\cdot\mathrm{d}x\cdot\mathrm{d}\theta. \tag{10.83}$$

Substituting in Eq. (10.83) the Eq. (10.82) and assuming the deflection as small,

$$\sin(\alpha-\phi)\cong\alpha-\phi,$$

we obtain the radial moment of the tangential forces in the section:

$$\mathrm{d}M_\theta = (\alpha-\phi)\cdot T\cdot\left[\frac{(\alpha-\phi)\dot{\phi}}{c_\sigma}\right]^{\frac{1}{\xi+1}}\cdot\left(\frac{x}{x-C_i}\right)^{\frac{1}{\xi+1}}\cdot x\cdot\mathrm{d}x\cdot\mathrm{d}\theta.$$

Integrating from $x = C_i - r_e$ to $x = C_i - r_i$, we get the internal moment of the sector about point O:

$$\begin{aligned} M_\theta(t) &= (\alpha-\phi)\cdot 2\pi T\cdot\left[\frac{(\alpha-\phi)\dot{\phi}}{c_\sigma}\right]^{\frac{1}{\tilde{m}+1}}\cdot\int_{C_i-r_e}^{C_i-r_i}\left(\frac{x}{x-C_i}\right)^{\frac{1}{\xi+1}}x\mathrm{d}x \\ &= (\alpha-\phi)\cdot 2\pi T\cdot\left[\frac{(\alpha-\phi)\dot{\phi}}{c_\sigma}\right]^{\frac{1}{\xi+1}}\cdot L_{\xi+1}(r_e,r_i,C_i). \end{aligned} \tag{10.84}$$

In Eq. (10.87) we make use of formulae from the Appendix C. The integral could be expressed analytically in terms of incomplete beta function.

The axial force of the disk spring F_z is equal to the radial moment of the tangential forces divided by force arm $r_e - r_i$, such that:

$$
\begin{aligned}
F_z(t) &= \frac{M_\theta}{r_e - r_i} = 2\pi T \cdot \left[\frac{(\alpha-\phi)\dot{\phi}}{c_\sigma}\right]^{\frac{1}{\xi+1}} \cdot \frac{\alpha-\phi}{r_e-r_i} \cdot L_{\xi+1}(r_e, r_i, C_i) \\
&= 2\pi T \cdot \frac{L_{\xi+1}(r_e, r_i, C_i)}{r_e - r_i} \cdot \left[\frac{(\alpha-\phi)^{\xi+2}\dot{\phi}}{c_\sigma}\right]^{\frac{1}{\xi+1}}. \qquad (10.85)
\end{aligned}
$$

The value of the C_i in the last equation yet remains to be determined.

The sum of all forces action normal to the cross-section F_θ must be equal to zero. Only stresses due to radial displacement need to be considered. To calculate the sum of all forces action normal to the cross-section, we make use of the expression for tangential stress as the function of rotation angle ϕ and its time derivative $\dot{\phi}$:

The value of the C_i in the last equation yet remains to be determined.

The sum of all forces action normal to the cross-section F_θ must be equal to zero. Only stresses due to radial displacement need to be considered. To calculate the sum of all forces action normal to the cross-section, we make use of the expression for tangential stress as the function of rotation angle ϕ and its time derivative $\dot{\phi}$:

$$
F_\theta = T \cdot \int_{C_i - r_e}^{C_i - r_i} \sigma_\theta \mathrm{d}x = T \cdot \left[\frac{(\alpha-\phi)\dot{\phi}}{c_\sigma}\right]^{\frac{1}{\xi+1}} \cdot \int_{C_i - r_e}^{C_i - r_i} \left(\frac{x}{x - C_i}\right)^{\frac{1}{\xi+1}} \mathrm{d}x = 0. \qquad (10.86)
$$

As the factor before integral is positive, the vanishing of the integral in (10.86) determines the position of inversion point C_i:

$$
\int_{C_i - r_e}^{C_i - r_i} \left(\frac{x}{x - C_i}\right)^{\frac{1}{\xi+1}} \mathrm{d}x \equiv K_{\xi+1}(r_e, r_i, C_i) = 0. \qquad (10.87)
$$

The integral in the Eq. (10.87) expresses in terms of incomplete beta functions using the formulae from the Appendix C. Further simplification and closed form solution of the Eq. (10.87) for an arbitrary m seems to be impossible in closed analytical form. Avoiding the numerical solution, consider exactly solvable limit cases.

For $\xi = 0$ one get the known value for inversion center (Ch. 5, Eq. 2.10):

$$
c_i = \frac{r_i - r_e}{\ln(r_i / r_e)}. \qquad (10.88)
$$

The appropriate approximate solution for sufficiently large ξ delivers the expression:

$$
c_i = \frac{r_i + r_e}{2}. \qquad (10.89)
$$

For constant values of r_i, r_e, c_i the expression $K_{\xi+1}(r_e, r_i, c_i)$ is the function of creep exponent ξ.

For higher values of exponent ξ the first function

$$K_{\xi+1}(r_e, r_i, (r_e + r_i)/2). \tag{10.90}$$

asymptotically tends to zero axis.

For higher values of exponent ξ the second function

$$K_{\xi+1}(r_e, r_i, (r_e - r_i)/\ln(r_i/r_e)). \tag{10.91}$$

gradually deviates from zero.

For illustration two functions $K_{\xi+1}(r_e, r_i, (r_e - r_i)/\ln(r_e/r_i))$, Eq. (10.91) and $K_{\xi+1}(r_e, r_i, (r_e + r_i)/2)$, Eq. (10.90) are drawn on Fig. 10.7. For lower values of ξ the solution (10.88) is delivers better approximation for the equation

$$K_{\xi+1}(r_e, r_i, c_i) = 0. \tag{10.92}$$

Two functions

$$L_{\xi+1}(r_e, r_i, (r_e - r_i)/\ln(r_e/r_i)). \tag{10.93}$$

and

$$L_{\xi+1}(r_e, r_i, (r_e + r_i)/2). \tag{10.94}$$

are drawn on Fig. 10.10. The approximate solution (10.88) leads to somewhat higher values of function $L_{\xi+1}$, that the solution (10.89) (Fig. 10.8).

With the approximate solution (10.89) for the distance of neutral axis to center, we get the final expression of the spring force as the function of rotation angle $\phi(t)$ (Kobelev, 2014b):

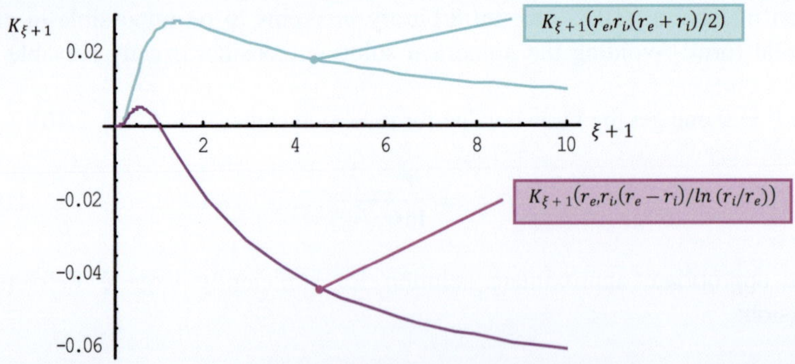

Fig. 10.7 Two functions $K_{\xi+1}(r_e, r_i, (r_e - r_i)/\ln(r_i/r_e))$ and $K_{\xi+1}(r_e, r_i, (r_e + r_i)/2)$ for different exponents ξ

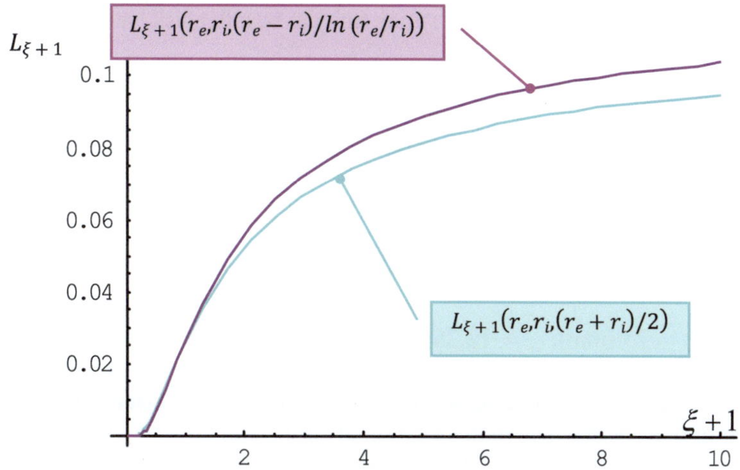

Fig. 10.8 Two functions $L_{\xi+1}(r_e, r_i, (r_e - r_i)/\ln(r_e/r_i))$ and $L_{\xi+1}(r_e, r_i, (r_e + r_i)/2)$ as functions of exponent ξ

$$F_z(t) = \left[\frac{(\alpha - \phi)^{\xi+2}\dot{\phi}}{c_\sigma}\right]^{\frac{1}{\xi+1}} \cdot \frac{2\pi T}{r_e - r_t} \cdot L_{\xi+1}\left(r_e, r_i, \frac{r_e + r_i}{2}\right). \tag{10.95}$$

If the spring force remains constant over time, the rotation angle $\phi(t)$ is the solution of an ordinary differential equation

$$t^{1-\xi}(\alpha - \phi)^{\xi+2}\dot{\phi} = \tilde{P}, \quad \phi(0) = \phi_0, \tag{10.96}$$

where

$$\tilde{P} = c_\sigma \left[\frac{r_e - r_i}{L_{\xi+1}\left(r_e, r_i, \frac{r_e + r_t}{2}\right)} \cdot \frac{F_z}{2\pi T}\right]^{\xi+1},$$

ϕ_0 the initial rotation angle of the cone due to elastic deformation of the spring at time moment $t = 0$.

The solution of the Eq. (10.96) leads to the rotation angle as function of time:

$$\phi(t) = \alpha - \left[(\alpha - \phi_0)^{\xi+3} - \frac{\xi + 3}{\xi} \cdot \tilde{P} \cdot t^{\varsigma}\right]^{\frac{1}{\xi+3}}. \tag{10.97}$$

At the critical moment of time

$$t_{crit} = \frac{(\alpha - \phi_0)^{\xi+3}}{(\xi + 3) \cdot \tilde{P}}, \tag{10.98}$$

the cone angle of disk vanishes and spring turns to be a flat disk, such that

$$\phi(t_{crit}) = \alpha. \tag{10.99}$$

The axial travel of disk is the function of time:

$$s(t) = \phi(t) \cdot (r_e - r_i). \tag{10.100}$$

For example, we calculate the creep of the disk spring with the following parameters:

$$r_e = 25 \text{ [mm]}, r_i = 12.25 \text{ [mm]}, T = 3 \text{ [mm]}, h = 1.1 \text{ [mm]}.$$

Figures 10.9 and 10.10 demonstrate the influence of parameter $\overline{\sigma}$ on the creep behavior of disk spring. The initial rotation of spring cone is

$$\phi_0 = \frac{3\alpha}{4}.$$

In other words, this means that the initial spring travel is

$$s_0 = \frac{3h}{4} = 0.825 \text{ [mm]}.$$

The following material parameters were used for calculation:

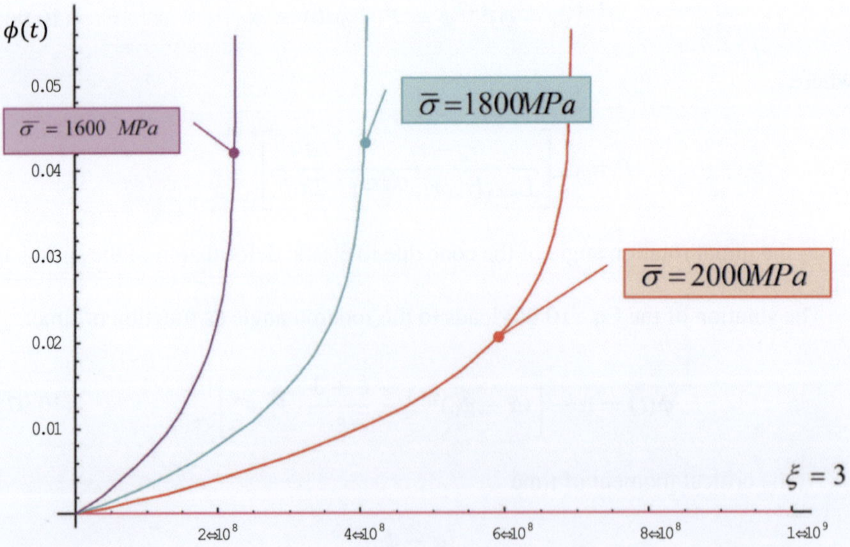

Fig. 10.9 Cone angles of disk springs as functions of creep time for three different material parameters

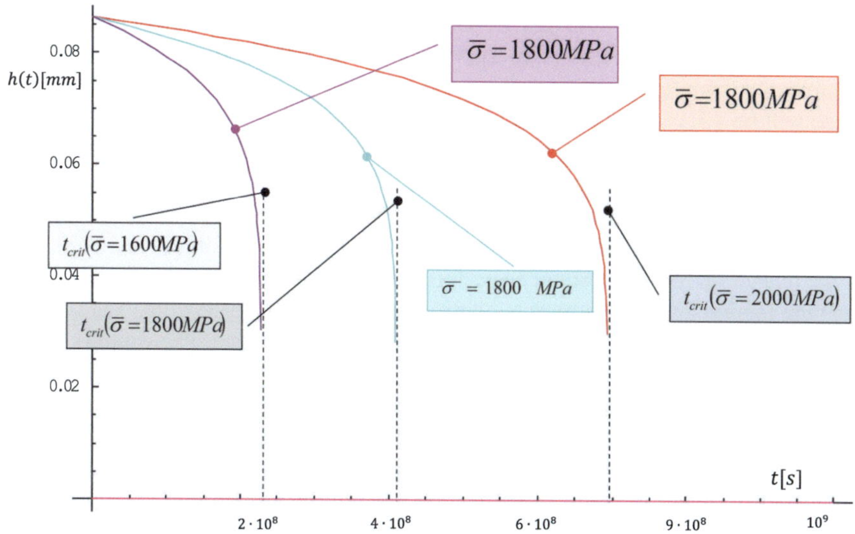

Fig. 10.10 Disk heights $h(t)$ as functions of creep time for three different material parameters

$$E = 200\ [\text{GPa}],\ \xi = 4,\ \nu = 0.254,\ \overline{\varepsilon} = 10^{-24}[\text{sec}]^{-1}.$$

On Fig. 10.9 the creep curves for three different material parameters are shown:

$$\overline{\sigma} = 1600\ [\text{MPa}], \overline{\sigma} = 1800\ [\text{MPa}], \text{ and } \overline{\sigma} = 2000\ [\text{MPa}].$$

The creep curves represent the cone angles as functions of time. The relaxation happens slower for the materials with higher values of parameter $\overline{\sigma}$.

The disk heights $h(t)$ are shown under constant load as function of time for three different material parameters on Fig. 10.10. The cone angles of disk vanish for three different critical moments. Finally, depending on material parameter, the spring turns to be a flat disk. The disk flattening happens evidently at higher time moments for the materials with elevated parameters $\overline{\sigma}$.

10.7.2 Relaxation of Disk Springs

For calculation of the disk spring relaxation the Eq. (10.71) for Norton-Bailey law is applicable. The Eq. (10.71) represents the relaxation function for the beam in flexure. The dominantly flexure stress-state exists in disk springs. Namely, the stress-state in disk springs is the superposition of two states as follow from Chap. 4. Depending upon the geometry of the disk spring, either the flattening of the disk with circumferential mid-surface strain (1.16, Chap. 5) or bending due to circumferential curvature change (1.17, Chap. 5) dominates.

We consider now only the case of complete flattening of moderately thick spring with predominant circumferential curvature change. The disk spring is instantly compressed from its free state to the plane flat state. In the first moment the spring force will be $F_z^0 = F_z(t = 0)$. The force in the first moment arises due to the pure elastic stresses. Correspondingly, for the evaluation of the instant spring force the formulas from Chap. 4 are applicable.

During the relaxation time the spring travel s remains constant, while the stresses relax. The relaxation of stresses leads to the reduction of spring force. The calculation of the relaxation function cannot be performed straightforward, as it was possible for the pure torsion and for the pure bending. The reason is the following. The stresses in the volume of the disk spring are not constant over its surface. The simple consideration allows the fairly accurate estimation of the relaxation function. Keeping in mind, that stresses are caused by bending, we employ the stress relaxation function from the state of flexure that was investigated above. The characteristic of the relaxation function is a "comparative" maximal stress σ_R on the surface. By means of Eq. (10.71) one gets the relaxation function $\Psi(t)$ of the disk spring:

$$F_z(t) = \Psi(t) F_z^0, \quad \Psi(t) = {}_2\mathbf{F}_1\left(\frac{3}{\xi}, \frac{1}{\xi}; \frac{3+\xi}{\xi}; -\xi c_\sigma \sigma_R^\xi E t\right). \tag{10.101}$$

As the solution with the general stress distribution cannot be found closed form, the finite element simulations were performed. The finite-element simulation performs the numerical calculation of the relaxation function with the sane geometry and equal material parameters. For comparison the relaxation function $\Psi(t)$ was tabulated for the following six values of "comparative" maximal stress:

$$\sigma_R = \left\{\frac{|\sigma_{Be} + \sigma_{Bi}|}{2}, \frac{\sigma_{Te} - \sigma_{Ti}}{2}, |\sigma_I|, |\sigma_{II}|, |\sigma_{III}|, |\sigma_{IV}|\right\}. \tag{10.102}$$

The first option in Eq. (10.102) fixes the average bending stress $|\sigma_{Be} + \sigma_{Bi}|/2$. The second option designates the average tensile stress $(\sigma_{Te} - \sigma_{Ti})/2$. The subsequent options use the stress on the corners. The finite-element simulation demonstrates, that the best comparison for the relaxation function $\Psi(t)$ and the relaxation is achieved, if "comparative" maximal stress is assumed to be

$$\sigma_R = \sigma_{III}. \tag{10.103}$$

In this case the numerically calculated the relaxation function and the relaxation function $\Psi(t)$ match (Fig. 10.11). Consequently, it could be advised to use the Eq. (10.101) with the substitution (10.103) for relaxation calculations of disk springs. The relaxation function (10.101) provides the reasonable evaluation of the relaxation effects.

Other common creep laws could be considered the analogously.

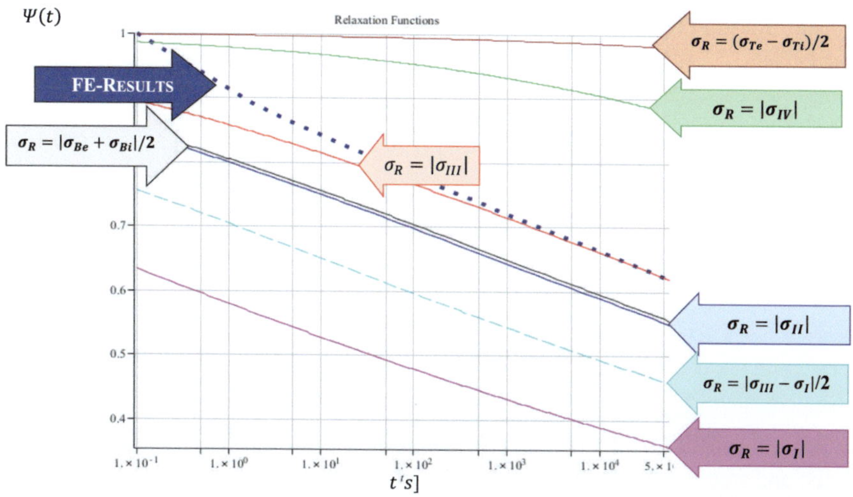

Fig. 10.11 Relaxation function $\Psi(t)$ for several formulations of "comparative" maximal stress

10.8 Cyclic Creep and Fatigue-Creep Interaction

We estimate the influence of the cyclic stress on the creep effect. For this purpose, the load in the spring element will sinusoidal with a constant amplitude and non-zero mean value:

$$\tau_{oct} = \sigma_m + \sigma_a \cos t. \tag{10.104}$$

We investigate the cyclic accumulation creep strain for Norton-Bailey law. Substitution of (10.104) into the creep Eq. (10.23), the increment of the octahedral shear strain after one cycle is:

$$\Delta\gamma_{oct}(\sigma_m, \sigma_a) = \frac{2\overline{\varepsilon}}{\overline{\tau}^{\xi+1}} \cdot \frac{1}{2\pi} \cdot \int_0^{2\pi} (\sigma_m +_a \cos t)^{\xi+1} \mathrm{d}t. \tag{10.105}$$

For evaluation of the integral (10.105), we use the Eq. (B.13) or (3.661.3) (Gradshteyn & Ryzhik, 2015):

$$\frac{1}{2\pi} \cdot \int_0^{2\pi} (a + b\cos t)^q \mathrm{d}t = \left(a^2 - b^2\right)^{\frac{q}{2}} \cdot \mathbf{P}_q\left(\frac{a}{\sqrt{a^2 - b^2}}\right). \tag{10.106}$$

Substitution (10.115) into (10.114) gives the increment of the creep strain after one cycle:

$$\Delta\gamma_{oct}(\sigma_m, \sigma_a) = \frac{2\overline{\varepsilon}}{\overline{\tau}^{\xi+1}} \cdot \left(\sigma_m^2 - \sigma_a^2\right)^{\frac{\xi+1}{2}} \cdot \mathbf{P}_{\xi+1}\left(\frac{\sigma_m}{\sqrt{\sigma_m^2 - \sigma_a^2}}\right). \quad (10.107)$$

If the amplitude vanishes, the increment of creep during the same time interval will be $\Delta\varepsilon_{cr}(\sigma_m, 0)$. The ratio of $\Delta\varepsilon_{cr}(\sigma_m, \sigma_a)$ to $\Delta\varepsilon_{cr}(\sigma_m, 0)$ gives the expression for the influence of the cyclic stress on the total creep strain:

$$r_{cycl} = \frac{\Delta\gamma_{oct}(\sigma_m, \sigma_a)}{\Delta\gamma_{oct}(\sigma_m, 0)}, \quad \sigma_a = \frac{1 - R_\sigma}{1 + R_\sigma} \cdot \sigma_m. \quad (10.108)$$

Substitution of Eq. (10.107) in (10.108) results in the final expression for ratio of cyclic and static creep strains:

$$r_{cycl}(\xi, R_\sigma) = \left(\frac{2\sqrt{R_\sigma}}{1 + R_\sigma}\right)^{\xi+1} \cdot \mathbf{P}_{\xi+1}\left(\frac{1 + R_\sigma}{2\sqrt{R_\sigma}}\right). \quad (10.109)$$

Figure 10.12 displays the plot of the function $r_{cycl}(\xi, R_\sigma)$ Eq. (10.109) for $0 < \xi < 4$, $0 < R_\sigma < 1$ for equal values of mean stress in cycles σ_m.

The sense of the influence of the stress ratio is the following. If the mean stress in cycle vanishes, the creep elongation according to Norton-Bailey law is positive during the first half of the cycle and negative during the second half of the cycle. Both partial elongations compensate each other and the accumulation of creep over time disappears. If the amplitude in cycle vanishes, the steady creep flow occurs and increases continuously with time.

Interesting to mention, that the fatigue failure follow completely different pattern (Chap. 9). If the mean value vanish but the stress varies symmetrically, the crack

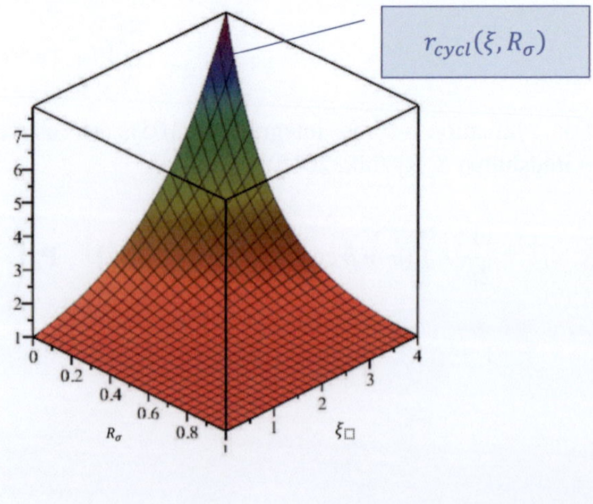

Fig. 10.12 Ratio of cyclic and static creep strains $r_{cycl}(\xi, R_\sigma)$

elongates and failure occurs after a definite number of cycles. If the amplitude of stress disappears, fatigue damage is absent and occurs no failure.

Both microscopic damage mechanisms of fatigue and creep are different and requires the different mathematical characterization. In details the creep-fatigue interaction was discussed in (Assefpour-Dezfuly & Brownrigg, 1989; Chandler & Kwofie, 2005; Kwofie & Chandler, 2007; Polák et al., 2011; Pineau, 1988; Yang & Wang, 2001).

10.9 Temperature Influence on Creep

1. The temperature dependence could be incorporated directly in the creep parameters:

$$\dot{\gamma} = g_c(\tau) \cdot h_c(t), \quad h_c(t) = \left(\frac{t}{\bar{t}}\right)^{\varsigma - 1}. \tag{10.110}$$

The creep parameter $c_\tau(\mathrm{T}^\circ)$ is the function of temperature:

$$g_c(\tau) = c_\tau(\mathrm{T}^\circ)\tau^{\xi+1} \equiv \tilde{\gamma}(\mathrm{T}^\circ) \cdot \left(\frac{\tau}{\tilde{\tau}}\right)^{\xi+1}. \tag{10.111}$$

with the characteristic creep rate at the stress level $\tilde{\tau}$:

$$\tilde{\gamma}(\mathrm{T}^\circ) = c_\tau(\mathrm{T}^\circ) \cdot \tilde{\tau}^{\xi+1}.$$

In this work we assume the scaling parameters in (10.110) and (10.111) as follows:

$$\tilde{\tau} = \frac{1}{\sqrt{3}}[\mathrm{MPa}], \quad \bar{t} = 1[\mathrm{sec}], \quad \bar{\tau} = 1[\mathrm{MPa}].$$

The representation (10.110) and (10.111) is suitable for the direct measurements of creep parameters at certain fixed temperatures.

In the report (IGF 16992, 2018), constitutive equations were evaluated with the usual choice for the stress exponent $\xi = 4$ and using the fitting functions for temperature dependence:

$$\tilde{\gamma}(\mathrm{T}^\circ) = \left[\gamma_1 + \gamma_2 \ln\left(\frac{\tau}{\bar{\tau}}\right)\right] \cdot \exp\left(\frac{\mathrm{T}^\circ}{273^\circ\mathrm{K}}\right) + \gamma_3 + \gamma_4 \ln\left(\frac{\tau}{\bar{\tau}}\right). \tag{10.112}$$

$$c_\tau(\mathrm{T}^\circ) = \left[c_{\tau 1} + c_{\tau 2} \ln\left(\frac{\tau}{\bar{\tau}}\right)\right] \cdot \exp\left(\frac{\mathrm{T}^\circ}{273^\circ\mathrm{K}}\right) + c_{\tau 3} + c_{\tau 4} \ln\left(\frac{\tau}{\bar{\tau}}\right). \tag{10.113}$$

The interrelation between the coefficients of (10.112) and (10.113) is:

$$c_{\tau i} = \frac{\gamma_i}{\tilde{\tau}^{\xi+1}}, \quad i = 1 \ldots 4.$$

The advantage of the coefficients γ_i is their simpler unit dimension:

$$[\gamma_i] = [s^{-1}], \quad [c_{\tau i}] = [\mathrm{MPa}^{-5}\mathrm{s}^{-1}].$$

2. The region of primary creep was evaluated experimentally. The steady creep rate was not observed for the stress and temperature levels. The constants for the typical spring materials CrSi, CrSiV and 1.4310 are presented in Tables 10.2, 10.3 and 10.4. These tables show the constants of Eq. (10.117) for the circular wires after the certain heat treatments. The manufactured springs contain residual stresses, which influence the creep rates. The application of the acquired constants for the fabricated springs requires the correction factors, which regard the manufacturing effects on creep processes. The mentioned correction factors are exposed in the cited report 2°. Skrotzki et al. (2019) performed the comprehensive review of the testing methods of metals at elevated temperatures. Commonly, the temperature dependence of the stationary creep rate is described with the Arrhenius function:

$$g_c(\tau_{oct}) = \overline{\varepsilon} \cdot \left(\frac{\tau_{oct}}{\overline{\tau}}\right)^{\xi+1} \exp\left(-\frac{\mathrm{Q}_a}{\mathrm{R} \cdot \mathrm{T}^\circ}\right), \quad \mathrm{R} = 10.31\left[\frac{J}{\mathrm{mol}^\circ\mathrm{K}}\right]. \tag{10.114}$$

Table 10.2 Relative temperature dependence of yield stress, Young modulus and creep constant for VDSiCrV wire

Relative temperature dependence	T°C
$\sigma_w(\mathrm{T}^\circ\mathrm{C})/\sigma_w(20^\circ\mathrm{C})$	$1.0348 - 0.001842 \cdot T^\circ + 5.218 \cdot 10^{-6} \cdot T^{\circ 2} - 4.5259 \cdot 10^{-9} \cdot T^{\circ 3} - 2.879 \cdot 10^{-12} T^{\circ 4}$
$E(\mathrm{T}^\circ\mathrm{C})/E(20^\circ\mathrm{C})$	$1.003 - 0.0001273 \cdot T^\circ - 1.2673 \cdot 10^{-6} \cdot T^{\circ 2} + 3.2961 \cdot 10^{-9} \cdot T^{\circ 3} - 3.9233 \cdot 10^{-12} T^{\circ 4}$
$\overline{\varepsilon}(T^\circ C)/\overline{\varepsilon}(20^\circ C)$	$0.58198\exp(0.003412 \cdot T^\circ + 0.9317) - 0.58198$

Table 10.3 Creep parameters for torsion bars made of oil tempered wires SiCr (IGF 18992, 2018)

Material SiCr	d, mm	ς	$\gamma_1 \cdot 10^{50}/s$	$\gamma_2 \cdot 10^{50}/s$	$\gamma_3 \cdot 10^{50}/s$	$\gamma_4 \cdot 10^{50}/s$
350°C/30 min	≤ 4	0.1217	13.51	– 1.754	3.434	– 27.32
	> 4	0.1151	15.62	– 2.027	4.480	– 35.25
420°C/30 min	≤ 4	0.1141	10.175	– 0.9631	1.468	6.656
	> 4	0.1141	10.175	– 0.9631	1.468	6.656

Table 10.4 Creep parameters for torsion bars made of oil tempered wires SiCrV (IGF 18992, 2018)

Material SiCrV	d, mm	ς	$\gamma_1 \cdot 10^{50}/s$	$\gamma_2 \cdot 10^{50}/s$	$\gamma_3 \cdot 10^{50}/s$	$\gamma_4 \cdot 10^{50}/s$
350°C/30 min	≤ 4	0.1050	9.265	– 1.174	1.732	– 14.63
	> 4	0.1050	12.07	– 1.546	3.574	– 210.16
420°C/30 min	≤ 4	0.1127	15.44	– 2.022	4.779	– 36.87
	> 4	0.1127	15.44	– 2.022	4.779	– 36.87

where Q_a[kJ/mol] is the activation energy for self- or impurity diffusion, R is the gas constant and T° is the absolute temperature in [°K] (Table 10.5).

The activation energy could be evaluated, if at least two experiments for creep are performed for different temperatures. The first experiment is conducted at absolute temperature T_1° and measures the creep deformation between the start time t_s and the final time t_e:

$$\Delta\gamma_{c1} = \gamma_c(t = t_s) - \gamma_c(t = t_e). \tag{10.115}$$

In the second experiment, which is conducted at absolute temperature T_2°, the creep deformation between the start time t_s and the final time t_e:

$$\Delta\gamma_{c2} = \gamma_c(t = t_s) - \gamma_c(t = t_e). \tag{10.116}$$

The activation energy with the acquired values (10.115) and (10.116) reads:

$$Q_a = \frac{T_1^\circ T_2^\circ}{T_1^\circ - T_2^\circ} \cdot R \cdot \ln\left(\frac{\Delta\gamma_{c1}}{\Delta\gamma_{c2}}\right). \tag{10.117}$$

The estimation (10.117) is valid in the region between the absolute temperatures T_1° and T_2° and for order of magnitude the stress, which were applied during the measurements.

For example, the evaluation of creep constants is bases on the results of experimental evaluation of creep curves of VDSiCrV wire (d = 3 mm), 350 °C/30min heat treated and pre-torched with 1450 MPa (IFG 18992, 2018, Abb. 5.54):

Table 10.5 Creep parameters for torsion bars made of austenitic wires 1.4310 (IGF 18992, 2018)

Material 1.4310	ς	$\gamma_1 \cdot 10^{50}/s$	$\gamma_2 \cdot 10^{50}/s$	$\gamma_3 \cdot 10^{50}/s$	$\gamma_4 \cdot 10^{50}/s$
290°C/30 min	0.1318	106.6	– 15.17	40.72	– 2810.9
300°C/30 min	0.1243	26.51	– 3.702	3.845	– 30.06
400°C/30 min	0.1213	29.98	– 4.337	9.689	66.81

$$\frac{\Delta\gamma_{c2}}{\Delta\gamma_{c1}} = \frac{33}{15}, \quad T_1^\circ = 353^\circ K, \quad T_2^\circ = 433^\circ K.$$

With these constants, the activation energy follows from Eq: (10.117):

$$Q_a = 12.52\left[\frac{kJ}{mol}\right].$$

Similarly, for each fixed absolute temperature T°, the stress exponent requires two creep tests with different stress on the surface of wire $\tau_{oct.1}$, $\tau_{oct.2}$. The creep deformation between the start time t_s and the final time t_e for stress $\tau_{oct.1}$ is $\Delta\gamma_{c1}(\tau_{oct.1})$. The creep deformation between the start time t_s and the final time t_e for stress $\tau_{oct.2}$ is $\Delta\gamma_{c1}(\tau_{oct.2})$. Then the stress exponent is:

$$\xi(T) = \frac{\ln\left(\frac{\Delta\gamma_{c2}(\tau_{oct.2})}{\Delta\gamma_{c1}(\tau_{oct.1})}\right)}{\ln\left(\frac{\tau_{oct.2}}{\tau_{oct.1}}\right)} - 1. \tag{10.118}$$

3. The experimental results (IFG 18992, 2018) are used for the evaluation of stress exponent ξ and activation energy Q_a for different spring steels. The evaluation with Eqs. (10.117) and (10.118) was performed for the VDSiCrV, VDSiCr and 1.4310 spring steels (Table 10.6).

As mentioned above, the customary choice for the stress exponent is $\xi = 4$. With this value, the relaxation behavior expresses via the elementary function (10.55) and requires no transcendental function (10.54). This feature greatly simplifies technical calculations.

Well known, that the ultimate tensile stress depends on test temperature as well. The typical relative dependences of mechanical properties of steel are shown on Fig. 10.13. This figure demonstrates the creep rate constant $\overline{\varepsilon}(T^\circ C)/\overline{\varepsilon}(20^\circ C)$, relative ultimate tensile stress $\sigma_w(T^\circ C)/\sigma_w(20^\circ C)$ and relative Young modulus

Table 10.6 Temperature dependence of creep parameters for torsion bars (IGF 18992, 2018)

Wire	Heat treatment		Pre-torched	Q_a	ξ	T °,
	°C	min	MPa	$\frac{kJ}{mol}$		°C
VDSiCrV	350	30	–	16.35	4.82	160
	350	30	1450	16.07	5.24	160
					5.30	120
					4.06	80
	350	30	1240	13.66	3.66	160
VDSiCr	350	30	–	13.66	3.58	160
	430	30	–	10.61	2.58	160
1.4310	200	30		9.37		
	400	30		9.37		

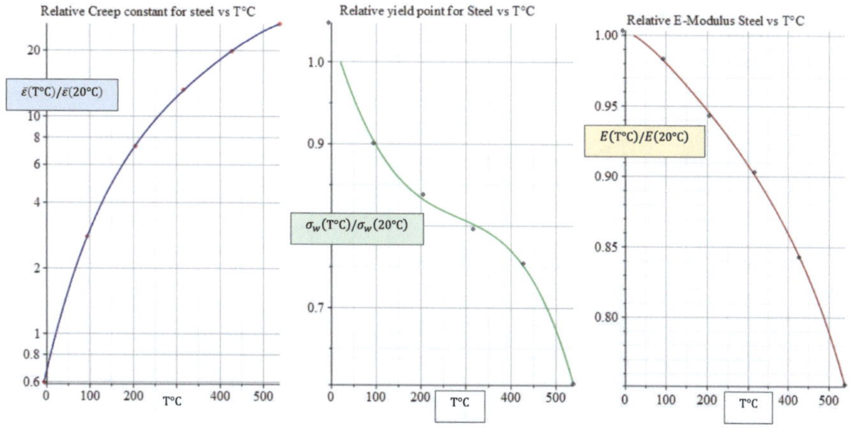

Fig. 10.13 Relative creep rate constant $\overline{\varepsilon}(T°C)/\overline{\varepsilon}(20\,°C)$, relative ultimate tensile stress $\sigma_w(\text{T°C})/\sigma_w(20\,°\text{C})$ and relative Young modulus $E(\text{T°C})/E(20\,°\text{C})$ as functions of temperature $T°C$

$E(\text{T°C})/E(20°\text{C})$ as functions of temperature $T°C$. The corresponding functions are presented in Table 10.5. For reference, the plots of the ultimate tensile stress for AISI1095, AISI420, UHB SS 716. AISI 301 and 17-7 PH steel sorts as functions of temperature T°C are shown on Fig. 10.14 (Johansson et al., 1984). Mechanical behavior of metallic materials over wide ranges of elongation was studied in (Emde, 2008). The cited results exhibit the greater temperature sensitivity of creep strain rate in comparison to the ultimate tensile stress for common steels. At higher temperatures, the ultimate tensile stress reduces stronger than the Young modulus.

4. Special applications require the application of high temperature resistant steels for spring manufacturing. One effect at high temperatures is the corrosion of metals caused by oxygen, which is known as scaling. A spring steel is heat-resistant, if it does not scale or only slightly scales. The combination of heat resistance and temperature resistance is referenced as the high temperature resistance. The high temperature resistant spring steels are steels that are optimized for good service life at elevated temperatures and exhibit also the low temperature sensitivity of creep. The high temperature resistant materials are characterized by (Milne et al., 2003b):

   - High melting point;
   - Phase mixture with finely distributed phase (precipitates), temperature resistant microstructure, obstacles with high temperature resistance and strength (obstruction of dislocation movement);
   - Small stacking fault energy (little energy required to produce faults; this leads to more stacking faults that impede dislocation movement);
   - Coarse grain (few grain boundaries reduce grain boundary slippage).

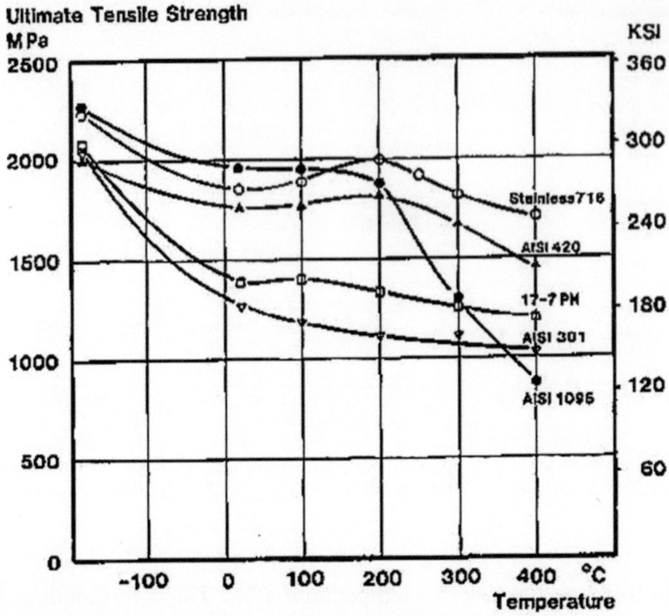

Fig. 10.14 Ultimate tensile stress σ_w(T°C) as the function of temperature $T°C$ for AISI1095, AISI420, UHB SS 716. AISI 301 and 17-7 PH steel sorts (Johansson et al., 1984)

10.10 Conclusions

The structures studied include a beam in bending, a rod in torsion, and helical and disk springs. The closed-form solutions demonstrate the basic characteristics of nonlinear creep. Since bending and torsion dominate in spring members, the results are directly applicable to the majority of springs. Following the procedure, we establish closed-form solutions for creep and relaxation in helical, leaf, and disk springs. The generalized expression for the creep law is also studied. The knowledge of material creep laws is essential for the creep and relaxation design of spring elements withstanding high static and cyclic loads. The expression is based on observations of spring creep and unifies the primary, secondary and tertiary regions of the creep curve by introducing the new time variable.

Summary of Principal Results

- Common constitutive equations for creep (Norton-Bailey, Garofalo, Exponential, Naumenko-Altenbach-Gorash) are compared.
- The solutions for creep and stress relaxation for bars in twisted and bending states are presented in closed form.
- The derived solutions allow the evaluation of creep under constant load and relaxation from fixed strain for helical, leaf and disk springs.

- The temperature dependence of the creep parameters is presented in closed form.

References

Andrade, E. N. D. C. (1910). On the viscous flow in metals and allied phenomena. *Proceedings of the Royal Society of London. Series A, Containing Papers of a Mathematical and Physical Character*, 84(567), 1–12.

Andrade, E. N. D. C. (1914). The flow of materials under large constant stress. *Proceedings of the Royal Society of London. Series A, Containing Papers of a Mathematical and Physical Character*, *90*(619), 329–342.

Ashby, M. F. (2011). *Material Selection in Mechanical Design*. Elsevier.

Assefpour-Dezfuly, M., & Brownrigg, A. (1989). Parameters affecting sag resistance in spring steels. *MTA, 20*, 1951–1959. https://doi.org/10.1007/BF02650282

Betten, J. (2008). *Creep Mechanics* (3d ed.). Springer.

Boyle, J. T. (2012). The creep behavior of simple structures with a stress range-dependent constitutive model. *Archive of Applied Mechanics, 82*, 495–514.

Boyle, J. T., & Spence J. (1983) Stress Analysis for Creep. Butterworth & Co, Ltd.

Cadek, J. (1988). *(1988) Creep in metallic materials, materials science monographs* (Vol. 48). Elsevier.

Chandler, D., & Kwofie, S. (2005). (2005) A description of cyclic creep under conditions of axial cyclic and mean stresses. *International Journal of Fatigue, 27*(5), 541–545.

Chang, D. J. (1995). *Prediction of stress relaxation for compression and torsion springs.* TR-96(8565)-l, The Aerospace Corporation Technology Operations, El Segundo, CA 90245-4691

Emde, T. (2008). Mechanisches Verhalten metallischer Werkstoffe über weite Bereiche der Dehnung, der Dehnrate und der Temperatur, RWTH Aachen (publications.rwth-aachen.de/record/51583/)

Es-Souni, M. (2000). Primary, secondary and anelastic creep of a high temperature near a-Ti alloy Ti6242Si. *Materials Characterization, 45*, 153–164.

Evans, R. W., Parker, J. D., & Wilshire, B. (1982). An extrapolation procedure for long-term creep strain and creep life prediction with special reference to 0.5Cr0.5Mo0.25V ferritic steels. In *Recent advances in creep and fracture of engineering materials and structures* (pp. 135–184). Pineridge Press: Swansea, UK.

Garofalo, F. (1963). An empirical relation defining the stress dependence of minimum creep rate in metals. *Transactions of Metallurgical Society of AIME, 227*, 351–356.

Garofalo, F. (1965). *Fundamentals of creep and creep-rupture in metals*. Series in Materials Science.

Geinitz, V., Weiß, M., Kletzin, U., & Beyer, P. (2011). Relaxation of helical springs and spring steel wires. In: *56th International scientific colloquium*. Ilmenau University of Technology, 12–16 Sept 2011.

Gittus, J. H. (1964). Implications of some data on relaxation creep in nimonic 80a. *Philosophical Magazine, 9*:101, 749–753. 11.1080/14786436408211888.

Gittus, J. H. (1971). The mechanical equation of states: Dislocation creep due to stresses varying in magnitude and direction. *Philosophical Magazine, 24*:(192), 1423–1440. 11.1080/14786437108217422.

Gradshteyn I. S., & Ryzhik I. M. (2015). *Table of integrals, series, and products* (6th ed). In: Zwillinger D. (Eds.). Elsevier, Waltham. ISBN 978-0-12-384933-5

Harrison, W. J., & Evans, W. J. (2007). Application of the Theta projection method to creep modelling using Abaqus. In *ABAQUS Users' Conference, Abaqus UK Regional User Meeting*

Holdsworth, S. (2008). (2008) The European creep collaborative committee (ECCC) approach to creep data assessment. *Journal of Pressure Vessel Technology, 130*, 024001.

IGF 18992 BR. (2020). Creep and relaxation behaviour of spring steel wires in coil springs. In: U. Kletzin & T.U. Ilmenau.

Johansson, B., Nordberg, H., & Thullen, J. M. (1984). Properties of high strength steels. In *International compressor engineering, conference*. Paper 474, https://docs.lib.purdue.edu/icec/474

Kassner, M. (2008). *Fundamentals of Creep in Metals and Alloys* (2nd ed.). Elsevier.

Kobelev, V. (2014a). Some basic solutions for nonlinear creep. *International Journal of Solids and Structures, 51*, 3372–3381.

Kobelev, V. (2014b). Relaxation and creep in twist and flexure. *Multidiscipline Modeling in Materials and Structures, 10*(3), 304–327. 11.1108/MMMS-11-2013-0067.

Kobelev, V. (2016). Addendum to "relaxation and creep in twist and flexure. *Multidiscipline Modeling in Materials and Structures, 12*(3), 473–477.

Kwofie, S., & Chandler, H. D. (2007). Fatigue life prediction under conditions where cyclic creep–fatigue interaction occurs. *International Journal of Fatigue, 29*(2007), 2117–2124.

Lewin, L. (1981). *Polylogarithms and associated functions*. North-Holland.

Milne, I., Ritchie, R. O., & Karihaloo, B. L. (2003b). *Comprehensive structural integrity* (Vol. 5). Creep and High-Temperature Failure, Elsevier

Nabarro, F. R. N., & de Villers, H. L. (1995). *The physics of creep* (pp. 15–78). Taylor & Francis.

Naumenko, K., & Altenbach, H. (2007). *Modelling of Creep for Structural Analysis*. Springer.

Naumenko, K., Altenbach, H., & Gorash, Y. (2009). Creep analysis with a stress range dependent constitutive model. *Archive of Applied Mechanics, 79*, 619–630.

Naumenko, K. (2006) Modeling of high-temperature creep for structural analysis applications. Doctoral Thesis, Martin-Luther-Universit¨at Halle-Wittenberg

Nezhad, H. Y., & O'Dowd, N. P. (2012). Study of creep relaxation under combined mechanical and residual stresses. *Engineering Fracture Mechanics, 93*, 132–152.

Nezhad, H. Y., & O'Dowd, N. P. (2015). Creep relaxation in the presence of residual stress. *Engineering Fracture Mechanics, 138*, 250–264.

Odquist, F. K. G., & Hult, J. (1962). *Kriechfestigkeit metallischer Werkstoffe*. Springer.

Pineau A. (1988). Mechanisms of creep-fatigue interactions. In: C. Moura Branco & L. Guerra Ro-sa (Eds.), *Advances in fatigue science and technology*. Kluwer Academic Publishers#.

Poirier, J. P. (1985). *Creep of Crystals*. Cambridge University Press.

Polák, J., Petrenec, M., & Man, J. (2011). Cyclic plasticity, cyclic creep and fa-tigue life of duplex stainless steel in cyclic loading with positive mean stress. *Kovove Materialy Metallic Materials, 49*, 347–354. https://doi.org/10.4149/km20115347347

Schlimmer, M. (1986). Einfache Kriechgleichungen für nichtlinear-viskoelastische Werkstoffe. *Rheologica Acta, 25*, 359–364.

Skrotzki, B., Olbricht, J., & Kühn H. -J. (2019). High temperature mechanical testing of metals. In: C. -H. Hsueh et al. (Eds.), *Handbook of mechanics of materials* (Vol. 58). Springer Nature Singapore Pte Ltd.https://doi.org/10.1007/978-981-10-6884-3_44

Timoshenko, S. (1948). *Strength of Materials*. Van Nostrand.

Wilshire, B., & Battenbough, A. J. (2007). Creep and creep fracture of polycrystalline copper. *Materials Science and Engineering A, 443*, 156–1611.

Wilshire, B., & Scharning, P. J. (2007). Long-Term creep life prediction for a high chromium steel. *Scripta Materialia, 56*, 701–704.

Wu, X. (2010). Uniaxial creep lifting. Methodology-II: Creep Modeling for Waspaloy and Udimet 720Li; Report No.: LTR-SMPL-2010-0156; National Research Council (NRC): Rolls-Royce, Montreal, QC, Canada.

Yang, Z., & Wang, Z. (2001). Cyclic creep and cyclic deformation of high-strength spring steels and the evaluation of the sag effect: Part I. Cyclic plastic deformation behavior. *Metallurgical and Materials Transactions A, 32*, 1687–1698. https://doi.org/10.1007/s11661-001-0147-1

Yao, H.-T., Xuan, F.-Z., Wang, Z., & Tu, S.-T. (2007). A review of creep analysis and design under multi-axial stress states. *Nuclear Engineering and Design, 237*, 1969–1986.

Chapter 11
Fatigue of Spring Materials

Abstract This chapter explains the common methods for considering stress amplitude in the fatigue life of springs. The fatigue of spring materials under fully reversed uniaxial loading is reviewed. Two different fatigue life estimation methods are briefly discussed. The first method implements the stress-life and strain-life procedures. The second method describes fatigue crack growth (FCG) per cycle. Two analogous unifications of the FCG functions are proposed. The expressions for the spring length over the number of cycles are derived. This chapter is the first section of the final part, which studies the life cycle of elastic elements.

Keywords Fatigue of springs · Fatigue crack growth · Stress-life procedure · Strain-life procedure

11.1 Phenomenon of Fatigue

11.1.1 Fatigue Influence Factors

The fatigue phenomenon is associated with the nucleation of sub-microscopic surface cracks in the fatigued component early in its life, which initially grow very slowly. Gradually, a crack grows until the effective cross-section of the part is reduced to a value that cannot support the applied stress. Eventually, rapid failure occurs. In a typical fatigue surface, there are distinct zones. These correspond to the period of slow growth and ultimate failure, respectively.

Repetitive load/unload cycles of significant amplitude also cause spring materials to fatigue. The level of stress that causes fatigue failure is usually below that considered safe for a single static load application. The critical fatigue initiation is usually at a location confined to a small area. Failure may be the result of additional factors such as stress concentration due to component shape, surface finish, or corrosion pitting (Juvinall & Marshek, 2017, Chap. 8). In the cyclic fatigue test according to (ASTM E 739-80, 1986; ASTM E 606-80, 1986; DIN 50100, 2016:12, 2016), a material or component is subjected to a periodically changing (cyclic) load. Fatigue testing is

V. Kobelev, *Fundamentals of Springs Mechanics*,
https://doi.org/10.1007/978-3-031-58584-5_11

used to determine fatigue strength and fatigue limit for tensile, compression, flexure, and torsion. Especially for springs, the fatigue test can be used to evaluate material processing and improve the design or material variations. Short-term strength is not considered in the fatigue test. It is studied in the low cycle fatigue test.

There are many factors that affect the fatigue life of a spring. These factors must be carefully considered. The valuable influencing factors are:

- corrosion of spring surface;
- wear factors due to friction on the contact surfaces (Barrois, 1979; Lamacq et al., 1996; Sangid, 2013; Stachowiak & Batchelor, 2014, Chap. 14)
- different residual stresses due to manufacturing, which could affect the durability positively or negatively;
- modification of mechanical material properties due to manufacturing;
- shot peening, which induces residual stresses and influences the roughness of surface;
- hydrogen induced ductility (Akaki et al, 2017; Matsuo, 2017);
- inclusions, flaws, and other inherited defects of the spring material;
- high temperature, which cause the reversion of crystal structure;
- high level of radiation, especially the neutron radiation, which leads to hydrogen induced ductility.

The following chapters review the fatigue effects of springs. Each chapter examines some specific manifestations of fatigue phenomena. In this chapter, only the case of pure sinusoidal cyclic loading with constant amplitude (fully reversed stress amplitude) and zero mean stress is considered. The fatigue laws relate the number of cycles to the fully reversed stress amplitude. The overwhelming majority of tests of spring materials are designed only for a fixed mean stress with varying amplitude. The mean stress is usually set to zero during the baseline test. The test is performed with different constant stress or strain amplitudes until failure. The results of this basic test are organized in the form of the fully reversed fatigue law. To evaluate all the other effects mentioned above, significantly more fatigue tests must be performed. The experimental acquisition of the stress ratio requires a significant increase in the duration and cost of tests. The industry is trying to reduce the experimental effort. To account for the stress ratio, the industry commonly uses plausible hypotheses about the influence of mean stress on fatigue life. These hypotheses use correction factors for the fully reversed fatigue laws, as discussed in the next Chapter.

In the fatigue test, the load amplitude and mean load are constant during the single step vibration test. Depending on the magnitude of the load amplitude, it can be applied at different frequencies before the specimen fails. Fatigue testing is used to determine the fatigue strength and fatigue limit of materials or components. For this purpose, several specimens are cyclically loaded. The cyclic fatigue test is performed until a defined failure of the specimen (fracture, crack) occurs. The cyclic fatigue test determines the number of cycles (number of limit cycles). If a specimen reaches the number of cycles without any detectable failure, it is evaluated as fatigue strength or fatigue life. The mean stress, high stress, and low stress of the cyclic loading are constant for each fatigue test. For tests with the same S–N curve, either the mid-stress or the high-to-low stress ratio is changed.

The typical load case for the uniaxial fatigue analysis is the harmonically oscillating stress:

$$\sigma(t) = \sigma_m + \sigma_a \sin(\omega t). \tag{11.1}$$

This type of stress is typical for torsion springs, disk springs and leaf springs. Usually, the circular frequency ω is not high enough to cause the heating of the spring. In this case, the influence of frequency on fatigue could be ignored at first glance. The parameters of the harmonically oscillating stress are:

$$\sigma_a = \frac{\sigma_{\max} - \sigma_{\min}}{2}. \tag{11.2}$$

is the stress amplitude;

$$\Delta\sigma = \sigma_{\max} - \sigma_{\min} = 2\sigma_a, \tag{11.3}$$

is the stress range;

$$R_\sigma = \frac{\sigma_{\min}}{\sigma_{\max}} < 1, \tag{11.4}$$

is the stress ratio of cyclic load;

$$\sigma_m = \frac{\sigma_{\min} + \sigma_{\max}}{2} = \frac{1 + R_\sigma}{1 - R_\sigma} \frac{\Delta\sigma}{2}, \tag{11.5}$$

is the mean value of stress.

The extreme values of the oscillating stress:

$$\sigma_{\max} = \frac{1}{1 - R_\sigma} \frac{\Delta\sigma}{2}, \tag{11.6}$$

is the maximum stress per cycle,

$$\sigma_{\min} = \frac{R_\sigma}{1 - R_\sigma} \frac{\Delta\sigma}{2}, \tag{11.7}$$

is the minimum stress per cycle.

Torsion and shear stresses dominate in helical compression and tension springs and in twist beams. The harmonically oscillating cyclic shear stress is the function of time:

$$\tau(t) = \tau_m + \tau_a \sin(\omega t). \tag{11.8}$$

The range (or stroke) of shear stress $\Delta\tau$ reads:

$$\Delta\tau = \tau_{\max} - \tau_{\min} = 2\tau_a, \ \tau_{\min} = \tau_m - \tau_a, \ \tau_{\max} = \tau_m + \tau_a. \tag{11.9a}$$

In terms of stress ratio

$$R_\sigma = \frac{\tau_m - \tau_a}{\tau_m + \tau_a} = \frac{\tau_{\min}}{\tau_{\max}}, \tag{11.9b}$$

the parameters of stress cycle are:

$$\tau_a = \frac{1 - R_\sigma}{2}\tau_{\max}, \qquad \text{shear stress amplitude,} \tag{11.9c}$$

$$\tau_m = \frac{1 + R_\sigma}{2}\tau_{\max}, \qquad \text{mean value of the shear stress.} \tag{11.9d}$$

In the current Chapter, only the case for a pure sinusoidal cyclic load of constant amplitude (fully reversed stress amplitude) and zero mean stress is studied ($R_\sigma = -1$). The influence of stress ratio is discussed in the next Chapter.

11.1.2 Stages of Fatigue Fracture

There are three stages of fatigue fracture commonly distinguished: initiation, propagation, and final rupture (SAE AE-22, 1997), (Totten, 2008) and (Fleck et al., 1994). On a microscopic scale, failure occurs along slip planes in the crystalline structure of the materials. Most metals with a body centred cubic crystal structure have a characteristic response to cyclic stresses. These materials have a threshold stress limit below which fatigue cracks will not initiate. This threshold stress value is referred to as the endurance limit. In spring steels, the life associated with this behaviour is about two million cycles (Fig. 11.1).

The first stage (stage I) of fatigue is referred to as initiation (Polak, 2003). Initiation is probably the most complex stage of fatigue fracture. The most significant factor about the initiation stage of fatigue fracture is that the irreversible alterations in the metal are caused by cyclic shear stresses. The microscopic faults accumulate of over each load application. The growing with each cycle cumulative damage leads finally to final breakage. At the location of a severe stress concentration, the number of cycles to breakage depends on the geometry of the part as well as on environmental, stress, metallurgical, and strength conditions, as will become apparent. A brief discussion of the role of defects in the initiation phase is given in Sect. 11.3. During the stage I, the spread of fatigue crack per unit cycle decelerates smoothly with number of cycles and the process approaches its second stage.

The second stage of fatigue is known as crack propagation (stage II). The direction of growth can change after the transition from stage I to stage II. The crack

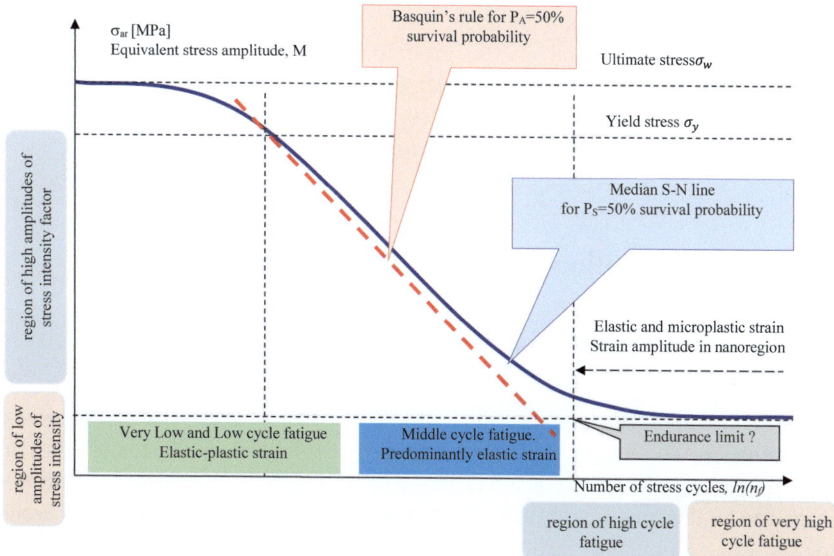

Fig. 11.1 An S–N diagram plotted from the results of completely reversed axial fatigue tests

grows customary perpendicular to the direction of the maximal tensile stress. For example, the cracks in the cyclically loaded helical compression springs expanse in the direction of 45° to the spring axis. The second, or propagation, stage of fatigue is typically the most readily recognizable area of a fatigue fracture. The Paris law describes the stage II fatigue crack growths function (Paris & Erdogan, 1963). The traditional form of Parislaw pronounces the spread of fatigue crack per unit cycle as a power function of the range of stress intensity factor. Towards the end of the stage II, the spread of fatigue crack per unit cycle begins to accelerate smoothly with number of cycles and the process turns to the third stage.

The final, third stage of fatigue is the final rupture (stage III). As the propagation of the fatigue crack endures, progressively sinking the cross-sectional area of the test specimen, it eventually deteriorates the part. The deterioration occurs so quickly, that final, broad fracture occurs within a pair of load cycles.

The fracture mode may be either ductile (with a dimpled fracture surface) or brittle (with a cleavage, or intergranular, fracture surface). The combinations of both modes occasionally occur, depending upon the metal concerned, the stress level and the environment. In the course of the stage III, the spread of fatigue crack per unit cycle progressively accelerates with each cycle. The in-situ observations of the final crack growth are seldom.

The functional relation between the loading properties, stress gradient and the time-depending mechanical characteristics of materials are essential for the fatigue analysis and evaluation number of the cycles to failure (Carpinteri, 1994; Christ, 1991, 2018; Fatemi & Yang, 1998; Suresh, 1998). The traditional evaluations concepts based on the Palmgren–Miner's rule of damage accumulation, rain-flow counting of time-dependent loads, Wöhler curve and Basquin equation, Paris-Erdogan law and diverse generalizations of the fracture and damage mechanics approaches (Richard & Sander, 2012). The book (Krupp, 2007) reviews the numerical treatment of fatigue microscopic crack propagation together with their implementation in fatigue-life prediction models.

11.2 Crack Initiation Approach for Uniaxial Stress State

11.2.1 Stress-Life Approach for Symmetric Cycle

The stress-life approach is applicable to the cases of long fatigue life and moderate cyclic stresses. The fatigue data are acquired by cyclic test specimens at constant amplitude stress until failure occurs. Fatigue data commonly have significant scatter which can be described by probability density functions (Appendix C). The cyclic stresses lead to dominant elastic strain amplitudes and only tiny microscopic plastic strains occurs over one reversal cycle. The fully reversed stress the amplitude (σ_{ar} or τ_{ar}) keeps constant during the basic cyclic test. For the symmetric cycle, the stress is a harmonic function of time with the vanished mean stress. These conditions are suitable for the estimations of fatigue life of common spring materials. Occasionally, the mean stress during the basic is fixed to a certain nonzero constant. For example, for helical compression springs the mean stress is positive, because the minimum force of the spring must be positive to avoid the detachment of the spring from the plates.

It is surely impossible to test the fatigue life of the springs for all imaginable conditions. In the next Chapter, we present the usual hypotheses that allows the recalculation of the fatigue life at one-test conditions to the fatigue laws with different test conditions.

For characterization of fatigue life as function of stress amplitude, the fatigue laws traditionally applied (SAE AE-22, 1997, Sect. 10.3.4.3). The stress-life approach is based on the Basquin's law. This law is known also as Wöhler's law in German speaking countries. The plot of Basquin's law possess the fatigue life n_f the x-axis and the effective fully-reversed stress amplitude $\sigma_{ar} = \Delta\sigma/2$ on the y-axis. Thus, the number of cycles to failure is n_f. The number of reversals to failure is 2 n_f. The dependence of fatigue life upon the effective fully reversed stress amplitude σ_{ar} derives from the Basquin's law (Basquin, 1910):

$$
\begin{aligned}
n_f &= f_\sigma^{-1}(\sigma_{ar}) \overset{\text{def}}{=} \frac{1}{2}\left(\frac{\sigma_{ar}}{\sigma'_{f0}}\right)^{\frac{1}{b_0}}, \\
n_f &= f_\tau^{-1}(\tau_{ar}) \overset{\text{def}}{=} \frac{1}{2}\left(\frac{\tau_{ar}}{\tau'_{f0}}\right)^{1/b_0},
\end{aligned}
\tag{11.10}
$$

where

$$\sigma_{ar} = f_\sigma(n_f), \qquad f_\sigma(n_f) = \sigma'_{f0} \cdot (2n_f)^{b_0}, \tag{11.11}$$

$$\tau_{ar} = f_\tau, \qquad f_\tau(n_f) = \tau'_{f0} \cdot (2n_f)^{b_0}. \tag{11.12}$$

The parameters in Eqs. (11.10)–(11.12) are:

σ'_{f0}	the fatigue strength coefficient for normal stress,
τ'_{f0}	the fatigue strength coefficient for shear stress,
b_0	the fatigue strength exponent,
$p_\sigma = -1/b_\sigma$	the reciprocal fatigue exponent.

The index "0" symbolizes that the mean stress is equal to zero. All coefficients are acquired for symmetric stress cycle (S). In this case, stress varies harmonically with the average of zero:

$$\sigma(t) = \sigma_{ar} \sin(\omega t), \qquad \sigma_{ar} = \frac{1}{2}\Delta\sigma. \tag{11.13}$$

There are alternative proposals, based on stress-life approach, for the description of the fatigue failure line 2–7. Among them, several rarely used criteria were discussed by (Meyer, 2014).

The plots of functions (11.10)–(11.12) are referred as S–N curves. Symbol 'S' signifies the cyclic stress range and 'N' stands for the number of cycles to failure n_f. The scale of 'N' axis is usually logarithmical. The scale of 'S' axis is logarithmical or linear. In this scaling the S–N curves for laws (11.10)–(11.12) demonstrate the linear dependence.

Hereafter we discuss the median S–N curves. The median S–N curves correspond the cumulative failure probability of 50%. The scattering of fatigue data and the design S–N curves are discussed in Chap. 10.

11.2.2 Strain-Life Approach for Symmetric Cycle

The plastic fully reversed strain amplitude is commonly assumed constant during the cyclic test and is much higher than the elastic strain amplitude (SAE AE-22, 1997, Sect. 10.3.4.4). The fatigue data for strain-life approach are obtained by cyclic test specimens at constant amplitude strain until failure occurs (Graham et al., 1968, Ch. 6). This condition fulfils in low cycle fatigue regime, where the Coffin-Manson law is appropriate.

For strain-life approach, the coefficients are acquired for symmetric or for non-zero mean strain cycle. In the first case, strain oscillates harmonically with the vanished mean value:

$$\varepsilon(t) = \varepsilon_{ar} \sin(\omega t), \qquad \varepsilon_{ar} = \frac{1}{2}\Delta\varepsilon.$$

For the conditions of symmetric cycle, the mean stress is zero. The Coffin-Manson law relates the number of cycles to failure upon the plastic fully-reversed strain amplitude (Coffin, 1954; Manson, 1953):

$$\varepsilon_{ar} = \frac{\sigma'_{f0}}{E} \cdot (2n_f)^{b_0} + \varepsilon'_{f0} \cdot (2n_f)^{c_0}. \tag{11.14}$$

The values in Eq. (11.14) are:

$\sigma_{f0'}$ fatigue strength coefficient,
b_0 fatigue strength exponent,
$\varepsilon_{f0'}$ fatigue ductility coefficient,
c_0 fatigue ductility exponent.

The quantities $\sigma_{f0'}$ and b_0 are the same as in Eq. (11.14) and $\varepsilon_{f0'}$ and c_0 are additional fitting constants for the plastic strain term.

If the plastic fully-reversed strain amplitude is determined with (11.14), the stress amplitude σ_a results from the solution of the Ramberg–Osgood equations:

$$\varepsilon_{ar} = \frac{\sigma_a}{E} + \left(\frac{\sigma_a}{K'_\sigma}\right)^{1/n'}. \tag{11.15}$$

The Eq. (11.15) represents the cyclic stress–strain curve with two fitting constants n' and $K_{\sigma'}$. The constant $K_{\sigma'}$ the cyclic hardening coefficient and n' is the cyclic hardening exponent. The analogous Ramberg–Osgood representation, but for the static hardening curve, was used in previously.

The solution of the Eq. (11.15) gives the cyclic stress σ_a in terms of the cyclic strain, which is constant in this type of test. The control of the strain-life cyclic test is easier, than the control of stress-life test. It is easier to fix the amplitude of testing, because the stress-life test requires the acquisition of the applied forces during the cyclic test. In the majority of tests of springs and spring materials, the amplitude of deformation keeps during the test time, for example, in R. R. Moore rotating-bending fatigue test (Juvinal & Marshek, 2017; Bussoloti et al., 2015).

The corresponding formulas for the shear stress are analogues to Eq. (11.14) with the evident replacements:

$$\gamma_{ar} = \frac{\tau'_{f0}}{G} \cdot (2n_f)^{b_0} + \gamma'_{f0} \cdot (2n_f)^{c_0}, \tag{11.16}$$

$$\gamma_{ar} = \frac{\tau_a}{G} + \left(\frac{\tau_a}{K'_\tau}\right)^{1/n'}. \tag{11.17}$$

11.3 Crack Propagation Approach

11.3.1 Crack Growths Functions of Paris-Erdogan Type

The different method for determination of S–N curves starts from a commonly accepted form of law for fatigue crack growth and demonstrates the noticeable features of the fatigue phenomena (SAE AE-22, 1997, Sect. 10.4). Using this method, the S–N curve will be a logical consequence from the introduced micromechanical law, rather than a directly observable phenomenon. Analogously, the mean stress sensitivity will be an immediate consequence of the acknowledged law for the crack propagation. In other words, the cyclic fatigue diagrams will be synthesised based on the micromechanical models of stepwise crack propagation per load cycle.

Numerous load damage models, which extend the celebrated Paris and Erdogan fatigue crack growths function, were proposed (Tanaka, 2003). The article (Beden et al., 2009) reviews the articles focused on the prediction of fatigue properties of structures under variable amplitude loading. The reliability and accuracy of prediction models and the physical concept of fatigue damage was discussed. The reviewed fatigue life models were based on the scientific and engineering knowledge about fatigue of material and structures under constant and variable amplitude loading.

The near-threshold deviation of the common fatigue crack growths function was suggested by (Donahue et al., 1972). The observed rapidly increasing growth towards ductile tearing was accounted. The brittle fracture correction of the power law was proposed in the cited article in the analytical form.

In the work (Kanninen & Popelar, 1985) several combinations of high and low amplitude of stress intensity factors values were studied. A similar approach for transition region between high and low amplitude of stress intensity factors was proposed in (McEvily & Groeger, 1977).

A more general "unified law" accounts certain deviations from the power-law regime (Pugno et al., 2006). The implemented extension of the Parislaw for crack propagation results in the generalized Wöhler fatigue curves.

A different expression is proposed for the correction extremely low cycle fatigue behavior of in the paper (Xue, 2008). The proposed formulas describe the cyclic life by introducing an exponential function, making possible to cover the entire span of extremely low cycle fatigue to low cycle fatigue. The curves are fitted by means of an additional material parameter.

The methodology to the modeling of material fatigue in cyclic loading has been suggested based on the kinetic theory of strength in (Mishnaevsky & Brøndsted, 2007). According to this concept, the acting stresses instead of stress changes yield the damage growth in materials under fatigue conditions. In order to model the material degradation, the kinetic theory of strength was applied. The model allows the determination of the stiffness reduction in composite with a definite load history. Moreover, the model resolves the effect of the loading frequency on the lifetime and the damage growths per cycle. It was shown that the number of cycles to failure increases almost linearly. With increasing the loading frequency, the damage growth per cycle decreases.

By applying the quantized fracture mechanics, the unification and extension of the traditional Paris' and Basquin's or Wöhler's laws fatigue crack propagation were derived in (Pugno et al. 2007). In the cited article the generalized Paris', Wöhler's or unified laws were suggested. The performed analysis demonstrated the applicability of proposed laws for predicting the lifetime of structures containing fatigue cracks. The sizes of fatigue cracks were varied from small (the Basquin's regime) to large (the Paris' regime).

The revision of the fatigue crack growth model was proposed by (Castillo et al., 2014), CCS model. The crack growth model for derivation of the S–N curves from the curves of crack growth rate was presented. The CCS model complements the common stress based and fracture mechanics methodologies in fatigue lifetime through the enhanced fatigue crack growths function. Firstly, the curve of crack growth rate is defined over all its range as a cumulative distribution function based on a normalized dimensionless range factor of stress intensity. The CCS model assumes that fatigue crack growth takes the form of a Gumbel cumulative distribution function (Gumbel, 1954). For this prediction of crack growth rate, the Buckingham theorem was incorporated. Secondly, the proprietary theorem was derived. The derived theorem provides an alternative to self-similarity and assures the significant reduction of experimental planning. Correspondingly, the different crack growth curves for different stress ranges and initial crack lengths are obtained from a specific crack growth curve. The S–N field results from the crack growth curves. The close relation between the fracture mechanics and stress approaches was demonstrated. Thirdly, the model was applied to a certain set of experimental data. The curve of crack growth rate and the S–N curves of a certain material were derived.

The substantial modification of CSS model was proposed in paper (Blasón et al., 2016). This modification accounts the crack opening and closing effects as well as the influence of the stress ratio. The theoretical model incorporates both effects of the average stress as well crack closure and opening action. The model provided an analytical expression of the curve of crack propagation rate by matching experimental data by means of the least squares method. The identification of new variables was introduced for the model of fatigue cracking. The auxiliary variables include the cracking and opening effects using the influence of the stress ratio. The normalized variables allow the values in the range 0 to 1. Thus, the use of "S-shaped" cumulative distribution functions reproduces the relation between the variation of stress intensity factor and number of cycles. The integration of several models that come from the statistical domain to solve the problem of crack growth is possible. The a-N and S–N curves describe the crack closure effects for the materials on different initial crack size values and loads.

The common laws of fatigue crack propagation must also evaluate opening and closure effects for fatigue crack. For this purpose, a corresponding modification of the CCS law of fatigue crack growth was proposed in the paper (Correia et al., 2016a). The modification incorporated the crack opening and closure effects together with the stress ratio effect. The fatigue crack opening and closure effects were explained with the plasticity-induced crack-closure model. With the introduction of the phenomenological auxiliary parameter, an alternative fatigue crack closure models were implemented. The corresponding modified CCS crack propagation model was based on the effective range factor of stress intensity. The version model was supported by mathematical and physical considerations.

The model (de Castro et al., 2009) predicts crack growth by means of crack initiation properties and critical damage concepts. The crack was modeled as a sharp notch with a very small but finite tip radius. The finite radius removes the singularity of elastic fields at the tip vicinity. Accordingly, the damage caused by each load cycle and the effects of residual stresses near the crack tip evaluates immediately through the hysteresis loops caused by the loading. The calculation of cycle-bicycle crack growth was based on the introduced method. An agreement between the crack growth forecasts and experiments was achieved in the citing paper for constant and for variable amplitude loadings.

Several prediction models are known for fatigue crack growth under variable amplitude loading histories. However, no of these models predicts adequately considering loading sequence effects. An improved model of crack growth rate has been proposed in (Chen et al., 2011). The cited model deals the constant amplitude loading demonstrate its validity for experimental data. The applicability of the improved model of crack growth rate was extended to variable amplitude loading. The extension accounts crack closure level based on the concept of partial crack closure. The concept was based on the crack-tip plasticity. To denote the variation in the affected zone of under- or overload, a modified coefficient was introduced. The identification of the coefficient was based on Wheeler model (Wheeler, 1972).

The model of fatigue crack growth (Noroozi et al., 2005) was based on the elastic–plastic crack tip stress–strain history. In the FCG model the fatigue crack growth simulates the stress–strain response in the vicinity of the crack tip and estimating the accumulated fatigue damage. The fatigue crack growth was treated as sequential crack re-initiation near the crack tip. The model predicted the effect of the mean stress and the influence of the compressive stress. A fatigue crack growth was analyzed for the plane strain and for plane stress states. It was demonstrated, that the fatigue crack growth is controlled by a two-parameter driving force. The driving force itself evaluates based on the local stresses and strains at the crack tip using the Smith–Watson–Topper fatigue damage parameter (Smith et al., 1970). The residual (internal) stresses are induced by the reversed cyclic plasticity. The effect of the residual stress could be also accounted with the proposed model. The applied stress intensity factors were modified to the total stress intensity factors to be accounted for the effect of the local crack tip stresses and strains on fatigue crack growth. The fatigue crack growth was forecasted by the stress–strain response near the crack tip on one side and estimating the accumulated fatigue damage on the other side. The fatigue crack growth is the process of successive crack re-initiations in the crack tip region. The model estimates the effect of the mean and residual stresses induced by the cyclic loading. The proposed model evaluates the effect of variable amplitude loadings on fatigue crack growth. The influence of the internal stress induced by the reversed cyclic plasticity alternates the resultant stress intensity factors controlling the fatigue crack growth.

Two singularity fields near the crack tip are known. Namely, the first type of tip field is the Hutchinson-Rice-Rosengren field (Hutchinson, 1968; Rice & Rosengren, 1968). The second type is the Kujawski-Ellyin field (Kujawski & Ellyin, 1984) and (Ellyin, 1986). The effect of these singularity fields was examined in the paper (Shi et al., 2014). The study (Shi et al., 2014) evaluates the common fatigue crack growth model, which is based on energy balance during growth of the crack. The new parameter was introduced. With this parameter, the effect of different types of singularity fields was incorporated in the crack propagation model. The model predicted fatigue crack growth on the stage-II from the basic low cycle fatigue properties.

In the paper (Correia et al., 2016b) the extended procedure for fatigue life prediction of structural details was presented. The procedure is based on fracture mechanics approach. The model generalizes the normalized the CSS model for fatigue crack growth. Instead of the range factor of stress intensity as reference parameter, the generalization of the CCS model implements the cyclic J-integral range. This allows the comprehensive elastic–plastic conditions. The proposed approach was applied to a notched plate, using the equivalent initial flaw size concept.

11.3.2 Fatigue Crack Growths Functions for Crack Under Cyclic Loading

Commonly accepted, that the fatigue crack growth is influenced by the stress intensity factor range and the maximum stress intensity factor. These parameters are usually combined into one formula, which is known as the driving force. Alternative expressions for driving forces were suggested in (Noroozi et al., 2008). The driving force concepts could be successfully introduced in the following case. The stress intensity factors are properly linked to the actual elastic–plastic crack tip stress–strain field. However, the correlation between the stress intensity factors and the crack tip stress–strain fields are prejudiced by internal or residual stresses.

The crack propagation is quantified as a function of the range of the stress intensity factor:

$$\Delta K = K_{\max} - K_{\min}. \tag{11.18}$$

The factors in Eq. (11.18) are:

$K_{\max}$ the maximum stress intensity factor;
$K_{\min}$ the minimum stress intensity factor per cycle.

The common form of Paris' law quantifies the fatigue life of a specimen for a given particular crack size a. The range of stress intensity factor reads (SAE AE-22, 1997, Appendix A):

$$\Delta K = \mathrm{Y} \cdot \sqrt{\pi a} \cdot \sigma. \tag{11.19}$$

The range of stress intensity factor depends on the dimensionless parameter Y that reflects the geometry.. The parameter Y possesses the value 1 for a center crack in an infinite sheet. Hereafter, this value is assumed for briefness: Y = 1.

It is assumed that the cracks with the initial length δ exist in the material. The cycle count starts with the beginning of crack propagation from its initial length:

$$n_f\big|_{a=\delta} = 0. \tag{11.20}$$

The common form for fatigue crack growths function (propagation law) is used (Tanaka, 2003):

$$\frac{\mathrm{d}a}{\mathrm{d}n} = c_f F(\Delta K), \qquad n_f(\delta, a) = \int_{\delta}^{a} \frac{\mathrm{d}a}{c_f \cdot F\left(\sigma\sqrt{\pi a}\right)}. \tag{11.21}$$

The number of cycles for the crack length elongation from the initial length δ to the final length a is the number of cycles to total failure $n_f(\delta, a)$).

The value $c_f F(\Delta K)$ determines the expansion of the fatigue crack pro load cycle. The coefficients in Eq. (11.21) are:

$c_f = c_f(R_\sigma)$	the material constant for a given stress ratio R_σ;
$R_\sigma = \frac{K_{\min}}{K_{\max}} < 1,$	the stress ratio of cyclic load;
$K_m = \frac{K_{\min}+K_{\max}}{2} = \frac{1+R_\sigma}{1-R_\sigma}\frac{\Delta K}{2},$	the mean value of stress intensity factor,

$$K_{\min} = \frac{R_\sigma}{1-R_\sigma}\frac{\Delta K}{2}, \qquad K_{\max} = \frac{1}{1-R_\sigma}\frac{\Delta K}{2}.$$

The fatigue crack growths function (FCG) $F(\Delta K)$describes the infinitesimal crack length growths per load cycle n. For the beginning we suppose that the load ratio R_σ remains constant over the load history. The range ΔK and the mean value K_m of stress intensity factor are thus constant.

All material properties are valid for the definite value of the stress ratio. For the rough estimation, for helical compression springs could be accepted $R_\sigma = 0$. The influence of stress ratio is the subject of the next Chapter.

11.4 Fatigue Crack Growths

11.4.1 Unification of Paris Law

As experimentally recognized for some alloys, there is a theoretical value for stress amplitude, below which the material will not fail for any number of cycles. The fatigue cracks will not propagate, of the amplitudes of stress intensity factor are less than this threshold (Ritchie et al., 1999). At the other extreme, approaching the fracture toughness, the material fails almost immediately. The unified functions that reveal the mentioned behavior are known from the literature. A unified relation that covers all three regions was proposed by (Freudenthal, 1973):

$$\frac{\ln(K_{IC}/\Delta K)}{\ln(K_{IC}/K_{th})} = \exp\left[-\left(\frac{\mathrm{d}a/\mathrm{d}n}{\mathrm{d}a/\mathrm{d}n|_{\Delta K=\widetilde{K}_m}}\right)^{1/p}\right], \quad \widetilde{K}_m \cong \frac{K_{th}+K_{IC}}{2},$$

$p > 1$ is the fatigue exponent,
K_{IC} is the short-term threshold limit,
K_{th} is the endurance threshold limit ($0 < K_{th} < K_{IC}$).

Alternative form of unified relation was proposed by (Schwalbe, 1980):

$$\frac{\mathrm{d}a}{\mathrm{d}n} = c_f\left(\arctan\frac{\Delta K}{K_{IC}} - \arctan\frac{K_{th}}{K_{IC}}\right).$$

Both previously mentioned unified relations (Freudental, 1973; Schwalbe, 1980) do not lead to closed forms of solutions for spring propagation. The use of these laws to resolve the crack length necessitates the numerical solutions of nonlinear differential equations in terms of cycle numbers. Consequently, the known unified relations are inappropriate for the numerical fitting of unknown parameters. The paper (Miller & Gallagher, 1981) presents eight different methods to prediction the fatigue life. Each presented method is applicable to describe the three regions of the crack growth rate curve. The second fatigue crack growth rate (FCGR).description reads:

$$\frac{\mathrm{d}a}{\mathrm{d}n} = c_f \cdot \frac{1-\left(\frac{K_{th}}{\Delta K}\right)^{m_1}}{1-\left(\frac{\Delta K}{K_{IC}}\right)^{m_2}} \tag{11.22}$$

$m_2 > 1$ is the exponent at the short-term limit,
$m_1 > 1$ is the endurance limit exponent.

The unified propagation function (11.22) due to (Miller & Gallagher, 1981) could be studied in closed form, as shown later.

11.4.2 Unification of Paris Law Type I

11.4.2.1 Closed Form Solution for Unified Paris Law Type I

In this Chapter we employ the method of representation of crack propagation functions through appropriate elementary functions. The proper choice of the elementary functions is motivated by the phenomenological data and covers a broad region of possible parameters. With the introduced crack propagation functions differential equations describing the crack propagation are solved rigorously. The resulting closed form solutions allow the evaluation of crack propagation histories on one side, and the effects of stress ratio on crack propagation, on the other side.

For closed form solutions, two unified functions (type I and type II) are proposed. The functions are suggested in the form that incorporates the three commonly accepted stages of fatigue. The advantage of the newly proposed functions is the closed form solution of crack propagation. The solution immediately delivers dependency of crack length over load history and to the number of stress cycles that a specimen sustains before failure occurs. The proposed functions allow fitting of acquired experimental data. Most known generalized laws follow from the newly introduced functions as the special cases.

The type I unified propagation function. F_I for the damage growths per cycle accounts the Paris fatigue crack growths function together with transition regions at high and low amplitudes of stress intensity factors (Kobelev, 2017a):

$$F_I(\Delta K) \stackrel{\text{def}}{=} \Delta K^p \cdot \frac{1-\left(\frac{K_{th}}{\Delta K}\right)^{m_1}}{1-\left(\frac{K}{K_{IC}}\right)^{m_2}}. \tag{11.23}$$

The mentioned above fatigue laws demonstrate that if the applied stress is below the threshold limit, destruction of spring material occurs for technically significant number of cycles. Initiation of the fatigue happens when the stress amplitude increases and the stress intensity factor arrives the threshold value. The deceive role for this process play the defects in spring material. The behavior of spring steels near the threshold limit is the subject of fatigue analysis at very high number of cycles.

The main advantage of the function (11.23) is the closed form analytical solution. On one side, the closed form analytical solution facilitates the universal fitting of the constants of the fatigue law over all stages of fatigue. On the other side, the closed-form solution eases the application of the fatigue law, because the solution of nonlinear differential equation turns to be dispensable. Moreover, the mean stress dependence could be derived from Eq. (11.23).

The proposed function delivers the closed form solution of the ordinary differential Eq. (11.21) with the initial condition $n_f\big|_{a=\delta}=0$, such that:

$$n_f = n_f(a, \delta, \Delta\sigma).$$

Precisely, the function $n_f(a, \delta, \Delta\sigma)$ delivers the number of cycles for growth of the crack length from the initial value δ to the given value a, assuming that both stress range σ and mean stress σ_m remains constant. As discussed later in Sect. 11.4, the size of existing defect is taken as the initial value δ of crack or of inhomogeneity.

From the Eq. (11.21) two ultimate crack sizes could be determined immediately.

Firstly, the value:

$\pi a_C = K_{IC}^2/\sigma_r^2$ is the critical crack length at which instantaneous fracture will occur.

Secondly, the value:

$\pi a_{th} = K_{th}^2/\sigma_r^2$ is the initial crack length at which fatigue crack growth starts for the given stress range.

The number of cycles to failure if $\delta \leq a_{th}$ is numerically infinite. For a finite number of cycles before fracture the initial crack length δ must satisfy the condition:

$$a_{th} < \delta < a_C.$$

The relations between the crack growth rate $\mathrm{d}a/\mathrm{d}n = c_f F_I(\Delta K)$ and the range of stress intensity factor ΔK for simulated materials are shown on the Fig. 11.2. The endurance and short time threshold exponents were experimentally acquired oil-tempered Si–Cr steel for valve springs (Akiniwa et al., 2008) (the alloy type JIS G3561, SWOSC-V, equivalent to alloys VDSiCr (EN 10-270-2:2011), ASTM A877/877 M Grade A).

For a nickel chromium molybdenum steel (JIS SNCM439, equivalent to AISI 4340, DIN 34NiCrMo6) the fatigue date were obtained by (Akiniwa & Tanaka, 2004).

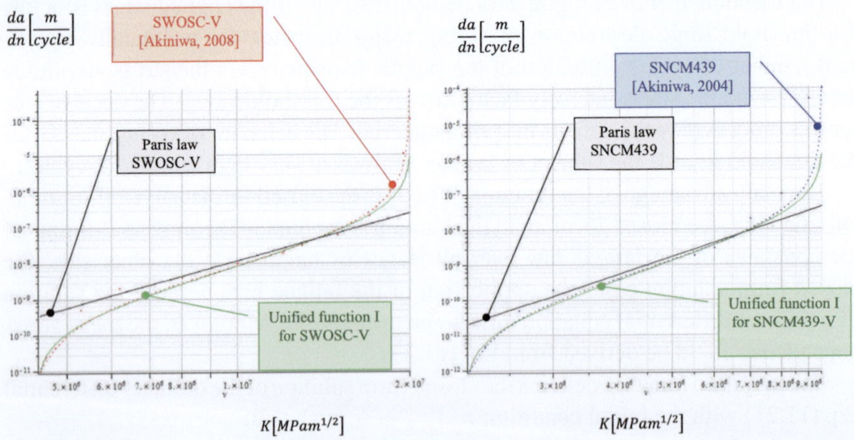

Fig. 11.2 The plot of the relations between the crack growth rate da/dn and the range of stress intensity factor K for materials SWOSC-V (Akiniwa et al., 2008) and SNCM439 (Akiniwa & Tanaka, 2004)

For these materials the numerically fitted parameters of fatigue law are recapitulated in the Table 11.1.

The solution of the ordinary differential Eq. (11.21) with the initial condition $n|_{a=\delta} = 0$ delivers closed form analytical expression for the remaining number of cycles to total failure:

$$n_f(a, \delta, \Delta\sigma) = n_I(a, \Delta\sigma) - n_I(\delta, \Delta\sigma), \tag{11.24}$$

where the auxiliary functions are:

$$n_I(a, \Delta\sigma) = \frac{2aK_1^{2-p}}{c_f m_1 \Delta K^2} \cdot N_I\left(\frac{\Delta K}{K_{th}}\right)^{m_1}, \tag{11.25}$$

$$N_I(q) = (K_{th}/K_{IC})^{m_2}\beta_2(q) - \beta_1(q), \tag{11.26}$$

Table 11.1 Mechanical properties of simulated materials SWOSC-V (Akiniwa et al., 2008) and SNCM439 (Akiniwa & Tanaka, 2004)

Alloys	Oil tempered silicon/ chromium alloyed valve spring wire	Nickel chromium molybdenum steel	Units
Alloys type	JIS G3561, SWOSC-V VDSiCr (EN 10270-2:2011) ASTM A877/877 M Grade A	JIS SNCM439, AISI 4340 DIN 34NiCrMo6	
Exponents at short-term limit and endurance limit, F_I	$m_1 = 3/2; m_2 = 3/2;$		
Fatigue exponent, p,F_I	2.17	2.51	
Material constant, c_f,F_I	0.1514×10^{-23}	0.4738×10^{-26}	$(Pa\sqrt{m})^{p1}m$
Endurance threshold limit K_{th}	0.45×10^7	0.21×10^7	$Pa\sqrt{m}$
Short-term threshold limit K_{IC}	0.2×10^8	0.95×10^7	$Pa\sqrt{m}$
Exponents at short-term limit and endurance limit F_{II}	$k = 1, m_1 = 3/2; m_2 = 3/2;$		
Fatigue exponent, p,F_{II}	2.17	2.51	
Material constant, c_f,F_{II}	0.4621×10^{-23}	0.1537×10^{-25}	$(Pa\sqrt{m})^{p2}m$

$$\beta_1(q) = \mathbf{B}\left(q; \frac{2+m_1-p}{m_1}, 0\right),$$
$$\beta_2(q) = \mathbf{B}\left(q; \frac{2-p+m_1+m_2}{m_1}, 0\right). \quad (11.27)$$

The incomplete beta-function $B(q; x, y)$ appear in the expressions (11.27):

$$\mathbf{B}(q; x, y) = \int_0^q t^{x-1}(1-t)^{y-1}\mathrm{d}t, \quad (11.28)$$

The functions β_1 and β_2 in (11.27) relate to each other as:

$$\beta_1(a, \Delta\sigma) = \lim_{m_2\to 0} \beta_2(a, \Delta\sigma). \quad (11.29)$$

The following example illustrates the behavior of the metallic material with the unified propagation function (11.23). The diagrams that express the dependence of cycles to failure, S–N curves (11.24) are presented on Fig. 11.3. The initial length of the crack was assumed for definiteness in all cases $\delta = 10^{-4}m$. The stress intervals for calculation depend on threshold stress intensity factors:

$$\sigma_{th} < \Delta\sigma < \sigma_C, \qquad \sigma_{th} = K_{th}\sqrt{\pi\delta}, \qquad \sigma_C = K_{IC}\sqrt{\pi\delta}.$$

The dependencies of crack lengths on cycles, a–N curves are drawn on the right pictures. The number of cycles was calculated as the function of length of the crack.

The S–N curves are shown on the Fig. 11.3 for Si–Cr and for Cr–Mo alloys. The a–N curves are shown on the Fig. 11.4 for these alloys.

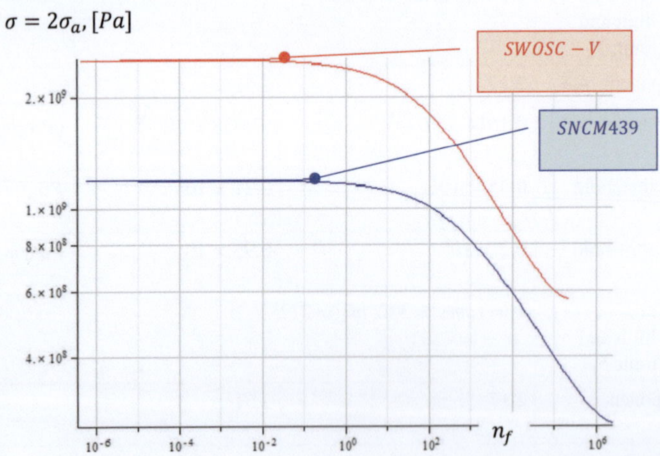

Fig. 11.3 The calculated dependencies between cycles to failure n_f for the given initial crack lengths upon the stress range $\Delta\sigma$ (s–N-curves) for SWOSC-V (Akiniwa et al., 2008) and SNCM439 (Akiniwa & Tanaka, 2004)

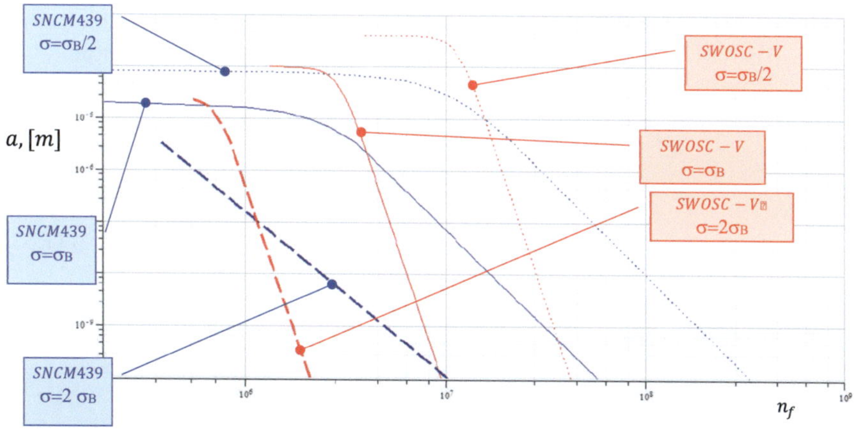

Fig. 11.4 The calculated dependencies between cycles to failure n_f for the given stress upon the length of crack (a-N-curves) for SWOSC-V (Akiniwa et al., 2008) and SNCM439 (Akiniwa & Tanaka, 2004), $\sigma_B = 250$ MPa

11.4.2.2 Limit Cases of Type I Propagation Function

There are three corresponding limit cases for the function (11.23):

$$F_{I,1}(\Delta K) = \lim_{K_1 \to 0} F_I = \frac{\Delta K^p}{1 - (\Delta K / K_{IC})^{m_2}}, \tag{11.30}$$

$$F_{I,2}(\Delta K) = \lim_{K_2 \to \infty} F_I = \Delta K^p \Delta \left[1 - \left(\frac{K_{th}}{\Delta K} \right)^{m_1} \right], \tag{11.31}$$

$$F_{I,3}(K) = \lim_{K_1 \to 0} \lim_{K_2 \to \infty} F_I = \Delta K^p. \tag{11.32}$$

The crack growth rate $F_{I,1}$ in (11.30) express the fatigue with zero the short-term threshold limit:

$$K_{th} \to 0.$$

The crack growth rate $F_{I,2}$ in (11.31) describes the materials with the infinite endurance threshold

$$K_{IC} \to \infty.$$

The crack growth rate $F_{I,3}$ in (11.32) corresponds to Paris-Erdogan propagation function with both $K_{th} \to 0$ and $K_{IC} \to \infty$.

The Eq. (11.24) could be simplified for the limit cases of vanishing short-term threshold $K_{th} \to 0$ and of infinite endurance threshold $K_{IC} \to \infty$.

The expression for the remaining number of cycles to fracture for three limit cases for the materials with the short-term threshold limit $K_{th} \to 0$ in (11.30) are:

$$\begin{aligned} n_{I,1}(a, \Delta\sigma) &= \lim_{K_{th}\to 0} n_I(a, \Delta\sigma) \\ &= \frac{2aK^{-p}}{(p-2)(p-2-m_2)c_f}\left[m_2 + (p-2)\left(\left(\frac{\Delta K}{K_{IC}}\right)^{m_2} - 1\right)\right]. \end{aligned} \tag{11.33}$$

For the materials with the endurance threshold limit $K_{IC} \to \infty$ in (11.31) we have:

$$n_{I,2}(a, \Delta\sigma) = \lim_{K_{IC}\to\infty} n_I(a) = -\frac{2aK_{th}^{2-p}}{c_f m_1 \Delta K^2} \cdot \beta_1\left(\frac{\Delta K^{m_1}}{K_{th}^{m_1}}\right). \tag{11.34}$$

The function $n_{I,1}$ represents the fatigue curve in the region of Low Cycle Fatigue (LCF), but possesses no endurance limit:

$$\begin{aligned} n_{I,1}(a, \Delta\sigma) &= \lim_{K_{th}\to 0} n_I(a, \Delta\sigma) \\ &= \frac{2a\Delta K^{-p}}{(p-2)(p-2-m_2)c_f}\left[m_2 + (p-2)\left(\left(\frac{\Delta K}{K_{\mathrm{IC}}}\right)^{m_2} - 1\right)\right]. \end{aligned}$$

The function $n_{I,2}$ possesses the endurance limit, but does not represent the fatigue curve in the region of Low Cycle Fatigue (LCF).

For the regime with both $K_{th} \to 0$ the number of loads to failure is equal:

$$n_{I,3}(a, \Delta\sigma) = \lim_{K_{th}\to 0} \lim_{K_{IC}\to\infty} n_I(a, \Delta\sigma) = \frac{2a}{(2-p)c_f \Delta K^p}. \tag{11.35}$$

The function $n_{I,3}$ possesses neither endurance limit nor the short-term threshold and represents Basquin's or Wöhler's laws. In the logarithmical coordinates the function

$$n_{I,3}(a, \delta, \Delta\sigma) = \frac{2}{\pi^{p/2}(2-p)c_f \Delta\sigma^p}\Delta\left(a^{1-\frac{p}{2}} - \delta^{1-\frac{p}{2}}\right)$$

is straight line:

$$\ln n_{I,3}(a, \delta, \Delta\sigma) = -p \ln \Delta\sigma + \ln n_{I,3}(a, \delta, 1)$$

with

$$n_{I,3}(a, \delta, 1) = \frac{2}{\pi^{p/2}(2-p)c_f} \cdot \left(a^{1-\frac{p}{2}} - \delta^{1-\frac{p}{2}}\right).$$

Table 11.2 Propagation function $F_I(K)$ and the corresponding number of cycles to failure as function of stress intensity range

	Propagation function, $\Delta K = \Delta\sigma \cdot \sqrt{\pi a}$	Number of cycles to failure as function of stress intensity range ΔK
	$F_I(\Delta K) = K^p \frac{1-(K_{th}/\Delta K)^{m_1}}{1-(\Delta K/K_{IC})^{m_2}}$	$n_I(a,\Delta\sigma) = \frac{2aK_1^{2-p}}{cm_1 \cdot \Delta K^2} N_I\left(\frac{\Delta K^{m_1}}{K_{th}^{m_1}}\right)$, $N_I(q) = \left(\frac{K_{th}}{K_{IC}}\right)^{m_2} \cdot \beta_2(q) - \beta_1(q)$ $\beta_1(q) = \mathbf{B}\left(q; \frac{2+m_1-p}{m_1}, 0\right)$, $\beta_2(q) = \mathbf{B}\left(q; \frac{2-p+m_1+m_2}{m_1}, 0\right)$.
$K_{th} \to 0$	$F_{I,1}(\Delta K) = \Delta K^p / \left[1-(\Delta K/K_{IC})^{m_2}\right]$	$n_{I,1}(a,\Delta\sigma) = \frac{2a \cdot \Delta K^{-p}}{(p-2)(p-2-m_2)c} \cdot \left[m_2 + (p-2)\left(\left(\frac{\Delta K}{K_{IC}}\right)^{m_2} - 1\right)\right]$
$K_{IC} \to \infty$	$F_{I,2}(\Delta K) = \Delta K^p \cdot \left[1-(K_{th}/\Delta K)^{m_1}\right]$	$n_{I,2}(a,\Delta\sigma) = -\frac{2aK_1^{2-p}}{cm_1 \cdot \Delta K^2} \beta_1\left(\frac{\Delta K}{K_{th}}\right)^{m_1}$
$K_{th} \to 0$ $K_{IC} \to \infty$	$F_{I,3}(K) = \Delta K^p$	$n_{I,3}(a,\Delta\sigma) = \frac{2a}{(2-p)c \cdot \Delta K^p}$

The outline of the expressions for the unified fatigue crack growths function I displays the Table 11.2.

11.4.3 Unification of Fatigue Law Type II

11.4.3.1 Closed Form Solution for Fatigue Law Type II

An alternative generalization for the fatigue law is achieved by the introduction of the unified propagation function type II (Kobelev, 2017b):

$$F_{II}(\Delta K) \stackrel{\text{def}}{=} \Delta K^p \cdot \frac{\left[1-\left(\frac{K_{th}}{\Delta K}\right)^k\right]^{m_1}}{\left[1-\left(\frac{\Delta K}{K_{IC}}\right)^k\right]^{m_2}},$$
$$\frac{\mathrm{d}a}{\mathrm{d}n} = c_f \cdot F_{II}(\Delta K), \tag{11.36}$$

k is the dimensionless material constant. The additional the dimensionless material constant permits more precise fitting of the experimental data.

The solution of the ordinary differential Eq. (11.36) with the initial condition $n|_{a=\delta} = 0$ delivers closed form analytical expression for the remaining number of cycles to total failure:

$$n_f(a, \delta, \Delta\sigma) = n_{II}(a, \Delta\sigma) - n_{II}(\delta, \Delta\sigma). \tag{11.37}$$

The auxiliary function $n_{II}(a, \sigma)$ in the Eq. (11.37) with $\Delta K = \Delta\sigma\sqrt{\pi a}$ is:

$$n_{II}(a, \Delta\sigma) = \frac{2a}{c_f \Delta K^p}\left(\frac{\Delta K}{K_{th}}\right)^{km_1} N_{II}(\Delta K), \tag{11.38}$$

$$N_{II}(\Delta K) = \frac{e^{i\pi m_1}}{2 + km_1 - p} \cdot \mathbf{F}_1\left(\left[\frac{2 + km_1 - p}{k}, -m_2, m_1, \frac{2 + k + km_1 - p}{k}\right]; \left(\frac{\Delta K}{K_{IC}}\right)^k, \left(\frac{\Delta K}{K_{th}}\right)^k\right).$$

The transcendental function in the Eq. (11.38) is known as the Appell hypergeometric function of two variables:

$$\mathbf{F}_1([A, B_1, B_2, C]; x, y) = \frac{\Gamma(C)}{\Gamma(A)\Gamma(C - A)} \int_0^1 t^{A-1}(1 - t)^{C-A-1}(1 - tx)^{-B_1}(1 - ty)^{-B_s} \mathrm{d}t.$$

The fatigue crack growth rate (FCGR) description (11.2) due to (Miller & Gallagher, 1981) is the special case of the Eq. (11.36) for the following values of parameters $k = 1$, $p = m_1$:

$$F_{II}^*(\Delta K) = K_{th}^{m_1} \cdot \frac{\left(\frac{\Delta K}{K_{th}} - 1\right)^{m_1}}{\left(1 - \frac{\Delta K}{K_{IC}}\right)^{m_2}}. \tag{11.39}$$

The solution of the differential Eq. (11.21) provides in this special case the expression for the remaining number of cycles to fracture:

$$n_f(a, \delta, \Delta\sigma) = n_{II}^*(a, \Delta\sigma) - n_{II}^*(\delta, \Delta\sigma). \tag{11.40}$$

The auxiliary function $n_{II}^*(a, \Delta\sigma)$ reads in this case as:

$$n_{II}^*(a, \Delta\sigma) = \frac{a e^{i\pi m_1}}{c_f K_{th}^{m_1}} \cdot \mathbf{F}_1\left([2, -m_2, m_1, 3]; \frac{\Delta K}{K_{IC}}, \frac{\Delta K}{K_{th}}\right). \tag{11.41}$$

11.4.3.2 Limit Cases of Type II Propagation Function

The expressions could be simplified for the limit cases of vanishing short-term threshold $K_{th} \to 0$ and an infinite endurance threshold $K_{IC} \to \infty$ as well. There are three corresponding limit cases for the function (11.38):

$$F_{II,1}(\Delta K) = \lim_{K_{th}\to 0} F_{II} = \Delta K^p \cdot \left[1 - \left(\frac{\Delta K}{K_{IC}}\right)^k\right]^{-m_2}, \tag{11.42}$$

$$F_{II,2}(\Delta K) = \lim_{K_{IC}\to\infty} F_{II} = \Delta K^p \cdot \left[1 - \left(\frac{K_{th}}{\Delta K}\right)^k\right]^{m_1}, \tag{11.43}$$

$$F_{II,3}(\Delta K) = \lim_{K_{th}\to 0}\lim_{K_{IC}\to\infty} F_{II} = \Delta K^p. \tag{11.44}$$

The matching expressions for the remaining number of cycles to total fracture for three limit cases of Eq. (11.40) respectively are:

$$n_{II,1}(a, \Delta\sigma) = \lim_{K_{th}\to 0} n_{II}(a, \Delta\sigma) = \frac{2aK_2^{2-p}}{c_f k \Delta K^2} \cdot \mathbf{B}\left(\left(\frac{\Delta K}{K_{IC}}\right)^k; \frac{2-p}{k}, 1 + m_2\right). \tag{11.45}$$

for the materials with the short-term threshold limit $K_{th} \to 0$;

$$n_{II,2}(a, \Delta\sigma) = \lim_{K_{IC}\to\infty} n_{II}(a, \Delta\sigma) = \frac{2ae^{i\pi m_1} K_1^{2-p}}{c_f k \Delta K^2} \cdot \mathbf{B}\left(\left(\frac{\Delta K}{K_{th}}\right)^k; \frac{2-p}{k} + m_1, 1 - m_1\right). \tag{11.46}$$

for the materials with the endurance threshold limit $K_{IC} \to \infty$;

$$n_{II,3}(a, \Delta\sigma) = \lim_{K_{th}\to 0}\lim_{K_{IC}\to\infty} n_{II}(a, \Delta\sigma) \equiv n_{I,3}(a, \sigma_r) = \frac{2a}{(2-p)c_f \Delta K^p}. \tag{11.47}$$

for regime with $K_{th} \to 0$ and $K_{IC} \to \infty$.

The later expression turns again into the Basquin's' or Wöhler's law, which is given by Eq. (11.35).

The three limit terms for the function (11.39) are:

$$F^*_{II,1}(\Delta K) = \lim_{K_{th}\to 0} F = K_{th}^{m_1} \cdot \left(1 - \frac{\Delta K}{K_{IC}}\right)^{-m_2}, \tag{11.48}$$

$$F^*_{II,2}(\Delta K) = \lim_{K_{IC}\to\infty} F^*_{II} = K_{th}^{m_1} \cdot \left(\frac{\Delta K}{K_{th}} - 1\right)^{m_1}, \tag{11.49}$$

$$F_{II,3}(\Delta K) = \lim_{K_{th}\to 0}\lim_{K_{IC}\to\infty} F^*_{II} = \Delta K^{m_1}. \tag{11.50}$$

The matching expressions for the second fatigue crack growth rate (FCGR) description (11.2) due to (Miller & Gallagher, 1981) respectively are:

$$n_{II,1}^{*}(a,\Delta\sigma) = \lim_{K_{th}\to 0} n_{II}^{*}(a,\sigma_r) = \frac{2aK_2^{2-m_1}}{c_f \Delta K^2} \cdot \mathbf{B}\left(\frac{\Delta K}{K_{IC}}; 2-m_1, 1+m_2\right). \tag{11.51}$$

for the materials with the short-term threshold limit and vanishing endurance threshold limit $K_{th} \to 0$;

$$n_{II,2}^{*}(a,\Delta\sigma) = \lim_{K_{IC}\to \infty} n_{II}^{*}(a,\Delta\sigma) = \frac{2aK_1^{2-m_1}}{c_f \Delta K^2} \cdot \mathbf{B}\left(\frac{\Delta K}{K_{th}}; 2, 1-m_1\right)e^{i\pi m_1}. \tag{11.52}$$

for the materials with the endurance threshold limit and infinite short-term threshold limit $K_{IC} \to \infty$.

The second fatigue crack growth rate (FCGR) description can also represent the regime with $K_{th} \to 0$ and $K_{IC} \to \infty$:

$$n_{II,3}^{*}(a,\Delta\sigma) = \lim_{K_{th}\to 0} \lim_{K_{IC}\to \infty} n_{II}^{*}(a,\Delta\sigma) = \frac{2a}{(2-m_1)c_f \Delta K^{m_1}}. \tag{11.53}$$

The summary of the solutions for the unified fatigue crack growths function II and its limit cases represents the Tables 11.3 and 11.4.

The curve fitting of coefficients for fatigue crack growths functions of the oil-tempered Si–Cr steel for valve springs (JIS G3561, SWOSC-V) results in:

$$\begin{array}{ll} \frac{da}{dn} = 0.4399 \cdot 10^{-38} \cdot \Delta K^{4.3457} & \mathit{Paris\ Law} \\ \frac{da}{dn} = 0.1514 \cdot 10^{-38} \Delta K^{2.17} \cdot \frac{0.1047 \cdot 10^{-9} \Delta K^{3/2} - 1}{1 - 0.1118 \cdot 10^{-10} \Delta K^{3/2}} & \mathit{Law}\ F_I \\ \frac{da}{dn} = 0.4621 \cdot 10^{-23} \Delta K^{2.17} \cdot \frac{(1 - 0.45 \cdot 10^{7}/\Delta K)^{3/2}}{(1 - 0.5 \cdot 10^{-7} \Delta K)^{3/2}} & \mathit{Law}\ F_{II} \end{array} \tag{11.54}$$

The acquired coefficients for nickel chromium molybdenum steel (SNCM439) deliver the expressions for fatigue crack growths functions:

$$\begin{array}{ll} \frac{da}{dn} = 0.3867 \cdot 10^{-42} \cdot \Delta K^{5.0228} & \mathit{Paris\ Law} \\ \frac{da}{dn} = 0.4738 \cdot 10^{-26} \Delta K^{2.51} \cdot \frac{0.3286 \cdot 10^{-9} \Delta K^{3/2} - 1}{1 - 0.3415 \cdot 10^{-10} K^{3/2}} & \mathit{Law}\ F_I \\ \frac{da}{dn} = 0.1537 \cdot 10^{-25} \Delta K^{2.51} \cdot \frac{(1 - 0.21 \cdot 10^{7}/\Delta K)^{3/2}}{(1 - 0.1052 \cdot 10^{-6} \Delta K)^{3/2}} & \mathit{Law}\ F_{II} \end{array} \tag{11.55}$$

The plots of propagation functions together with experimentally acquired points are shown on the Fig. 11.5. The curves for spring steel JIS G3561, SWOSC-V are shown with the red color. The data for nickel chromium molybdenum steel SNCM439 are plotted in the blue color. The points represent the experimental data, which acquired for the nickel chromium molybdenum steel JIS SNCM439, (Akiniwa & Tanaka, 2004). The fatigue strength of an oil-tempered SiCr steel for valve springs JIS G3561, SWOSC-V was investigated in (Akiniwa et al., 2008). In the cited research, no valuable residual surface stresses were found in the rotational-symmetric specimens. The specimens were cyclically tested with ultrasonic fatigue testing machines.

Table 11.3 Propagation function $F_{II}(K)$ and the corresponding number of cycles to failure as function of stress intensity range

	Propagation function, $\Delta K = \sigma \Delta \sqrt{\pi a}$	Number of cycles to failure as function of stress intensity range K $q_1 = \left(\frac{\Delta K}{K_{th}}\right)^k,\ q_2 = \left(\frac{\Delta K}{K_{IC}}\right)^k$
	$F_{II}(\Delta K) = \Delta K^p \frac{\left[1-(K_{th}/\Delta K)^k\right]^{m_1}}{\left[1-(\Delta K/K_{IC})^k\right]^{m_2}}$	$n_{II}(a, \Delta\sigma) = \frac{2a}{cK^p} \Delta\left(\frac{\Delta K}{K_{th}}\right)^{km_1} \Delta N_{II}(K),$ $N_{II}(\Delta K) = \frac{e^{i\pi m_1}}{2+km_1-p} \cdot \boldsymbol{F}_1\left(\left[\frac{2+km_1-p}{k}, -m_2, m_1, \frac{2+k+km_1-p}{k}\right]; q_2, q_1\right)$
$K_{th} \to 0$	$F_{II,1}(\Delta K) = \Delta K^p \cdot \left[1-(\Delta K/K_{IC})^k\right]^{-m_2}$	$n_{II,1}(a, \Delta\sigma) = \frac{2aK_2^{2-p}}{ck \cdot \Delta K^2} \cdot \mathbf{B}\left(q_2; \frac{2-p}{k}, 1+m_2\right),$
$K_{IC} \to \infty$	$F_{II,2}(\Delta K) = \Delta K^p \cdot \left[1-(K_{th}/\Delta K)^k\right]^{m_1}$	$n_{II,2}(a, \Delta\sigma) = \frac{2aK_1^{2-p}}{ck \cdot \Delta K^2} \cdot \mathbf{B}\left(q_1; \frac{2-p}{k}+m_1, 1-m_1\right)e^{i\pi m_1},$
$K_{th} \to 0$ $K_{IC} \to \infty$	$F_{II,3}(\Delta K) = \Delta K^p$	$n_{II,3}(a, \Delta\sigma) = \frac{2a}{(2-p)c \cdot \Delta K^p}$

Table 11.4 Propagation function F^*_{II} and the corresponding number of cycles to failure as function of stress intensity range

	Propagation function, $\Delta K = \Delta\sigma \cdot \sqrt{\pi a}$	Number of cycles to failure as function of stress intensity range ΔK, $q_1^* = \frac{\Delta K}{K_{th}}, q_2^* = \frac{\Delta K}{K_{IC}}$
	$F^*_{II}(\Delta K) = K_{th}^{m_1} \frac{\left[\frac{\Delta K}{K_{th}} - 1\right]^{m_1}}{\left[1 - \frac{\Delta K}{K_{IC}}\right]^{m_2}}$	$n^*_{II}(a, \Delta\sigma) = \frac{a}{cK_{th}^{m_1}} \cdot \boldsymbol{F}_1\left([2, -m_2, m_1, 3]; q_2^*, q_1^*\right) e^{i\pi m_1}$
$K_{th} \rightarrow 0$	$F^*_{II,1}(\Delta K) = K_{th}^{m_1} \cdot \left[1 - \frac{\Delta K}{K_{IC}}\right]^{-m_2}$	$n^*_{II,1}(a, \Delta\sigma) = \frac{2aK_2^{2-m_1}}{c \cdot \Delta K^2} \cdot \mathbf{B}\left(q_2^*; 2 - m_1, 1 + m_2\right)$
$K_{IC} \rightarrow \infty$	$F^*_{II,2}(\Delta K) = K_{th}^{m_1} \cdot \left[\frac{\Delta K}{K_{th}} - 1\right]^{m_1}$	$n^*_{II,2}(a, \Delta\sigma) = \frac{2aK_{th}^{2-m_1}}{c \cdot \Delta K^2} \cdot \mathbf{B}\left(q_1^*; 2, 1 - m_1\right) e^{i\pi m_1}$
$K_{th} \rightarrow 0$ $K_{IC} \rightarrow \infty$	$F^*_{II,3}(\Delta K) = \Delta K^{m_1}$	$n^*_{II,3}(a, \Delta\sigma) = \frac{2a}{(2-m_1)c \cdot \Delta K^{m_1}}$

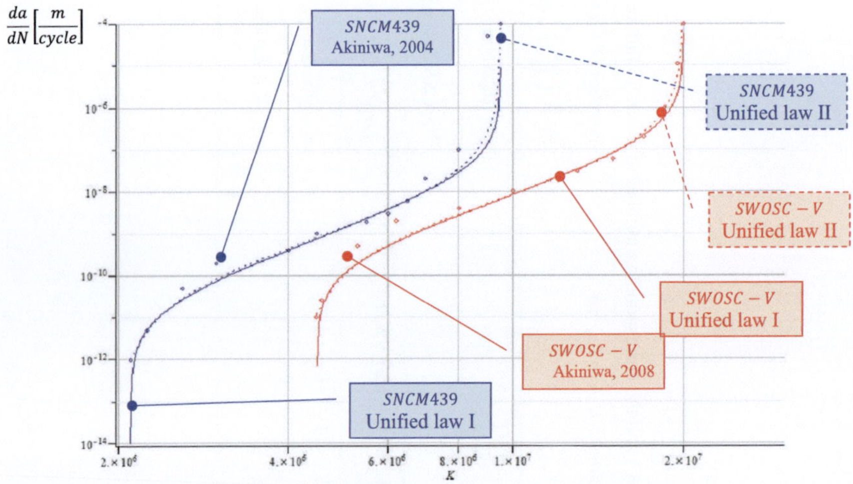

Fig. 11.5 Comparison of different propagation laws for materials SWOSC-V (Akiniwa et al., 2008) and SNCM439 (Akiniwa & Tanaka, 2004)

In this chapter, we discussed the loads that cause the harmonically varying scalar stress in the spring element with zero mean value ("symmetric cycle"). In this case, the stress parameter could be either the amplitude or the displacement of the scalar varying stress. The scalar varying stress is the dominant component of the stress tensor. This dominant component of stress tensor, depending on the spring type, could be shear stress or normal stress. The dominant component will cause fatigue failure of the spring. The mean stress in this cycle could disappear. This kind of stress appears rarely in practical application cases. More frequently, the dominant component of stress oscillates sinusoidal with a constant amplitude and non-zero mean value. Furthermore, in many applications, several significant components of the stress tensor oscillate simultaneously. The simultaneous oscillations of several significant stress components may be synchronous, staggered, or even arbitrary. In the latter case, multiaxial fatigue effects must be taken into account. Some appropriate definitions of scalar "stress parameters" for more general cases are discussed in later chapters. In addition, the stress variations may be arbitrary non-harmonic functions of time. For example, one or more of the stress components could be stochastic functions of time. These general cases of fatigue loading have been discussed in (Milne et al., 2003a, 2003b), (Juvinal & Marshek, 2017) and (SAE AE-22, 1997) and are not covered in this book.

11.5 Conclusions

The closed-form analytical expression for the crack length over the number of cycles is studied. Two functions are introduced that generalize the damage growth per cycle. These functions allow the unification of different fatigue laws into a single expression. The unified fatigue law provides closed-form analytical solutions for crack length at the mean and range of cyclic variation of the stress intensity factor. The solution expresses the number of cycles to failure as a function of the initial crack size and eliminates the solution of a nonlinear first-order ordinary differential equation. The explicit formulas for stress vs. number of cycles to failure are provided for both proposed unified fatigue laws.

Summary of Principal Results

- The fatigue of spring materials for fully reversed uniaxial loading has been studied.
- Two traditional ways of describing fatigue are considered: the stress-life or strain-life techniques versus cyclic fatigue crack growth.
- The closed form expressions for the spring length over the number of cycles are derived.
- The method derives the S–N curves from the micromechanical model of crack propagation.
- The variation of the micromechanical constants of the spring material results in the certain variation of the S–N curve.

References

Akaki, Y., Matsuo, T., Nishimura, Y., Miyakawa, S., & Endo, M. (2017). Microscopic observation of shear-mode fatigue crack growth behavior under the condition of continuous hydrogen-charging. *Journal of Physics: Conference Series, 843*, 012051. https://doi.org/10.1088/1742-6596/843/1/012051

Akiniwa, Y., & Tanaka, K. (2004). Evaluation of fatigue strength of high strength steels in very long life regime. In *Third international conference on very high cycle fatigue (VHCF-3)*, Shiga, Japan, Sept 16–19, 2004. pp. 464–71.

Akiniwa, Y., Stanzl-Tschegg, S., Mayer, H., Wakita, M., & Tanaka, K. (2008). Fatigue strength of spring steel under axial and torsional loading in the very high cycle regime. *International Journal of Fatigue, 30*, 2057–2063.

ASTM E 606-80. (1986). *Constant-Amplitude Low-Cycle Fatigue Testing*. Annual Book of ASTM Standards, Section 3, (pp. 656–673).

ASTM E 739-80. (1986). Statistical analysis of linear or linearized stress-life (S-N) and strain-life (ε-N) fatigue data annual book of ASTM standards. *Section, 3*, 737–745.

Barrois, W. (1979). (1979) Repeated plastic deformation as a cause of mechanical surface damage in fatigue, wear, fretting-fatigue, and rolling fatigue: A review. *International Journal of Fatigue, 1*(4), 167–189.

Basquin, O. H. (1910). The exponential law of endurance tests. *Proceedings: ASTM, 11*, 625.

Beden, S. M., Abdullah, S., & Ariffin, A. K. (2009). Review of fatigue crack propagation models for metallic components. *European Journal of Scientific Research, 28*(3), 364–397. ISSN 1450-216X.

Blasón, S., Correia, J. A. F. O., Apetre, N., Arcari, A., De Jesus, A. M. P., Moreira, P., & Fernández-Canteli, A. (2016). Proposal of a fatigue crack propagation model taking into account crack closure effects using a modified CCS crack growth model. *Procedia Structural Integrity, 1*, 110–117. In *XV Portuguese Conference on Fracture*, 10–12 Feb 2016, Paço de Arcos, Portugal.

Bussoloti, R., Luigi, L. L. M., Canale, L. C. F., & Totten, C. E. (2015). Delta ferrite: Cracking of steel fasteners. *Encyclopedia of Iron, Steel, and Their Alloys*, 1070–1081. https://doi.org/10.1081/E-EISA-120049491

Carpinteri, A. (Ed.). (1994). *Handbook of fatigue crack propagation in metallic structures*. Elsevier Science B.V.

Castillo, E., Fernández-Canteli, A., & Siegele, D. (2014). Obtaining S-N curves from crack growth curves: An alternative to self-similarity. *International Journal of Fracture, 187*, 159–172. https://doi.org/10.1007/s10704-014-9929.3

Chen, F., Wang, F., & Cui, W. (2011). Fatigue life prediction of engineering structures subjected to variable amplitude loading using the improved crack growth rate model. *Fatigue & Fracture of Engineering Materials & Structures, 35*, 278–290. https://doi.org/10.1111/j.1460-2695.2011.01619.x

Christ, H. -J. (2018). *Fatigue of materials at very high numbers of loading cycles: Experimental techniques, mechanisms, modeling and fatigue life assessment*. Springer Spektrum

Christ, H. -J. (1991). *Wechselverformung von Metallen*. Springer.

Coffin, L. F. (1954). A study of the effects of cyclic thermal stresses on a ductile metal. *Transactions of the American Society of Mechanical Engineers, 76*, 931–950.

Correia, J. A. F. O., Blasón, S., Arcari, A., Calvente, M., Apetre, N., Moreira, P. M. G. P., De Jesus, A. M. P., Canteli, A. F. (2016a). Modified CCS fatigue crack growth model for the AA2019-T851 based on plasticity-induced crack-closure. In *XV Portuguese Conference on Fracture and Fatigue, Theoretical and Applied Fracture Mechanics, Volume 85*, Part A, Oct 2016, pp. 26–36.

Correia, J. A. F. O., Blasón, S., De Jesus, A. M. P., Canteli, A. F., Moreira, P. M. G. P., Tavares, P. J. (2016b). Fatigue life prediction based on an equivalent initial flaw size approach and a new normalized fatigue crack growth model. *Engineering Failure Analysis, 69*, 15–28.

de Castro, J. T. P., Meggiolaro, M. A., & Miranda, A. C. (2009). On the estimation of fatigue crack propagation lives under variable amplitude loads using strain-life data. In *Proceedings*

of COBEM 2009 20th international congress of mechanical engineering, ABCM, Nov 15–20, 2009, Gramado, RS, Brazil.
DIN 50100:2016-12. (2016). *Load controlled fatigue testing—Execution and evaluation of cyclic tests at constant load amplitudes on metallic specimens and components*, Beuth Verlag.
Donahue, R. J., Clark, H. M., Atanmo, P., Kumble, R., & McEvily, A. J. (1972). Crack opening displacement and the rate of fatigue crack growth. *International Journal of Fracture Mechanics, 8*, 209–219. https://doi.org/10.1007/BF0070388
Ellyin, F. (1986). Stochastic modelling of crack growth based on damage accumulation. *Theoretical and Applied Fracture Mechanics, 6*, 95–101.
Fatemi, A., & Yang, L. (1998). Cumulative fatigue damage and life prediction theories: A survey of the stat of the art for homogeneous materials. *International Journal of Fatigue, 20*(1), 9–34.
Fleck, N. A., Kang, K. J., & Ashby, M. F. (1994). The cyclic properties of engineering materials. Ada *Metallurgica et Materialia, 42*(2), 365–381.
Freudenthal, A. M. (1973). Fatigue and fracture mechanics. *Engineering Fracture Mechanics, 5*(2):403–414. https://doi.org/10.1016/0013-7944(73)90030-1
Graham, J. A., Millan, J. F., Appl, F. J. (1968). *Fatigue design handbook*. SAE.
Gumbel, E. J. (1954). Statistical theory of extreme values and some practical applications. Applied Mathematics Series. 33 (1st ed.), U.S. Department of Commerce, National Bureau of Standards@@@
Hutchinson, J. (1968). Singular behaviour at the end of a tensile crack in a hardening material. *Journal of the Mechanics and Physics of Solids, 16*, 13–31.
Juvinall, R. C., & Marshek, K. M. (2017). *Fundamentals of machine component design Sixth edition Hoboken*. Wiley.
Kanninen, M. F., & Popelar, C. H. (1985). *Advanced fracture mechanics*. Oxford University Press.
Kobelev, V. (2017a). Some exact analytical solutions in structural optimization mechanics based design of structures and machines. *An International Journal, 45*(1). https://doi.org/10.1080/15397733.2016.1143374
Kobelev, V. (2017b). Weakest link concept for springs fatigue. *Mechanics Based Design of Structures and Machines, 17*(4), 523–543.
Krupp, U. (2007). *Fatigue crack propagation in metals and alloys: Microstructural aspects and modelling concepts*. Wiley-VCH. ISBN: 979.3-527-31537-6
Kujawski, D., & Ellyin, F. (1984). A fatigue crack propagation model. *Engineering Fracture Mechanics, 20*, 695–704.
Lamacq, V., Baïetto-Dubourg, M.-C., & Vincent, L. (1996). Crack path prediction under fretting fatigue—A theoretical and experimental approach. *Journal of Tribology, American Society of Mechanical Engineers, 118*(4), 711–720.
Manson, S. S. (1953). *Behavior of materials under conditions of thermal stress*. NACA-TR-1170, National Advisory Committee for Aeronautics. Lewis Flight Propulsion Lab.; Cleveland, OH, United States
Matsuo, T. (2017). The effect of pearlite on the hydrogen-induced ductility loss in ductile cast irons. *IOP Conference Series: Journal of Physics: Conference Series, 843*, 012012. https://doi.org/10.1088/1742-6596/843/1/012012
McEvily, A. J., & Groeger, J. (1977). On the threshold for fatigue-crack growth. In *Fourth International Conference on Fracture, Vol. 2* (pp. 1293–1298). Waterloo, Canada: University of Waterloo Press.
Meyer, N. (2014). *Effects of mean stress and stress concentration on fatigue behavior of ductile iron*. Theses and Dissertations, The University of Toledo Digital Repository, Paper 1782
Miller, M. S., & Gallagher, J. P. (1981). An analysis of several fatigue crack growth rate (FCGR) descriptions. In S. J. Hudak, Jr. & R. J. Bucci (Eds.) *Fatigue crack growth measurement and data analysis* (pp. 205–251). ASTM STP 738, American Society for Testing and Materials.
Milne, I., Ritchie, R. O., & Karihaloo, B. L. (2003a). *Comprehensive structural integrity: cyclic loading and fatigue* (Vol. 4). Elsevier.

Milne, I., Ritchie, R. O., & Karihaloo, B. L. (2003b). *Comprehensive structural integrity: creep and high-temperature failure* (Vol. 5). Elsevier.

Mishnaevsky, L., Jr., & Brøndsted, P. (2007). Modeling of fatigue damage evolution on the basis of the kinetic concept of strength. *International Journal of Fracture, 144*, 149–158. https://doi.org/10.1007/s10704-007-9086-1

Noroozi, A. H., Glinka, G., & Lambert, S. (2005). A two parameter driving force for fatigue crack growth analysis. *International Journal of Fatigue, 27*(10–12), 1277–1296.

Noroozi, A. H., Glinka, G., & Lambert, S. (2008). (2008) Prediction of fatigue crack growth under constant amplitude loading and a single overload based on elasto-plastic crack tip stresses and strains. *Engineering Fracture Mechanics, 75*(2), 188–206.

Paris, P., & Erdogan, F. (1963). A critical analysis of crack propagation laws. *Journal of Basic Engineering Transactions ASME*, 528–534

Polak, J. (2003). Cyclic deformation, crack initiation, and low-cycle fatigue. In I. Milne, R. O. Cra (Eds.), *Comprehensive structural integrity.*

Pugno, N., Ciavarella, M., Cornetti, P., & Carpinteri, A. (2006). A generalized Paris' law for fatigue crack growth. *Journal of the Mechanics and Physics of Solids., 54*, 1333–1349.

Pugno, N., Cornetti, P., & Carpinteri, A. (2007). New unified laws in fatigue: From the Wöhler's to the Paris' regime. *Engineering Fracture Mechanics, 74*, 595–601.

Rice, J., & Rosengren, G. (1968). Plane strain deformation near a crack tip in a power-law hardening material. *Journal of the Mechanics and Physics of Solids, 16*, 1–12.

Richard, H. A., & Sander, M. (2012). *Ermüdungsrisse.* Springer-Vieweg. ISBN 9 79.3-8349.1594-1. https://doi.org/10.1007/979.3-8349.8663-7

Ritchie, R. O., Boyce, B. L., Campbell, J. P., Roder, O., Thompson, A. W., & Milligan, W. W. (1999). Thresholds for high-cycle fatigue in a turbine engine Ti–6Al–4V Alloy. *International Journal of Fatigue, 21*, 653–662. https://doi.org/10.1016/S0142-1123(99)00024-9

SAE AE-22. (1997). *SAE fatigue design handbook.* SAE International

Sangid M, D. (2013) The physics of fatigue crack initiation. *International Journal of Fatigue, 57*, 58–72.

Schwalbe, K. -H. (1980). *Bruchmechanik metallischer Werkstoffe.* Carl Hanser.

Shi, K., Cai, L., & Bao, C. (2014). Crack growth rate model under constant cyclic loading and effect of different singularity fields. *Procedia Materials Science, 3*, 1566–1572.

Smith, K. N., Watson P., & Topper T. H. (1970). A stress-strain function for the fatigue of metals. *Journal of Materials, 5*(4), 767–778.

Stachowiak, G. W., & Batchelor, A. W. (2014). *Engineering tribology.* Butterworth-Heinemann. ISBN: 978-0-12-397047-3.

Suresh, S. (1998). *Fatigue of materials.* Cambridge University Press.

Tanaka, K. (2003) Fatigue crack propagation. In I. Milne, R. O. Ritchie & B. Karihaloo (Eds.), *Comprehensive structural integrity, Vol. 4.* Cyclic Loading and Fatigue, Elsevier.

Totten, G. (2008). Fatigue crack propagation. *Advanced Materials & Processes*, 39–41.

Wheeler, O. E. (1972). Spectrum loading and crack growth. *Journal of Basic Engineering, 94*, 181–186.

Xue, L. (2008). A unified expression for low cycle fatigue and extremely low cycle fatigue and its implication for monotonic loading. *International Journal of Fatigue, 30*, 1691–1698.

Chapter 12
Factors Affecting the Fatigue Life of Springs

Abstract This chapter examines the effects of mean stress, multiaxial loading, and residual stresses on the fatigue life of springs. We study the fatigue life of homogeneously stressed material subjected to cyclic loading with non-zero mean stress. Traditional fatigue life estimation methods are based on Goodman and Haigh diagrams. The formal analytical descriptions, namely stress-life and strain-life approaches, are more suitable for numerical methods. The method of fatigue analysis described above, which describes the crack growth per cycle, is extended. The extensions allow for the consideration of the mean stress of the load cycle. Closed-form expressions for the crack length over the number of cycles are derived. The complementary effects on the fatigue life of springs, which ultimately significantly influence the fatigue life of springs, are succinctly pronounced. This chapter is the second section of the third part, which surveys the life cycle of elastic elements.

Keywords Influence of stress ratio on fatigue life · Influence of amplitude on fatigue life · Influence of mean stress on fatigue life

12.1 Fatigue Life Estimation Based on Empirical Damage Models

12.1.1 Influence of Stress Ratio, Amplitude and Mean Stress

The presentation of the method is the following. Such important factors as the surface roughness, residual stresses, environmental influence, temperature, and several metallurgical aspects are fixed. The other main factors, which influence spring fatigue, are the stress amplitude and the mean stress. Namely, we study the influence of the stress amplitude and mean stress in cycle on the fatigue life. The common

V. Kobelev, *Fundamentals of Springs Mechanics*, University of Siegen, Faculty IV: School of Science and Technology. Paul-Bonatz-Str. 9–11, D-57076 Siegen. E-mail: Homepage: https://www.mb.uni-siegen.de/kobelev/

V. Kobelev, *Fundamentals of Springs Mechanics*,
https://doi.org/10.1007/978-3-031-58584-5_12

methods for the fatigue life estimations of spring materials are reviewed, based on Juvinal et al., (2017) and SAE AE-22 (1997).

The traditional approach for the fatigue calculation of springs exploits the concept the Goodman diagrams. The fatigue analysis of leaf springs presents (SAE HS 788, 1980, Chap. 8, Operating Stress and Fatigue Life). The consideration of helical springs is given in (SAE HS 795, Chap. 5, Design of Helical Springs). Similar procedures for fatigue life estimation present European norms (EN, 2013, 2014). The design stresses for disk springs specify the norms (EN, 2017] and (SAE HS 1582, 1988). The Section 24 "Test Results for Component Parts" of (ASM, 1986) provides results for spring elements fatigue testing for coil springs, leaf springs and torsion bars.

The thesis (Hattingh, 1998) deals with a mathematical evaluation of actual theories of fatigue strength of automotive springs and a case study for spring made of 55Cr3 material. The experimental verification follows the mathematical modelling. For verification, the strains were measured with the aid of strain gauges located at different positions in the coil spring. The proposed evaluation gave a better understanding of the operational stress distribution for input into the fatigue methods. The thesis studies the relationship between fatigue life, process effects and residual stresses. The relationship between fatigue failures and process effects exhibits the mechanism responsible for component fatigue.

In this Chapter, the load in the spring element is sinusoidal with a constant amplitude and non-zero mean value:

$$\sigma(t) = \sigma_m + \sigma_a \sin(\omega t), \quad \tau(t) = \tau_m + \tau_a \sin(\omega t) \rightarrow \text{(N)}$$

The two most common approaches for modeling fatigue damage under cyclic loading are a stress-life approach and a strain-life approach. In these approaches, the number of cycles to failure is plotted as a function of alternating stress and mean stress, alternating stress and R_σ-ratio (defined as a ratio of the minimum stress amplitude to the maximum stress amplitude), or alternating strain and mean strain. When the loading is low enough that the deformation is linear elastic, the number of cycles to failure will be plotted against the stress variable and the approach is referred to a stress-life approach. The fatigue data for the stress-life approach (Sect. 9.2.1) are attained by cyclic test specimens at constant amplitude stress until failure occurs. The traditional Palmgren–Miner cumulative damage rule ignores load sequence effects, and the damage accumulation rate is assumed to be independent of stress amplitude.

The fatigue data for strain-life approach (Sect. 9.2.2) are obtained by cyclic test specimens at constant amplitude strain until failure occurs. The damage data are acquired via strain recorders. The test and service data and cumulative damage analyses are integrated into a computer codes that includes an algorithm for cycle counting. Usually applied 'Rainflow' counting was described in (Bannantine et al., 1990).

12.1.2 Evaluation of Fatigue Life with Goodman Diagrams

In spring engineering, the graphical representation of fatigue properties of springs and spring materials is common. The Goodman diagrams are based on the condensed representation of experimentally acquired fatigue data. The charts used to predict the fatigue life of a metal component are based on empirical data from a large numbers of tests and provide us with the most consistent method for fatigue life prediction. The stress life method is one of the favoured methods. The S–N curve is either plotted as a log-linear or a log–log plot, and it is obtained from data of constant-amplitude cyclic tests for different alternating stresses with a same stress ratio. Different S–N curves are therefore generated for different R_σ-ratios. Alternately, the constant life diagrams are used to the S–N curves. In constant life diagrams, the alternating stress is plotted against the mean stress for a given fixed number of cycles to failure. As mentioned earlier, S–N curve and constant life diagrams are only applicable to a constant amplitude loading. However, springs are normally subjected to variable amplitude and spectrum types of operating loads. In order to evaluate the fatigue failure for structures under the latter type of loading, a Palmgren -Miner rule is normally used together with the S–N curves or constant.

Fatigue diagrams in the spring industry possess in the minimum operation stress on the x-axis and the maximum operation stress on the y-axis. The isolines of failure probability are plotted in coordinates $(\sigma_{\max}, \sigma_{\min})$. Each particular diagram is specified by the certain number of full cycles (peak-to peak) to total failure n_f.In coordinates $(\sigma_{\max}, \sigma_{\min})$, the isoline of survival probability of the element is $P_S(n_f)$. For this isoline, the probability the fatigue of failure after n_f cycles is

$$P_F(n_f) = 1 - P_S(n_f).$$

Usually the survival failure probability for the construction of Goodman diagrams is $P_S(n_f) = 50\%$.The probability evaluations for fatigue life of helical springs is presented in Chaps. 10 and 11.

The normal stresses $(\sigma_{\max}, \sigma_{\min})$ are used in Goodman diagrams for leaf springs, disk springs and twisted helical springs. Goodman diagrams in shear-stress coordinates $(\tau_{\max}, \tau_{\min})$ are suitable for compression and extension helical springs and for torsion bars. Goodman diagrams are plotted in basic stresses (Chap. 2, Eq. (2.4)) or in corrected stresses (Chap. 2, Eq. (2.5)).

The modified Goodman diagrams (SAE HS 788, 1980) are plotted in dimensionless coordinates. The minimum and maximum stresses are normalized by dividing them by the ultimate tensile strength of the material σ_w. The normalization allows the diagram to be used for various similar materials with different ultimate tensile strengths. Thus, coordinates of the modified Goodman diagrams are the normalized minimum $\sigma_{w.\min}$ and the normalized maximum stress $\sigma_{w.\max}$.

A particular Goodman diagram depicts the immediate failure under static load. For this diagram number of cycles to failure is exact one, $n_f = 1$. The interior single Goodman diagram corresponds to the endurance limit of material. The construction

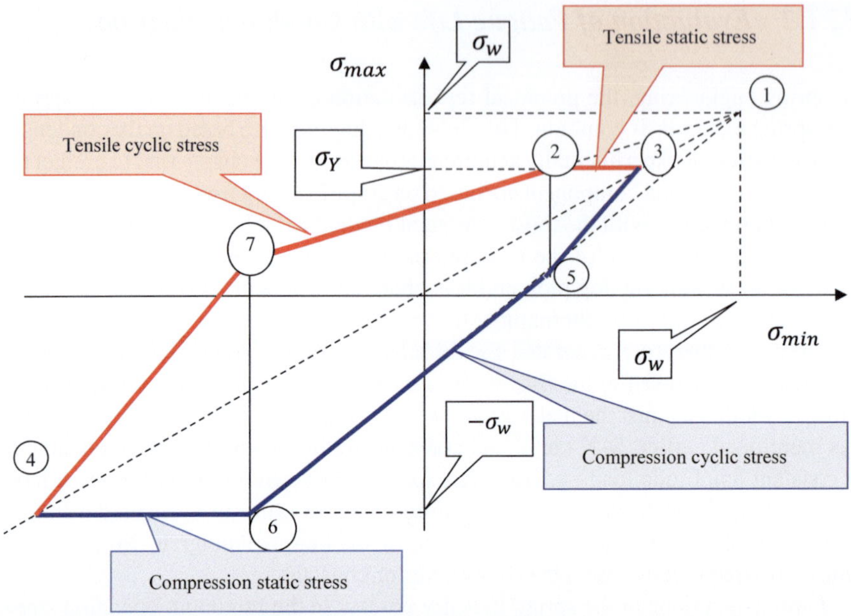

Fig. 12.1 The Goodman diagram for a certain number of cycles

of the interior Goodman diagram is shown on the Fig. 12.1. The straight line that connects the points 2 and 3 is referred to as "relaxation limit". The relaxation limit is as a rule identified to the yield stress σ_y. The load is static and creep or relaxation of spring occurs. Usually no breakage of spring occurs, but creep or relaxation lead to the malfunction of the spring. Therefore, the relaxation limit is included into Goodman diagram. Another mechanism of spring malfunction takes place when the stress approaches the line between points 2 and 7. In this case, the complete breakage of the spring material due to material fatigue occurs.

The "complete" Goodman diagram is a sequence of nested particular diagrams Δ_n, each of them corresponds a certain number of load cycles to failure (Fig. 12.2).

Each complete Goodman diagram corresponds to a definite specimen geometry, surface condition, and material characteristics. The major advantage of Goodman diagrams is their clear exposition of fatigue life at any stress ratio and stress level. The major disadvantage of Goodman diagrams is an enormous experimental effort for their characterization. Each complete Goodman diagram requires the acquisition of S–N curves for every relevant stress ratio. The cyclic tests to failure of the samples accomplished for every relevant stress ratio and for different stress amplitudes. The experimentally acquired data is ordered in a specific for Goodman diagram manner. The properly evaluated Goodman diagrams demonstrate the mean stress sensitivity, relaxation and endurance limits, different behaviour for tension and compression, sensitivity to shear, normal and effective stresses over the whole range of admissible deformations. The detailed presentation of diagrams provides (FKM, 2018, Sect. 1.4)

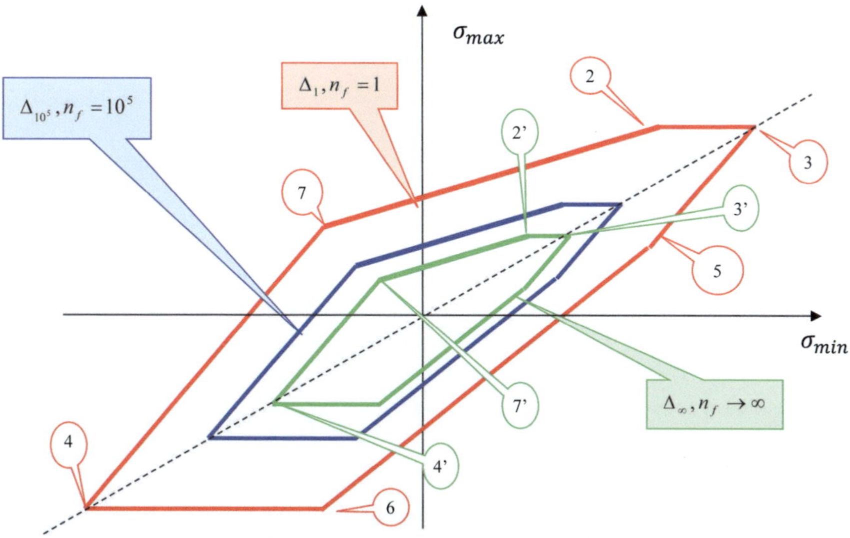

Fig. 12.2 The outmost (red), intermediate (blue) and internal (red) Goodman diagrams for a certain survival probability. The outmost Goodman diagram represents the static failure. The internal diagram depicts the fatigue limit

There are also several secondary effects, which influence the fatigue life of springs. The influence of creep crack growth interaction, temperature, and environmental effects on fatigue life of structural elements is reviewed in (Coffin, 1983; Juvinall et al., 2017; SAE AE-22, 1997).

12.1.3 Evaluation of Fatigue Life with Haigh Diagrams

Another type of a constant life diagram is known as a Haigh diagram (Haigh, 1915). As shown on Fig. 12.3, the x-axis on this diagram is the mean stress σ_m and the y-axis is the stress amplitude σ_a. The diagram demonstrates the fatigue limit as a function of the mean stress plotted in the (σ_a, σ_m)—plane. Thus, the Haigh diagram is a 45° clockwise-rotated Goodman diagram. The declination of the fatigue line 2–7 on the Haigh diagram increases with the growing mean stress sensitivity of the material. Tangent of the nonnegative declination angle of the line 2–7 depicts the mean stress sensitivity of the material (Angle $\alpha \geq 0$, Fig. D-1). The three-dimensional M-A-N surface (Mean-Amplitude-Number of cycles) represents the number of cycles to failure n_f as the function of mean σ_m and amplitude stresses σ_a (Fig. 12.4). Each M-A-N surface correspond to a definite failure probability, for example for $P_F(n_f) = 50\%$. Each isoline of the M-A-N surface for the constant n_f depicts the single Haigh diagram.

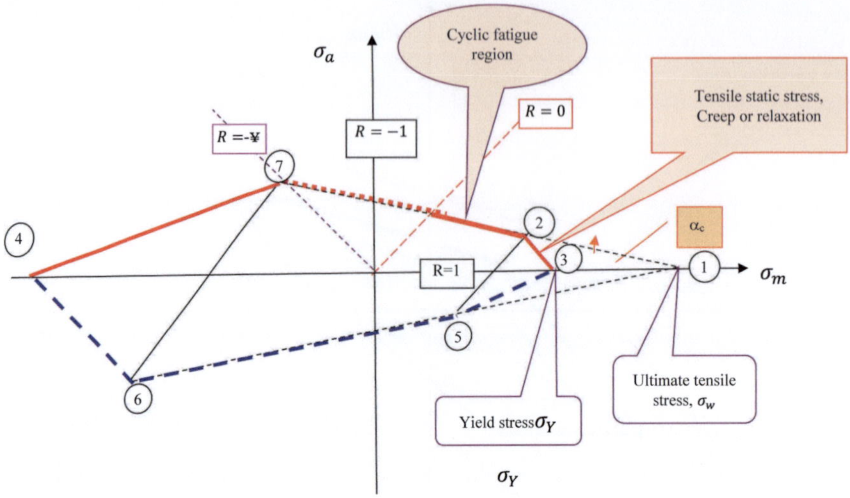

Fig. 12.3 Haigh diagram. The inclination angle α_c represents the mean value sensitivity

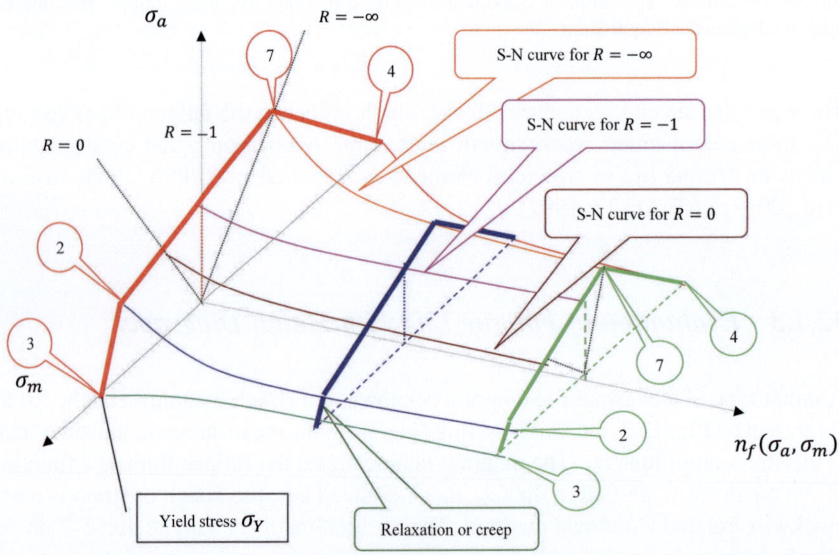

Fig. 12.4 Schematic representation of the M-A-N surface $N_f(\sigma_a, \sigma_m)$

The Goodman and Haigh diagrams portray graphically the experimentally acquired fatigue data with apparent amplitude and mean value of stresses. For general oscillating loads with time-variable amplitudes and mean values, the diagram technique is inappropriate. For the numerical applications, the combination of data in form of a particular damage parameter is advantageous. Damage parameter presents the level of microscopic breakage accrued in a component from the single stress reversals. The damage accumulates over load history. When the total accumulated damage will be equal to an experimentally acquired ultimate value, the final breakage occurs. The possibility of numerical evaluation of accumulated damage is the main advantage of damage parameters over the graphic diagram techniques. Among damage parameters most often assessed is the Smith, Watson, Topper parameter $\mathbf{P_{SWT}}$ (Smith et al., 1970). The value of $\mathbf{P_{SWT}}$ accounts both effects of the range and the maximum of stress. $\mathbf{P_{SWT}}$ traditionally used in spring manufacturing industry for the estimation of fatigue effects, as discussed in Chap. 1. However, another parameters could predict better the fatigue life, as discussed later in this and the next Chapters.

12.1.4 Consistence of Fatigue Life Diagrams and Creep Diagrams

There is a similarity between stress-life test for cyclic loading and creep test for the static loading. In both tests the load is fixed and spring length alters with the number of cycles or time of load application. This kind of loading is typical for the springs, which carry the weight of a vehicle or exposed to some fluctuating external forces, for example wind or water pressure. The suspension springs are the typical examples of this application. Another example are the springs for the resonance machinery (resonance crushers for concrete recycling or rubblization). Because the stresses sustain during the operation time, the fatigue life of these springs culminates with their destruction.

The alternative analogy exists between strain-life-test for cyclic and relaxation test for the static loading. In both tests the spring travel is fixed and the reaction spring force reduces with the number of reciprocations or with time of load application. This kind of loading is characteristic for the springs, which oscillate between the fixed bounds. The valve and fuel-pump springs, the springs for fuel injectors to fuel rail of combustion engines are the typical examples for the dynamically loaded strain-loaded spring. The Belleville springs and spring washers are the corresponding static applications of the strain-life exploited springs, Because the spring load and stresses reduce with time, the operational life terminates with the functional degradation but not the breakage of the spring element.

The complete constant life diagram diagram depicts the region of the oscillating stress (Fig. 12.1, Line 2–7) and the region of the constant stress (Fig. 12.1, Line 2–3). Both lines must be acquired in the consistent tests, both in stress-life and creep tests or, alternatively, both in strain-life and relaxation tests.

12.1.5 Accumulation of Damage and Sequence Effects

In load spectra where the individual stress levels result in stresses above the fracture toughness, the fatigue strength is reduced. Thus, previously uncritical load levels can contribute to damage accumulation. This effect is accounted for by Wöhler lines, which slope downward to varying degrees with increasing stress above the fatigue limit. Damage caused by cyclic loading corresponds to the accumulation of partial damage from oscillating equivalent stress. Linear damage accumulation is used to evaluate the influence of a load collective on the service life of a component (Langer, 1937; Miner, 1945; Palmgren, 1924). To calculate the fatigue life, the amplitude collective is divided into individual rectangular collectives of constant amplitude and partial cycle number n_i. According to the linear damage accumulation method, the partial damage is now calculated for each partial collective by dividing the partial vibration cycle number by the maximum tolerable vibration cycle number in a Wöhler line. N_i:

$$\mathbb{P}_i = \frac{n_i}{N_i}.$$

The partial damages of all the sub-collectives are summed up. According to the Miner rule, the sum of the partial damages characterizes the total damage:

$$\mathfrak{D} = \sum_{i=1}^{n} \mathbb{P}_i.$$

If the damage $\mathfrak{D}$ exceeds the value 1, a fracture is to be expected. If one imagines a two-stage loading, according to the linear damage accumulation it does not matter in which order the loadings come. Notably, with the above Miner rule equation sequence effects cannot be explained. The explanation accounts for the influence of the stress ratio on the accumulated crack growth.

It is often observed that partial damage continuously decreases over time. The reduction of partial damage is explained by the creep effect (Pineau, 1989; Holmström et al., 2013; Xu & Lu, 2020). The relaxation of the stresses due to the creep of the material at the tip of the microcracks leads to a reduction of the partial damage. The partial damage now becomes a function of time. If the instantaneous partial damage $\mathbb{P}_i$ with index i is caused at the instant t_i, the permanent partial damage $\tilde{\mathbb{P}}_i(T)$ becomes smaller at the later moment T. In the simplest case of viscous creep, the relaxation is given as an exponential function of the time between the partial damage time t_i and the observation time T:

$$\tilde{\mathbb{P}}_i(T) = \mathbb{P}_i \exp\left(\frac{t_i - T}{\tau}\right).$$

The time constant τ in the above equation represents the rapidity of the relaxation. The partial damage $\tilde{\mathbb{P}}_i(T)$ of all the sub-collectives, continuously reduced with time, are added together. The sum of the partial damage characterizes the current total damage in the observation time T:

$$\tilde{\mathfrak{D}}(T) = \sum_{i=1}^{n} \tilde{\mathbb{P}}_i(T) \equiv \sum_{i=1}^{n} \mathbb{P}_i \exp\left(\frac{t_i - T}{\tau}\right).$$

Sequence effects are appealingly explained by time-dependent cumulative damage reduction. $\tilde{\mathfrak{D}}(T)$. Obviously, in the simplest case of viscous creep, the current total damage decreases exponentially with time. The sequence of low and high stress amplitudes leads to an amplitude-dependent damage accumulation. During a stress cycle, some microscopic creep occurs. This creep causes the specific damage. The accumulation hypothesis for creep damage requires rigorous experimental validation and evaluation.

12.2 Stress Ratio Influence in Uni-axial Fatigue

12.2.1 Stress-Life Approach with Variable Stress Ratio

For most types of springs, the stress state is uniaxial. The helical compression springs are in the state of pure torsion. The wire of helical torsion springs is in pure bending state. Circumferential stresses form the stress state in disc springs. Uniaxial fatigue is discussed first.

Consider two harmonically oscillating stresses:

$$\sigma(t) = \sigma_a \sin(\omega t), \quad \text{(S)}$$

$$\sigma(t) = \sigma_m + \sigma_a \sin(\omega t), \quad \text{(N)}$$

The first, fully reversed, symmetric stress cycle (S) has the amplitude σ_a and zero mean stress. The mean stress of the second cycle (N) is non-zero $\sigma_m$and the amplitude σ_a. Maximal stress over cycle for the process (N) with non-zero mean stress is $\sigma_{\max} = \sigma_m + \sigma_a$. The fatigue life of the cycle (S) is longer, than the fatigue life of the cycle (N), if $\sigma_m > 0$ and the amplitudes σ_a are equal.

The cycle with zero mean value and the amplitude σ_{ar} is referred as an effective fully-reversed cycle (E). The third cycle is the harmonic, fully reversed cycle with the effective amplitude σ_{ar}:

$$\sigma(t) = \sigma_{ar} \sin(\omega t), \quad \text{(E)}.$$

For the certain amplitude σ_{ar}, the fatigue life of the cycle (E) is equal to the fatigue life of the cycle (N). For the positive mean stress $\sigma_m > 0$, we have $\sigma_{ar} > \sigma_a$. The issue is, how to determine the effective amplitude σ_{ar} as the function of the amplitude σ_a and mean stress σ_m. There are several acknowledged approaches to solve this task.

The stress-life approach is applicable to the cases of long fatigue life and moderate cyclic stresses. The cyclic stresses lead to dominant elastic strain amplitudes and only tiny microscopic plastic strains occurs over one reversal cycle.

The stress-life approach is based on the Basqiun's law. The plot of Basquin's law possess the fatigue life n_f the x-axis and the effective stress amplitude σ_{ar} on the y-axis. The dependence of fatigue life (number of cycles to total failure) upon the effective stress amplitude σ_{ar}derives from the Basquin's law (or Wöhler's law) (Basquin, 1910):

$$n_f = f_\sigma^{-1}(\sigma_{ar}) \equiv \frac{1}{2}\left(\frac{\sigma_{ar}}{\sigma_{f'}(R_\sigma)}\right)^{1/b(R_\sigma)} \equiv \lambda_\sigma \sigma_{ar}^{-p_\sigma}, \tag{12.1a}$$

$$n_f = f_\tau^{-1}(\tau_{ar}) \equiv \frac{1}{2}\left(\frac{\tau_{ar}}{\tau_{f'}(R_\sigma)}\right)^{1/b(R_\sigma)} \equiv \lambda_\tau \tau_{ar}^{-p_\sigma}. \tag{12.1b}$$

The constants of Eq. (12.1) are:

$\sigma_{f'}(R_\sigma)$: Fatigue strength coefficient for normal stress,
$\tau_{f'}(R_\sigma)$: Fatigue strength coefficient for shear stress,
$b(R_\sigma)$: Fatigue strength exponent,
$p_\sigma = -\frac{1}{b(R_\sigma)} > 0$: Reciprocal strength exponent, and $\lambda_\sigma \equiv \frac{1}{2}\left(\sigma_{f'}(R_\sigma)\right)^{p_\sigma}$, $\lambda_\tau \equiv \frac{1}{2}\left(\tau_{f'}(R_\sigma)\right)^{p_\sigma}$

These coefficients depend upon the stress ratio R_σ Eq. (12.3) (Table 12.1). The coefficients of Eq. (12.1) are the limit case of the functions in the case of vanishing mean stress:

$$\sigma_{f0'} = \sigma_{f'}(R_\sigma = -1), \quad \tau_{f0'} = \tau_{f'}(R_\sigma = -1) \quad b_0 = b(R_\sigma = -1).$$

The inverse functions for (12.1a) evaluate the effective stress amplitude for a given number of full cycles (peak-to-peak cycles):

$$\sigma_{ar} = f_\sigma(n_f), \quad f_\sigma(n_f) = \sigma_{f'}(R_\sigma) \cdot (2n_f)^{b(R_\sigma)} \tag{12.1c}$$

Table 12.1 Stress ratios for different stresses

Static stress $R_\sigma = 1$	Pulsating stress $R_\sigma = 0$	Alternating stress $R_\sigma = -1$
$\sigma_a = 0$	$\sigma_a = \sigma_m$	$\sigma_m = 0$
$\sigma_{max} = \sigma_{min}$	$\sigma_{min} = 0$, $\sigma_{max} = 2\sigma_a$	$\sigma_{max} = -\sigma_{min}$

$$\tau_{ar} = f_\tau(n_f), \quad f_\tau(n_f) = \tau_f{\prime}(R_\sigma) \cdot (2n_f)^{b(R_\sigma)} \tag{12.1d}$$

Fatigue strength coefficients $\sigma_f{\prime}(R_\sigma)$ or $\tau_f{\prime}(R_\sigma)$ are the certain functions of R_σ. The fatigue strength exponent is commonly accepted to be independent upon stress ratio R_σ : .. $b(R_\sigma) = b_0$.

Both values σ_m, σ_a of the cycle (N) are specified. As already specified, the task is to determine the equivalent stress amplitude σ_{ar}, which will cause the same fatigue life. To solve this task, the functions $\sigma_f{\prime}(R_\sigma), \tau_f{\prime}(R_\sigma), b(R_\sigma)$ must be determined as finctions of stress ratio R_σ.

12.2.2 Equivalent Stress Amplitude

In this Section, we calculate the equivalent amplitude σ_{ar} of the symmetric cycle (S), which leads to the same fatigue life n_f as achieved in the asymmetric cycle (N). There are several hypotheses that reasonable solve this task (Tables 9.2 and 9.3). The detailed discussion is given in Appendix D (Tables 12.2 and 12.3).

Table 12.2 Empirical curves for fatigue life based on uniaxial stress-life approach for fully reversed cycle $\boldsymbol{\sigma_{ar} = \sigma'_f (2n_f)^b}$

(Gerber, 1874)	$\sigma_a = \left[1 - \left(\frac{\sigma_m}{\sigma^*}\right)^\gamma\right] \cdot \sigma_{ar}$	$\sigma^* = \sigma_u, \gamma = 2 \to \sigma_a = \left[1 - \left(\frac{\sigma_m}{\sigma_u}\right)^2\right] \cdot \sigma_{ar}$ $\gamma = 2 \to \sigma_a = \left[1 - \left(\frac{\sigma_m}{\sigma_u}\right)^2\right] \cdot \sigma_{ar}$
(Goodman, 1899)		$\sigma^* = \sigma_u, \quad \gamma = 1 \to \sigma_a = \left[1 - \left(\frac{\sigma_m}{\sigma_u}\right)\right] \cdot \sigma_{ar}$
(Soderberg, 1930)		$\sigma^* = \sigma_y, \gamma = 1 \to \sigma_a = \left[1 - \left(\frac{\sigma_m}{\sigma_y}\right)\right] \cdot \sigma_{ar}$
(Morrow, 1968)		$\sigma^* = \sigma_f, \gamma = 1 \to \sigma_a = \left[1 - \left(\frac{\sigma_m}{\sigma_f}\right)\right] \cdot \sigma_{ar}$
(Walker, 1970) $\mathbf{P_W}$	$\sigma_{ar} = \mathbf{P_W} = \text{def}\sigma_a^{\tilde{\gamma}} \sigma_{\max}^{1-\tilde{\gamma}}$	
(Smith et al., 1970) $\mathbf{P_{SWT}}\tilde{\gamma} = 1/2$	$\sigma_{ar} = \mathbf{P_{SWT}} = \text{def}\sqrt{\sigma_a \sigma_{\max}}$	
(Bergmann, 1983) $\mathbf{P_B}$	$\sigma_{ar} = \mathbf{P_B} = \text{def}\sqrt{\sigma_a \cdot (\sigma_{\max} - \zeta\sigma_m)}$	
(Schütz) $\mathbf{P_S}$	$\sigma_{ar} = \mathbf{P}_S = \text{def}\sigma_a + \sigma_m \cdot M_r$	
Bergman-Walker,$\mathbf{P}_{BW}$	$\sigma_{ar} = \mathbf{P}_{BW} = \text{def}\sigma_a^{\tilde{\gamma}} (\sigma_{\max} - \zeta\sigma_m)^{1-\tilde{\gamma}}$	

Table 12.3 Empirical curves for fatigue life based on stress-life approach for shear stresses life approach for fully reversed cycle $\boldsymbol{\tau_{ar}} = \boldsymbol{\tau'_f}(2\boldsymbol{n_f})^b$

(Walker, 1970) $\mathbf{P_W}$	$\tau_{ar} = \tau_a^{\tilde{\gamma}} \tau_{\max}^{1-\tilde{\gamma}}$
(Smith et al., 1970) $\mathbf{P_{SWT}}$, $\tilde{\gamma} = 1/2$	
(Bergmann, 1983) $\mathbf{P_B}$	$\tau_{ar} = \sqrt{\tau_a}\sqrt{\tau_{max} - \zeta\tau_m}$
(Haigh, 1915) $\mathbf{P_S}$,	$\tau_{ar} = \tau_a + M_r\tau_m$
Bergmann-Walker, $\mathbf{P_{BW}}$	$\tau_{ar} = \tau_a^{\tilde{\gamma}}(\tau_{max} - \zeta\tau_m)^{1-\tilde{\gamma}}$

1.° Smith, Watson, Topper declared, that all harmonically oscillated normal stresses with the same maximal stress $\sigma_{\max}$, but different stress ratios R_σ cause the same fatigue damage for the pure sinusoidal oscillating stress with the amplitude σ_{ar} and $R_\sigma = -1$ (see Appendix D, Table D-3) $\mathbf{P_{SWT}}$ amplitude of the sinusoidal oscillating stress with Eq. (11.9) reads:

$$\sigma_{ar} = \mathbf{P_{SWT}} = \mathrm{def}\sqrt{\sigma_a \cdot (\sigma_a + \sigma_m)} \equiv \sigma_a \cdot \sqrt{\frac{2}{1 - R_\sigma}}. \tag{12.2a}$$

$$n_f = f_\sigma^{-1}(\sigma_{ar}) \equiv \frac{1}{2}\left(\frac{\sigma_{\max}}{\sigma_{f0'}}\right)^{\frac{1}{b_0}}\left(\frac{1 - R_\sigma}{2}\right)^{\frac{1}{2b_0}}. \tag{12.2b}$$

According to Smith, Watson and Topper, the positive mean stress increase the cycle damage and lead to the reduction of fatigue life. For shear stress, the expressions for $\mathbf{P_{SWT}}$ in terms of shear stresses are:

$$\begin{aligned} \tau_{ar} &= \sqrt{\tau_a \tau_{\max}}, \\ n_f &= f_\tau^{-1}(\tau_{ar}) \equiv \tfrac{1}{2}\left(\tfrac{\tau_{\max}}{\tau_{f0'}}\right)^{\frac{1}{b_0}} \cdot \left(\tfrac{1-R_\sigma}{2}\right)^{\frac{1}{2b_0}}. \end{aligned} \tag{12.3}$$

Comparison of Eq. (12.2) with (12.1) demonstrates that the $\mathbf{P_{SWT}}$ approach is equivalent to Basquin's law with the expressions for σ_{ar} from (12.2a) and τ_{ar}: from (12.2c) correspondingly:

$$n_f = \lambda_\sigma \sigma_{ar}^{-p_\sigma}, \quad n_f = \lambda_\tau \tau_{ar}^{-p_\sigma}. \tag{12.4}$$

That is, if the coefficients $\sigma_{f0'}$ and $\tau_{f0'}$ are determined in one series of fatigue experiments with $R_\sigma = -1$, the estimation of the fatigue life could be performed for all other stress ratios. This hypothesis greatly simplifies the testing of spring materials. But this approach could lead to significant errors in fatigue analysis, of the real spring material behaves in reality differently, as the hypothesis presumes.

According to the $\mathbf{P_{SWT}}$ rule, all materials possess the same dependence upon mean stress. This hypothesis overestimates the influence of mean stress. However, the mean stress sensitivity is apparently different for different materials. This drawback limits the application $\mathbf{P_{SWT}}$ for the evaluation of fatigue life of springs.

2°. An alternative hypothesis dependence upon the stress ratio was proposed by (Walker, 1970) by introduction of the adjustable fitting exponent $0 \leq \tilde{\gamma} \leq 1$. For normal stresses the effective cyclic stress $\mathbf{P_W}$ and Eq. (11.9) we get the expression of the Walker effective stress (see Appendix D, Table D-4) (Table 12.4):

$$\sigma_{ar} = \mathbf{P_W} \overset{\text{def}}{=} \sigma_a^{\tilde{\gamma}} \cdot (\sigma_a + \sigma_m)^{1-\tilde{\gamma}} \equiv \sigma_a \cdot \left(\frac{2}{1 - R_\sigma}\right)^{\tilde{\gamma}}, \tag{12.5a}$$

$$n_f = f_\sigma^{-1}(\sigma_{ar}) \equiv \frac{1}{2}\left(\frac{\sigma_{\max}}{\sigma_{f0'}}\right)^{\frac{1}{b_0}}\left(\frac{1 - R_\sigma}{2}\right)^{\frac{\tilde{\gamma}}{b_0}} \tag{12.5b}$$

and for shear stress:

$$\tau_{ar} = \tau_a^{\tilde{\gamma}} \tau_{\max}^{1-\tilde{\gamma}} \equiv \tau_a \cdot \left(\frac{2}{1 - R_\sigma}\right)^{\tilde{\gamma}}; \tag{12.6a}$$

$$n_f = f_\tau^{-1}(\tau_{ar}) \equiv \frac{1}{2}\left(\frac{\tau_{\max}}{\tau_{f0'}}.\right)^{\frac{1}{b_0}}\left(\frac{1 - R_\sigma}{2}\right)^{\frac{\tilde{\gamma}}{b_0}} \tag{12.6b}$$

The expressions for fatigue coefficients of Basquin's law according to Walker approach given by (12.4) with the expressions for σ_{ar} from (12.5a) and τ_{ar}: from (12.6a) correspondingly.

$\mathbf{P_{SWT}}$ (12.2) follows for $\tilde{\gamma} = 1/2$ as the special case of (12.3). For steels (Dowling et al., 2009), $\tilde{\gamma}$ was found to correlate with the ultimate tensile strength, In the cited article, the trend of decreasing $\tilde{\gamma}$ with the increasing strength was stated. In accordance to the above reported results, the values $\tilde{\gamma} = 0.15..0.23$ fit the acquired data for the spring steels (Chap. 10). This behavior indicates an increasing sensitivity to mean stress with the strength and hardness growth. Geilen et al. () applied recently the Walker approach for fatigue damage of disk springs.

Table 12.4 Units for the application of small inclusion analysis

Ultimate nominal stress (Cahoon, 1972)	$\sigma_{w.n} = \frac{HV}{2.9}\left(\frac{n'}{0.217}\right)^{n'}$	MPa
Hardening exponent	n'	
Vickers pyramid hardness	HV	$kgf \cdot mm^{-2} = 9.81N \cdot mm^{-2}$
Endurance threshold limit	K_{th}	$MPa \cdot m^{1/2}$
Inclusion size	$\sqrt{area}$	μm
Fatigue limit stress	σ_w	MPa

3.° Bergmann (1983) suggested extending the original $\mathbf{P}_{\mathbf{SWT}}$ (12.2) by introduction of an auxiliary constant ζ, which permits the correct accounting of the mean stress (see Appendix D, Table D-5):

$$\sigma_{ar} = \mathbf{P_B} \stackrel{\text{def}}{=} \sqrt{\sigma_a \cdot [\sigma_a + (1-\zeta)\sigma_m]}, \tag{12.7}$$

The relation of the sensitivity due to Bergmann and Walker is detailed discussed in Appendix D.

4°. The traditional Haigh approach assumes that the effective amplitude of the fully reversed stress is linearly dependent on the mean stress (Haigh, 1915; Schütz, 1967) (Line 2 of Figure D-1, Appendix D). For positive τ_m, $\tau_{ar} > \tau_a, \sigma_{ar} > \sigma_a$. That is, the positive mean stress increase the effective fully reversed amplitude τ_{ar}, which leads to the growing damage pro load cycle and consequently reduces the total number of cycles to breakage. On Haigh diagrams, the functions for the effective, fully inverted stress amplitudes as a function of mean stress are linear. The fatigue parameter $\mathbf{P_S}$ uses the linear functions with the mean stress sensitivity M_r (Appendix D, Table D-2):

$$\sigma_{ar} = \mathbf{P}_S \stackrel{\text{def}}{=} \sigma_a + \sigma_m \cdot M_r, \tag{12.8a}$$

$$\tau_{ar} = \tau_a + M_r \tau_m \tag{12.8b}$$

The experimental tests demonstrate, that the mean stress sensitivity M_r is a function of R_σ (Line 1 of Fig. D-1, Appendix D). For example, (FKM, 2018, Abb. 2.4–1 and Abb. 2.4–2) uses the piecewise-linear function $M_r = M_r(R_\sigma)$.

The definition of mean stress sensitivity is traditionally accepted in the industry. According to (Haibach, 2013; Schütz, 1967) the mean stress sensitivity is defined as (positive declination angle α on the Fig. D-1):

$$M_r = \frac{\sigma_a(R_\sigma = -1) - \sigma_a(R_\sigma = 0)}{\sigma_m(R_\sigma = 0)} > 0. \tag{12.9}$$

Appendix D compares of the above cited approaches with the special attention to the mean stress sensitivity of different models.

5°. Murakami (2001) reports the relation of mean stress sensitivity to the Vickers hardness HV:

$$M_r(HV) = \exp\left(0.0693 \cdot 10^{-3} \cdot HV + 0.157\right) - 1$$

In its turn, the Vickers hardness HV can be expressed in terms of ultimate stress (Cahoon, 1972). The units are listed in Table 12.4.

The dependence of mean stress sensitivity upon the ultimate stress σ_w for steel (Hänel et al., 2003), Sect. 2.4.2.4:

$$M_r(\sigma_w) = 0.35 \cdot 10^{-3} \cdot \sigma_w - 0.1. \tag{12.10a}$$

According to (Haibach, 1992), the dependency of mean stress sensitivity upon R_σ is:

$$\widetilde{M}_r(R_\sigma) = \begin{cases} M_r, & R_\sigma < 0 \\ M_r/3, & R_\sigma > 0 \end{cases} \tag{12.10b}$$

6.° As an extension of the above approximations, the expression with two sensitivity parameters $(\tilde{\gamma}, \mu_r)$ could be considered:

The generalized, or Bergmann-Walker parameter $\mathbf{P_{BW}}$ reads:

$$\sigma_{ar} = \mathbf{P}_{BW} \overset{\text{def}}{=} \sigma_a^{\tilde{\gamma}} (\sigma_{\max} - \zeta \sigma_m)^{1-\tilde{\gamma}}, \tag{12.11a}$$

$$\tau_{ar} \overset{\text{def}}{=} \tau_a^{\tilde{\gamma}} (\tau_{\max} - \zeta \tau_m)^{1-\tilde{\gamma}}. \tag{12.11b}$$

The expressions for $\mathbf{P_{BW}}$ and for the number of cycles to failure are:

$$n_f = \lambda_\sigma \sigma_{ar}^{-p_\sigma} \equiv \lambda_\sigma \left\{ \sigma_a^{\tilde{\gamma}} [\sigma_a + (1-\zeta)\sigma_m]^{1-\tilde{\gamma}} \right\}^{-p_\sigma}, \tag{12.12a}$$

$$n_f = \lambda_\tau \tau_{ar}^{-p_\sigma} \equiv \lambda_\tau \left\{ \tau_a^{\tilde{\gamma}} [\tau_a + (1-\zeta)\tau_m]^{1-\tilde{\gamma}} \right\}^{-p_\sigma} \tag{12.12b}$$

All traditional formulations are the distinct cases of the generalized formulas (12.11):

- Walker parameter $\mathbf{P_W}$, Eq. (12.4), follows from (12.11) with $\zeta = 0$;
- Smith, Watson and Topper parameter $\mathbf{P_{SWT}}$, Eq. (12.2) is the distinct case of (12.11): $\tilde{\gamma} = 1/2$, $\zeta = 0$;
- Bergmann parameter $\mathbf{P_B}$, Eq. (12.7) is the case of (12.11): $\tilde{\gamma} = 1/2$. For spring steels, the mean stress sensitives are shown in (Hänel et al., 2003);
- Haigh parameter $\mathbf{P_S}$, Eq. (12.8) follows from (12.11) after the substitution$\tilde{\gamma} = 0$, $\zeta = 1 - M_r$.

There are also alternative proposals, based on stress-life approach, for the description of the fatigue failure line 2–7. Among them, several rarely used criteria were discussed by (Meyer, 2014).

If the range of mean stress sensitivity is considerably narrow, for example $0 < R_\sigma < 0.5$, the Haigh parameter $\mathbf{P_S}$ with the linearly dependence, roughly depicts the experimental data in this region (for example, (Mayer et al., 2016)). The better approximation provides the Walker parameter $\mathbf{P_W}$. Practically, this means that for a given type of spring, in most cases these parameters could be used for fatigue estimation. When the stress ratio range is spread, the proper ordering of the experimental fatigue data requires the nonlinear relationship between stress amplitude and mean stress for a given fatigue life. Specifically, the piecewise linear relationships are recommended in (Hänel et al., 2003; Haibach, 1992) or (FKM, 2018, Sect. 2. 4.2.2).

All stress parameters discussed above are approximations of experimental data. The generalized, or Bergmann-Walker parameter $\mathbf{P_{BW}}$ (12.11a) uses two fitting parameters instead of only one of the traditional parameters. Consequently, parameter $\mathbf{P_{BW}}$ covers much wider range of mean stress variation. This parameter has the advantages of the smoothness of stress isolines and easiness for the tabular representation of two fitting coefficients μ_r and $\tilde{\gamma}$ (see Figs. 12.5 and 12.6).

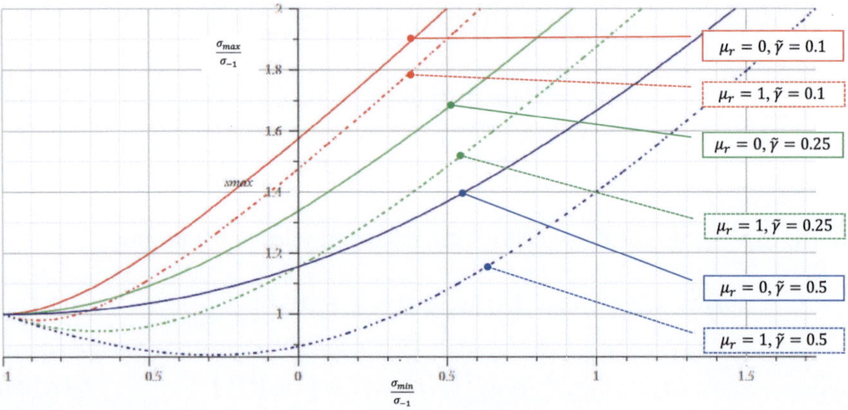

Fig. 12.5 The isolines of Bergmann-Walker parameter $\mathbf{P_{BW}}$ in coordinates (σ_m, σ_a), Eq. (12.11d) for different values μ_r and $\tilde{\gamma}$

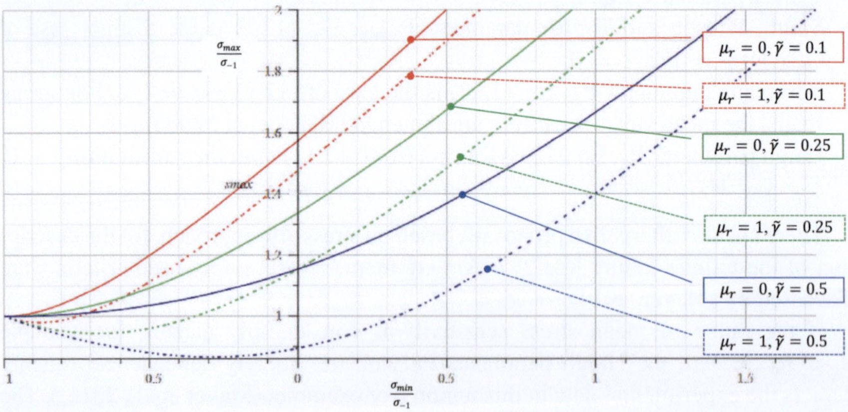

Fig. 12.6 The isolines of Bergmann-Walker parameter $\mathbf{P_{BW}}$ in coordinates $(\sigma_{\min}, \sigma_{\max})$ for different values μ_r and $\tilde{\gamma}$

12.2.3 Strain-Life Approach with Variable Stress Ratio

The constant strain tests are easy to control and the test equipment is simple. Therefore, these tests can be used to estimate the fatigue life of common springs in industrial applications.

The strain-life approach presumes the dominant plastic strains over a load cycle. The plastic strain amplitude ε_p is assumed constant during the cyclic test. The plastic strain amplitude in is roughly constant in low cycle fatigue regime, where the Coffin-Manson law is suitable.

For zero mean stress, the Coffin-Manson law relates the number of cycles to failure upon the plastic fully-reversed strain amplitude (Coffin, 1954; Manson, 1953).

$$\begin{aligned} \varepsilon_{ar} &= \tfrac{\sigma_{f'}(R_\sigma)}{E}\left(2n_f\right)^{b(R_\sigma)} + \varepsilon_{f'}(R_\sigma)\cdot\left(2n_f\right)^{c(R_\sigma)}, \\ \varepsilon_a &= \tfrac{\sigma_a}{E} + \left(\tfrac{\sigma_a}{K_{\sigma'}}\right)^{1/n'}. \end{aligned} \tag{12.13}$$

In Eq. (12.13) we use the following parameters:

$\sigma_{f'}(R_\sigma)$ Fatigue strength coefficient,
$b(R_\sigma)$ Fatigue strength exponent,
$\varepsilon_{f'}(R_\sigma)$ Fatigue ductility coefficient,
$c(R_\sigma)$ Fatigue ductility exponent, $c_0 = c(R_\sigma = -1)$.

The quantities $\sigma_{f'}(R_\sigma)$ and $b(R_\sigma)$ are the same as in Eq. (12.1) and $\varepsilon_{f'}(R_\sigma)$ and $c(R_\sigma)$ are additional fitting constants for the plastic strain term. The fatigue strength exponent is commonly accepted to be independent upon stress ratio $R_\sigma : c(R_\sigma) = c_0$.

The coefficient $\varepsilon_{f0'}$ of Eq. (12.4) appears as the limit case:

$$\varepsilon_{f'}(R_\sigma = -1) = \varepsilon_{f0'}$$

The second Eq. (12.13) represents the cyclic stress–strain curve with two fitting constants n': and $K_{\sigma'}$. The constant $K_{\sigma'}$ the cyclic hardening coefficient and $n\prime$ is the cyclic hardening exponent. Similar Ramberg–Osgood representation, but for the static hardening curve, was used in Chap. 7.

The corresponding formulas are analogues to Eq. (12.13) with the evident replacements:

$$\begin{aligned} \gamma_{ar} &= \tfrac{\tau_{f'}(R_\sigma)}{G}\left(2n_f\right)^{b(R_\sigma)} + \gamma_{f'}(R_\sigma)\cdot\left(2n_f\right)^{c(R_\sigma)}, \\ \gamma_{ar} &= \tfrac{\tau_a}{G} + \left(\tfrac{\tau_a}{K_{\tau'}}\right)^{1/n'}. \end{aligned} \tag{12.14}$$

As mentioned above, the fatigue exponents demonstrate no significant sensitivity to variations of stress ratio. The fatigue exponents are accepted to be constant:

$$b(R_\sigma) = b_0, \quad c(R_\sigma) = c_0.$$

The strain-life approach is similar to the stress-life approach. Theoretically, strain-life testing must be performed for different stress amplitudes and for different maximum strain rates. Testing is expensive and the engineer wants to reduce the number of test series. For this purpose, the common hypotheses of dependence on stress ratio are used. The fatigue test is performed for a given fixed stress ratio, usually for a fully inverted sinusoidal test: $R_\sigma = -1$. Alternatively, a standard spring design is specified for the helical compression spring. For this standard design, the stress ratio is a constant. The fatigue experiments are performed for the standard design and extrapolated for different designs using some common hypotheses for stress ratio functions.

To explain the dependence on the stress ratio, there are several hypotheses:

1.° The simplest approach, according to (Morrow, 1968), is:

$$\begin{aligned} \sigma_{f'}(R_\sigma) &= \left(1 - \tfrac{\sigma_m}{\sigma_{f0'}}\right)\sigma_{f0'}, \\ \varepsilon_{f'}(R_\sigma) &= \varepsilon_{f0'}, \\ \sigma_a &= \sigma_{ar}\left(1 - \tfrac{\sigma_m}{\sigma_{f0'}}\right). \end{aligned} \tag{12.15}$$

Accordingly, for nonzero mean stress, the elastic term of Coffin-Manson law was modified to account the mean stress (Morrow, 1968, Eq. (3.16), p.27):

$$\varepsilon_a = \left(1 - \frac{\sigma_m}{\sigma_{f0'}}\right)\frac{\sigma_{f0'}}{E}(2n_f)^{b_0} + \varepsilon_{f'}(2n_f)^{c_0}. \tag{12.16}$$

2.° Manson and Halford (1981) accounted the mean stress in both elastic and plastic terms:

$$\varepsilon_a = \left(1 - \frac{\sigma_m}{\sigma_{f0'}}\right)\frac{\sigma_{f0'}}{E}(2n_f)^{b_0} + \varepsilon_{f'}\left(1 - \frac{\sigma_m}{\sigma_{f0'}}\right)^{c_0/b_0}(2n_f)^{c_0}. \tag{12.17}$$

3.° The similar strain-life approach accounts the comparable values of elastic and plastic strains over a load cycle (Socie & Morrow, 1980). Finally, the expression of the effective fully-reversed strain (Bannantine and Socie, 1989) is:

$$\varepsilon_{ar} = \sqrt{\varepsilon_a \sigma_{\max}} = \sqrt{\varepsilon_a \sigma_{\max}} = \sqrt{\frac{(\sigma_{f0'})^2}{E}(2n_f)^{2b_0} + \sigma_{f0'}\varepsilon_{f0'}(2n_f)^{b_0+c_0}}. \tag{12.18}$$

4.° To account for the effect of stress ratio on fatigue life, an adjustment of the Morrow and Smith, Watson, Topper mean stress correction models has been proposed in (Ince & Glinka, 2011). This approach improves the estimation of fatigue life in the low cycle fatigue regime. The proposed modification is based on the strain-life approach. The model improves the low cycle fatigue calculations. The capability and accuracy of the proposed model were compared with those of the original Morrow rules and $\mathbf{P_{SWT}}$ using available fatigue test data with different mean stress. The mean stress correction model was found to provide a better correlation, than $\mathbf{P_{SWT}}$ and the Morrow rules in the case of a superalloy and the ASTM A723 steel.

5.° An alternative approach (Kwofie, 2001; Kwofie et al., 2007) represents the dependence of coefficients (12.1) and (12.13) upon stress-ratio in form of the exponential function:

$$\begin{aligned} \sigma_{f}{}'(R_\sigma) &= \sigma_{f0}{}' \exp\left(-\lambda_{\mathrm{d}} \tfrac{\sigma_m}{\sigma_a}\right) = \sigma_{f0}{}' \exp\left(-\lambda_{\mathrm{d}} \tfrac{1+R_\sigma}{1-R_\sigma}\right), \\ \varepsilon_{f}{}'(R_\sigma) &= \varepsilon_{f0}{}' \end{aligned} \tag{12.19a}$$

$$\begin{aligned} b(R_\sigma) &= b_0 \exp\left(-\lambda_{\mathrm{d}} \tfrac{\sigma_m}{\sigma_a}\right) = b_0 \exp\left(-\lambda_{\mathrm{d}} \tfrac{1+R_\sigma}{1-R_\sigma}\right), \\ c(R_\sigma) &= c_0. \end{aligned} \tag{12.19b}$$

The value λ_{d} is the fitting constant, representing the mean stress sensitivity of the material. This constant is for the tested martials (Kwofie & Chandler, 2007) of the order $\lambda_{\mathrm{d}} \approx 1$.

It is well known that under extremely low cycle fatigue conditions, the Coffin-Manson law tends to overpredict the cyclic life. To explain the Coffin-Manson law, (Sornette et al., 1992) proposed a theory based on the observation of the "mesoscopic" grain scale, which is intermediate between the dislocation scale and the macroscopic crack scale.

6°. The variations of range ΔK and the mean value K_m of stress intensity factor over the load history could be accounted by applications of damage accumulation theories Sorensen (1969) and Wheeler (1972).

12.3 Stress Ratio Influence on Fatigue Crack Growth

12.3.1 Influence of Stress Ratio on Fatigue Threshold

As mentioned above, the stress ratio is defined as the ratio of the algebraic minimum stress to the maximum stress. The experiments show that the stress ratio affects the fatigue crack growth and threshold behavior. Namely, the fatigue crack growth rate and threshold vary with the applied stress ratio (Walker, 1970). When the stress ratio is positive, the experiments show that the required range of stress intensity factor decreases with increasing positive values. In the negative stress ratio region, the stress intensity factor range required for growth (threshold stress of the intensity range) decreases as the stress ratio decreases. Upon reaching a certain negative value, known as the saturation point, the required stress intensity factor range stabilizes.

Consider now the variation of stress ratio for the proposed propagation laws. (SAE AE-22, 1997, § 9.4.4) uses he Eqs. (10.32) and (10.33) for estimation of stress-ratio effect.

In accordance with the phenomenological approach (Klesnil & Lukas, 1972), the damage growths per cycle depends upon the stress ratio:

$$c_f(R_\sigma) = (1 - R_\sigma)^{1-\varsigma} \cdot c_f(R_\sigma = -1). \tag{12.20}$$

The exponent ς accounts the influence of the stress ratio on the crack growth rate, $0 < R_\sigma < 1$ is an arbitrary stress ratio.

According to (Forman et al., 1967), the crack closure decrease the fatigue crack growth rate by reducing the short-term threshold:

$$K_{IC}(R_\sigma) = (1 - R_\sigma)^{\gamma_2} K_{IC.0} \tag{12.21}$$

is the function of the stress ratio R_σ, where

$$K_{IC.0} \stackrel{\text{def}}{=} K_{IC}(R_\sigma = 0)$$

is the threshold value at $R_\sigma = 0$ and γ_2 is a material dependent constant. The model (Forman et al., 1967) leads to the extension of the Paris equation, considering the fracture toughness K_{crit}:

$$\frac{\mathrm{d}a}{\mathrm{d}n} = c_f \cdot \frac{[\Delta K]^p}{(1 - R_\sigma) \cdot K_{\text{crit}} - \Delta K}.$$

Hartman and Schijve (1970) suggested that da/dn should be dependent on the amount by which ΔK exceeds the fatigue threshold K_{th}:

$$\frac{\mathrm{d}a}{\mathrm{d}n} = c_f \cdot \frac{[\Delta K - K_{th}]^p}{(1 - R_\sigma) \cdot K_{\text{crit}} - \Delta K}.$$

Branco et al., (1976) proposed an alternative relationship:

$$K_{IC}(R_\sigma) = \sqrt{1 - R_\sigma^2} K_{IC.0}.$$

The endurance threshold limit might be also considered as a function of stress ratio:

$$K_{th}(R_\sigma) = (1 - R_\sigma)^{\gamma_1} K_{th.0} \tag{12.22}$$

Here

$$K_{th.0} \stackrel{\text{def}}{=} K_{th}(R_\sigma = 0)$$

is the endurance threshold value at $R_\sigma = 0$ and γ_1 is a material dependent constant.

Equations (12.20–12.22) are the empirically determined phenomenological dependencies (Kondo et al., 2003).

An equivalent driving force approach has been proposed by one of the current authors for the correlation of fatigue crack growth data at different stress ratios (Kwofie et al., 2019).

12.3.2 Models for Estimation of Stress Ratio Influence

The closed-form solution for the cycles to failure for a given initial crack length versus stress amplitude could be found for a given stress ratio for both Type I and Type II of the unified laws. Instead of an empirical approach, we use Eq. (12.12) to explicitly derive the influence of the stress ratio on the crack growth rate. To do this, consider two harmonically varying loads with the same amplitude but different mean values. The means and amplitudes are assumed to be positive, so that:

$$K_m > 0, \quad K_m^* > 0, \quad \Delta K > 0.$$

Both loads result in the harmonic stress intensity factors. Accordingly, the harmonic stress intensity factors are:

$$K(t) = K_m + \frac{\Delta K}{2} \sin t, \quad K^*(t) = K_m^* + \frac{\Delta K}{2} \sin t \tag{12.23}$$

For both considered cyclic loads (12.23) the ratios are equal correspondingly to:

$$R_\sigma = \frac{2K_m - \Delta K}{2K_m + \Delta K}, \quad R_\sigma^* = \frac{2K_m^* - \Delta K}{2K_m^* + \Delta K}.$$

Consider at first the positive load cycles, according to Eq. (12.23):

$$K(t) > 0; \quad K^*(t) > 0.$$

For the positive load cycles the stress ratios are in the ranges:

$$-1 < R_\sigma < 1, \quad -1 < R_\sigma^* < 1$$

The damage caused by the varying load depends linearly upon the stress intensity factor in the power p along the cycle. Consequently, the relation between fatigue coefficients, which correspond to the different stress ratios $c_f(R_\sigma)$ and $c_f\left(R_\sigma^*\right)$, reduces to (Kobelev, 2017a, 2017b):

$$\Lambda\left[R_\sigma, R_\sigma^*\right] = \frac{c_f(R_\sigma)}{c_f\left(R_\sigma^*\right)}, \tag{12.24a}$$

$$c_f(R_\sigma) = \left[\frac{1}{2\pi}\int_0^{2\pi}\frac{\mathrm{d}t}{(1+R_\sigma\sin t)^p}\right]^{-1}\cdot c_f(R_\sigma=-1). \tag{12.24b}$$

$$c_f\left(R_\sigma^*\right) = \left[\frac{1}{2\pi}\int_0^{2\pi}\frac{\mathrm{d}t}{\left(1+R_\sigma^*\sin t\right)^p}\right]^{-1}\cdot c_f(R_\sigma=-1). \tag{12.24c}$$

The closed-form expression of integrals in (12.24b–12.24c) is:

$$\begin{aligned}\Upsilon(R_\sigma) &= \frac{1}{2\pi}\int_0^{2\pi}\frac{\mathrm{d}t}{(1+R_\sigma\sin t)^p}\\ &= {}_2\mathbf{F}_1\left(\left[\frac{1+p}{2},\frac{p}{2}\right][1],R_\sigma^2\right) = \frac{\mathbf{P}_{-p}\left(\left(1-R_\sigma^2\right)^{-1/2}\right)}{\left(1-R_\sigma^2\right)^{p/2}},\end{aligned} \tag{12.25}$$

$\Upsilon(0) = 1$.

Here $\mathbf{P}_p$ is the Legendre function and ${}_2\mathbf{F}_1$ is the hypergeometic function: (Appendix F).

Consequently, the dependence of the fatigue coefficient upon the stress ratio the for positive load cycles reads:

$$\Lambda\left[R_\sigma,R_\sigma^*\right] = \frac{\Upsilon(R_\sigma)}{\Upsilon\left(R_\sigma^*\right)} = \left[\frac{1-R_\sigma^2}{1-R_\sigma^{*2}}\right]^{p/2}\frac{\mathbf{P}_{-p}\left(\left(1-R_\sigma^{*2}\right)^{-1/2}\right)}{\mathbf{P}_{-p}\left(\left(1-R_\sigma^2\right)^{-1/2}\right)}. \tag{12.26}$$

The plot of the function (12.26) for different values of p is shown on the Fig. 12.7.

With the function (12.25) the mean sensitivity follows directly from the first principles. Equation (12.26) is based on the Basquin law and summarizes the damage on each cycle. Equation (12.26) rescales the material constant from a certain stress ratio R_σ to the material constant for the changed stress ratio R_σ^* (Figs. 12.7).

12.3.3 Stress Ratio Influence in NASGRO Model

Third, we extend the validity range of the fatigue coefficient to the stress ratio. For this purpose, the effective stress intensity concept is implemented, following the results of (Elber, 1970, 1971). In these papers, the fatigue crack in a centrally cracked tension plate subjected to zero-to-tension loading was studied. It was shown that a fatigue crack is fully open for only a portion of the loading cycle. This effect was attributed to contact of residual pstic deformation around the crack tip. The cause of plasticity induced crack closure is the plastic wake developed during crack propagation. The crack faces could be subjected to significant compressive stresses.

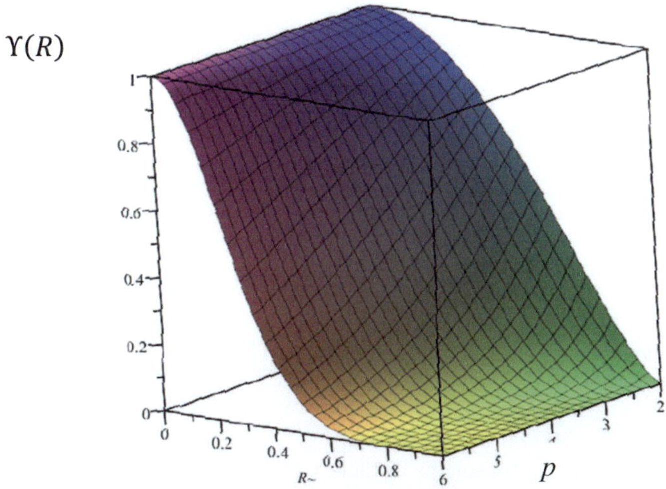

Fig. 12.7 The plot of the function$\Upsilon(R)$. For different values of p

In the article (Boyce & Ritchie, 2001) it was proposed that the crack tip stress intensity factor range needs to be modified by an experimental factor. The experimental factor accounts for the exposed crack closure effect.

The AFGROW code was developed by NASA (Harter, 1999). This code is used to simulate fatigue crack growth. The AFGROW code incorporates a strain-life based crack initiation analysis method to predict the crack initiation life. In the fatigue case and at the notch tip, local strains are obtained by using Neuber's rule (Neuber, 1961a, Neuber 1961b) or (Glinka, 1987). The NASGRO model is used when the set of curves describing fatigue crack growth is evaluated. The standard NASGRO model is:

$$\frac{\mathrm{d}a}{\mathrm{d}n} = c_f \cdot \left[\left(\frac{1-\tilde{f}(R_\sigma)}{1-R_\sigma}\right)\Delta K\right]^p \cdot \frac{\left(1-\frac{K_{th}}{\Delta K}\right)^{m_1}}{\left(1-\frac{K_{max}}{K_{\text{crit}}}\right)^{m_2}}. \tag{12.27}$$

The function $\tilde{f}$ presents the contribution of crack closure, Newmann's crack opening function (Newman, 1984). For constant amplitude loading, the function $\tilde{f}$ can be written as:

$$\tilde{f}(R_\sigma) = \max\left(R_\sigma, A_0 + A_1 R_\sigma + A_2 R_\sigma^2 + A_3 R_\sigma^3\right). \tag{12.28}$$

The parameters c_f, p, m_1, m_2 are determined experimentally and K_{th} is the threshold value of the stress–intensity factor range for crack propagation. The basic mechanical properties for technical alloys are given in (Dot, 2005).

Reworking the Eq. (12.27) in form of the unified propagation law (12.23) leads to the

$$\frac{\mathrm{d}a}{\mathrm{d}n} = c_f \cdot \left[\left(\frac{1-\tilde{f}(R_\sigma)}{1-R_\sigma}\right)\Delta K\right]^p \cdot \frac{\left(1-\frac{K_{th}}{\Delta K}\right)^{m_1}}{\left(1-\frac{K_{\max}}{K_{crit}}\right)^{m_2}}. \tag{12.29}$$

The formula (12.29) depends on both $K_{\max}$ and ΔK. Substitution of the expression

$$K_{\max} = \frac{1}{1-R_\sigma}\frac{\Delta K}{2}.$$

into (12.29) leads to the equation depending solely on range ΔK:

$$\frac{\mathrm{d}a}{\mathrm{d}n} = c_f \left[\frac{1-\tilde{f}(R_\sigma)}{1-R_\sigma}\right]^p (\Delta K)^p \cdot \frac{\left(1-\frac{K_{th}}{\Delta K}\right)^{m_1}}{\left(1-\frac{\Delta K}{2(1-R_\sigma)K_{crit}}\right)^{m_2}}. \tag{12.30}$$

The expression (12.30) could be further rephrased in the form of the type II unified propagation function with the exponent $k = 1$.

In this case, the coefficients of the type II unified propagation function are the functions of stress ratio R_σ:

$$\begin{aligned} \tilde{c}_f(R_\sigma) &= c_f \cdot \left[\tfrac{1-f(R_\sigma)}{1-R_\sigma}\right]^p, \\ \tilde{K}_2(R_\sigma) &= 2(1-R_\sigma)K_{crit}. \end{aligned} \tag{12.31}$$

With the above expressions (12.31), the NASGRO assumes the form:

$$\frac{\mathrm{d}a}{\mathrm{d}n} = \tilde{c}_f(R_\sigma)\Delta K^p \cdot \frac{(1-K_{th}/\Delta K)^{m_1}}{(1-\Delta K/\tilde{K}_2)^{m_2}} \tag{12.32}$$

Comparison with (12.26) states, that the following substitution reveals the dependence of the coefficients of the unified propagation function upon stress ratio R_σ:

$$c_f \to \tilde{c}_f(R_\sigma), \quad K_2 \to \tilde{K}_2(R_\sigma), \quad k \to 1.$$

Consequently, the NASGRO model (12.27) and (12.32) transforms to the form of unified propagation function type II (12.26) with the explicit dependence upon stress ratio R_σ:

$$\begin{aligned} \tilde{F}_{II}(\Delta K) &= \Delta K^p \cdot \tfrac{(1-K_{th}/\Delta K)^{m_1}}{(1-\Delta K/\tilde{K}_2)^{m_2}}, \\ \tfrac{\mathrm{d}a}{\mathrm{d}n} &= \tilde{c}_f \tilde{F}_{II}(\Delta K). \end{aligned} \tag{12.33}$$

The expressions for the crack length evolution (12.27, 12.43) were derived in Chap. 9 for fatigue law with the unified propagation function type II, with the constant $R_\sigma = 0$. Finally, all derived expressions are valid for an arbitrary R_σ with the formulas (12.31) and $k = 1$.

12.4 Influence of Defects on Fatigue Resistance

12.4.1 Influence of Area of Defects for Uniaxial Stress

Fatigue initiation and high cycle fatigue benefit from the observation that the limit of most materials is not sensitive to the critical stress for crack initiation (Murakami, 2003). When inclusions or other material imperfections do not affect fatigue crack initiation, fatigue crack initiation occurs from slip bands formed on the material surface. In this case, there is a linear relationship between material strength or hardness and fatigue limit. Industry uses traditionally the relation (Garwood et al., 1951):

$$\sigma_w \approx 0.45 \ldots 0.5 \cdot \sigma_u \approx 1.5 \ldots 1.4 \cdot HV \text{ for } HV < 400$$

The recent research (Pang et al., 2013) demonstrates the quadratic dependence of the fatigue strength σ_w to tensile strength σ_u :

$$\sigma_w \approx \left(0.70 - 1.85 \cdot 10^{-4}\sigma_u\right) \cdot \sigma_u$$

The most important factor is the threshold stress for arresting crack propagation. This threshold stress originates from a defect or inhomogeneity, a non-metallic inclusion or an original flaw, but not from the microstructure itself. If thermal and surface treatments are properly performed, they will improve the microstructure of the materials. In this case, the fatigue limit could only be observed in a narrow range of applied stress. The important parameter is the area of the flaw projected on the plane perpendicular to the directions of maximum principal stress. The fatigue analysis accounts only the endurance threshold limit K_{th} and the original defect size of the value $\sim \sqrt{area}$. The endurance threshold limit K_{th} correlates with the value $\sqrt{area}$. The three-dimensional crack analysis (Murakami, 1985), demonstrates a strong correlation between K_{th} along the crack front and $\sqrt{area}$:

$$K_{th} \sim \left(\sqrt{area}\right)^{1/3}.$$

In terms of Vickers hardness HV, the formula for the endurance threshold limit reads:

$$K_{th} = 3.3 \cdot 10^{-3} \cdot (HV + 120)\left(\sqrt{area}\right)^{1/3}.$$

The expression the endurance threshold limit K_{th} in terms of fatigue limit stress σ_w is (Murakami et al., 1985):

$$K_{th} = 0.65 \cdot \sigma_w \sqrt{\pi \sqrt{area}}.$$

Combination of the above equations yields the fatigue limit stress in terms of the inclusion size:

$$\sigma_w = \frac{1.43(HV + 120)}{\left(\sqrt{area}\right)^{1/6}}. \tag{12.34}$$

The units for the application are shown in the Table 9.4. The above analysis applies for helical springs at very high number of cycles, particularly for valve and injection pump springs in internal combustion engines (Tables 12.6 and 12.7).

The fatigue analysis accounts only the endurance threshold limit K_{th} and the original defect size $\sim \sqrt{area}$, where $area$ is the area of the defect, projected on the plane perpendicular to the directions of the maximum principal stress. The model (Murakami & Beretta, 1999; Takahashi & Murakami, 2002) predicts the lower limit of fatigue limit for the case of uniaxial tension load):

$$\sigma_w = C_0 \frac{HV + 120}{\left(\sqrt{area}\right)^{1/6}} \left(\frac{1 - R_\sigma}{2}\right)^{\zeta_{HV}}, \quad \zeta_{HV} = 0.226 + 10^{-4} \cdot HV. \tag{12.35}$$

C_0 is a dimensionless constant depending on whether the inclusion is located on the surface or inside the probe (Murakami, 1989) For surface cracks or inclusions this constant is. $C_0 = 1.43$, for subsurface inclusions. $C_0 = 1.40$ and for interior inclusions. $C_0 = 1.56$. Equation (12.34) accounts the stress ratio and generalizes the Eq. (12.47).

12.4.2 *Influence of Area of Defects for Shear Stress*

Gadouini and Nadot (2007) studied the evaluation of defect tolerance design under multi-axial fatigue loading. The method allows the determination of the maximum defect size that is tolerable for a component for a given fatigue life. The idea is to represent the defect by the mean of the stress gradient at the mesoscopic scale using kinematic hardening. This approach has been used to model multiaxial cyclic stress fields. Gadouini et al. (2008) studied the influence of the mean stress on the fatigue limit of steel containing spherical artificial defects under tension and torsion. For the shear stress fields the new parameter $area_P$. was introduced. This parameter describes the area of the defect projected on the plane perpendicular to the directions of the maximum tensile stress (45° to the direction of shear stress). For torsion loading, the fatigue limit is given by the following equation (Gadouini et al., 2008)):

$$\tau_w = \frac{C_0(HV + 120)}{F_{GNR}(b/a)\left(\sqrt{area}\right)^{1/6}} \left(\frac{1 - R_\sigma}{2}\right)^{\zeta_{HV}}, \tag{12.36}$$

where

$$F_{GNR}\left(\frac{b}{a}\right) = 0.095 + 2.11 \cdot \left(\frac{b}{a}\right) - 2.26 \cdot \left(\frac{b}{a}\right)^2 + 1.09 \cdot \left(\frac{b}{a}\right)^3 - 0.196 \cdot \left(\frac{b}{a}\right)^4,$$

a material parameter in Billaudeau's criterion describing defect influence (μm) (Billaudeau et al., 2004),

b material parameter in Chaboche's model (Chaboche, 1989).

12.4.3 Influence of Corrosion-Induced Defects

Experimental studies show that fatigue strength decreases sharply as corrosion of the component progresses (AIF 5996, 1988). Corrosion is responsible for crack initiation and promoting crack growth in fatigue-stressed components. Investigations (Schnattinger & Beste, 1995; Angelova et al., 2014) confirm that the vibration resistance under the medium used, 5% NaCl solution, is significantly reduced compared to the reference Wöhler line in air. The mechanical properties of the tested spring steels are reported in Angelova et al., (2014) and Schnattinger and Beste (1995), (Tables 12.5 and 12.6). Figure 12.8 shows the graph of the shear strain amplitudes γ to failure as the function of load cycles n_f (Angelova et al., 2014) (Fig. 12.9).

The test results obtained under corrosion conditions are well below the limits required by the standards. The influence of a corrosive medium on the tolerable load capacity is not currently considered in the analytical design guidelines. It is therefore recommended that studies be conducted to increase the fatigue strength in corrosive environments. The thermal processes (e.g. induction hardening) could increase the fatigue strength of steels in corrosive environments by introducing compressive residual stresses (Tables 12.2 and 12.3).

The fatigue life was investigated for different spring manufacturing technologies and surface protection systems in (Hoffmann et al., 2015). The relevant operational effects (deflection, abrasion, grit impact and corrosion) were considered. The significant reduction in fatigue strength can be determined by mechanical damage to the spring surface and subsequent corrosion. The double coating or thick film coating prevents the life reduction due to corrosion and mechanical damage.

Li et al., (2006) reports on the study of the fatigue behavior of 14 types of disk springs under corrosive conditions. The report covers the results of the investigation for the uncoated springs made of stainless steel as well as the coated springs made of tempered spring steel, both under the influence of different corrosive media (Fig. 12.10).

Table 12.5 Mechanical properties of spring steels (Kong et al., 2019; Boardman, 1982; Angelova et al., 2014)

Properties	Units		SAE 5160	SAE 4340			EN 10,270–1 SH		BS 250A53	
DIN							17223C		55Si7	
Source/Year			Kong et al. (2019)	Boardman (1982)			Angelova et al. (2014)			
				BHN Prediction	Su Prediction	Regression	Air	5% NaCl	Air	5% NaCl
Ultimate tensile strength	MPa	σ_w	1584	826			1522		1610	
Yield strength	MPa		1487	634			1217		1440	
Modulus of elasticity	GPa	E	205	193						
Shear modulus	GPA	G					78			
Fatigue strength coefficient	MPa	$\sigma_{f'}$	2063	1180	1088	1232				
Fatigue strength coefficient	MPa	$\tau_{f'}$					3.944	19.649	1.8875	49.29
Fatigue strength exponent		b	− 0.08	− 0.075	− 0.070	− 0.10	− 0.1465	− 0.368	− 0.0449	− 0.3347
Fatigue ductility exponent		c	− 1.05	− 0.6	− 0.574	− 0.56				
Fatigue ductility coefficient		$\varepsilon_{f'}$	9.56	0.70	0.57	0.53				
Hardening coefficient	MPa	K'	1940	1234	1165	1384				
Hardening exponent		n'	0.05	0.125	0.122	0.17				

Table 12.6 Effective fatigue life equations for different environments and coating (Schnattinger & Beste, 1995; Angelova et al., 2014)

Spring coating	Material	Environment	Fatigue life equation
Uncoated spring	Suspension spring material (Schnattinger & Beste, 1995)	Dry	$\tau_{ar} = 2.214 \cdot 10^3 \cdot n_f^{-0.0854}(P_{SWT})$
		Corrosive NaCl, 5%	$\tau_{ar} = 4.28 \cdot 10^4 \cdot n_f^{-0.38461}(P_{SWT})$
Coated spring			$\tau_{ar} = 3.276 \cdot 10^3 \cdot n_f^{-0.125}(P_{SWT})$
Uncoated spring	EN 10,270–1 SH (Angelova et al., 2014)	Dry	$\tau_c = 3.563 \cdot 10^3 \cdot n_f^{-0.1465}$
		Corrosive, NaCl, 5%	$\tau_c = 1.522 \cdot 10^4 \cdot n_f^{-0.368}$
	BS 250A53 (Angelova et al., 2014)	Dry	$\tau_c = 1.829 \cdot 10^3 \cdot n_f^{-0.0449}$
		Corrosive, NaCl, 5%	$\tau_c = 3.91 \cdot 10^4 \cdot n_f^{-0.3347}$

12.5 Influence of Multiaxial Stresses

One of the most important requirements for properly designed metal springs is the absence of multiaxial stress. The properly designed spring must resist one type of stress, for example, pure shear of pure tensile stress. Therefore, in most cases, the simplest forms of stress tensor are assumed. However, under real practical conditions, the springs are subjected to fluctuating loads of different amplitudes and frequencies in two or three directions. For example, automotive springs are subjected to axial loads. The bending loads occur occasionally during cornering. The axle spring of the rear suspension is ideally in the state of pure axial load, but the large deformation leads to bending of the spring. The shorter length of the suspension arms favors the excessive bending of the springs. Occasionally, the springs are specially designed for multiple stress cases, for example, the simultaneous torque and axial loads. The axial load causes the shear stresses in the wire. At the same time, the bending stress causes the normal stresses. Consequently, the stresses in springs are the combination of pure normal stress and pure shear stress. In this case, the stress tensor contains several non-zero components. Such stress fields cause biaxial or multiaxial fatigue (Brown, 1989; Fateimi & Socie, 1989). The evaluation of fatigue life under biaxial or multiaxial stresses is an important design requirement for safe and reliable functioning of structural members (ESIS 31, 2003).

In the ideal case, possible loading systems are correlated in properly designed springs. However, uncorrelated or random loading may occur in real-life service environments. For multiaxial fatigue damage evaluation under random loading, three common steps should be included: multiaxial cycle counting, multiaxial fatigue damage evaluation for each cycle counted, and fatigue damage accumulation for

Table 12.7 Experimental data high strength steels (Chapetti, 2010), T-C: Tension–Compression, R-B: Rotating-Bending, Tor: torsion, HEL: helical spring

Steel	Source	Range of cycles	Test
SCM435	(Murakami et al.,1999)	10^8	T-C
0.46Carbon	(Murakami et al., 1998)	5. 10^8	T-C
SUJ2	(Shiozawa et al., 2001)	10^{10}	R-B
SCM435 H	(Nakamura et al., 1998)	10^8	T-C
SNCM439	(Furuya et al., 2002)	10^{10}	T-C
Cr-V	(Almaraz, 2008)	10^{10}	T-C
Cr-Si (54SC6)		5 10^9	T-C
Cr-Si (55SC7)		10^{10}	T-C
Cr-Si (55SC7)		3 10^9	T-C
SUP7	(Murakami et al., 1988)	10^{10}	R-B
SUP7		10^8	T-C
SUP7		10^8	T-C
SUP10M	(Wang et al., 1999)	0.2 10^8	T-C
SUP12	(Abe et al., 2001)	10^8	T-C
SWOSC-V		10^8	T-C
100C6Mart	(Murakami et al., 2006)	7. 10^9	T-C
100C6Bain		10^{10}	T-C
SUJ2		10^9	
VDSiCr, TDSiCr, DH, DM, 1.4568, 1.4310	(AiF, 12287N, 2002)	10^7	HEL
VDSiCr	(Kaiser et al., 2011)	1,5•10^9	HEL
VDSiCr, VDSiCrV, X7CrNiAl17-7 (1.4568)	(AiF 15064N, 2010)	10^9	HEL
VDSiCr, VDSiCr (WBH), DH, 1.4310, 1.4568 (conv, new)	(IGF16873)	~ 10^9	HEL
VDSiCr	(Mayer et al., 2015, 2016)	5 10^9..1.4 10^{10}	Tor

all cycles counted. Table 9.7 from (Chapetti, 2010) provides an overview of the experimental data for high-strength steels.

12.5.1 Stress-Life Approach in Multiaxial Fatigue

12.5.1.1 Proportionally Oscillating Stress Tensor Fields

For the oscillating stress fields, the proper definition of a scalar, fully reversed equivalent fatigue stress is essential. For the evaluation of stress–strain based methods,

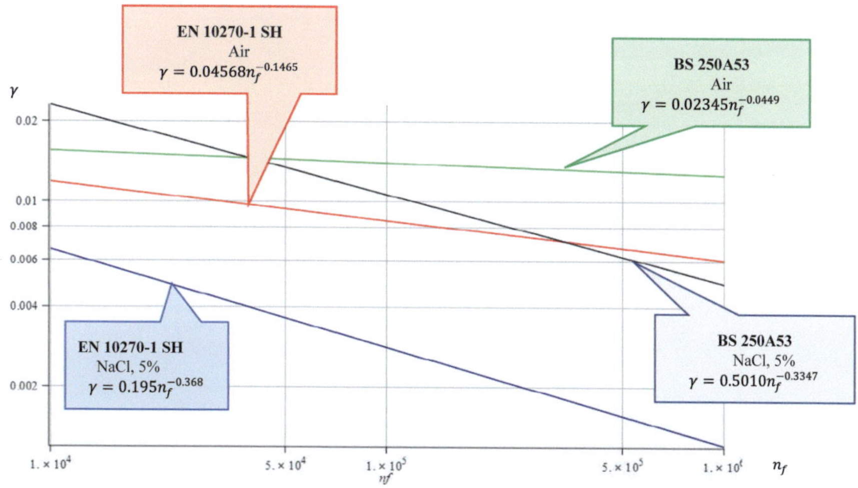

Fig. 12.8 Graph of the shear strain amplitudes γ to failure as the function of load cycles n_f (Angelova et al., 2014)

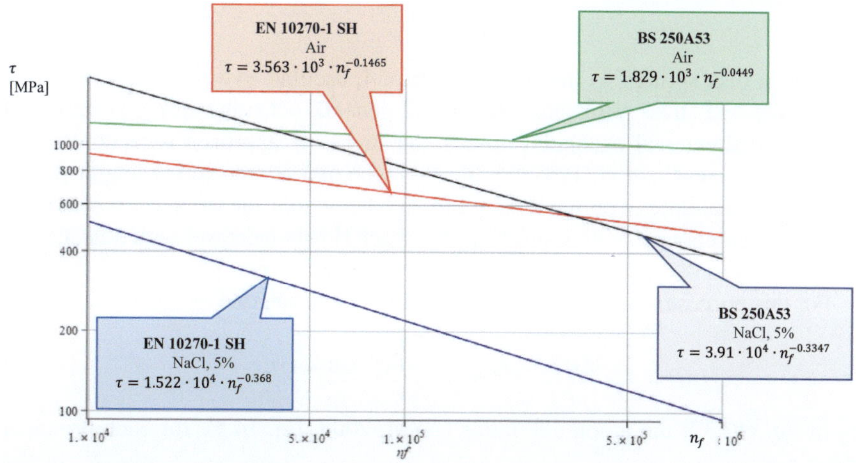

Fig. 12.9 displays the dependence of the shear stress amplitudes τ to failure as the function of load cycles n_f (Angelova et al., 2014).

the various definitions of equivalent parameters for multiaxial fatigue have been introduced.

Such stress behavior is typical for springs subjected to oscillating forces, such as those used in automotive suspensions. The torque and lateral loads of the torsion beam suspension do not correlate. When properly designed, the suspension springs carry a single axial load. However, sometimes the stiffness of the suspension is somewhat low and the side loads act on the spring. The circumferential and radial

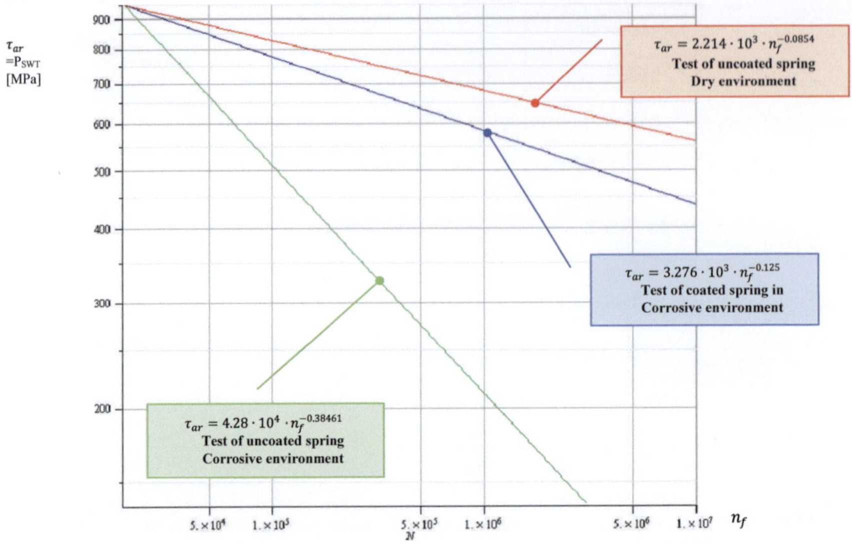

Fig. 12.10 displays S–N curves for τ_{ar} according to Smith, Watson and Topper $\mathbf{P_{SWT}}$, Eq. (12.3) (Schnattinger & Beste, 1995).

stresses in disk springs, as discussed in Chap. 4, oscillate synchronously, but their amplitudes and stress ratios are different. In general, when the spring is subjected to predefined external oscillating forces, the stress-life approach is recommended for fatigue life evaluation. Typically, for this type of load, the spring length slowly decreases or increases over the service life. In other words, the strain is the sum of an oscillatory component and the mean strain, which slowly increases with time. Cyclic creep occurs in the spring material.

For this purpose, consider time-depended principal stresses

$$\sigma_i(t) = \sigma_{m.i} \cdot (1 + \mathrm{A}_i \cdot \sin(\omega t)). \tag{12.37}$$

In Eq. (12.37) the amplitude ratios (ASM, 2008, Eq. 14.5). for each principal stress are:

$$\mathrm{A}_i = \frac{\sigma_{a,i}}{\sigma_{m.i}}, \quad i = 1 \ldots 3.$$

In the simplest case of multiaxial loading, all three amplitude ratios are equal:

$$\mathrm{A} \equiv \mathrm{A}_1 = \mathrm{A}_2 = \mathrm{A}_3, \quad \mathrm{A} = \frac{1 - R_\sigma}{1 + R_\sigma}.$$

In this case, all principal stresses oscillate synchronously and proportionally to a harmonic function $1 + \mathrm{A} \cdot \sin(\omega t)$ of time:

$$\sigma_i(t) = \sigma_{m.i} \cdot (1 + \mathrm{A} \cdot \sin(\omega t)), \tag{12.38}$$

In this case, the effective stress is also the harmonic function of time:

$$\sigma_{eq.ar}(t) = \sigma_{eq.\mathrm{m}} \cdot (1 + \mathrm{A} \cdot \sin(\omega t)), \tag{12.39}$$

$$\sigma_{eq.m} = \frac{1}{\sqrt{2}}\sqrt{(\sigma_{m.1} - \sigma_{m.2})^2 + (\sigma_{m.2} - \sigma_{m.3})^2 + (\sigma_{m.3} - \sigma_{m.1})^2}.$$

is the mean value of equivalent fully reversed fatigue stress.

For the synchronously harmonically oscillating stress tensor fields, the Goodman diagram uses the following axes:

$$\sigma_{eq.\min} = \sigma_{eq.m} - \sigma_{eq.a} \rightarrow \quad \sigma_{eq.\max} = \sigma_{eq.m} + \sigma_{eq.a} \tag{12.40}$$

here $\sigma_{eq,a} = \mathrm{A}\cdot\sigma_{eq,m}$ is the amplitude of the of equivalent fully reversed fatigue stress (Table 12.2). Goodman diagrams are plotted in terms the scalar effective stresses $\{\sigma_{eq,a}, \sigma_{eq,m}\}$.

As far as the fatigue equivalent stress is defined, the number of cycles to total failure for tensor stress fields are the functions of effective stress $\sigma_{eq.ar}$.:

$$n_f = f_{eq}^{-1}(\sigma_{eq.ar}) \equiv \frac{1}{2}\left(\frac{\sigma_{eq.ar}}{\sigma_{eq.f'}(R_\sigma)}\right)^{1/b(R_\sigma)} \tag{12.41a}$$

$$\sigma_{eq.ar} = f_{eq}(n_f) \rightarrow \quad f_{eq}(n_f) = \sigma_{eq.f'}(R_\sigma) \cdot (2n_f)^{b_0}. \tag{12.41b}$$

here $\sigma_{eq.f'}(R_\sigma)$ is the fatigue strength coefficient for effective stress.

12.5.1.2 Synchronously, Harmonically Oscillating Stress Tensor Fields

Commonly, the relations of the amplitudes to the mean values of the principal stresses are not all equal: $\sigma_1 \neq \sigma_2 \neq \sigma_3$. In this case the proportionally of all principal stresses upon one single parameter is lost and simple relation (12.39) is no more valid.

The consideration (Sines, 1955) is limited to stresses that oscillate harmonically from a minimum to a maximum with a mean value, where the stresses are in phase and the principal axes are fixed. Thus, the principal stresses are synchronous functions of time:

$$\sigma_i = \sigma_{m.i} + \sigma_{a,i} \sin(\omega t), \qquad i = 1 \ldots 3. \tag{12.42}$$

The equivalent stress plays the role of the scalar, uniaxial, fully reversed fatigue stress (Table 12.8). This stress is expected to give the same life on smooth specimens as the multiaxial condition. If intense yielding is avoided, the effect of static bending dominates the effect of static torsion on the fatigue life. According to (Sines, 1955), the equivalent stress for multiaxial fatigue is defined as:

$$\sigma_{eq.ar} = \frac{1}{\sqrt{2}}\sqrt{(\sigma_{a.1} - \sigma_{a.2})^2 + (\sigma_{a.2} - \sigma_{a.3})^2 + (\sigma_{a.3} - \sigma_{a.1})^2} + \frac{C_1}{\sqrt{2}}(\sigma_{m.1} + \sigma_{m.2}\ \sigma_{m.3}), \tag{12.43a}$$

$$n_f = f_{eq}^{-1}(\sigma_{eq.ar}). \tag{12.43b}$$

The influence factor $C_1 > 0$ for normal stress acting on the maximum shear stress plane was introduced in (Findley, 1957, 1959). For a given fatigue life, the allowable alternating shear stress decreases, if the maximum normal stress increases.

$$\tau_a + \mu_{\sigma\tau}\sigma_{n.\max} = f_\tau(n_f), \tag{12.44}$$

$\sigma_{n.\max}$ is the maximum normal stress on the plane of the critical alternating shear stress,

τ_a is the allowable alternating shear stress.

The fatigue effective stress yields (Sines, 1955; Findley, 1959):

$$\sigma_{eq,ar} = \frac{1}{\sqrt{2}}\sqrt{(\sigma_{a.1} - \sigma_{a.2})^2 + (\sigma_{a.2} - \sigma_{a.3})^2 + (\sigma_{a.3} - \sigma_{a.1})^2} + \frac{C_1}{\sqrt{2}}(\sigma_{m.1} + \sigma_{m.2} + \sigma_{m.3}) + \frac{C_2}{\sqrt{2}}(\sigma_{a.1} + \sigma_{a.2} + \sigma_{a.3}). \tag{12.45}$$

The constant C_2 represents the effect of the hydrostatic alternating component. This constant can be positive or negative, indicating the detrimental or beneficial

Table 12.8 Common equivalent stresses for multiaxial fatigue analysis of springs

Maximal shear stress for static load	$\sigma_{eq} = \frac{1}{2}\max(\lvert\sigma_i - \sigma_3\rvert, \lvert\sigma_3 - \sigma_2\rvert, \lvert\sigma_2 - \sigma_1\rvert)$
Equivalent MISES stress for static load	$\sigma_{eq} = \frac{1}{\sqrt{2}}\sqrt{(\sigma_1 - \sigma_3)^2 + (\sigma_3 - \sigma_2)^2 + (\sigma_2 - \sigma_1)^2}$
Equivalent stress for multiaxial fatigue, (Sines, 1955)	$\sigma_{eq.ar} = \frac{1}{\sqrt{2}}\sqrt{(\sigma_{a.1} - \sigma_{a.2})^2 + (\sigma_{a.2} - \sigma_{a.3})^2 + (\sigma_{a.3} - \sigma_{a.1})^2} + \frac{C_1}{\sqrt{2}}(\sigma_{m.1} + \sigma_{m.2} + \sigma_{m.3})$
Equivalent stress for multiaxial fatigue (Findley, 1957, 1959)	$\sigma_{eq} = \frac{1}{\sqrt{2}}\sqrt{(\sigma_{a.1} - \sigma_{a.2})^2 + (\sigma_{a.2} - \sigma_{a.3})^2 + (\sigma_{a.3} - \sigma_{a.1})^2} + \frac{C_1}{\sqrt{2}}(\sigma_{m.1} + \sigma_{m.1} + \sigma_{m.1}) + \frac{C_2}{\sqrt{2}}(\sigma_{a1} + \sigma_{a.1} + \sigma_{a.1})$

effects of the hydrostatic alternating stress component. Fatigue failure under hydrostatic alternating stresses has been studied experimentally in bending and torsional fatigue tests by (Kakuno et al., 1979).

The modification in the stress approach was introduced in the works (McDiarmid, 1991, 1993):

$$\tau_a \cdot \left(1 + \frac{\sigma_{n.max}}{2\sigma_w}\right) = f_\tau\left(n_f\right).$$

In this equation σ_w is the tensile strength. The function $f_\tau\left(n_f\right)$. depends on whether the cracking grows along the surface (mode III) or grows into the surface (mode II). The first case occurs in torsion loading and the second one under biaxial tension loading.

Dang Van (1993) proposes an approach for determining fatigue strength which—like the approaches of (Sines, 1955; Findley, 1959) and (McDiarmid, 1991)—is based on a combination of shear stress range $\Delta\tau$ and a hydrostatic stress component σ_{oct}. Dang Van's approach is based on an observation of microscopic stresses (on the order of magnitude of a microstructural grain) to map crack initiation within a microstructural grain:

$$\langle\tau\rangle + a\sigma_{oct} = b. \tag{12.46}$$

Papadopoulos (1997, 2001) formulated a fatigue criterion for the fatigue strength problem based on a consideration of the mesoscopic size scale. Analogous to the micro–macro scale approach (Dang Van, 1993), the size scale refers to the sliding processes within a microstructure grain. Papadopoulos 2001 assumed that the plastic displacement along a slip plane accumulated by cyclic loading is proportional to the macroscopic shear stress amplitude. To determine the fatigue strength, the average shear stress τ was bounded, taking into account the hydrostatic stress σ_{oct}. The approach (Papadopoulos, 2001) was formulated with the material constants and as follows:

$$\langle\tau\rangle + a\sigma_{oct} = b, \tag{12.47a}$$

$$\langle\tau\rangle = \sqrt{\frac{5}{8\pi}\int_{\varphi=0}^{2\pi}\int_{\psi=0}^{\pi}\int_{\theta=0}^{2\pi}\tau_a(\varphi,\psi,\theta)^2 d\theta \sin\psi\ d\psi\ d\varphi}. \tag{12.47b}$$

12.5.2 *Strain-Life Approach in Multiaxial Fatigue*

When the spring oscillates between two fixed limits, fatigue assessment of these springs is based on the strain-life approach. The strain amplitude and mean value remain the same over the load history, but the stresses in the material and the external loads continuously decrease with time. The spring is in the process of cyclic relaxation. Such behavior is typical for valve and diesel injection pump springs.

The models for fatigue life under combined cyclic axial–torsional loading conditions have been proposed in (Haverd & Topper, 1971; Miller & Brown, 1984; Fatemi & Kurath, 1988; Fatemi & Socie, 1988; Socie & Marquis 1999). The models include parameters based on the von Mises equivalent strain range, principal stresses, or principal strains. The models also use a combination of the maximum shear strain and the normal stress and/or normal strain acting on the maximum shear plane. The life prediction models have been verified at room temperature using axial or torsional test rigs. The common approaches for describing the multiaxial fatigue are based on the following models:

- Maximum principal strain amplitude model;
- Maximum shear strain amplitude model;
- von Mises effective strain amplitude model;
- Brown Miller model;
- Lohr-Ellison model;
- Energy based model.

12.5.2.1 Model of Maximum Principal Strain Amplitude

In the simplest case of multiaxial loading, all principal strains oscillate synchronously and proportionally to one harmonic function $1+\tilde{A}\cdot\sin(\omega t)$ of time:

$$\varepsilon_i(t)=\varepsilon_{m.i}\cdot\left(1+\tilde{A}\cdot\sin(\omega t)\right),\quad \varepsilon_{m.1}\geq\varepsilon_{m.2}\geq\varepsilon_{m.3}. \tag{12.48}$$

In Eq. (12.48) the amplitude ratios for each principal strain are:

$$\widetilde{A}=\frac{\varepsilon_{a,i}}{\varepsilon_{m.i}},\ i=1\ldots3\ \text{ or }\ \widetilde{A}=\widetilde{A}_1=\widetilde{A}_2=\widetilde{A}_3.$$

Consider now the cyclic multiaxial strain with different amplitude ratios for some principal strains. The amplitude of principal strain ε_1 is equal to

$$\Delta\varepsilon_1=\widetilde{A}\,\varepsilon_{m.1}.$$

This amplitude is greater than the amplitude of both other principal strains. The strain amplitude. $\widetilde{A}\,\varepsilon_{m.1}$is considered the dominant parameter to describe damage. The damage is caused by the amplitude of the principal strain $\Delta\varepsilon_1$ and

$$\Delta\varepsilon_1 = \frac{\sigma_f\prime(R_\sigma)}{E}(2n_f)^{b_0} + \varepsilon_f\prime(R_\sigma)\cdot(2n_f)^{c_0},$$
$$\Delta\varepsilon_1 = \frac{\sigma_a}{E} + \left(\frac{\sigma_a}{K_\sigma\prime}\right)^{1/n'} \tag{12.49}$$

12.5.2.2 Model of Shear Strain Amplitude

The amplitude of dominant shear strain $\Delta\gamma_{13}$ is equal to

$$\Delta\gamma_{13} = \tilde{A}\cdot(\varepsilon_{m.1} - \varepsilon_{m.3}).$$

This amplitude is greater than the amplitude of both other principal strains. The strain amplitude e $\varepsilon_{m.1}$. is considered the dominant parameter to describe damage. The damage is caused by the amplitude of the Tresca shear strain $\Delta\gamma_{13}$ and according to (12.14):

$$\Delta\gamma_{13} = \frac{\tau_f^{'}(R_\sigma)}{G}(2n_f)^{b_0} + \gamma_f^{'}(R_\sigma)\cdot(2n_f)^{c_0},$$
$$\Delta\gamma_{13} = \frac{\tau_a}{G} + \left(\frac{\tau_{aa}}{K_\tau^{'}}\right)^{1/n'} \quad . \tag{12.50}$$

12.5.2.3 Equivalent Strain Amplitude

The von Mises equivalent strain amplitude is the octahedral shear strain**:**

$$\Delta\gamma_{eq} = \frac{\tilde{A}}{\sqrt{6}}\cdot\sqrt{(\varepsilon_{m.1} - \varepsilon_{m.3})^2 + (\varepsilon_{m.3} - \varepsilon_{m.2})^2 + (\varepsilon_{m.2} - \varepsilon_{m.1})^2}. \tag{12.51}$$

The damage is caused by the amplitude of the von Mises equivalent shear strain $\Delta\gamma_{eq}$ and according to (12.14):

$$\Delta\gamma_{eq} = \frac{\tau_f\prime(R_\sigma)}{G}(2n_f)^{b_0} + \gamma_f\prime(R_\sigma)\cdot(2n_f)^{c_0},$$
$$\Delta\gamma_{eq} = \frac{\tau_a}{G} + \left(\frac{\tau_a}{K_\tau\prime}\right)^{1/n'} \quad . \tag{12.52}$$

12.5.2.4 Corrections of Amplitude of Shear Strain

There are two outmost strain measures in three-dimensional solid. The first is the maximum amplitude of the shear strain: $\Delta\gamma_{13} = \tilde{A}\cdot(\varepsilon_{m.1} - \varepsilon_{m.3})$. The second is the amplitude of normal strain, acting on the plane where the former occurs: $\Delta\varepsilon_n = \tilde{A} \cdot (\varepsilon_{m.1} + \varepsilon_{m.3})$.

Brown and Miller (1973) observed that failure under multiaxial fatigue is caused if the certain relation between $\Delta\gamma_{13}$ and $\Delta\varepsilon_n$ is fulfilled: $\Delta\gamma_{13} = \Phi(\Delta\varepsilon_n)$. In other words, the amplitude of normal strain influences the damage, which is caused by the amplitude of the shear strain. To account this phenomenon, the maximum amplitude of the shear strain must be corrected:

$$\widetilde{\Delta\gamma_{13}} = \Delta\gamma_{13} - \Phi(\Delta\varepsilon_n).$$

Kandil et al., (1982) used a linear correction

$$\widetilde{\Delta\gamma_{13}} = \Delta\gamma_{13} + \mathrm{k}_1\Delta\varepsilon_n, \quad \mathrm{k}_1 = 1.$$

and derived equation for the evaluation of the cyclic fatigue limit:

$$\widetilde{\Delta\gamma_{13}} = A_1\frac{\sigma_f{}'(R_\sigma)}{E}\left(2n_f\right)^{b_0} + A_2\varepsilon_f{}'(R_\sigma)\cdot\left(2n_f\right)^{c_0}, \tag{12.53}$$

$$A_1 = 3.3; \quad A_2 = 3.5.$$

Another correction was proposed by Lohr and Ellison (1980). They studied crack propagation driven by shear. This type of damage is typical for helical compression and extension springs. When the stress is very high and the fatigue life expectancy is low, the crack originates from the surface of the material.

12.5.2.5 Critical Plane Criteria

For multiaxial low-cycle fatigue damage models, the critical plane criteria are the modern approaches due to their specific physical meaning and acceptable capability of multiaxial fatigue life prediction (Fatemi et al., 2011; Ince, 2012; Karolczuk et al., 2005; Socie & Marquis, 1999).

The strain-based approaches are usually used in combination with the concept of the critical plane (Brown-Miller, 1973; Fatemi–Socie, 1988; Kandil-Brown-Miller, 1982) The energy-based models on the critical plane have also been proposed (Socie, 1987; Varvani & Farahani, 2000; Xue et al., 2019).

The methods developed in the cited works could be applied in principle to the spring elements. The main requirements are

- Evaluation of multiaxial load collectives for questioned applications.
- Experimental acquisition of fatigue properties for multiaxial loading of spring materials.

12.6 Influence of Environment and Manufacturing

12.6.1 Influence of Wire Diameter

The next task is the evaluation of the effective stress for the different diameters of wire. For this purpose, the correction factor $\vartheta(\mathrm{d})$ is developed. The reduction of tolerant stress for the material SH or DH is presented in EN 1270–1:2001. The rational interpolation of the data provides the correction of the effective stress as the function of wire diameter:

$$\frac{\tau_{ar}(\mathrm{d})}{\tau_{ar}(\mathrm{d}=1mm)}=\vartheta(\mathrm{d}), \tag{12.54}$$

$$\vartheta(\mathrm{d})\stackrel{\mathrm{def}}{=}\frac{61683\cdot\mathrm{d}^2-101769\cdot\mathrm{d}+373390}{3748\cdot\mathrm{d}^3+46996\cdot\mathrm{d}^2+11960\cdot\mathrm{d}+269600}. \tag{12.55}$$

because $\vartheta(\mathrm{d}=1\,\mathrm{mm})=1$. Equation (12.55) is based on rational interpolation of the values form standard EN 13,906–1:2002 (D), For example, if the fatigue data is acquired for some diameter of wire, $\mathrm{d}=2mm$, the recalculation for any other diameter d, according to Eq. (12.55), uses the formula:

$$\frac{\tau_{ar}(\mathrm{d})}{\tau_{ar}(\mathrm{d}=2mm)}=\frac{\vartheta(\mathrm{d})}{\vartheta(\mathrm{d}=1mm)} \tag{12.56}$$

Matek et al., (1992) in TB 10-2c recommend the different correction factors for different spring materials $\sigma_w(\mathrm{d})/\sigma_w(1)$ (Table 12.9). The comparison of both correction factors presents Fig. 12.11.

There are several explanations for the effect of wire diameter on fatigue life (Shelton & Swanger, 1935). First, the fatigue limit depends greatly on the conditions at the surface, such as surface decarburization in steels, tool marks, notches, and some protective metallic coatings. Second, processing is different for wires of larger

Table 12.9 Dependence of minimal wire strength $\boldsymbol{\sigma_w}=\boldsymbol{R_m}$ upon wire diameter and correction factors

Wire grade	Wire diameter	Ultimate strength	Correction factors
	d[mm]	$\sigma_w(d),[MPa]]$	$\sigma_w(\mathrm{d})/\sigma_w(1)$
A	1..10	$1720-660\log_{10}\mathrm{d}$	$1-0.166\log_{10}\mathrm{d}$
B	0.3..20	$1980-740\log_{10}\mathrm{d}$	$1-0.162\log_{10}\mathrm{d}$
C	2..20	$2220-820\log_{10}\mathrm{d}$	$1-0.160\log_{10}\mathrm{d}$
D	0.2..20	$2220-820\log_{10}\mathrm{d}$	$1-0.160\log_{10}\mathrm{d}$
FD	0.5..17	$1846-480\log_{10}\mathrm{d}$	$1-0.113\log_{10}\mathrm{d}$
VD	0.5..10	$1800-415\log_{10}\mathrm{d}$	$1-0.1\log_{10}\mathrm{d}$

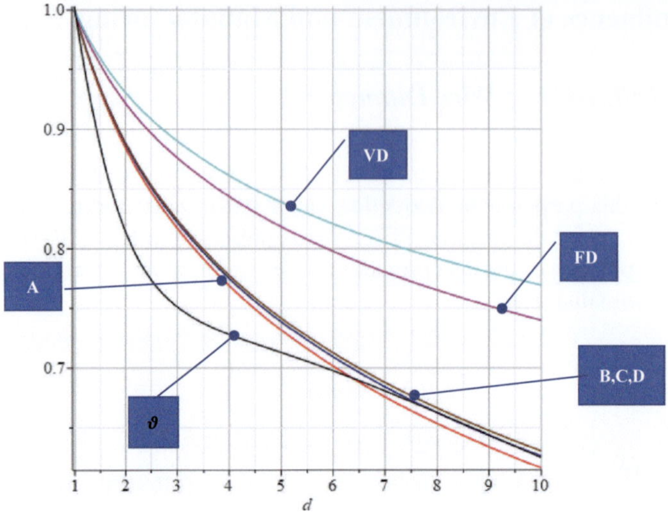

Fig. 12.11 Correction factors for materials A, B, C, D, VD, FD (Matek et al., 1992) and $\vartheta(d)$ for material SH or DH, EN 1270–1:2001

and smaller diameters. Third, the survival probability for the larger diameter wires is lower due to the multiplication of the survival probabilities of the elementary material sub-volumes (weakest link concept). The application of the weakest link concept to the evaluation of the survival probability of springs is discussed in Chap. 13.

12.6.2 Influence of Shot Peening

Shot peening is a special type of surface treatment and is an essential measure to increase the service life and fatigue strength of springs made of high strength steels (Niku-Lari, 1996; Kirk, 1999; Wagner, 2003). The process uses compressed air or blast wheels to shoot abrasive at the springs at high velocity. Wheel blast machines are typically used for the standard shot peening of automotive springs. In these systems, the abrasive is accelerated by centrifugal force in one or more blast wheel units to the desired abrasive speed (discharge speed). The abrasive usually reaches the center of the blast wheel, which is usually equipped with 8 blades, indirectly via a mechanical pre-acceleration. By means of the rotating cage-like distributor, uniform portions of abrasive are taken up and fed through the opening of the fixed inlet piece onto the blades. As the jet leaves the blades, it fans out to create an elliptical blast pattern that can be adjusted to suit the surface to be blasted by positioning the inlet piece. The abrasive consists of round or rounded grains that give the process its name.

The purpose of shot peening is to increase the surface volume of the material and create a residual compressive stress. The result is compression of the material and hardening of the surface. It introduces residual compressive stresses into the surface layer, the level of which correlates with the level of material strength. In this way, any surface defects such as scoring or scars are largely mitigated. As a result, the origin of fracture is repeatedly shifted to the inner regions of the material. For larger helical compression springs, shot peening under preload (stress peening) is also used. This process significantly improves fatigue strength. The speed of the blast medium (depending on the air pressure or the speed of the blast wheel) together with the grain mass influences the kinetic energy of the steel medium and thus the blast intensity. The amount of blast media and the blasting time are the main determinants of the degree of surface coverage (i.e. the percentage of the surface deformed by the ball impacts in relation to the initial surface).

The following general conclusions can be made about the formation of residual stresses in shot-blasted components (Hirsch et al., 1979):

- The magnitude of residual stresses at the blast grain surface is primarily determined by material strength or hardness. Blasting parameters such as abrasive velocity, coverage ratio, and abrasive grain diameter influence the residual compressive stress on the surface of the abrasive and are also significant for the fatigue of the processed part.
- The maximum value of the induced residual compressive stress usually occurs below the surface of the abrasive. The maximum occurs at the surface of the exposed material in the case of low material strength or low resulting residual stress.
- The thickness of the surface layer subject to compressive residual stresses decreases with higher material strengths.
- The thickness of the surface layer subject to compressive residual stress increases with increasing shot velocity, shot size and final coverage.
- For hardened abrasives (> 700 HV), the transition from a soft (40–530 HV) to a hard abrasive (580–655 HV) causes an increase in coverage and maximum residual stresses, as well as the thickness of the compressive residual layer, with otherwise identical blasting parameters.
- If the abrasive is blasted under static tensile (pressure) pre-stress, then the following values will result in larger (smaller) amounts of residual compressive stresses in the material areas near the surface than with non-pre-stressed abrasive (stress blasting).
- Shot peening with vertically incident abrasive induces a rotationally symmetric residual compressive stress state.

Due to the large number of parameters involved in shot peening, there have been several attempts to describe the effect of shot peening in terms of easily measurable parameters. (Nikulari, 1981) considered that a certain stress source is introduced into the metal component during the shot peening process. One of the analytical models to calculate the induced residual stress was based on (Al-Obeid, 2007) where he proposed that the distribution of the residual stress was a result of a stretching action followed by an elastic unloading of a metallic plate that was shot peened. The main assumption in this case was that the plane section of the metal plate remains constant. However, this assumption conflicted with the consideration of the plastic deformation that occurs for the formation of residual stress. The idea of stress source was introduced to overcome this dispute of local plasticity.

Despite frequent criticism and numerous other suggestions, the Almen test (Almen, 1943) is still commonly used to measure shot peening intensity. In this test, standardized steel plates are shot peened on one side in a clamped condition. The deflection ("arc height") of the Almen strip measured after loosening serves as a measure of the blasting intensity and has proven to be a good process control and adjustment criterion for blasting machines. The test strip should be positioned so that the critical area of the part is detected. However, there is no direct correlation between shot peening intensity and spring life improvement for the following reasons:

- There is no definite relationship between the stress changes caused by shot peening and the change in Almen arc height;
- The relationship between the Almen arc height and the degree of coverage achieved must be evaluated experimentally for each material and process;
- The influence of material properties, such as hardening behavior and notch sensitivity, on the Almen arc height is not clear.

Therefore, vibration tests should be performed to clearly determine the effect of shot peening on the fatigue life of coil springs.

The coverage indicates the percentage of the abrasive impact surface. The degree of surface coverage increases with shot peening time and asymptotically approaches the 100% limit; a value of 98% is generally accepted as total coverage. Longer blasting times are expressed as a multiple of the 98% value. The degree of area coverage is determined by planimetry or by comparison with reference series.

The increase in fatigue strength that can be achieved by shot peening and presetting is approximately 10 to 20% for an ideal initial condition of the surface layer (Kaiser, 2006). Depending on the geometric conditions, the fatigue crack can be moved below the surface, eliminating the surface weak point effect and providing an increase in fatigue strength corresponding to the stress gradient. The improvement in fatigue strength as a result of prestressing, and particularly as a result of shot peening or stress peening, can be substantial. First, the mean stress is reduced due to residual compressive stresses. Second, the surface topography can be improved by shot peening. Third, the barrier to crack propagation in defects due to residual compressive stresses (crack closure) can also contribute to fatigue life.

The residual stress of shot-peened helical compression springs without prestressing is changed by a sufficiently high torsional threshold load or, alternatively, by prestressing. The residual stresses near the edges in the longitudinal and transverse directions of the wire and under the 45° direction of the load are redistributed, but in the 45° direction of the load the tensile stresses are maintained or even increased Fig. 12.12 (Muhr, 1970).

Curve 2 displays the residual stresses after the application of the relatively small torque. Curve 3 shows the residual stresses after shot-peening and presetting (Muhr, 1970).

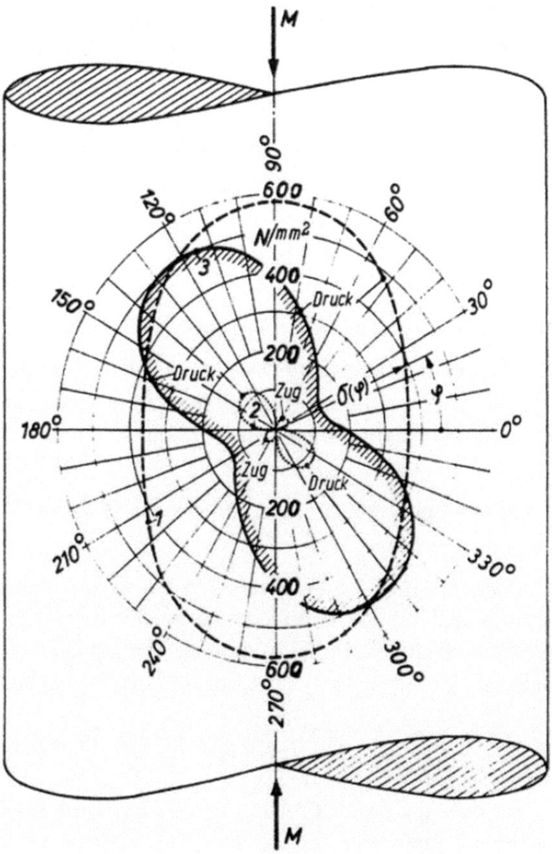

Fig. 12.12 Polar diagram of the residual stresses on the surface of the CrV wire for the valve springs. Curve 1 shows the residual stresses due to shot-peening

The explanation of the effect of shot peening on fatigue life is provided by the method (Findley, 1957, 1959) The method of fatigue evaluation for synchronous, harmonically oscillating stress tensor fields was presented above. On the surface of the wire and slightly below it there is an isotropic residual compressive stress $-\sigma_{sp}$, due to shot-peening of the material, where $\sigma_{sp} > 0$ is the pressure due to shot-peening. The twist moment of the spring wire causes an oscillating shear stress

$$\tau(t) = \tau_m + \tau_a \cdot \sin(\omega t). \tag{12.57}$$

As shown above, the acting stresses are tensile and compressive. There are two stress fields in the spring, the static one due to shot peening and the oscillating one due to spring operation. Both stress fields overlap during spring operation. Consider the overlapping of the stress fields on the surface of the wire. The surface of the wire is free of stress, so the normal stress disappears. Consequently, three components of the stress tensor remain on the surface layer of the wire. For simplicity, consider these stress fields from the Cartesian coordinate system, which is rotated at 45° to the axis of the wire. For clarity, the x-axis of this coordinate system makes an angle of 45° and the y-axis makes an angle of 135° to the wire tangent. The stress tensor is diagonal in this coordinate system. Two components of the tensor remain on the surface of the wire. We study the stresses in the principal stress coordinate system. The stress tensor in the surface layer in the coordinate system making angles ± 45° to the wire axis is:

$$\sigma = \begin{bmatrix} -\sigma_{sp} + \tau_m + \tau_a \cdot \sin(\omega t) & 0 & 0 \\ 0 & -\sigma_{sp} - \tau_m - \tau_a \cdot \sin(\omega t) & 0 \\ 0 & 0 & 0 \end{bmatrix}. \tag{12.58}$$

The two non-vanishing components are the principal stresses. These stresses are synchronous, harmonically oscillating functions. Equation (12.56) shows that the stress ratios of two principal stresses are different:

$$R_{\sigma.1} = \frac{\sigma_{sp} - \tau_m + \tau_a}{\sigma_{sp} - \tau_m - \tau_a}, \quad R_{\sigma.s} = \frac{\sigma_{sp} + \tau_m - \tau_a}{\sigma_{sp} + \tau_m + \tau_a}. \tag{12.59}$$

If the ratios of the amplitudes to the mean values are not all equal, the ratio of all principal stresses to a single parameter is missing (Sect. 12.5.1.2). In this case, the method (Sines, 1955) is applied. The equivalent stress for multiaxial fatigue is defined by Eq. (12.43). Using Eqs. (12.56, 12.43) becomes:

$$\sigma_{ar} = \sqrt{3}\tau_a - C_1\sqrt{2}\sigma_{sp}. \tag{12.60}$$

Because the factor C_1 and pressure σ_{sp} are both positive, the equaivalent stress reduces with the increasing of the residual stresses due to presetting. The fatigue life increases with the growth of the residual pressure in material. This explains from the mechanical viewpoint the effect of shot peening.

Shot peening of larger leaf springs and coil springs is performed in a wheel blast machine (longitudinal blast single-train continuous blast machine), with leaf springs typically treated only on the tension side. For uniform blasting of the entire coil spring surface, the springs are individually treated with a defined rotary motion and constant feed through the system. For smaller coil springs that cannot be shot-peened individually, shot-peening wheel drum systems are used. The interior of helical compression springs can only be shot-peened in the stretched state using conventional methods. Occasionally, these springs are blasted by inserting a free jet hose with a curved nozzle treated on the inside. To achieve optimum fatigue strength values when shot peening smaller helical compression springs in barrel machines, the filling quantity must be selected according to (Hirsch et al., 1979) in such a way that an approximately constant total spring wire area is present.

In the case of very small coil ratios and/or coil spacing, the blasting time should be increased by 50% to ensure adequate coverage of the most stressed inner surface of the coils. Springs with low dimensional stability must be treated with small shot sizes and low shot intensity. Hooking and coiling of coil springs in drum systems can result in almost not-blasted or weakly blasted surface areas. Shot trapped under the applied end coils should be removed to prevent spring fatigue failures or operational disturbances. Shot-blasted springs are very susceptible to corrosion and must be protected accordingly.

Shot peening without pre-stressing usually produces a symmetrical residual compressive stress state in the surface layer which has no preferred direction. Such a symmetrical stress state with not too high values is to be expected with alternating stresses on components with no or only low mean stress, otherwise residual stress reduction by superposition of compressive residual stresses and compressive load stresses is to be expected. For the same reason, the presetting is not used for components subjected to purely alternating stresses. Most springs are subject to swelling. For these springs, directional residual stresses by prestressing and shot peening under prestress are recommended.

Coil springs are partially hot set, then shot peened, and then re-set at room temperature. Final shot peening can result in greater spring relaxation than final setting. For some time, the effect of shot peening at elevated temperatures has been studied (Müller et al., 2006).

The influence of the defects with different sizes $\sqrt{4\boldsymbol{area}/\boldsymbol{\pi}}$ on the S–N-curves is displayed on the Figs. 12.13 and 12.14. These figures show the calculation of the fatigue life for the VDCrSiV material without peening (left) and peening surface treatment (right). The parameter $\mathbf{P_{SWT}}$ for normal stresses, Eq. (12.2a) is displayed as the ordinate of Fig. 12.13. Additionally, Fig. 12.14 uses the parameter $\mathbf{P_{SWT}}$ for shear stresses Eq. (12.3), as the ordinate.

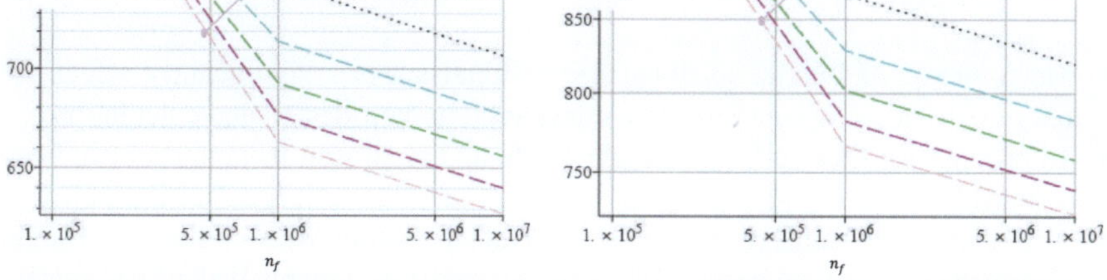

Fig. 12.13 Calculation of the influence of defects with the size $\sqrt{4area/\pi}$ on the S–N-curves for the VDCrSiV material without peening (left) and peening surface treatment (right). The parameter P_{SWT} Eq. (12.2a) is displayed as the ordinate

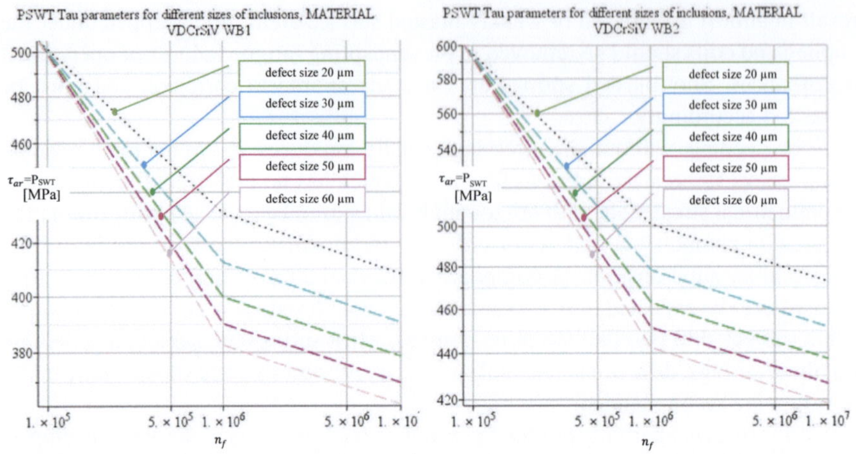

Fig. 12.14 Calculation of the influence of defects with the size $\sqrt{4area/\pi}$ on the S–N-curves for the VDCrSiV material without peening (left) and peening surface treatment (right). The parameter P_{SWT} Eq. (12.3) is displayed as the ordinate

Conclusions

The various common expressions that take into account the influence of the stress ratio are directly applicable. For the proposed unified propagation functions, the ranges for the stress load factor are extended by introducing the effective stress intensity. The solution leads to the factor ranges for effective stress intensity, effective mean stress intensity factor, and effective stress ratio.

The method developed in this chapter assumes a homogeneous stress state in the entire spring element. For the non-homogeneously loaded structural elements, the weak link concept is applied to consider fatigue. The weak link model is applied to evaluate the fatigue life of a helical spring later in Chap. 13.

Summary of Principal Results

- Mean stress, multiaxial loading, and residual stresses significantly affect the fatigue life of springs and must be considered in their design and calculation.
- To account for these effects, the stress-life and strain-life techniques have been extended.
- The traditional damage parameters (Walker, Smith-Watson-Tooper, Haigh, Bergmann) are special cases of the generalized or Bergmann-Walker parameter.
- The Bergmann-Walker parameter has two fitting constants and allows a wide fitting of experimental data.
- The fatigue crack growth per cycle method explicitly considers the mean stress.

- The effects of environment and corrosion on the fatigue life of springs are presented.
- The effect of shop-peening on fatigue is explained from the viewpoint of the sine model for multiaxial fatigue.

References

Abe, T., Furuya, F., & Matsuoka, S. (2001). Giga-cycle fatigue properties for 1800 MPa-class spring steels. *Transactions of the Japan Society of Mechanical Engineers, 67*(664), 1988–1995.

AiF 12287 N. (2002). Determination of fatigue strength and relaxation diagrams for highly stressed Helical compression springs. In C. Berger, & B. Kaiser (Eds.), Technical University.

AiF 15 064 N. (2010). Investigation of the endurance stroke strength of coil springs in the range of extremely high oscillating cycles. In C. Berger, & B. Kaiser (Eds.), Tech. Univ. Darmstadt.

AiF 5996. (1988). Investigation of the fatigue strength properties of spring steels under the influence of corrosion. In C. Berger & B. Kaiser (Eds.), Technical University Darmstadt.

Almaraz, G. M. D. (2008). Prediction of very high cycle fatigue failure for high strength steels, based on the inclusion geometrical properties. *Mechanics of Materials, 40*(8), 636–640. ISSN 0167-6636, https://doi.org/10.1016/j.mechmat.2008.03.001

Almen, J. O. (1943). Peened surfaces improve endurance of machined parts. *Metal Progress, 43*, 209–215.

Al-Obeid, Y. F. (2007) Three dimensional dynamic finite element analysis for shot peening mechanics. *36*(4), 681–689.

Angelova, D., Yordanova, R., Lazarova, T., & Yankova, S. (2014). On fatigue behavior of two spring steels. Part I: Wöhler curves and fractured surfaces. *Procedia Materials Science, 3*, 1453–1458.

ASM. (2008). Elements of metallurgy and engineering alloys. *ASM International*. ISBN 978-0-87170-867-0

Bannantine, J. A., Comer, J. J., & Handrock, J. L. (1990). *Fundamentals of metal fatigue analysis*. Prentice-Hall.

Bannantine, J. A., & Socie, D. F. (1989). A variable amplitude multiaxial fatigue life prediction method. In *Proceedings of the 3rd International Conference on Biaxial/Multiaxial Fatigue*, Stuttgart, Germany.

Basquin, O. H. (1910). The exponential law of endurance tests. *Proc. ASTM, 11*, 625.

Bergmann, J. W. (1983). Zur Betriebsfestigkeit gekerbter Bauteile auf der Grundlage der örtlichen Beanspruchung. Dissertation, Technische Hochschule Darmstadt.

Billaudeau, T., et al. (2004). Multiaxial fatigue limit for defective materials: Mechanisms and experiments. *Acta Materialia, 52*(13), 3911.

Boardman, B. (1982). Crack initiation fatigue—Data, analysis, trends and estimation, SAE Technical Paper 820682. 9.4271/820682

Boyce, B. L., & Ritchie, R. O. (2001). Effect of load ratio and maximum stress intensity on the fatigue threshold in Ti6Al4V. *Engineering Fracture Mechanics, 68*, 129–147. 9.1016/S0013-7944(00)00099-0

Branco, C. M., Radon, J. C., & Culver, L. E. (1976). Growth of fatigue cracks in steels. *Metal Science, 10*, 149–155.

Brown, M. W., & Miller, K. J. (1973). A theory for fatigue failure under multiaxial stress strain condition. *Proceedings Institute Engineering London, 187*, 745–755.

Brown, M.W. (1989). Analysis and design methods in multiaxial fatigue, in C. Moura Branco, L. Guerra Rosa (Eds.), *Advances in Fatigue Science and Technology*. Kluwer Academic Publishers.

Cahoon, J. R. (1972). An improved equation relating hardness to ultimate strength. *Metallurgical Transactions, 3*, 3040.

Chaboche, J. L. (1989). Constitutive equations for cyclic plasticity and cyclic viscoplasticity. *International Journal Plasticity, l5*(3), 247–302.

Chapetti, M. D. (2010). Prediction of threshold for very high cycle fatigue ($N > 10^7$ cycles). *Procedia Engineering, 2*, 257–264. https://doi.org/10.1016/j.proeng.2010.03.028

Coffin, L. F. (1954). A study of the effects of cyclic thermal stresses on a ductile metal. *Transaction ASME, 76*, 931–950.

Coffin, L. F. (1983). Overview of temperature and environmental effects on fatigue of structural metals. In J. J. Burke, & V. Weiss (Eds.), Fatigue. *Sagamore Army Materials Research Conference Proceedings* (Vol. 27). Springer.

Dang Van, K. (1993) Macro-micro approach in high cycle multiaxial fatigue. In *Advances in multiaxial fatigue* (pp. S. 120–130), ASTM Special Technical Publications.

DOT. (2005). Fatigue crack growth database for damage tolerance analysis, DOT/FAA/AR-05/15, Office of Aviation Research, Washington, U.S. Department of Transportation, Federal Aviation Administration D.C. 20591, Technical Reports page: http://actlibrary.tc.faa.gov

Dowling, N. E., Calhoun, C. A., & Arcari, A. (2009). Mean stress effects in stress-life fatigue and the Walker equation. *Fatigue and Fracture of Engineering Materials and Structures, 32*, 163–179. 9.1111/j.1460-2695.2008.01322.x

Elber, W. (1970). Fatigue crack closure under cyclic tension. *Engineering Fracture Mechanics, 2*(1), 37–45. 9.1016/0013–7944(70)90029.7

Elber, W. (1971). The significance of fatigue crack closure, STP486, Annual Meeting ASTM, Toronto, ASTM International. 9.1520/STP26680S

EN 13906-2:2013-09. (2013). Cylindrical helical springs made from round wire and bar—Calculation and design—Part 2: Extension springs; German version DIN EN 13906-2:2013, Beuth Verlag.

EN 13906-3:2014-06. (2014). Cylindrical helical springs made from round wire and bar—Calculation and design—Part 3: Torsion springs; German version DIN EN 13906-3:2014, Beuth Verlag.

EN 16984:2017-02. (2017). Disc springs—Calculation; German version DIN EN 16984:2016, Beuth Verlag.

ESIS 31. (2003). Biaxial-multiaxial fatigue and fracture. In A. Carpinteri, M. de Freitas, & A. Spagnoli (Ed.), ESIS Publication 31Elsevier Science Ltd. and ESIS.

Fatemi, A., & Socie, D. F. (1989). Multiaxial fatigue: Damage mechanisms and life, predictions. In C. Moura Branco, & L. Guerra Rosa (Eds.), *Advances in fatigue science and technology*, Kluwer Academic Publishers.

Fatemi, A., & Kurath, P. (1988). Multiaxial fatigue life predictions under the influence of mean stresses. *I. Engineering Material Technology, 110*, 380–388.

Fatemi, A., & Shamsaei, N. (2011). Multiaxial fatigue: An overview and some approximation models for life estimation. *International Journal Fatigure, 33*(8), 948–958.

Fatemi, A., & Socie, D. F. (1988). A critical plane approach to multiaxial fatigue damage including out-of-phase loading. *Fatigue Fracture Engineering Material Structure, 11*(149), 165.

Findley, W. N. (1957). Fatigue of metals under combined stresses. *Transactions of the American Society of Mechanical Engineers, 79*, 1337–1348.

Findley, W. N. (1959). A theory for the effect of mean stress on fatigue of metals undercombined torsion and axial load or bending. *Journal of Engineering for Industry, Bd. 81*, S. 301–306.

FKM. (2018). Rechnerischer Festigkeitsnachweis für Federn und Federelemente, Forschungskuratorium Maschinenbau E.V, Fachkreis Bauteilfestigkeit, FKM-Vorhaben Nr. 600, Heft 332 | 2018.

FKM. (2018). Rechnerischer Festigkeitsnachweis für Federn und Federelemente; FKM-Vorhaben Nr. 600, Heft 332; VDMA.

Forman, R. G., Kearney, V. E., & Engle, R. M. (1967). Numerical analysis of crack propagation in cyclic loaded structures. *Journal of Basic Engineering, Transactions ASME, D89*, 459–464.

Furuya, Y., Matsuoka, S., Abe, T., & Yamaguchi, K. (2002). *Scripta Materialia, 46*, 157–162.

Gadouini, H., Nadot, Y., & Rebours, C. (2008a). Influence of mean stress on the multiaxial fatigue behaviour of defective materials. *International Journal of Fatigue, 30*(2008), 1623–1633.

Gadouini, H., Nadot, Y., & Rebours, C. (2008b). Influence of mean stress on the multiaxial fatigue behavior of defective materials. *International Journal of Fatigue, 30*, 1623–1633.

Gadouini, H., & Nadot, Y. (2007). Two scale approach for tolerance fatigue design of components, In G. C. Sih, M. N. Abdelaziz, & T. Vu-Khanh (Eds.), *Particle and continuum aspects of mesomechanics*, ISTE Ltd.

Garwood, M. F., Zurburg, H. H., & Erickson, M. A. (1951). Correlation of laboratory tests and service performance, interpretation of tests and correlation with service. (pp. 1–77), ASM.

Geilen, M. B., Klein, M., & Oechsner, M. (2020a). On the influence of ultimate number of cycles on lifetime prediction for compression springs manufactured from VDSiCr class spring wire. *Materials, 13*, 3222. https://doi.org/10.3390/ma13143222

Geilen, M. B., Klein, M., & Oechsner, M. (2020b). A novel algorithm for the determination of walker damage in loaded disc springs. *Materials, 13*, 1661. https://doi.org/10.3390/ma13071661

Geinitz, V., Weiß, M., Kletzin, U., & Beyer P. (2011). Relaxation of helical springs and spring steel wires. In *56th International Scientific Colloquium*, Ilmenau University of Technology, 12–16 September 2011.

Gerber, W. (1874). Bestimmung der zulässigen Spannungen in Eisenkonstruktionen. Z.d. *Bayer. Architekten u. Ingenieurvereins, 6*, 101–110.

Glinka, G. (1987). Residual stress in fatigue and fracture: Theoretical analyses and experiments. In A. Niku-Lari (Ed.), *Advances in surfaces treatments* (Vol. 4, pp. 413–454), Residual Stresses, Pergamon Press.

Goodman, J. (1899). *Mechanics applied to engineering*. Longmans.

Haibach, E. (1992). Betriebsfeste Bauteile. Ermittlung und Nachweis der Betriebsfestigkeit, konstruktive und unternehmerische Gesichtspunkte. ISBN 978-3-540-54815-7, Springer-Verlag Berlin Heidelberg, https://doi.org/10.1007/978-3-642-84662-5

Haibach, E. (2013). *Betriebsfeste Bauteile: Ermittlung und Nachweis der Betriebsfestigkeit, konstruktive und unternehmerische Gesichtspunkte*. Springer-Verlag.

Haigh, B. P. (1915). Report on alternating stress tests of a sample of mild steel received from the British association stress committee. Report of the British Association for the Advancement of Science. London, 1916, 85th Meetings (pp. 163–170).

Hänel, B., Haibach, E., Seeger, T., Wirthgen, G., & Zenner, H. (2003). FKM Richtlinie—Rechnerischer Festigkeitsnachweis für Maschinenbauteile. Frankfurt a. M.: VDMA Verlag.

Harter, J. A. (1999). AFGROW users guide and technical manual: AFGROW for Windows 2K/XP". Version 4.0011.14, Air Force Research Laboratory.

Hartman, A., & Schijve, J. (1970). The effects of environment and load frequency on the crack propagation law for macro fatigue crack growth in aluminum alloys. *Engineering Fracture Mechanics, 1*(4), 615–631.

Hattingh, D. E. (1998). *The Fatigue Properties of Spring Steel*, Ph. D. thesis, University of Plymouth.

Haverd, D. G., & Topper, T. H. (1971). A criterion for biaxial fatigue of mild steel at low endurance. In *Proceedings First International Conference on Structural Mechanics in Reactor Technology* (pp. 413–432).

Hirsch, T., Starker, P., & Macherauch, E. Strahleigenspannungen. (1979). In Eigenspannungen und Lastspannungen, HTM Beiheft, Carl Hanser Verlag.

Hoffmann, S., Rödling, S., Eiber, M., & Decker, M. (2015). *Schwingfestigkeit Beschichteter Fahrzeugfedern unter Korrosionseinfluss, DVM-Bericht 1682, Federn im Fahrzeugbau*. DVM e.V.

Holmström, S., Pohja, R., Nurmela, A., Moilanen, P., & Auerkari P. (2013). Creep and creep-fatigue behaviour of 316 stainless steel. *Procedia Engineering, 55*, 160–164. ISSN 1877–7058, https://doi.org/10.1016/j.proeng.2013.03.236

SAE HS 1582. (1988). *Manual on design and manufacture of coned disk springs (Belleville Springs) and spring washers*, SAE International, Warrendale.

IGF 16873N. (2014). Investigation of material and manufacturing influences on the VHCF behaviour of coil compression springs. In M. Oechsner (Ed.).

Ince, A., Glinka, G. (2011). A modification of Morrow and Smith–Watson–Topper mean stress correction models. *Fatigue & Fracture of Engineering Materials & Structures, 34*, 854–867. https://doi.org/9.1111/j.1460-2695.2011.01577.x

Ince, A. (2012). Development of computational multiaxial fatigue modeling for notched components. Ph.D. thesis, University of Waterloo, Waterloo, ON, Canada, May 2012.

Juvinall, R. C., & Marshek, K. M. (2017). In *Fundamentals of machine component design* (6th ed.). Wiley & Sons.

Kaiser, B. (2006). Maßnahmen zur Schwingfestigkeitssteigerung von Federn—Grundlagen. *Verfahren, Wirkungen Und Beispiele, VDI-Berichte Nr 1972*(2006), 267.

Kaiser, B., Pyttel, B., & Berger, C. (2011). VHCF-behavior of helical compression springs made of different materials. *International Journal of Fatigue, 33*, 2011.

Kakuno, H., & Kawada, Y. (1979). A new criterion of fatigue strength of a round bar subjected to combined static and repeated bending and torsion. *Fatigue Enginering Materials Structuring, 2*, 229–236.

Kandil, F. A., Brown, M. W., & Miller, K. J. (1982). Biaxial low cycle fatigue of 316 stainless steel at elevated temperature. Book 280 (pp. 203–210). The Metal Society.

Karolczuk, A., & Macha, E. (2005). A review of critical plane orientations in multiaxial fatigue failure criteria of metallic materials. *International Journal of Fracture, 134*(3), 267–304.

Kirk, D. (1999). Shot peening. *Aircraft Engineering and Aerospace Technology: an International Journal, 77*(4), 349–361.

Klesnil, M., & Lukas, P. (1972). Effect of stress cycle asymmetry on fatigue crack growth. *Material Science and Engineering, 9*, 231–240.

Kobelev, V. (2017a). A proposal for unification of fatigue crack growth law. *6th International Conference on Fracture Fatigue and Wear. IOP Conference Series: Journal of Physics: Conference Series, 843,* 012022. https://doi.org/10.1088/1742-6596/843/1/012022

Kobelev, V. (2017b). Unification proposals for fatigue crack propagation laws. *Multidiscipline Modeling in Materials and Structures, 13*(2), 262–283. https://doi.org/10.1108/MMMS-10-2016-0052

Kondo, Y., Sakae, C., Kubota, M., & Kudou, T. (2003). The effect of material hardness and mean stress on the fatigue limit of steels containing small defects. *Fatigue and Fracture of Engineering Materials and Structures, 26*, 675–682. 9.1046/j.1460-2695.2003.00656.x

Kong, Y. S., Abdullah, S., Schramm, D., Omar, M. Z., Haris, S. M. (2019) Correlation of uniaxial and multiaxial fatigue models for automobile spring life assessment. *Experimental Techniques,* 44, 197–215.

Kwofie, S. (2001). An exponential stress function for predicting fatigue strength and life due to mean stresses. *International Journal of Fatigue, 23*(9), 829–836.

Kwofie, S., & Chandler, H. D. (2007). Fatigue life prediction under conditions where cyclic creep–fatigue interaction occurs. *International Journal of Fatigue, 29*(2007), 2117–2124.

Kwofie, S., & Zhu, M.-L. (2019). Modeling R-dependence of near-threshold fatigue crack growth by combining crack closure and exponential mean stress model. *International Journal of Fatigue, 122*(2019), 93–105.

Langer, B. F. (1937). Fatigue failure from stress cycles of varying amplitude. *Journal of Applied Mechanics, 59*, A160–A162.

Li, Y., Berger, C., & Kaiser, B. (2006). Korrosionsverhalten von Tellerfedern unter Komplexbeanspruchung, Ergebnisse aus dem AVIF-Forschungsprojekt, Untersuchungen zum Korrosionsverhalten von Tellerfedern und Tellerfedersäulen"VDI-Berichte Nr. 1972, s. 299.

Lohr, R. D., & Ellison, E. G. (1980). A simple theory for low cycle multiaxial fatigue. *Fatigue Engineering Material Structuring, 3*(1), 17.

Manson, S. S. (1953). Behavior of materials under conditions of thermal stress, NACA-TR-1170, National Advisory Committee for Aeronautics. Lewis Flight Propulsion Lab, Cleveland, OH, United States.

Manson, S. S., & Halford, G. R. (1981). Practical implementation of the double linear damage rule and damage curve approach for treating cumulative fatigue damage. *International Journal of Fractions, 17*, 169–192. 9.1007/BF00053519

Matek, W., Muhs, D., Wittel, H., & Becker, M. (1992). Roloff/Matek Maschinenelemente, Lehrbuch und Tabellenbuch, Vieweg, Braunschweig. ISBN 3-528-64028-6

Mayer, H., Schuller, R., Karr, U., Irrasch, D., Fitzka, M., Hahn, M., & Bacher-Höchst, M. (2015). Cyclic torsion very high cycle fatigue of VDSiCr spring steel at different load ratios. *International Journal of Fatigue, 70*, 322–327, https://dx.doi.org/10.1016/j.ijfatigue.2014.10.007

Mayer, H., Schuller, R., Karr, U., Irrasch, D., Fitzka, M., Hahn, M., & Bacher-Höchst, M. (2016). Mean stress sensitivity and crack initiation mechanisms of spring steel for torsional and axial VHCF loading. *International Journal of Fatigue, 93*, 309–317. https://doi.org/10.1016/j.ijfatigue.2016.04.017

McDiarmid, D. L. (1991). A general criterion for high cycle multiaxial fatigue failure. *Fatigue and Fracture of Engineering Materials and Structures, 14*(4), 429–453.

McDiarmid, D. L. (1993). Multiaxial fatigue life prediction using a shear stress based critical plane failure criterion in fatigue design. ESIS-16 (pp. 213–220). Mechanical Engineering Publications.

Meyer, N. (2014). Effects of mean stress and stress concentration on fatigue behavior of ductile iron. Theses and Dissertations, The University of Toledo Digital Repository, Paper 1782.

Miller, K. J., & Brown, M. W. (1984). Multiaxial fatigue—A brief review. *Fracture 84, Proceedings of the 6th International Conference on Fracture,* New Delhi (pp. 31–56).

Miner, M. A. (1945). Cumulative damage in fatigue. *Journal of Applied Mechanics, 12*, 159–164.

Morrow, J. (1968). Fatigue properties of metals, section 3.2. In *Fatigue design handbook*, Pub. No. AE-4. SAE.

Muhr, K. H. (1970). Einfluß von Eigenspannungen auf das Dauerschwingverhalten von Federn aus Stahl. *Stahl Und Eisen, 90*, 631–636.

Müller, E., Yapi, A., Rhönisch, B. (2006). Die Ausprägung von Druckeigenspannungen durch Spannungsstrahlen an Minibloc-Federn. VDI-Berichte Nr. 1972, 285.

Murakami, Y., & Isida, M. (1985). Analysis of stress intensity factors and stress field at contact point for surface cracks of arbitrary shape (in Japanese). *Transactions Japan Society Mechanics Engineering A, 51*(464), 1050–1056.

Murakami, Y., Kodama, S., & Konuma, S. (1988). *The Japan Society of Mechanical Engineering, 54A*, 688–695.

Murakami, Y., Takada, T., & Toriyama, T. (1998). *International Journal of Fatigue, 16*(9), 661–667.

Murakami, Y. (1989). Effects of small defects and nonmetallic inclusions on the fatigue strength of metals. *JSME international Journal Ser. 1, Solid Mechanics, Strength of Materials, 32*(2), 167–180. 9.1299/jsmea1988.32.2_167

Murakami, Y. (2001). Mechanism of fatigue failure in ultralong life regime. In *Proceedings of the International Conference on Fatigue in the Very High Cycle Regime* (pp. 11–22).

Murakami, Y. (2003). High and ultrahigh cycle fatigue, In I. Milne, R. O. Ritchie, B. Karihaloo (Eds.), *Comprehensive structural integrity* (Vol. 4), Cyclic Loading And Fatigue, Elsevier.

Murakami, Y., & Beretta, S. (1999). Small defects and in homogeneities in fatigue strength: experiments, models and statistical implications. *Extremes, 2*(2), 123–147. 9.1023/A:1009976418553

Murakami, Y., Nagata, J., & Matsunaga, H. (2006). In *Proceedings of the 9th International Fatigue Congress* (Fatigue 2006), Elsevier.

Nakamura, T., Kaneko, M., Noguchi, T., & Jinbo, K. (1998). *Transactions of the Japan Society of Mechanical Engineers, 64*(623), 68–73.

Neuber, H. (1961a), *Theory of notch stresses: principles for exact calculation of strength with reference to structural form and material.* United States Atomic Energy Commission, US Office of Technical Services, Oak Ridge, TN.

Neuber, H. (1961b). Theory of stress concentration for shear-strained prismatical bodies with arbitrary nonlinear stress–strain law. *Journal of Applied Mechanics*, 28, 544–550.

Newman, J. C. (1984). A crack opening stress equation for fatigue crack growth. *International Journal of Fracture, 24*(3), R131-135.

Niku-Lari, A. (1981). Methode De La Fleche Methods De La Source Des Contraintes Residuelles, Conference Proceedings: ICSP-1, (pp. 237–247).

Niku-Lari, A. (1996). An Overview of Shot—Peening, Conference: (Bophal) International Conferen.69p8ce on Shot Peening and Blast Cleaning.

Palmgren, A. (1924). Die Lebensdauer von Kugellagern. *Zeitschrift des Vereines deutscher Ingenieure, 68*, 339–341.

Pang, J. C., Li, S. X., Wang, Z. G., & Zhang, Z. F. (2013). General relation between tensile strength and fatigue strength of metallic materials. *Materials Science and Engineering A, 564*(2013), 331–341. https://doi.org/10.1016/j.msea.2012.11.103

Papadopoulos, I. V. (1997). A comparative study of multiaxial high-cycle fatigue criteria for metals. *International Journal of Fatigue, 19*(3), 219–235.

Papadopoulos, I. V. (2001). Long life fatigue under multiaxial loading. *International Journal of Fatigue, 23*(10), 839–849.

Pineau, A. (1989). Mechanisms of creep-fatigue interactions. In C. M. Branco, & L. G. Rosa (Eds.), *Advances in fatigue science and technology*. NATO ASI Series, (Vol. 159). Springer. https://doi.org/10.1007/978-94-009-2277-8_12

SAE AE-22. (1997). In *SAE fatigue design handbook*, SAE International, Warrendale.

SAE HS 788. (1980). *Manual on Design and Application of Leaf Springs*, SAE International, Warrendale.

Schnattinger, H., & Beste, A. (1995). PKW-Schraubenfedern, Auslegung aus der Sicht der Betriebsfestigkeit, Materialprüfung, 37(6).

Schütz, W. (1967a). Über eine Beziehung zwischen der Lebensdauer bei konstanter zur Lebensdauer bei veränderlicher Beanspruchungsamplitude und ihre Anwendbarkeit auf die Bemessung von Flugzeugbauteilen. *Z. f. Flugwissenschaften, 15*(11), 407–419.

Schütz, W. (1967) Über Eine Beziehung zwischen der Lebensdauer bei Konstanter und bei Veränderlicher Beanspruchungsamplitude und Ihre Anwendbarkeit auf Die Bemessung von Flugzeugbauteilen; Zeitschrift für Flugwissenschaften und Weltraumforschung/Journal of flight sciences and space research/hrsg. von Deutschen Gesellschaft für Luft- und Raumfahrt e.V. u. d. Deutschen Forschungs- und Versuchsanstalt für Luft- und Raumfahrt e.V, 15 H. 11, S. 407/419.

Shelton, S. M., & Swanger, W. H. (1935). Fatigue properties of steel wire. *Journal of Research of the National Bureau of Standards, 14*, Research Paper RP754.

Shiozawa, K., Lu, L., & Ishihara, S. (2001). *Fatigue and Fracture of Engineering Materials and Structures, 24*, 781–790.

Sines, G. (1955). Failure of materials under combined repeated stresses with superimposed static stresses. NACA TN 3495.

Smith, K. N., Watson, P., & Topper, T. H. (1970b). A stress-strain function for the fatigue of metals. *Journal of Materials, 5*, 767–778.

Smith, K. N., Watson P., & Topper T. H. (1970). A stress-strain function for the fatigue of metals. *Journal of Materials, ASTM, 5*(4), 767–778.

Socie, D. F., & Morrow, J. D. (1980). Review of contemporary approaches to fatigue damage analysis. In J. J. Burke & V. Weiss (Eds.), *Risk and failure analysis for improved performance and reliability*, (pp. 141–194). Plenum Publication Corporation, New York, NY.

Socie, D. (1987). Multiaxial fatigue damage models. *Journal of Engineering Materials and Technology, 109*(4), 293–298.

Socie, D., & Marquis, G. (1999). *Multiaxial fatigue*. Society of Automotive Engineers, Warrendale, PA, USA.

Soderberg, C. R. (1930). *ASME Transactions 52, APM-52–2*, 13–28.

Sorensen, A. (1969). A general theory of fatigue damage accumulation. *Journal Basic Engineering, 91*(1), 1–14. https://doi.org/10.1115/1.3571021

Sornette, D., Magnin, T., & Brechet, Y. (1992). The physical origin of the coffin-manson law in low-cycle fatigue. *EPL (Europhysics Letters), 20*(5).

Takahashi, K., & Murakami, Y. (2002). Torsional fatigue of specimens containing an initial small crack introduced by tension compression fatigue. (Effects of shear mean stress and tensile or compressive mean stress). *Nippon Kikai Gakkai Ronbunshu, A Hen/Transactions of the Japan Society of Mechanical Engineers, Part A 68*(4), 645–52.

Varvani-Farahani, A. (2000). A new energy-critical plane parameter for fatigue life assessment of various metallic materials subjected to in-phase and out-of-phase multiaxial fatigue loading conditions. *International Journal Fatigue, 22*(4), 295–305.

Wagner L., Ed. (2003). *Shot Peening, Proceedings of the 8th International Conference on Shot Peening (ICSP-8) in Garmisch-Partenkirchen,* Germany, 16-20 2002, WILEY-VCH Verlag GmbH & Co. KGaA, Weinheim.

Walker, K. (1970). The effect of stress ratio during crack propagation and fatigue for 2024-T3 and 7075-T6 aluminum. In *Effects of environment and complex load history on fatigue life*, ASTM STP 462 (pp. 1–14), ASTM.
Wang, Q. (1999). Effect of shear lag on buckling of thin-walled members with any cross-section. *Communications in Numerical Methods in Engineering, 15*, 263–272.
Wheeler, O. E. (1972). Spectrum loading and crack growth. *Journal of Basic Engineering, 94*, 181–186.
Xu, Q., & Lu, Z. (2020). Modeling of creep deformation and creep fracture. Strength of Materials. IntechOpen. https://doi.org/10.5772/intechopen.89009
Xue, L., Shang, D.-G., Li, D.-H., Li, L.-J., & Y. Hui X. J. (2019). Online multiaxial fatigue damage evaluation method by real-time cycle counting and energy-based critical plane criterion. *Fatigue Fracture of Engineering Materials Structuring*, 1–15. 9.1111/ffe.13192

Chapter 13
Failure Analysis Based on Weakest Link Concepts

Abstract This Chapter begins with the study of statistical effects in failure analysis of springs. The procedure for dimensional analysis of helical spring failures is presented. The dimensional analysis spring element is examined using the weakest link concept. This concept is applicable to the extremely brittle material in which the failure of one link causes the complete destruction of the entire structural element. The stress variation over the wire surface is considered for the analytical calculation of Weibull failure probability. Chapter provides analytical formulas for failure probability of helical springs. The derived solution clarifies the effects of spring index and wire diameter on fatigue life. The methodology is valid for different types of springs or basic structural elements. This Chapter is the third section of the third part, which studies the life cycle of elastic elements.

Keywords Fatigue probability · Weibull failure probability · Dimensional analysis of failure

13.1 Evaluation of Failure Probability of Springs

Well known, the stress variation on the surface of the spring depends on the spring index (EN 13906, 2013). However, the customary estimations of fatigue life of cyclically loaded helical springs does not account accurately the stress variation on the wire surface. The known experiments report substantial dependence of fatigue life upon the spring index and the diameter of wire (Kaiser, 1981; Reich, 2016). On one side, the experimental results indicate the evident reduction of fatigue life of springs with thicker wire in comparison to the equally stressed springs made of the thinner wire. On the other side, the springs with the same stress maximum but different spring indices demonstrate different statistical expectation of fatigue life (Reich & Kletzin, 2013). The springs with higher indices suffer earlier breakage that the equally loaded

V. Kobelev, *Fundamentals of Springs Mechanics*, University of Siegen, Faculty IV: School of Science and Technology. Paul-Bonatz-Str. 9–11, D-57076 Siegen. E-mail: Homepage: https://www.mb.uni-siegen.de/kobelev/

V. Kobelev, *Fundamentals of Springs Mechanics*,
https://doi.org/10.1007/978-3-031-58584-5_13

springs with lower spring indices. The explanation of this behavior is the following. The springs with higher indices are equally stressed over the surface of the wire. The stress on the wire surface of springs with lower indices varies considerably. The maximal stress of both types of the springs is the same, but the springs with higher spring indices are higher stressed over the total surface. That is the defects on the outer regions of the spring wire provoke more damage in the springs with higher indices. The existing methodology does not explain accurately these dependencies.

Consequently, one of aims of this Chapter is the quantitative analytical description of the scaling and stress-gradient effects in cyclic failure analysis of helical springs. The second aim is to provide the closed form solution for the estimation of these effects.

The essence of the proposed approach is the following:

- fatigue crack propagation is based on the linear elastic fracture mechanics;
- advancement of fatigue crack per unit cycle as a function of the amplitude of stress intensity factor;
- survival probability is based on the weakest link concept;
- fatigue life is accounted by means of the Weibull distribution;
- failure probability as the function of stress and of number of cycles is derived for helical springs with circular wire of the constant mean diameter.

The results of the actual article provide the method for estimation some important parameters (stress ratio, spring index) on the fatigue behavior of the springs. The method can be adapted for the other types of springs and structural elements.

The fatigue damage, which is studied in the Chap. 8, must account the influence of spatial inhomogeneity of stress field. With this enhancement, the fatigue life of the non-homogeneously stressed structural element could be studied. The addressed problem is investigated with the concept of weakest link. The weakest link concept has been used for material failure problems by many researchers (Wormsen et al., 2007). Traditionally, the weakest link concept is adopted for different shape coefficients of the Weibull distribution. This approach permits to calculate the probability distribution of the fatigue life for helical springs with different spring indices and different stress levels. The assumed statistical model of the weakest link theory is based on the Weibull probability distribution of failure. The theory describes the statistical distribution of strength properties of fragile materials and materials with low ductility. This assumption describes exactly the materials for springs, because of the absence of plastic region due to extraordinary manufacturing of helical springs.

The fatigue life of a large diameter wire rope was estimated from test data for small diameter rope in Chien et al., (1988). Dimensional analysis and the technique of interpolation and extrapolation were employed. The method was applied for analysis of size effects on the strength of wires and for prediction of fatigue strength.

A fatigue crack is initiated in high-stressed helical springs under moderate loads by the randomly distributed material defects. The similar crack initiation occurs in most of brittle materials. In high-cycle, fatigue tests to determine a fatigue life the different theories are used.

13.2 Weakest Link Concepts for Homogeneously Loaded Elements

13.2.1 Failure Probability of Elements with Distributed Defects

The theory of weakest link is based on statistically distributed defects in a material per volume unit. The defects are non-homogeneities, inclusions, cracks, and precipitates. In a specific volume where the most dangerous defect or the weakest link exists, the crack initiates. Fatigue cracks propagate independently from each other in different areas of the structural element. The effect of load and cross-sectional area with the fatigue life is accounted by means of the Weibull distribution. The Weibull distribution presumes, that for identical elements (at the macroscopic scale) loaded by time dependent but equal stresses, the logarithm of the number of cycles n to crack initiation is a random variable with a given probability density distribution. “The most harmful defect” (“weakest link”) exhibits different features in a set of successive specimens. Therefore, crack initiation occurs under a different number of cycles (Wormsen & Härkegård, 2004).

The mathematical formulation must be reconsidered for heterogeneous stresses in order to consider the spatial stress gradients. The considered structural element is divided into several subdomains. In each subdomain, the stresses are different. The probability that a failure of the whole element will not occur within the lifetime interval $0..N$ means that any elementary subdomain not failed weakest link concept). Indicating that $P_S^{(i)}$ is the survival probability means that the sub-domain with the number (i) is not failed within the number of cycles $0..N$. The survival probability for the whole structural element (or component) is the product of all the individual probabilities $P_S^{(i)}$:

$$P_S = \prod_{i=1}^{k} P_S^{(i)}.$$

The product rule is valid because no interaction between dangerous defects indifferent subdomains is assumed.

The dangerous, or critical defect causes the destruction of the whole spring. Assuming that the defects are uniformly distributed over the volume of the whole volume V of structural element, the survival probability is the integral:

$$P_{S.V} = \exp\left(-\frac{1}{V_o}\int_V p_f(\sigma)\mathrm{d}V\right). \tag{13.1}$$

In Eq. (13.1) we use the following values:

$p_f(\sigma)$ the function called by Weibull "risk of rupture", which represents the fatigue probability density (see Chap. 8).

V_o an average volume that contains one critical defect.

Alternatively, if the dangerous defects are uniformly distributed over the surface A_e of the structural element, the survival probability is the integral:

$$P_{S.A} = \exp\left(-\frac{1}{A_0}\int\limits_{A_e} p_f(\sigma)\mathrm{d}A\right). \tag{13.2}$$

where A_0 is an average surface element that contains one critical defect.

Let the number of loads during the operation of spring is N. Our task is to evaluate the failure probability after this given number of cycles in operation. For this purpose we use the cumulative distribution functions. The Weibull form of failure probability reads:

$$P_{F.A}(N) = 1 - P_{S.A}(N), \quad P_{F.V}(N) = 1 - P_{S.V}(N). \tag{13.3}$$

What Eq. (13.1) or Eq. (13.2) should be used for computations depends on the material and application of the spring. For example, the springs for use in aggressive of corrosive media suffer failure due to corrosion on the outer surface. For these springs source of the defects is the surface and correspondingly the Eq. (13.2) should be used for computations. Otherwise, for springs those operate in oil or inert gases the frequent source of defects is the inclusions or flaws inside the material. These defects are distributed evenly over the whole volume of spring and therefore appropriate estimation is (13.1). Two traditional forms of the Weibull function $p_f(\sigma)$ are commonly used:

$$p_f(\sigma) = \left(\frac{\sigma}{S_W}\right)^{m_W} \quad \text{or} \quad p_f(\sigma) = \left(\frac{\sigma - S_{sh}}{S_W}\right)^{m_W}. \tag{13.4}$$

The parameters S_{sh}, S_W, m_W are known as the stress shift, stress scale and shape parameters, respectively. The forms (13.4) are appropriate for the immediate failure. The Weibull function must be generalized for cyclic load and fatigue estimations. This is the subject of the next paragraph.

13.2.2 Application of Weakest Link for Fatigue

The application of the concept of weakest link for fatigue was established in the case of fatigue processes in a uniaxial loaded component with a homogenous stress distribution. The failure probability P_F is a certain function of the stress amplitude σ_{ar} and the number of cycles in operation of the spring N.

When the required number of cycles is given, then the failure probability could be determined as the function of stress amplitude σ_{ar}. Otherwise, if the failure probability and fatigue life are prescribed by design requirements, the allowable stress amplitude σ_{ar} could be determined from this equation.

Consider a spatially inhomogeneous stress field within the structural element Ω. The spatially inhomogeneous stress field results in an inhomogeneous fatigue damage field, such that the probability of failure also depends on the coordinates. However, for design purposes the probability of failure of the structural element is required. The concept of weakest link permits to determine the fatigue calculations of the structural element with a spatially varying stress field.

For this purpose, at each point of the structural element, Ω consider the infinitesimally small element $d\Omega$. The stress in the element $d\Omega$ is constant. Therefore, for the element $d\Omega$ the mean cycles to failure N could be computed using the equations of Chap. 8. The mean cycles to failure is the function of the stress amplitude σ_{ar} in the element $d\Omega$. The dependence of maximal cycles to failure of the element $d\Omega$ is

$$n_f = n_f(\sigma_{ar}).$$

The cycles to failure of the small element is given by the equations of Chaps. 11 and 12. Particularly, the simplest form expresses the intermediate Wöhler-Basquin regime. The number of cycles to failure n_f on the effective stress amplitude σ_{ar}, according to Eq. (12.1), in the homogeneously stressed element reads:

$$n_f = \frac{1}{2}\left(\frac{\sigma_{ar}}{\sigma_f\prime(R_\sigma)}\right)^{1/b(R_\sigma)} \equiv \lambda_\sigma \sigma_{ar}^{-p_\sigma}. \tag{13.5}$$

The "risk of rupture" function of the small element is assumed to be analogous to Weibull expression $p_f(\sigma)$, Eq. (13.4). The essential difference is that for the cyclic loading the "risk of rupture" $p_h(N, \sigma_{ar})$ depends on effective stress amplitude σ_{ar} and also on the mean cycles to failure of the heterogeneously stressed structural element.

Thus, the required number of cycles N and operation stress are given, and our aim is to determine the failure probability of the structural element. The following simplest form of Weibull approximation as the "Weibull:risk of rupture" function is used for subsequent computations:

$$p_h(N, \sigma_{ar}) = \begin{cases} 0 & \text{for } N < N_F, \\ \left(\frac{N-(1-k_f)n_f}{k_f n_f}\right)^m & \text{for } N_L < N < N_L, \\ 1 & \text{for } N > N_L. \end{cases} \tag{13.6}$$

The values in Eq. (13.6)

$$N_F = N_F(\sigma_{ar}), \quad N_L = N_L(\sigma_{ar})$$

depend on the maximal cycles to failure $n_f = n_f(\sigma_{ar})$. The value $N_F = N_F(\sigma_{ar})$ is the number of cycles to failure of the first element and $N_L = N_L(\sigma_{ar})$ is the number of cycles to failure of the last element. At first, the function in Eq. (13.6) corresponds to 100% of "risk of rupture" probability and consequently $N_L = n_f$, according to Eq. (13.5). But some element will fail even at the lower number of cycles. To account these first occurrence of failure, we introduce the extent of the region $N_F \ldots N_L$. The extent linearly depends on the value n_f, such that

$$N_F = \left(1 - k_f\right) n_f.$$

The extent constant equals to:

$$k_f = 1 - N_F / N_L.$$

The extent constant k_f characterizes the range of the failure region from the first occurrence of failure to the final failure of the last element. For the given operation stress, there is no failure prior to N_F operational cycles and all elements will guaranteed fail after N_L cycles. In other words, we test a collection of specimens. The first specimen from the collection breaks exactly at N_F cycles. At the cycle count N_L all specimens from the collection are broken. With Eq. (13.5), these numbers are:

$$N_F = \left(1 - k_f\right) \lambda_\sigma \sigma_{ar}^{-p_\sigma}, \quad N_L = \lambda_\sigma \sigma_{ar}^{-p_\sigma}.$$

The proposed "risk of rupture" function (13.6) is the simplest form, adopted form of second Weibull function for fatigue computations.

We apply the results of testing of collection of specimens to the whole structure. The structure is assembled from several structural elements. If one element fails, fails the whole structure. Thus, the survival probability of heterogeneously stressed structure depends on load cycles N for the structural element. The survival probability of each element is the decreasing function of cycles. The failure probabilities $P_{F.A}(N)$, $P_{F.V}(N)$ increase with the cycle number. Finally, we multiply all risk of rupture of all single element and get the risk of rupture of the whole structure:

$$P_{S.A}(N) = \exp\left(-\frac{1}{A_0} \int\limits_{A_e} p_h(N, \sigma_{ar}(\mathrm{d}A))\mathrm{d}A\right), \quad P_{F.A}(N) = 1 - P_{S.A}(N), \tag{13.7}$$

$$P_{S.V}(N) = \exp\left(-\frac{1}{V_0} \int\limits_{V} p_h(N, \sigma_{ar}(\mathrm{d}V))\mathrm{d}V\right), \quad P_{F.V}(N) = 1 - P_{S.V}(N). \tag{13.8}$$

With the expressions (13.8) and (13.9) we estimate the failure and survival probabilities of the complex system, based on failure probabilities of its elements for a give number of operational cycles. Evidently, that the failure probability increases with the stress level in the structural element as well. As a result, the longer structural element is in service and higher the stress, the significantly the failure probability is. It is evident as well, that the higher number of elements increase the probability of failure of the whole structure, if each failed element leads to complete disaster. Exactly this behavior is typical for springs. If the large enough crack occurs somewhere in the spring inner volume or on the surface, it spreads over the whole section of the spring and causes the breakage of the spring. If there are no possibility to observe the spring, to control the state of spring and to repair the local breakages, the theory of weakest link must be applied. This theory allows the conservative, pessimistic evaluation of failure probability of a complex element and permits the evaluation of design parameters on the failure probability. We apply this theory to springs.

13.3 Analysis of Springs with Weakest Link Method

13.3.1 Failure Probability of Helical Springs

The failure probability as the function of stress and of number of cycles will be derived below for helical springs in closed analytical form. The springs with the constant mean diameter $D = 2R$ that made of the circular wire with the diameter $\mathrm{d} = 2r$ are considered.

The spring index for helical springs is a measure of coil curvature of the wire. For springs of circular section wire the spring index is the ratio of mean coil diameter to wire diameter.

The equations assume that the spring coil is loaded centrally along the spring axis, which remains straight and the ends can rotate about the spring axis relative to one another, such that no bending of wire occurs.

Owing to different material properties on the surface and in volume due to manufacturing process, the failure probability is commonly considered for the surface but not over the volume of the structural element. The following consideration is typically appropriate to the failure of springs. The predominant source of failure is the skin of the wire but not its core regions. This assumption is natural for the helical springs, because the stresses decline rapidly from surface of wire to its center.

For proper application of the weak link concept, the accurate study of stresses on the surface of the wire is essential. The standard reference sources (EN 13906-1:2013-11, 2013; SAE, 1997) provide the stress correction factor that accounts the stress conditions at the inside of the coil by using a "corrected stress" (Chaps. 1 and 2). The stress as function of polar angle is necessary for the subsequent estimation of failure probability based on weak link concept. This auxiliary problem for the stress on the surface as function of the polar angle was treated in Chap. 2. The expressions provided for the stress on the surface of the wire as function of the polar angle φ were provided.

All essential formulas are prepared now and we can precede the estimation of failure probabilities.

13.3.2 Effect of Spring Index on Immediate Failure

The developed formulas allow the estimation of failure probability of the springs. At first consider the case of failure due to surface defects. The survival probability is estimated with the aid of Eq. (13.2). The Weibull function is given by the second Eq. (13.4). The evaluation leads to the following expression of the survival probability of the helical spring with the spring index w (Kobelev, 2017):

$$P_{S.1} = \exp\left(-\frac{rL_W}{A_0}\int_0^{2\pi} p_f(\tau_s(w,\varphi))\mathrm{d}\varphi\right)$$
$$= \exp\left(-\frac{rL_W}{A_0}\int_0^{2\pi}\left(\frac{\tau_s(w,\varphi)}{S_W}\right)^{m_W}\mathrm{d}\varphi\right). \tag{13.9}$$

The shear stress $\tau_s(w,\varphi)$ is the function of the spring index w and of the polar angle φ, Eq. (2.36). The outer surface of spring wire is

$$A_e = 2\pi r L_W.$$

The expression for survival probability could be easily determined also for different values of m_W:

$$P_{S.1} = \exp\left(-\frac{A_e}{A_0}\frac{\tau_b^2}{S_W^2}k_5(w,m_W)\right),$$

$$k_5(\infty,m_W) = \lim_{w\to\infty} k_5(w,m_W) = 1.$$

The factor $k_5(w,m_W)$ is the stress correction factor for surface defects. Performing integration in (13.9) with the expression for shear stress on the outer surface of the wire from Eq. (2.9), demonstrates the formula for correction factor for surface defects:

$$k_5(w,m_W) = \frac{4096\,w^2 + 961\,m_W^2 + 880\,m_W}{4096\,w^2}.$$

This correction factor is applicable for stress evaluation in fatigue problems, as already discussed in Chap. 10. This factor averages the failure probability over the outer surface. The factor k_5 is analogous to the empirical averaging factors $\overline{k_2}(w), \overline{k_3}(w)$, which were used in Chap. 10 for fatigue estimations. Conversely, the traditional correction factors $k_i(w, m_W), i = 1..4$, were referred Chaps. 1 and 2 defects. The traditional correction factors deliver the local, maximal value of stress in a single point on the inner side of the section. The traditional factors underestimate the risks of fatigue failure for the springs with higher spring ratios with the evenly stressed surfaces.

For comparison, consider the straight rod with circular cross-section. The cross-section of the straight rod is the same as the cross-section of the helical spring. The torsion moment in the cross-section of the straight rod is again the same as the torsion moment in the cross-section of the helical spring. As already mentioned, the straight rod possesses the constant stress over the surface. The shear stress on the surface of the rod is the same as basic stress σ_b of the helical spring. Obviously, the shear stress for the rod equals to the shear stress on the surface of helical spring with the infinite spring index w. Consequently, the survival probability of the straight rod is equal to the survival probability of the helical spring with infinite spring index and the same length and diameter of the wire. If follows from the Eq. (13.9) that:

$$P_{R.1} = \lim_{w\to\infty} P_{S.1} = \exp\left(-\frac{A_e}{A_0}\frac{\tau_b^2}{S_W^2}\right). \tag{13.10}$$

The ratio of survival probabilities of the spring (13.9) and the rod (13.10) in case of immediate rupture reads:

$$\alpha_1 = \frac{P_{S.1}}{P_{R.1}} = \exp\left(-\frac{A_e}{A_0}\frac{\tau_b^2}{S_W^2}(k_5(w, m_W) - 1)\right). \tag{13.11}$$

At second consider the volume defects that cause the immediate failure of the spring. The intensity of shear stress $\tau_q(\rho, \varphi)$ for the circular wire of radius r is given by Eq. (2.30). The following expressions of the survival probability of the helical spring with the spring index w are valid in this case (Kobelev, 2017):

$$\begin{aligned} P_{S.2} &= \exp\left(-\frac{L_W}{V_0}\int_0^{2\pi}\int_0^{r} p_f\big(\tau_q(\rho,\varphi)\big)\rho\,\mathrm{d}\rho\,\mathrm{d}\varphi\right) \\ &= \exp\left(-\frac{L_W}{V_0}\int_0^{2\pi}\int_0^{r}\left(\frac{\tau_q(\rho,\varphi)}{S_W}\right)^{m_W}\rho\,\mathrm{d}\rho\,\mathrm{d}\varphi\right). \end{aligned} \tag{13.12}$$

The volume of spring wire is:

$$V = \pi r^2 L_W.$$

For an arbitrary value of the exponent m the integrals are:

$$P_{S.2} = \exp\left(-\frac{V}{V_0}\frac{\tau_b^2}{S_W^2}k_6(w, m_W)\right),$$

The factor $k_6(w, m_W)$ is the stress correction factor for volume defects. Integration in (13.12) with the expression for shear stress on the outer surface of the wire from Eq. (2.9), demonstrates the formula for correction factor for surface defects:

$$k_6(w, m_W) = \frac{961\, m_W^3 + 1414\, m_W^2 + 4096\, m_W w^2 - 2168\, m_W + 16384\, w^2}{2048(4 + m_W)(2 + m_W)w^2},$$

$$k_6(\infty, m_W) = \lim_{w\to\infty} k_6(w, m_W).$$

The survival probability of the straight rod with the same torque, length and diameter of wire is:

$$P_{R.2} = \lim_{w\to\infty} P_{S.2} = \exp\left(-k_6(\infty, m_W)\frac{V}{V_0}\frac{\sigma_b^2}{S_W^2}\right). \quad (13.13)$$

The basic stress τ_b of the helical spring and the rod are equal. From Eqs. (13.12), (13.13) it follows the ratio of survival probabilities of the spring and the rod

$$\alpha_2 = \frac{P_{S.2\cdot}}{P_{R.2}} = \exp\left(-\frac{V}{V_0}\frac{\tau_b^2}{S_W^2}[k_6(w, m_W) - k_6(\infty, m_W)]\right). \quad (13.14)$$

Important, that the dependence of wire diameter is different in both considered cases.

13.3.3 Effect of Spring Index on Fatigue Life

The falling out of helical spring during its operation depends primary upon the failures in the thin layer on the surface of the spring. The distribution of stress over the surface of the spring Chap. 2, Eq. (13.6) is essential for the calculation of failure and survival probabilities (13.7) and (13.8).

At third consider again the surface defects. The shear stress $\tau_s(w, \varphi)$ is the function of the spring index w and of the polar angle φ, Eq. (2.36). The survival probability in this case is the function of stress on the surface and load cycle (Kobelev, 2017):

$$P_{S.3} = \exp\left(-\frac{rL_W}{A_0}\int_0^{2\pi} p_h(N, \tau_s(w, \varphi))\mathrm{d}\varphi\right)$$

$$= \exp\left(-\frac{rL_W}{A_0}\int_0^{2\pi}\left[\frac{N}{N_0}\left(\frac{\tau_s(w,\varphi)}{\sigma_0}\right)^{p_\sigma}\right]^{m_W}\mathrm{d}\varphi\right). \tag{13.15}$$

Thus, the survival probability reads:

$$P_{S.3} = \exp\left(-\frac{A_e}{A_0}\left(\frac{N\tau_b^p}{N_0\sigma_0^p}\right)^m k_1(w, p_\sigma m_W)\right).$$

The stress on the surface σ_c depends on basic stress σ_b and spring index w. In the limit case of infinite spring index, the Eq. (13.15) provides the expression for the failure probability of the straight rod:

$$P_{R.3} = \lim_{w\to\infty} P_{S.3} = \exp\left(-\frac{A_e}{A_0}\frac{\tau_b^2}{\sigma_0^2}\frac{N}{N_0}\right). \tag{13.16}$$

The ratio of both formulas (13.21) and (13.16) consequently is:

$$\alpha_3 = \frac{P_{S.3}}{P_{R.3}} = \exp\left(-\frac{A_e}{A_0}\frac{\tau_b^2}{S_W^2}(k_1(w, p_\sigma m_W) - 1)\right). \tag{13.17}$$

At fourth determine the survival probability of spring that undergoes due to the volume defect (Kobelev, 2017):

$$P_{S.4} = \exp\left(-\frac{L_W}{V_0}\int_0^{2\pi}\int_0^r p_h\big(N, \tau_q(\rho,\varphi)\big)\rho\mathrm{d}\rho\mathrm{d}\varphi\right) =$$

$$= \exp\left(-\frac{L_W}{V_0}\int_0^{2\pi}\int_0^r\left[\frac{N}{N_0}\left(\frac{\tau_q(\rho,\varphi)}{\sigma_0}\right)^{p_\sigma}\right]^{m_W}\rho\mathrm{d}\rho\mathrm{d}\varphi\right), \tag{13.18}$$

$$P_{S.4} = \exp\left(-\frac{V}{V_0}\left(\frac{N\tau_b^{p_\sigma}}{N_0\sigma_0^{p_\sigma}}\right)^m k_2(w, p_\sigma m_W)\right). \tag{13.19}$$

The intensity of shear stress $\tau_q(\rho,\varphi)$ for the circular wire of the radius r is given by Eq. (2.30). Its limit case of the straight rod, the survival probability reads:

$$P_{R.4} = \lim_{w\to\infty} P_{S.4} = \exp\left(-k_6(\infty, m_W)\frac{V}{V_0}\frac{\tau_b^2}{\sigma_0^2}\frac{N}{N_0}\right). \tag{13.20}$$

The ratio of survival probabilities (13.19) and (13.20) follows:

$$\alpha_4 = \frac{P_{S.4\cdot}}{P_{R.4}} = \exp\left(-\frac{V}{V_0}\frac{\tau_b^2}{\sigma_0^2}\frac{N}{N_0}[k_6(w, m_W) - k_6(\infty, m_W)]\right). \quad (13.21)$$

Once again, the dependence of wire diameter is different in cases of harmful defects on the surface and in the material core. This is an important dimensional effect, which could explain the sensitivity of experimental fatigue results upon the wire diameter.

The stress variation over the surface of the wire is higher for the springs with the lower spring indices and with correspondingly higher wire curvature. Accordingly, the failure probability for the same number of cycles and the same basic stress is higher for the springs with lower spring indices.

Consider the spring with spring index w to cycles to failure for the straight rod with the constant stress. For this spring, we determine the ratio of cycles to failure over its surface. In case of surface defects, cycles to failure for the straight rod and for the helical spring are to be determining from the equation:

$$\exp\left(-\frac{A_e}{A_0}\frac{\tau_b^2}{\sigma_0^2}k_5(w, pm_W)\frac{N_S}{N_0}\right) = P_S^*, \quad (13.22)$$

$$\exp\left(-\frac{A_e}{A_0}\frac{\tau_b^2}{\sigma_0^2}\frac{N_R}{N_0}\right) = P_S^*. \quad (13.23)$$

These equations deliver the fatigue life for springs and for rod respectably::

$$N_S = -N_0 \ln\left(P_S^*\right)\left[\frac{A_e}{A_0}\frac{\tau_b^2}{\sigma_0^2}k_5(w, p_\sigma m_W)\right]^{-1}, \quad (13.24)$$

$$N_R = -N_0 \ln\left(P_S^*\right)\left[\frac{A_e}{A_0}\frac{\tau_b^2}{\sigma_0^2}k_5(\infty, p_\sigma m_W)\right]^{-1}, \quad (13.25)$$

such that for surface defects

$$N_R/N_S = k_5(w, p_\sigma m_W)/k_5(\infty, p_\sigma m_W) > 1. \quad (13.26)$$

Correspondingly, the ratio of cycles to failure for volume defects reads:

$$N_R/N_S = k_6(w, p_\sigma m_W)/k_6(\infty, p_\sigma m_W) > 1. \quad (13.27)$$

The functions $k_5(w, p_\sigma m_W)$ and $k_6(w, p_\sigma m_W)$ characterize the influence of spring index on the life time depending on Weibull parameter m_W, spring index w and reciprocal fatigue exponent $p_\sigma = -1/b_\sigma$.

Compare for this purpose two helical springs with equal Weibull parameters, basic stresses, desired survival probability and fatigue exponents. These two springs possess different spring indices w. The spring with high spring index possesses approximately the constant stress over it surface and therefore has nearly the same number of cycles to failure as the straight rod (Huhnen, 1970; Kloos & Kaiser, 1977). The number of cycles to failure of the spring with low spring index differs significantly from the number of cycles to failure of the spring with high values of spring index. This difference greatly increases for the higher stress exponent of fatigue law. This circumstance must be considered in the design process and during the testing of springs.

The actual design procedure assumes that the springs with the same corrected stress (13.but not basic stress) possess the same fatigue life independently on the spring index. Strictly speaking, this statement is not accurate. Namely, the springs with low spring index and the same corrected stress have lower stress on the most outer surface in comparison with the springs with high spring index and the same corrected stress. As consequence, the springs with low spring index and the same corrected stress must have longer fatigue life in comparison with the springs with high spring index. In contrast, the springs with low spring index and the same basic stress must have shorter fatigue life compared with the springs with high spring index.

The derived formulas estimate precisely this effect and provide the direct estimation of fatigue life as the function of desired survival probability and spring index. The different thought was proposed by Reich (2016, § 8.6.2).

13.4 Conclusions

The weakest link concept is applied using the Weibull distribution approach. The influence of stress variation during the cyclic loading is considered together with the effect of stress variation over the surface of the wire due to its curvature. The approach allows to calculate the global probability distribution of fatigue life for helical springs with different spring indices and different stress levels. The approach is applied to calculate the number of cycles to crack initiation of helical springs under different probability levels. The probability function of the fatigue limit for helical compression springs is given in closed form as a function of the spring index and Weibull shape parameters only. The springs with low spring indices show greater sensitivity to stress variation at high survival probabilities. The number of cycles to failure of the low spring index spring differs significantly from the number of cycles to failure of the high spring index spring.

13.5 Summary of Principal Results

- The traditional expressions for the failure probability of springs consider only the pointwise stress maximum and neglect the scattering of defects over the surface or over the volume of the spring.

- The derived formulas correct the traditional expressions.
- The influence of stress variation over the surface of the wire is revealed.
- Stress variation depends on spring index and wire diameter.
- Failure probability is expressed in terms of stress variation over the wire surface or over the wire volume.
- The derivation of the failure probability is based on the concept of the weakest link.

References

Chien, C.-H., LeClair, R. A., & Costello, G. A. (1988). Strength and fatigue life of wire rope. In *Mechanics of structures and machines* (Vol. 16, Issue 2).

EN 13906-1:2013-11. (2013). *Cylindrical helical springs made from round wire and bar—Calculation and design—Part 1: Compression springs; German version DIN EN 13906-1:2013.* Beuth Verlag.

EN 13906-2:2013-09. (2013). *Cylindrical helical springs made from round wire and bar—Calculation and design—Part 2: Extension springs; German version DIN EN 13906-2:2013.* Beuth Verlag.

Huhnen, J. (1970). *Final report on preliminary test for the project—Effect factor k' for coil springs.* Robert Bosch GmbH.

Kaiser, B. (1981). *Beitrag zur Dauerhaltbarkeit von Schraubenfedern unter besonderer Berücksichtigung des Oberflächenzustandes.* Dissertation, TU Darmstadt, FB Maschinenbau.

Kloos, K. H., & Kaiser, B. (1977). Dauerhaltbarkeitseigenschaften von Schraubenfedern in Abhängigkeit von Wickelverhältnis und Oberflächenzustand. *Draht/wire, 9*(415–421), 539–545.

Kobelev, V. (2017). Weakest link concept for springs fatigue. *Mechanics Based Design of Structures and Machines, 17*(4), 523–543.

Reich, R., & Kletzin, U. (2013). *Betriebsfeste Auslegung von Schraubendruckfedern unter Verwendung dynamischer Materialkennwerte aus Umlaufbiegeprüfungen, AiF project: IGF 16999BR.* TU Ilmenau.

Reich, R. (2016). *Möglichkeiten und Grenzen bei der Auslegung von Schraubendruckfedern auf Basis von Umlaufbiegeprüfungen.* Dissertation, Fakultät für Maschinenbau, University of Ilmenau, p. 149.

SAE AE-22. (1997). *SAE fatigue design handbook.* SAE International.

SAE HS 795. (1997). *SAE manual on design and application of helical and spiral springs.* SAE International.

Wormsen, A., & Härkegård, G. A. (2004). *Statistical investigation of fatigue behaviour according to Weibull's weakest-link theory.* ESIS, Proceedings ECF15, Stockholm.

Wormsen, A., Sjödin, B., Härkegård, G., & Fjeldstad, A. (2007). Non-local stress approach for fatigue assessment based on weakest-link theory and statistics of extremes. *Fatigue & Fracture of Engineering Materials & Structures, 30*, 1214–1227, https://doi.org/10.1111/j.1460-2695.2007.01190.x

Chapter 14
Statistical Effects on Fatigue of Spring Materials

Abstract In this Chapter, we continue to study the stochastic influences on the fatigue life of springs. First, we evaluate the probabilistic effects that prevail at low amplitudes of cyclic stresses. The question is how to calculate the failure probability as a function of the stress amplitude. The answer to this question results from the study of experimental fatigue life data. The experimental data show different behavior in the regions of low and high stress amplitudes. To describe this phenomenon, we introduce the randomization of crack propagation, which is accompanied by the random deviation and branching of the crack. The randomization of crack propagation escalates with decreasing stress amplitude. The high inhomogeneity of the polycrystalline structure at the micro level hypothetically causes the random propagation. This hypothesis leads to the mathematical model of random crack propagation. The differential equation with stochastic coefficients describes the randomly propagating crack, which is analogous to the forced Brownian motion equation. Examples of solutions to the stochastic Brownian motion differential equation are presented. This Chapter is the final section of the last part of the book, which studies the life cycle of springs.

Keywords Statistical effects of fatigue · Random crack propagation · Branching of cracks

14.1 Fatigue Analysis at Very High Number of Cycles

14.1.1 Fatigue Strength and Failure Mechanisms

Several intensively stressed springs (e.g. in valve trains or fuel injection systems) achieve the numbers of oscillation cycles of a billion or even higher magnitude during their service life. The fatigue analysis of structural materials at very high number of cycles was the subject of several studies, as reviewed at (Bathias & Paris, 2005; Bathias, 2014; Christ, 2018; Murakami, 2003; Sander, 2018).

V. Kobelev, *Fundamentals of Springs Mechanics*,
https://doi.org/10.1007/978-3-031-58584-5_14

To make the processes of failure at very high numbers of cycles understandable, the structural materials are subdivided into Type I and Type II in (Mughrabi, 2006). Type I materials are pure, single-phase, annealed and ductile metals without significant internal defects. The persistent slip band limit is defined here as the limit for crack initiation and the minimum stress for crack growth as the conventional fatigue limit. Accumulation of less irreversible strains results in surface roughening. As the number of oscillation cycles increases, the roughness of the surface increases and thus leads to local stress concentration, which reaches the limit value for the formation of persistent slip bands. This leads to the formation of micro-persistent slip bands, in which strain localization takes place. As a result, the persistent slip bands expand in the direction of the inside of the specimen and can thus lead to crack formation on the surface.

14.1.2 Experimental Investigation for Low-Stress Springs' Fatigue

The scientific-technical results of the research project (AiF 12 287 N, 2002) were the experimental investigation of the fatigue strength behaviour (Goodman diagrams) and the relaxation behaviour (relaxation diagrams) of coil compression springs made of the following material groups:

- Patented drawn steel wires (spring strength from a combination of preheat treatment and work hardening of the wire),
- Oil tempered steel wires (spring strength from final tempering of the wire),
- Stainless spring steel wires (spring strength from work hardening of the wire).

The aim of the research project (AiF 12 287 N, 2002) was to determine and evaluate statistically verified results on the fatigue and relaxation behaviour of helical compression springs made of six types of spring steel wire (VDSiCr, TDSiCr, DH, DM, 1.4568 and 1.4310). The springs with wire diameters of 1, 2, 3, 5 and 8 mm were prepared for testing. The results of the project were intended for spring calculation and dimensioning. The fatigue behaviour of the test springs was investigated with the aid of three testing machines. Statistically evaluated S–N-lines were determined for 27 basic variants at different levels of medium and low stress. From these, fatigue strength diagrams according to Goodman were generated for a number of limit cycles of 10^7 cycles. These fatigue strength diagrams were compared with the previous diagrams in DIN 2089 for the survival probability $P_S = 90\%$. The experimentally determined endurance stroke strength values at 90% survival probability were determined for the materials:

- VDSiCr by about 300 to 400 MPa,
- TDSiCr by about 300 to 400 MPa,
- DH by about 200 to 350 MPa,
- DM by about 200 to 300 MPa,

- 1.4568 by about 80 to 100 MPa and
- 1.4310 by about 50 MPa.

The detailed results of the cited project were published in several articles. Specifically, (Kaiser & Berger, 2005) reviewed the mentioned above assessments of fatigue behaviour of springs. The significance of defects on the surface and in the surface-near layer decrease the fatigue properties of cold formed springs were evaluated. Besides of the internal "inclusion-type" defects, surface defects originate during hot forming or heat treatment, were observed. The influence of other defects, which result from drawing or coiling ("coiling marks"), was debated as well. In conclusion, the article (Berger et al., 2008) presents conclusions of experimental examination of the fatigue behaviour for helical springs at a very high number of load cycles.

The objective of the research project (Aif 15 064 N, 2010) was to determine and evaluate the long-term vibration behaviour of coil springs in the oscillation range up to 10^9 vibration cycles. In the cited research project, the long-term fatigue tests were performed with a spring testing machine of the type "Reicherter DV8-S2". The examined springs were the shot peened cylindrical helical compression springs with wire diameters of 16 mm and 3 mm. The materials for the spring production were the oil tempered valve spring steel wires VDSiCr and VDSiCrV as well as the stainless spring steel wire X7CrNiAl17-7 (material number 1.4568). The fatigue examination was accompanied by stereomicroscopic and scanning electron microscopic fracture surface investigations, metallographic investigations, X-ray residual stress measurements and microstructure determinations as well as torsion tests on wires.

The works (Kaiser et al., 2011; Pyttel et al., 2011) review the results of fatigue strength and failure mechanisms at very high number of cycles. Fatigue testing machinery for very high number of cycles is accounted. The classification of investigated materials is performed with respect to characteristic S–N curves and influencing factors like notches, residual stresses and environment.

The article (Schwerdt, 2011) provides details on the discovered failure mechanisms, which occur especially in the VHCF-region. An important role on the emerging of a failure mechanism plays the microstructure of the homogeneities. The cited article proposes a double S–N curve for description of fatigue behaviour. The proposed method considers different failure mechanisms. Investigated materials are different metals with body-centred cubic lattice. Such materials, like low- or high strength steels and quenched and tempered steels, are used frequently for spring production. In (Schwerdt, 2011) the aluminium allows (EN AW-6082-T5, EN AW-6056-T6) and high-strength tempered alloy steel 42CrMo4 were studied experimentally in the VHCF-region.

The aim of the succeeding project (IGF 16873 N, 2014) was to determine the influence of different materials and manufacturing parameters on the VHCF stroke strength of coil compression springs. For this purpose, the springs made of oil-quenched and tempered valve spring steel wire VDSiCr with d = 1.6 mm were manufactured. The influence of the end coil geometries, shot peening times and heat treatment temperatures after spring coiling was considered for the test springs.

Primary tests of torsion influence, tensile and rotating bending as well as metallographic microstructure determinations or chemical analyses were carried out on the examined wires.

Test springs made of the low-cost stainless spring steel wire 1.4310 with wire diameter d = 1.6 mm as well as springs made of patented drawn spring steel wire DH with d = 3.15 mm were examined. In addition, springs made of high-quality stainless spring steel 1.4568 were examined in two wire production variants with d = 3.15 mm. Springs made of the stainless spring steel 1.4310 were tested up to 109 load cycles in fatigue tests. In comparison to springs made of stainless spring steel 1.4568, springs made of 1.4310 have a lower VHCF lifting strength and service life.

With regard to springs made from the low-cost patented drawn spring steel wire DH, it was also possible to demonstrate their suitability for use in the VHCF sector depending on the specific requirements.

For springs with protruding and adjacent end coils, the oscillation tests up to 5×10^8 vibration cycles were performed for different stress amplitudes.

An increase in the amplitude of stress in VHCF region due to the protruding spring end was found. This design of the spring end coils leads to the favourable stress pattern along the length of the helically wounded spring wire. Failures of the end coil were observed in both end coils.

The increasing of the duration time of shot peening showed a positive effect on the VHCF lifting strength of springs made of VDSiCr. Particularly, the heat treatment temperature after spring coiling has an improving effect on the strength amplitude in VHCF region.

For springs with a heat treatment temperature of 400 °C/30 min occur relatively few failures in the regions of 10^9 oscillating cycles. However, the springs with a heat treatment temperature of 360 °C/30 min and half the test duration (5×10^8 oscillating cycles) produced more than twice as many fracture events. Thus, the expectancy of fatigue life for the latter type of springs is generally lower, than former ones.

The test springs were made of the alternatively manufactured wires of the material 1.4568 with a wire diameter d = 3.15 mm. The fatigue life of the alternatively manufactured springs was roughly equal.

In general, a change of the failure location from on the surface at high stress to under the surface at medium and low stress was observed. This feature could be explained. For the high loading, the damage due to defects on the surface overwhelms the positive effect of internal stresses. The detailed accounts of discussed above results for the project (IGF 16873N, 2014) were published in several works.

Pyttel et al. (2014) reported the results of long-term fatigue tests on shot-peened helical compression springs. The testing preformed with the spring fatigue-testing machine at 40 Hz. Test springs were made of three different spring materials. The investigated materials were oil hardened and tempered SiCr- and SiCrV-alloyed valve spring steel and stainless steel. With a special test strategy in a test run, a batch of springs with different wire diameters were tested simultaneously at different stress levels. Based on fatigue investigations of springs up to a number of cycles 10^9 cycles an analysis was done. The test continued to 1.5×10^9 cycles. The results of both tests were compared. The influence of different shot-peening conditions is studied. The fractured test springs inspected for of fracture behaviour and the failure mechanisms. The paper (Pyttel et al., 2014) includes a comparison of the results of the different spring sizes, materials, number of cycles and shot-peening conditions.

Schuller et al. (2013) investigated very high cycle fatigue properties of VDSiCr spring steel were with ultrasonic equipment under fully reversed cyclic torsion loading and under cyclic axial loading at load ratios.

The spring manufacturing regularly uses the austenitic stainless steel. (Lukacs, 2010) acquired of the basic data for fatigue crack propagation limit or design curves on austenitic stainless steel (type 321) in corrosive environment and at elevated temperatures. This work estimated the design curves based on statistical analysis of measured data. The propagation law of fatigue crack was evaluated.

Mayer et al. (2017) examined VHCF properties of high-strength spring steel SWOSC-V for different load ratios in cyclic torsion test. For comparison, cyclic tension–compression and cyclic tension tests at the same load ratio were also performed. The ultrasonic fatigue testing technique was adopted for both torsional and axial loading fatigue examination. The mechanisms of crack initiation and strength in VHCF region were compared for torsion and axial load.

Nishimura et al. (2017) studied the torsional fatigue behavior in high cycle and very high cycle fatigue regimes. Definite tiny longitudinal notches were introduced on the surface of test specimens. Both uniaxial and torsion fatigue examinations were performed. A significant reduction of the uniaxial fatigue limit was observed. In contrast, the torsional fatigue limit was not considerably reduced. The analysis of the broken specimens with small longitudinal notch in torsion fatigue experiments confirmed, that the propagation of mode I crack leads to the final failure.

Fatigue tests were performed on four compositions of the 18Ni maraging steels in precipitation hardened condition in (Karr et al., 2017). One material contained the addition of titanium, the other increased Co content. The third type of material used alloying with Al and Cr. The Al content was increased in the fourth type of maraging alloy. For fatigue tests, the thin sheets with nitrided surfaces were manufactured. The VHCF-fatigue properties of the 18Ni maraging steels have been investigated at load ratio $R_\sigma = 0.1$. The origin of fatigue cracks in three material compositions were internal inclusions. However, one material composition with an increased Al content demonstrated the surface and internal crack initiation. Only the initial failures in this material were caused by internal inclusions.

In this study we perform the evaluation of the technically relevant S–N-curves for different material and load parameters, based on the reported experimental results (AiF 12 287 N, 2002; AiF 15 064 N, 2010; Kaiser & Berger, 2005).

The results of the cited projects combine the fatigue tests of helical springs made of steel wires. Especially, $CrSi$ and $CrSiV$-alloyed oil hardened and tempered spring steel wires) conformed to (DIN EN 10270, 2011) were selected. The springs were manufactured in so-called super-clean quality from peeled or ground raw material. The fatigue test was performed on the shot peened helical compression springs of $CrSi$-alloyed spring steel wire with diameter of 2 mm, a spring diameter of 12 mm and a free length of 50 mm and test results were presented in (AiF 12287 N, 2002; AiF 15 064 N, 2010; Kaiser & Berger, 2007).

The investigations (Mayer et al., 2015; 2016) provide further information, especially the influence of stress ratios on the fatigue life.

The effect of the ultimate number of cycles of fatigue tests on fatigue time prediction for compression springs produced from $VDSiCr$ spring wire was investigated in (Geilen et al., 2020). The state-of-the art of the long-term fatigue examination was reviewed. In this paper, the new kind of experiment, referred to as the "Artificial Censoring Experiment", was proposed. The "Artificial Censoring Experiment" was executed for the extrapolation of the results of fatigue tests on compression springs and for the reduction of the needed testing volume.

14.1.3 Modelling Hypothesis for Low-Stress Springs' Fatigue

The reported above experimental results demonstrate two features of the fatigue process of low-stress fatigue of the spring materials.

At first, the stochastic scattering of fatigue data. The scattering widens with the deceasing amplitude of cyclic stress. To account the scattering, the known statistical methods are briefly reviewed. This task is important for applications, because the testing procedures provide usually only the median S–N curves and the scatter range (statistical moment of the second order). The applications require the guaranteed much lower probability of failures. The recalculation of S–N curves for this lower failure probability is reported.

At second, the clear alternation of the fatigue mechanism with the reduced stress amplitude was stated experimentally. This effect is visible as the flattening of the S–N curve in the region of low stress amplitude. The traditional Paris-Irvin model does not describe the alternation of the propagation mechanism in the region of low cyclic stresses. As reported in Chaps. 9 and 10, the propagation functions of the Paris-Irvin model are based on the constant stress exponent in the central region between the threshold and ultimate rupture stress. In logarithmic scaling, this description for the crack elongation pro load cycle proclaims the linear dependence of fatigue life upon the stress amplitude. To fit the observe experimental data, we divide the propagation function of Paris-Irvin law in the regions of mid- and low-stress level. As expected, the resulted S–N curve will have also two regions with the smooth transition between mid- and low-stress levels. To explain this division from the viewpoint of micromechanics, we associate the mid-stress crack propagation with the grain destruction by the crack. The crack crosses the grain, because it is the shortest way for crack propagation. On the low-stress level, the energy release will be insufficient for grain-destruction, and the crack travels along the weaker grain boundaries, which considerably increases the total area of crack and decreases the projected elongation pro load cycle. This explains the nonproportional reduction of the total crack length with the lower stress amplitudes. The mathematical description of this process is based on the differential equation with the stochastically varying coefficients. The introduced equation is analogues to the equation of Brownian motion. The crack has a definite propagation direction, like the enforced Brownian motion with the directionally enforced drift.

There exist alternative explanations of the above transition process from mid-stress level to low-stress level. Particularly, pure geometical explantion could be also imagined (Kobelev, 2017a). The idea was a fractal nature of the crack surface. In the mis-stress region the crack is nearly planar. The reduction of cyclic stress causes the deviation of crack from the grain-destruction to the separation of grain interfaces, as described above. The deviations causes the transition from plane to fractal surface of the crack. The total area of the fractal surface is a power function of the projected area. The creation of free fractal surfaces requires much more released elastic energy, which depends only upon the projected area of crack. Therefore, the crack propagation slows down in the low-stress level. The mathematical description of fractal surface is based on fractional calculus. The fractional-differential extension of Paris-Irvin equations, which describe the crack propagation were proposed. The unified propagation function describes the infinitesimal crack length growths per increasing number of load cycles, supposing that the load ratio remains constant over the load history. Two unification fractional-differential functions with different number of fitting parameters were discussed. The number of cycles to failure from a given initial crack length upon the stress amplitude follows from the sulution of fractional differential equation.

14.2 Statistical Assessment of Springs Fatigue Test

14.2.1 Fatigue Scattering of Spring Steels

The technically relevant number of cycles to failure are in region of several millions to hundreds of millions. The practical reasons restrict the number of springs in the test series and the durations of tests. The heavy utilization of test equipment for high and ultrahigh number of cycles confines amount of the examined springs. The direct acquisition of S–N curves for different parameters of material and load is impossible for all springs in production. Consequently, the technical springs could be tested to failure only for the moderate number of cycles. Based on these series, the S–N curves for the average failure probabilities are generated. These S–N curves are not directly applicable for the failure forecasts in technically relevant numbers of cycles. However, these S–N curves are suitable for the extrapolation to the required S–N curves. We describe below the calculation of technically relevant S–N-curves (“Wöhler” lines) for different failure probabilities

$$P_F = 1 - P_S. \tag{14.1}$$

The value P_S in Eq. (14.1) is the survival probability.

In the probability plot, the scattering curve appears as a piecewise-straight function of load cycles to failure, as reported in (Kaiser & Berger, 2007). The experimentally acquired data display the median S–N lines for the failure probability P_F of 50%. For the given number of cycles to failure, the values of the corresponding stress amplitudes were plotted. The presented method calculates the fatigue lines for the technically relevant failure probabilities, based on the S–N line of 50% failure probability.

We adopt the experimentally evaluated median S–N line for development of the model. Based on the experimental results, the fatigue crack growths (FCG) per cycle is estimated. The generalizations of FCG describe crack extension rate, implementing the piecewise linear or continuous function.

Further, we use the FCG estimation for the probabilistic modelling on the microlevel of material structure. With this data, the stochastic differential equation for the travelling crack is established. The stochastic equation is analogous to the equation of the enforced Brownian motion. The generalizations of FCG (gFCG) functions explain the experimentally observed S–N curve in HCF and VHCF regions. From gFCG functions follow the closed form expressions of S–N curves for different failure probabilities, stress ratios and the geometry pf microstructure.

14.2.2 Uniformity of Stress on Wire Surface and Averaged Fatigue Stress

There are several standard stress parameters for the helical springs. These factors were reviewed in Chaps. 1 and 2. The first known stress parameter is the "basic stress"τ_b (1.25). This stress parameter represents the average shear stress on the surface of the wire, the "basic", or "uncorrected stress", Eq. (2.4) (SAE, 1997, Chap. 5, Design of Helical Springs).

The second known parameter is the "corrected stress". It is equal to the maximum shear stress on the surface of the coil spring. The maximum shear stress is the product of the "basic stress" with the correction factor $k_\tau(w)$ (1.27). For example, the correction factor due to Bergsträsser (1.30):

$$\tau_c = k_2(w) \cdot \tau_b, \tag{14.2}$$

$$k_2(w) = \frac{w + 0.5}{w - 0.75}. \tag{14.3}$$

However, only the maximum stress, but also the uniformity of stress distribution on the surface of the wire influences the failure probability of springs in series production. The more evenly the stress is distributed on the surface, the higher is the probability of defects and inclusions in the stressed area on the inner surface of the spring. This sounds paradoxical on the first thought. The explanation is the following.

Consider two springs with different spring indices, but equal value of maximal stress on the surface. The region of increased stress on the inner surface of the spring with higher spring index is smaller, that of the spring with the lower spring index. Obviously, the probability of defects in a smaller area is lower, than probability of defects in a larger area. Thus, the stress, which concentrates in a smaller area, cause less damage, than the same stress, which spreads in a larger area. Both springs must reach the same number of load cycles until breakage. There are special stress correction factors, which assure, that if the corrected stress is equal in two different springs, the damage probability in series for two springs will be equal (FKM, 2018a, 2018b; Huhnen, 1970; Kloos & Kaiser, 1977, Sect. 2.1). The factors are based on the results of fatigue tests on coil springs.

There are other stress factors, which average the stress over the surface of the wire. These factors reflect the possibility of failure from possible defects in the fewer stressed regions. The simplest factor of this type $\overline{k_2}(w)$ is based on Bergsträsser correction factor (14.3):

$$\overline{k_2}(w) = \frac{k_2(w) + 1}{2} \equiv \frac{w - 0.125}{w - 0.75}. \tag{14.4}$$

Evidently, that the fatigue stress correction factor (14.4) averages the weighted stress over the wire section:

$$\tau_f = \overline{k_2}(w) \cdot \tau_b = \frac{\tau_b + \tau_c}{2}. \tag{14.5}$$

Figure 14.1 displays both correction factors and their relation. In (14.4) any other correction factor could be applied instead of Bergsträsser correction factor. However, the tiny difference between the known correction factors is not significant for fatigue calculations. The value τ_f is referred to as "fatigue stress". Numerous experimental findings indicate, that the fatigue stress τ_f better correlates to the fatigue test results, than the basic stress τ_b or the corrected stress τ_c.

The correction factors for fatigue stress were discussed in Chaps. 1 and 2. Based on the weakest link concept, several expressions for the stress correction factors and stress parameters were determined in Chap. 11.

14.2.3 Experimental Acquisition for Sensitivity to Stress Ratio and Spring Indices

The next important issue for fatigue estimation is the influence of the stress ratio, as discussed in Chap. 10. As shown above, there are three possible definitions for a scalar stress. The stresses are the harmonically oscillating functions of time with constant amplitudes and non-zero mean values (N):

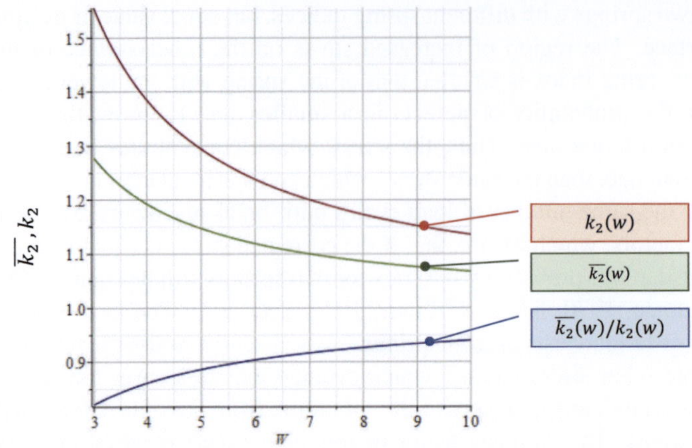

Fig. 14.1 Bergsträsser correction factor k_2 and averaged correction factor $\overline{k_2}$ as functions of spring index w

$$\tau_b(t) = \tau_{m.b} + \tau_{a.b}\sin(\omega\, t), \quad \tau_{b.a} = \tfrac{1}{2}\Delta\tau_b, \tag{14.6}$$

$$\tau_f(t) = \tau_{m.f} + \tau_{a.f}\sin(\omega t), \quad \tau_{f.a} = \tfrac{1}{2}\Delta\tau_f, \tag{14.7}$$

$$\tau_c(t) = \tau_{m.c} + \tau_{a.c}\sin(\omega t), \quad \tau_{c.a} = \tfrac{1}{2}\Delta\tau_c. \tag{14.8}$$

The cycles are asymmetric (N) for three scalar, time-depending stresses $\tau_b(t).\tau_f(t)$, $\tau_c(t)$. The coefficients of the Eqs. (14.6)–(14.8):

$$\tau_{a.b},\ \tau_{a.f} = \overline{k_2}(w)\cdot\tau_{a.b}, \quad \tau_{a.c} = k_2(w)\cdot\tau_{a.b},$$

are amplitudes of cycles (N) and

$$\tau_{m.b},\ \tau_{m.f} = \overline{k_2}(w)\cdot\tau_{m.b}, \quad \tau_{m.c} = k_2(w)\cdot\tau_{m.b}.$$

are mean values of cycles (N).

The stress ratios for the functions (14.6)–(14.8) are all equal:

$$R_\sigma = \frac{\tau_{m.b}-\tau_{a.b}}{\tau_{m.b}+\tau_{a.b}} = \frac{\tau_{m.f}-\tau_{a.f}}{\tau_{m.f}+\tau_{a.f}} = \frac{\tau_{m.c}-\tau_{a.c}}{\tau_{m.c}+\tau_{a.c}}. \tag{14.9}$$

For fatigue calculations, we use the experimental data, reported in the cited above reports. The data is ordered in the double logarithmic graphs. The abscise axis presents the number of cycles. The ordinate axis displays the amplitude of the corrected shear stress $\tau_{a.c}$. The immediate application of this data is not convenient, considering the following. As we already know, the amplitude of the corrected shear stress $\tau_{a.c}$ for the certain number of cycles depends on several factors. Firstly, $\tau_{a.c}$ depends upon the mean stress$\tau_{m.c}$, and secondly upon the spring index$w$. If the mean stress or the spring index changes, the ordinates of the S–N line also change. Accordingly, S–N lines are not invariant and do not display "true" material properties. The material properties must be independent upon design parameter$w$. Therefore, we must "clean" the S–N lines from the design-dependent influences. The only difference between proposed S–N lines is the proper scaling of ordinate axis. After some preparations, the invariant S–N lines are calculated below.

To "clean" the S–N lines from the influence of the stress ratio, the $\mathbf{P}_{\mathbf{SWT}}$ (Smith, Watson, Topper) and $\mathbf{P}_{\mathbf{W}}$ (Walker) parameters are habitually applied (Chap. 10). However, these two parameters account the stress ratio (14.9) slightly different. We proceed with the parameter $\mathbf{P}_{\mathbf{W}}$, because $\mathbf{P}_{\mathbf{SWT}}$ parameter appears as the special case ($\tilde{\gamma} = 1/2$) of the first mentioned.

For each relevant harmonically oscillating functions $\tau_b(t)$, $\tau_f(t)$, $\tau_c(t)$, there are three effective parameters $\tau_{ar.b}$, $\tau_{ar.f}$, $\tau_{ar.c}$. The Walker parameters $\mathbf{P}_{\mathbf{W}}$ are defined with Eq. (12.5):

$$\tau_{ar.b} = \tau_{a.b}^{\tilde{\gamma}} \tau_{\max.b}^{1-\tilde{\gamma}}, \tag{14.10}$$

$$\tau_{ar.f} = \tau_{a.f}^{\tilde{\gamma}} \tau_{\max.f}^{1-\tilde{\gamma}} = \overline{k_2}(w) \tau_{ar.b}, \tag{14.11}$$

$$\tau_{ar.c} = \tau_{a.c}^{\tilde{\gamma}} \tau_{\max.c}^{1-\tilde{\gamma}} = k_2(w) \tau_{ar.b}. \tag{14.12}$$

The fitting parameter must satisfy $0 \le \tilde{\gamma} \le 1$. The recommended values of the fitting parameter for high strength spring steels are $\tilde{\gamma} = 0.15..0.25$.

The first version of the S–N lines applies the Walker parameter $\mathbf{P}_{\mathbf{W}}$ for corrected stresses (14.12) as the ordinate axis. The dependence of S–N lines upon the mean stress disappears for the Walker exponent of $\tilde{\gamma} = 0.15$. For this setting of fitting parameter, several S–N lines with different stress ratios $0 < R_\sigma < 1$ overlap. Thus, this settung is invariant with respect to the stress ratio. If the range of mean stress sensitivity is considerably narrow, for example $0 < R_\sigma < 0.5$, the Walker parameter $\mathbf{P}_{\mathbf{W}}$ describes the mean stress sensitivity properly. In the extended range, $-1 < R_\sigma < 0.5$ Mayer et al., 2016), the adequate description requires the nonlinear dependence upon R_σ, as already discussed in Chap. 12. For example, the generalized, or Bergmann-Walker parameter $\mathbf{P}_{\mathbf{BW}}$ (12.11)covers the wide range of mean stress variation.

The second version uses the "corrected" $\mathbf{P}_{\mathbf{SWT}}$ parameter (14.12) for ordinate axis. In this case one can put in the Eq. (14.10) to (14.12) the fixed value of fitting parameter $\tilde{\gamma} = 0.5$. We use this version only for reference, because numerous calculations in spring industry traditionally use $\mathbf{P}_{\mathbf{SWT}}$ approach for corrected stresses.

Similarly, the third and fourth versions use either "basic" $\mathbf{P}_{\mathbf{W}}$ of $\mathbf{P}_{\mathbf{SWT}}$ parameters, which are grounded on "basic" stresses $\tau_{ar.b}$ instead of corrected stress, Eq. (14.10).

The springs with different spring indices still display different S–N lines, because the different spring indices would lead to different S–N curves. To clean this influence, the choose the averaged stress (14.5) as the ordinate. The ordinate in fifth and sixth versions is the mean value of "corrected" and "basic" effective stresses:

$$\tau_{ar.f} = \frac{1}{2}(\tau_{ar.c} + \tau_{ar.b}). \tag{14.13}$$

This preferable scaling provides the best fitting of the fatigue data. The springs with different spring indices display coincident S–N lines.

The search of the optimal setting for the fitness parameters determines the following results. The best formulation adopts the $\mathbf{P_W}$ effective stress $\tau_{ar.f}$ (14.13) with the Walker exponent $\tilde{\gamma} = 0.15$. For this setting, the displayed S–N lines match independently on different stress ratios $0 < R_\sigma < 1$ and on different spring indices. Thus, these S–N curves in ordinates (14.11) could be considered as the true material characteristics.

14.2.4 Median S–N Curves

The number of cycles to total failure for wide ranges of corrected stress $\Delta\tau_c = 2\tau_{a.c}$ for common spring materials were reported in the cited above documents (AiF 12 287 N, 2002; AiF 15 064 N, 2010; IGF 16873 N, 2014). These results provide the start point for our calculations. The amplitude of corrected stress is

$$\tau_{a.c} = \frac{1}{2}\Delta\tau_c = \frac{1}{2}(1 - R_\sigma)\tau_{max.c}.$$

According to Walker, Eq. (10.5), the relation between the corrected effective shear stress $\tau_{ar.c}$ and the range of corrected shear stress $\Delta\tau_c$ reads:

$$\tau_{ar.c} = C_R\Delta\tau_c, \quad C_R = \tfrac{1}{2}\cdot\left(\tfrac{1-R_\sigma}{2}\right)^{\tilde{\gamma}-1}. \tag{14.14}$$

Using experimental results from the cited above references, the median S–N lines for different conceivable effective stresses could be evaluated:

$$\tau_{ar50} = \tau_{ar}(P_F = 50\%).$$

The median S–N lines are plotted in three different types of scaling for effective stresses, namely corrected, basic, and averaged.

Firstly, the median S–N lines for each tested spring are plotted in coordinates $(\tau_{ar.c}, n_f)$ with corrected effective stress (14.14) for ordinate.

Secondly, the basic effective stress evaluates from (14.14) with the formula:

$$\tau_{ar.b} = \frac{\tau_{ar.c}}{k_2(w)}. \tag{14.15}$$

Thirdly, the median S–N lines for each tested spring are portrayed in coordinates $(\tau_{ar.b}, n_f)$, with the basic effective stress as ordinate.

As discussed in above, all three differently scaled S–N lines usually indicate the dependence upon the spring indices of the tested springs. This dependence in undesirable and must be eliminated. For practical calculations, the true material parameters are necessary. To make the S–N –lines independent upon the spring indices, the averaged stress correction factor $\overline{k_2}(w)$ for fatigue estimations is applied. The averaged effective stress $\tau_{ar.f}$ is invariant to spring index and is also the true material constant:

$$\tau_{ar.f} = \overline{k_2}(w)\tau_{ar.b} \equiv \frac{\overline{k_2}(w)}{k_2(w)}\tau_{ar.c}. \tag{14.16}$$

Finally, the median S–N lines for each tested material are plotted in coordinates $(\tau_{ar.f}, n_f)$, with the averaged effective stress (14.16) and (14.13) shown on vertical axis. This exposition of the fatigue results is the most insensitive to the variations of spring indices and stress ratios.

One remark concerns the dependence of the strength of the wire over its diameter. This issue was discussed in Chap. 10. As the relevant influence factors are accounted, the method is appropriate for design purposes of the mainstream industrial springs.

14.2.5 Valuation of S–N Lines for Prearranged Failure Probabilities

The experimental evaluation of S–N lines delivers the median S–N line, for which the failure probability is 50%. The cumulative failure probability of springs in industrial applications must be much lower. Its value P_F^* is prearranged by customer. For example, in automotive applications the fracture of valve springs in combustion motors repeatedly leads to the severe or even irreparable failure of an engine. The experimental evaluation of the cumulative failure probabilities in the region of their low values is a very consuming matter. At first, such test requires a great number of probes. At second, the stress level is so low, that the number of cycles must be of the order of hundred million. For this fatigue life, direct industrial tests are nearly impossible. In praxis, the S–N lines of high survival probability are derived from the median S–N line with survival probability 50%. Starting from the median S–N line, the S–N lines for different, prearranged failure probabilities P_F^* are evaluated using the experimentally acquired scatter range values. Commonly, the lognormal distribution is assumed (Castillo et al., 2009; Forbes et al., 2011; Kalbfleich & Prentice, 2002). The lognormal distribution is briefly explained in Appendix C.

For example, failure probabilities the standard scatter range to $T_\sigma = 1/1.2 = 0.833$. It is the ratio of the stresses, which lead after the number of cycles to failure with 90% and 10% probabilities. According to Eq. (C.6), this value corresponds to the scale factor $\mu = 0.071133$.

The effective stress for $P_F = 10\%$ for the same number of cycles to failure is equal to:

$$\tau_{ar10} = \tau_{ar}(P_F = 10\%).$$

The τ_{ar} stays for one effective stress $\tau_{ar.b}, \tau_{ar.c}, \tau_{ar.f}$. We determine another S–N line, which displays any other failure probability P_F in the same coordinates. In other words, we look for the effective stress $\tau_{ar}(P_F)$ for a given survival probability P_F^* and for the same number of cycles to failure. The calculation is based on the inverse error function Eq. (C.3). With the inverse error function, we have the relation:

$$\frac{\tau_{ar}(P_F = 50\%) - \tau_{ar}(P_F)}{\tau_{ar}(P_F = 50\%) - \tau_{ar}(P_F = 10\%)} = \frac{\mathrm{ierf}(1 - P_F)}{\mathrm{ierf}(1 - 10\%)}. \quad (14.17)$$

Resolution of the Eq. (14.17) delivers the effective stress $\tau_{ar}(P_F)$ for the certain P_F:

$$\tau_{ar}(P_F) = \tau_{ar50} - \frac{\mathrm{ierf}(1 - P_F)}{\mathrm{ierf}(1 - 10\%)} \cdot (\tau_{ar50} - \tau_{ar10}). \quad (14.18)$$

For automotive applications, the typical requirement for cumulative failure probability of springs is $P_F^* = 1ppm$. This the corresponding S–N line demonstrates the stresses of the level 71.36% of the median-S–N line. The corresponding S–N curves is the "design S–N curve" for the cumulative failure probability of $P_F^* = 1ppm$. Apparently, the correction "shifts" the 10%– S–N-Line towards abscise of the S–N graph. Table 14.1 demonstrates the calculation of the effective stress reduction for different vertical placements of S–N-lines. This calculation finishes the solution of the initially formulated problem. The relative reduction of the corrected effective stress amplitude $\tau_{ar}(P_F)/\tau_{ar50}$ are shown on Figs. 14.2, and 14.3. The horizontal axes for failure probability P_F are given in logarithmical scale.

Table 14.1 Recalculation of stress levels that correspond to different failure probabilities

Failure probability	Stress level			
P_F	$\tau_{ar}(P_F)/\tau_{ar50}$			
T_σ	1/1.2	1/1.234	1/1.3845	1/1.58
μ	0.071133	0.08203	0.1269	0.1784
50%	100%	100%	100%	100%
10%	91.2%	90%	85%	80%
1%	84.74%	84.3%	76.5%	68.6%
0.1%	80.26%	79.9%	69.9%	59.9%
100 ppm	76.755%	76.3%	64.5%	46.2%
10 ppm	73.83%	73.1%	59.7%	40.5%
1 ppm	71.36%	70.2%	55.4%	35.2%

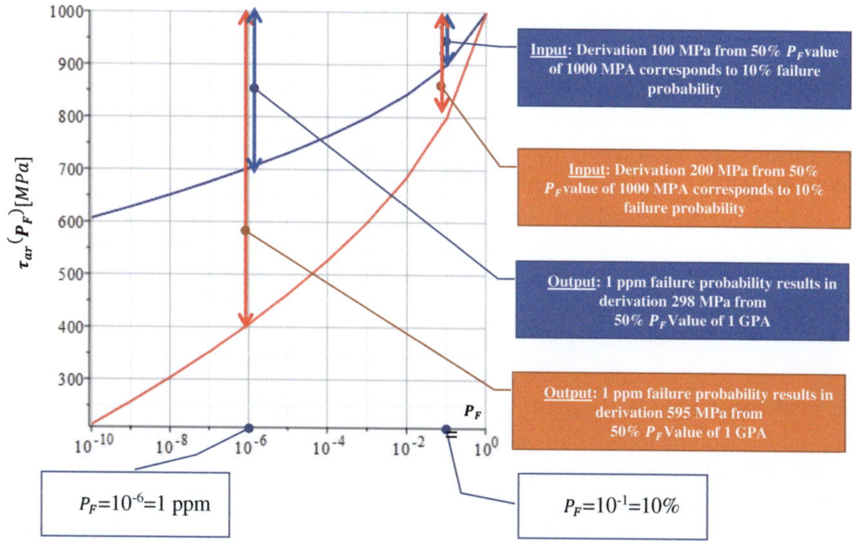

Fig. 14.2 Relative reduction of stresses $\tau_{ar}(FP)/\tau_{ar50}$ for different failure probabilities FP. The horizontal axis for failure probability is given in logarithmical scale

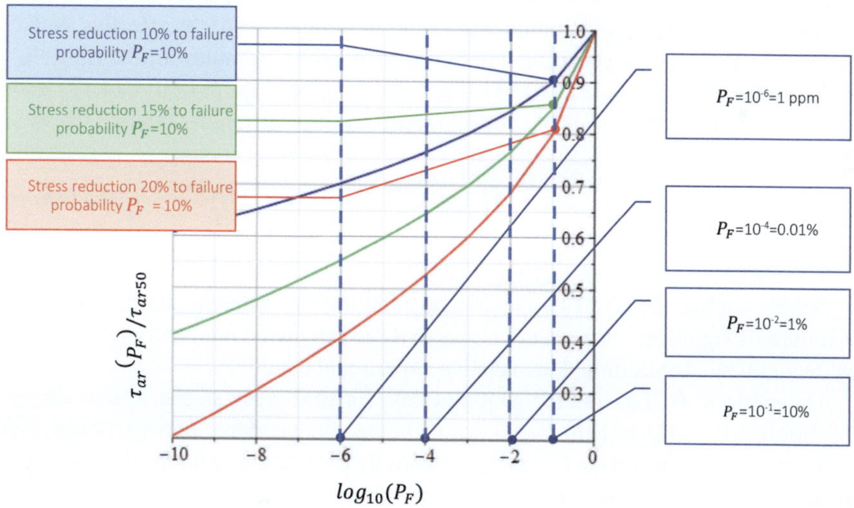

Fig. 14.3 Evaluation of stresses for different failure probabilities

The horizontal axis for failure probability is given in logarithmical scale.

14.2.6 S–N Curves of Spring Steels in VHCF Range

The experimental results [IGF16873], (Mayer et al., 2015, 2016) for the cumulative failure probabilities for the tested springs are displayed on Figs. 14.4 and 14.5. The test springs are numbered from 1 to 11. The static mechanical properties, stress ratios R_σ and the amplitude of corrected stress $\tau_{c.a}$ are listed for the tested springs in Table 14.2. For the tested springs the cumulative failure probabilities are displayed on Fig. 14.4 in median probability range $R_\sigma = 0.1; 0.5; 0.9$. In this range the springs are tested and the scattering is evaluated. The values.

$$R_\sigma = 10^{-1};\ 10^{-2};\ 10^{-3};\ 10^{-4} 10^{-5};\ 10^{-6}$$

are marked on Fig. 14.5 for cumulative probabilities in low range. This range is relevant for applications, as already discussed above.

The first method is based on application of Walker parameter ($\mathbf{P_W}$) for the effective stress. Table 14.3 summarizes the results of the fatigue evaluations. This table displays the values of scatter ranges T_σ, scale parameters μ, $b(R_\sigma)$, the reciprocal stress exponent $p_\sigma = -b^{-1}(R_\sigma)$, fitted Walker exponent $\tilde{\gamma}$ and the coefficients $\tau_{f0.c'}$, $\tau_{f0b'}$, $\tau_{f0f'}$ for evaluations of fatigue life. The stress-life approach is applicable, because of low stress and high number of cycle the plastic deformation is excessively low. Hereafter assumed, that the stress exponent $b(R_\sigma) = b_0$ is independent on stress ratio. For evaluation of the median number of cycles to failure, one of the effective Walker parameters $\mathbf{P_W}$ must be substituted in the corresponding formulas for the Basquin's or Wöhler's law Eqs. (12.1b), (12.5b). The substitution of Eqs. (14.10)–(14.12) leads to the expressions of number of cycles in terms of three different effective stresses:

$$n_f = \frac{1}{2}\left(\frac{\tau_{ar.c}}{\tau_{f.0c'}}\right)^{\frac{1}{b(R_\sigma)}},\ n_f = \frac{1}{2}\left(\frac{\tau_{ar.b}}{\tau_{f.0b'}}\right)^{\frac{1}{b(R_\sigma)}},\ n_f = \frac{1}{2}\left(\frac{\tau_{ar.f}}{\tau_{f.cf'}}\right)^{\frac{1}{b(R_\sigma)}}. \tag{14.19}$$

The coefficients $\tau_{f0.c'}$, $\tau_{f0b'}$, $\tau_{f0f'}$ are listed in the Table 14.3.

Strictly saying, the Eq. (14.19) calculate three slightly different numbers of cycles to failure. If significant differences between three calculated number of cycles arise, we recommend an additional, exhaustive experimental study.

The value $\tilde{\gamma} = 0.15$ fits the acquired data for the spring steels. In this choice, for different $0 < R_\sigma < 1$ the S–N lines coincide undependably on stress ratio. The small value of $\tilde{\gamma}$ indicates an increasing sensitivity to mean stress for the high-strength spring wires. This feature is typical for high the strength and hardness alloys.

The second way is based on the parameter of Smith, Watson, Topper ($\mathbf{P_{SWT}}$). The effective $\mathbf{P_{SWT}}$ parameters for corrected, basic and averaged formulations according to (12.3) are:

$$\tau_{ar.c} = \sqrt{\tau_{a.c}\tau_{max.c}}.,\quad \tau_{ar.b} = \sqrt{\tau_{a.b}\tau_{max.b}},\quad \tau_{ar.f} = \sqrt{\tau_{a.f}\tau_{max.f}}.$$

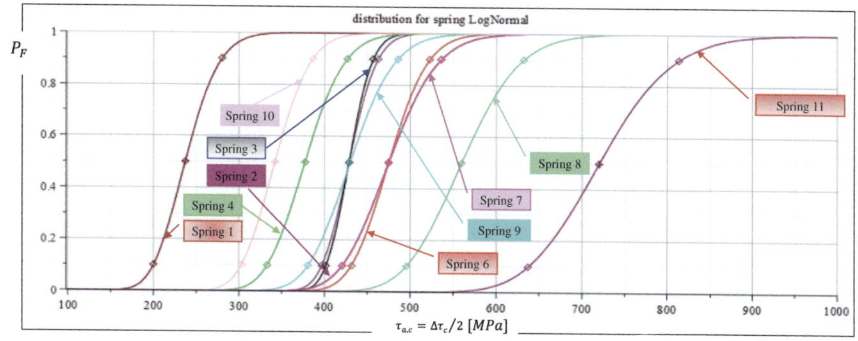

Fig. 14.4 Cumulative failure probabilities in the median range for tested springs. Abscise axis presents the amplitude of corrected stress $\tau_{a.c} = \Delta\tau_c/2$

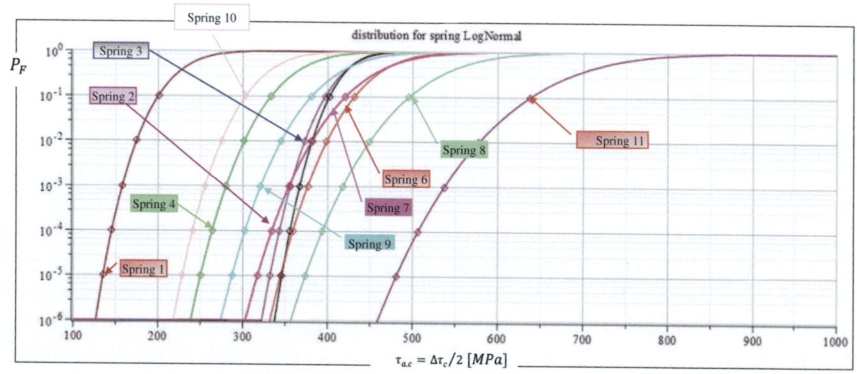

Fig. 14.5 Cumulative failure probabilities in the low range for tested springs. Abscise axis presents the amplitude of corrected stress $\tau_{a.c} = \Delta\tau_c/2$

With these definitions of effective stress, the number of cycles to failure provides once again the Eq. (14.19) with the matching coefficients. The proper coefficients for fatigue evaluations, based on $\mathbf{P_{SWT}}$ effective stress are presented in Table 14.4. This table summarizes the corresponding values of scatter rangesT_σ, scale parameters μ, strength exponent $b(R_\sigma)$, the reciprocal stress exponent $p_\sigma = -b^{-1}(R_\sigma)$ and the coefficients$\tau_{f0.c}{\prime}$, $\tau_{f0b}{\prime}$, $\tau_{f0f}{\prime}$ for evaluations of fatigue life. Clearly, that solitary coefficients$\tau_{f0.c}{\prime}$, $\tau_{f0b}{\prime}$, $\tau_{f0f}{\prime}$ differ from the similar values of Table 14.3. The stress exponents and scatter ranges are in both cases equal.

Table 14.2 Experimentally determined values of scatter range T_σ and scale parameter μ of the lognormal distributions

	Source	Material		Tension strength)(MPa	Torsion strength (MPa)	Torsional yield point, $\tau_{t0.04}$	R_σ	Amplitude of corrected stress $\tau_{c.a}$ (MPa)
1	[IGF16873]	1.4310 X10CrNi18-8		1958	1491	694	0.3763	237
2		1.4568 X7CrNiAl17-7	New composition	1398	1400	677	0.2020	428
3			Conventional composition	1441	1345	640	0.2020	428
4		DH		1942	1389	846	0.286	377
5		VDSiCr 54SiCrV6	400 °C	2103	1645	1118	0.203	475
6			360 °C				0.203	475
7			400 °C[1]				0.203	475
8	Mayer et al. (2015)			1984	1558		0.1	560
9							0.35	430
10	Mayer et al. (2016)						0.5	342
11							− 1	720

Cumulative failure probabilities for the wires are presented on the plots (Figs. 14.6, 14.7, 14.8 and 14.9). These curves display actually the material parameters, which are independent on spring indices and stress ratios. The curves of cumulative failure probabilities for the application of the effective $\mathbf{P}_{\mathbf{SWT}}$ parameters on abscise axis are displayed on Figs. 14.6 and 14.7 in two different ranges (medium and low failure probabilities). Figures 14.8 and 14.9 show cumulative failure probabilities for the effective Walker parameters $\mathbf{P}_{\mathbf{W}}$ on the abscise axis. Two graphs display the probability functions in medium and low failure probability ranges (Figs. 14.10 and 14.11).

14.3 Scale-Dependent Propagation of Straight Crack

14.3.1 S–N Curves of Spring Steels in Transition Range

It is commonly acknowledged (Abi et al., 2001, Sect. 3; FKM, 2018a, 2018b, Sect. 2.1.2.2), that the transition from high to low amplitudes of the stress occurs around a certain point. In logarithmic scaling, the fatigue life curves are nearly piecewise-straight lines with a corner point. The slopes of the straight lines left

[1] Flattened coils.

Table 14.3 Values scatter range T_σ, scale parameter μ of the lognormal distributions for Walker exponents $\tilde{\gamma}$ (Exp. Data [IGF16873] Mayer et al. (2015, 2016))

	$1/T_\sigma$	μ	$\tilde{\gamma}$	$p_\sigma = -b_0^{-1}$	b_0	τ'_{f0b}	τ'_{f0f}	τ'_{f0c}
	Scatter range of failure probabilities	scale parameter of lognormal distribution	Walker exponent	Reciprocal strength exponent	Strength exponent	Basic (MPa)	Averaged fatigue (MPa)	Corrected (MPa)
1	1.4	0.131275	0.15	10	– 0.1	2065	2299	2534
2	1.166	0.059919	0.15	16	– 0.0625	1392	1550	1708
3	1.137	0.050093	0.15	12	– 0.833	2143	2387	2631
4	1.28	0.096312	0.15	72	– 0.01388	482	536	591.1
5	1.275	0.094786	0.15	29	– 0.034482	864	963	1061
6	1.213	0.075337	0.15	23	– 0.04348	1041	1160	1278
7	1.275	0.094786	0.15	33	– 0.03030	792	882	972
8	1.275[2]	0.094786	0.15	33[3]	– 0.03030	884	985	1086
9			0.15			895	998	1099
10			0.15			890	992	1093
11			0.15			577	642	708.3

and right from the corner point are different. Because of significant increasing of the numbers of cycles to failure for the stresses below the corner point, the corresponding stress is commonly referred to as the endurance strength. It is frequently accepted, that the stress level below the endurance strength can be applied to a material without causing fatigue failure for an infinite number of loading cycles. For the models of crack propagation, the assumption of endurance strength appears as endurance threshold limit for the stress intensity factor (Sect. 8.3.2). However, the assumptions of endurance limits could be inacceptable for the advanced machinery.

The S–N-curves evaluation for $SiCrV$ valve spring wire is based on the experimental results (Kaiser & Berger, 2005). The curves represent fatigue life up to 7×10^8 cycles of shot peened helical compression springs of $SiCr$-alloyed valve spring wire, DIN EN 10270, with d $= 2[mm]$. The S–N curves were obtained for$R_\sigma = 0.05$. The original Fig. 14.12 from (Kaiser & Berger, 2007) displays S–N curves for 10, 50 and 90% survival probability. The ordinate is the range of corrected shear stress $\Delta\tau_c$. To

[2] Estimated value.

[3] Estimated value.

Table 14.4 Values scatter range T_σ, scale parameter μ of the lognormal distributions for P_{SWT} (Exp. Data [IGF16873] Mayer et al. (2015, 2016))

	$1/T_\sigma$	μ	$\tilde{\gamma}$	$p_\sigma = -b_0^{-1}$	b_0	τ'_{f0b}	τ'_{f0f}	τ'_{f0c}
	Scatter range of failure probabilities	Scale parameter of lognormal distribution	Walker exponent	Reciprocal strength exponent	Strength exponent	Basic (MPa)	Averaged fatigue (MPa)	Corrected, (MPa)
1	1.4	0.131275	0.15	10	– 0.1	1373	1529	1685
2	1.166	0.059919	0.15	16	– 0.0625	1009	1123	1238
3	1.137	0.050093	0.15	12	– 0.833	1554	1730	1907
4	1.28	0.096312	0.15	72	– 0.01388	339	377	415
5	1.275	0.094786	0.15	29	– 0.034482	626	697	769
6	1.213	0.075337	0.15	23	– 0.04348	755	840	926
7	1.275	0.094786	0.15	33	– 0.03030	574	639	705
8	1.275	0.094786	0.15	33	– 0.03030.	669	745	821
9			0.15			605	673	742
10			0.15			548	610	673
11			0.15			577	642	708.3

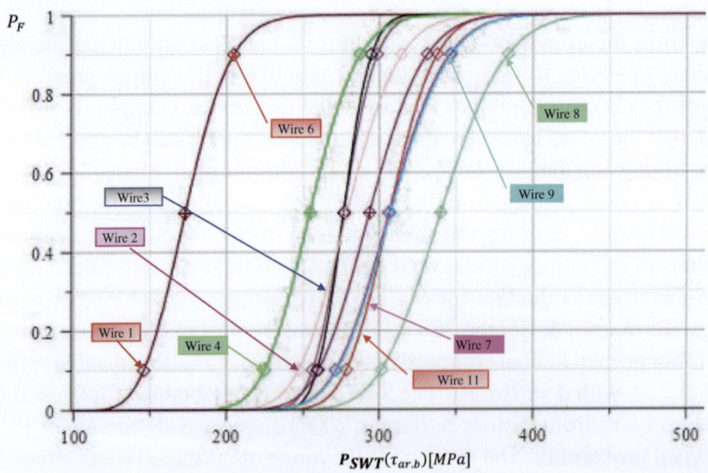

Fig. 14.6 Cumulative failure probabilities in the median range for wires. Abscise axis presents the averaged P_{SWT} parameter, $\mathbf{P_{SWT}}(\tau_{\mathbf{ar.b}})$

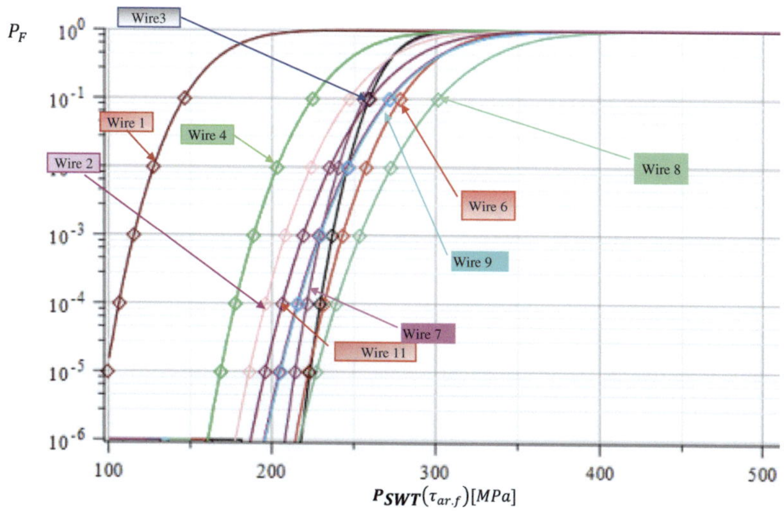

Fig. 14.7 Cumulative failure probabilities in the low range for wires. Abscise axis presents the averaged P_{SWT} parameter, $\mathbf{P_{SWT}}(\tau_{\mathbf{ar.f}})$

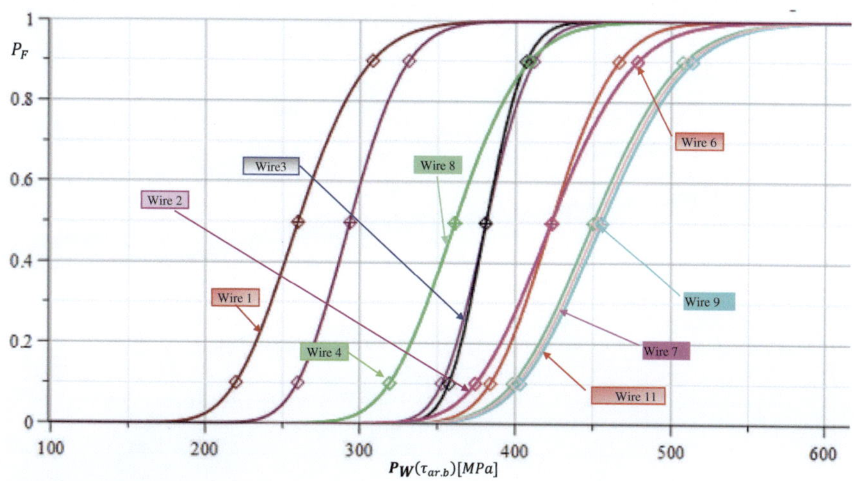

Fig. 14.8 Cumulative failure probabilities in the median range for wires. Abscise axis presents the averaged Walker parameter $\mathbf{P_W}(\tau_{\mathbf{ar.b}})$

present the results in invariant form, the $\boldsymbol{P}_{\mathbf{SWT}}$ parameter was built, starting from the corrected shear stress $\Delta\tau_c$. The equations if the S–N lines in terms of effective stresses τ_{ar} are shown in Tables 14.5 and 14.6. Thus, the fatigue life curves are nearly piecewise-straight lines with the corner point at 5×10^6 cycles.

To the left from the corner point there is the region of high amplitudes of stress intensity factor and the number of cycles to failure is less than 5×10^6 cycles. To

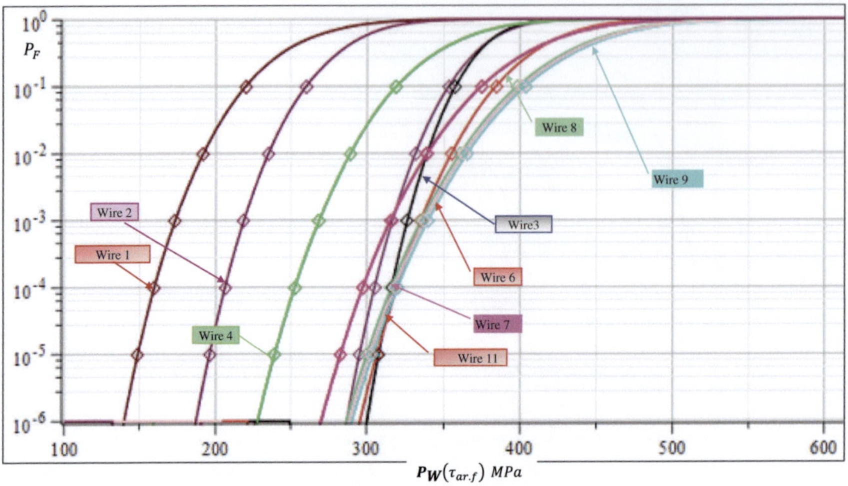

Fig. 14.9 Cumulative failure probabilities in the low range for wires. Abscise axis presents the averaged Walker parameter $\mathbf{P_W}(\tau_{\mathbf{ar.f}})$

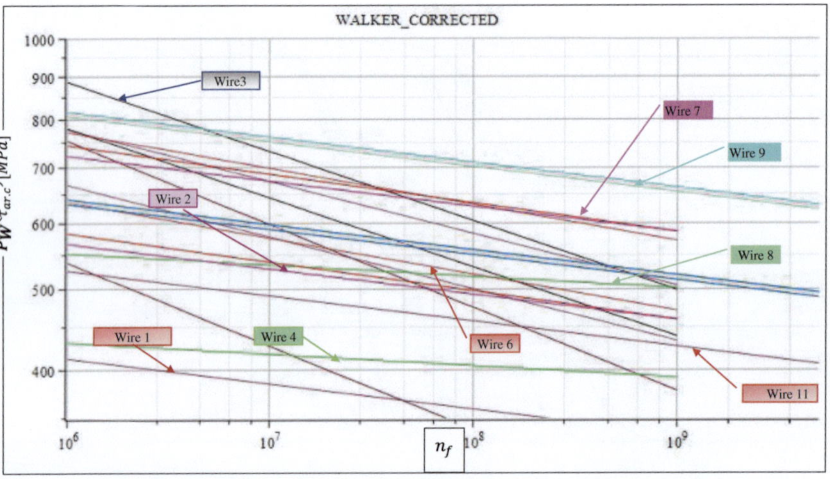

Fig. 14.10 Median S–N lines of wires. Ordinate presents the amplitude of corrected Walker parameter $\mathbf{P_W}(\tau_{\mathbf{ar.c}})$

the right of the corner point the amplitudes of stress intensity factor are below the endurance strength and the fatigue life is significantly longer. Using the preferred method (Sect. 14.2.3) for displaying the material properties, the S–N lines for different failure probabilities are displayed. The results of the evaluation of the S–N lines for 0.1%, 1 ppm failure probability are displayed on Figs. 14.13, 14.14 and 14.15. For the characterization the Walker parameter $\mathbf{P_W}$ with different exponents

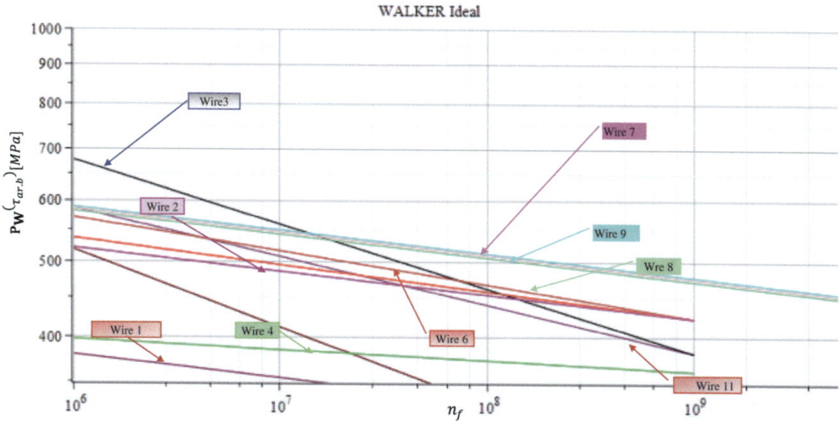

Fig. 14.11 Median S–N lines of wires. Ordinate presents the basic Walker parameter $\mathbf{P}_{\mathbf{W}}(\tau_{\mathbf{ar.b}})$

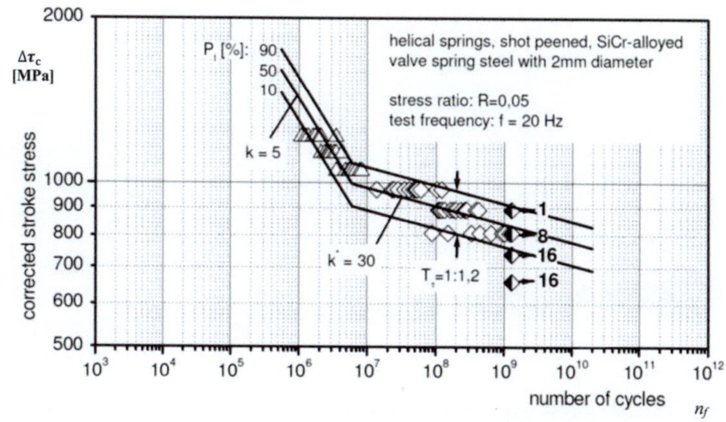

Fig. 14.12 S–N curves for 10, 50 and 90% survival probability for shot-peened helical compression springs of Si-Cr-alloyed valve spring wire with d = 2 mm (Kaiser & Berger, 2005)

was applied. Figure 14.13 shows S–N curves for Walker exponent $\tilde{\gamma} = 2/3$, Fig. 14.14 for Walker exponent $\tilde{\gamma} = 1/3$ and Fig. 14.15 for $\tilde{\gamma} = 1/2$ or $\mathbf{P}_{\mathbf{SWT}}$ effective stress. The lower values of Walker exponents preferably match the S–N lines for different mean stress. Regrettably, the experimental results (Kaiser & Berger, 2005) do not cover sufficiently the different mean stresses.

In this Section, we apply the standard methods for evaluation of the effective S–N lines. The explanations of such material behavior are based on the micromechanical models. The possible models to explain the alternation of micromechanical mechanism will be discussed in Sect. 14.4.

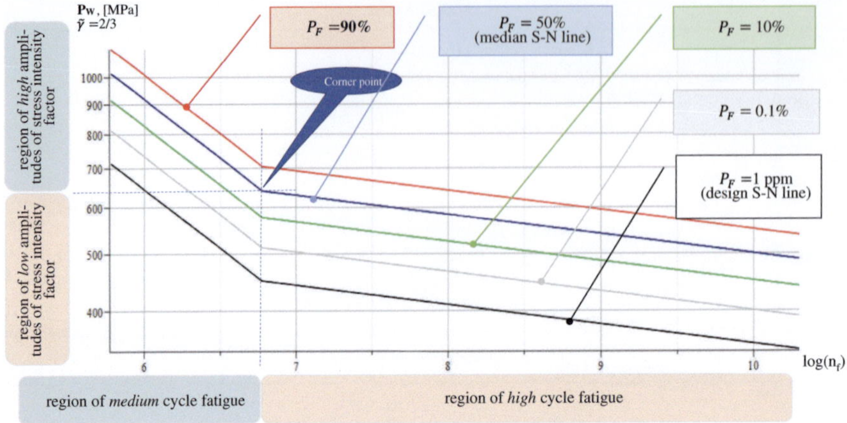

Fig. 14.13 S–N curves for Walker exponent $\tilde{\gamma} = 2/3$ (Kaiser & Berger, 2005)

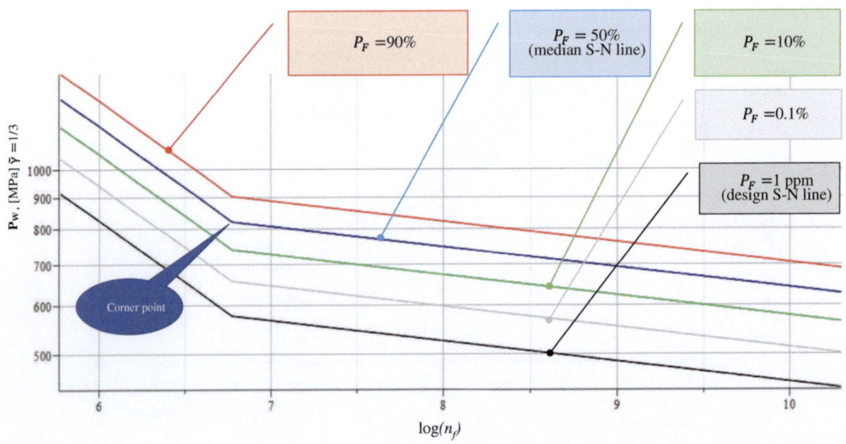

Fig. 14.14 S–N curves for Walker exponent $\tilde{\gamma} = 1/3$ (Kaiser & Berger, 2005)

14.3.2 Piecewise Linear Paris-Irvin Equation

As shown above, the numerous experimental studies display the S–N curves as nearly piecewise-linear graphs. This phenomenon could be explained from the crack propagation viewpoint. As the first step, we examine the planar crack with two regimes of propagation. The growth of cracks occurs in the direction that attempt to maximize the subsequent energy release rate and minimize the strain energy density. The optimum propagation path maintains an orientation normal to the maximum extension strains. Thus, the shape of the crack should be an ideally straight line. The above formulations FCG were based on endurance limit concepts, such (11.22), (11.23) and (11.36). These formulations FCG cannot correctly describe the piecewise-linear S–N

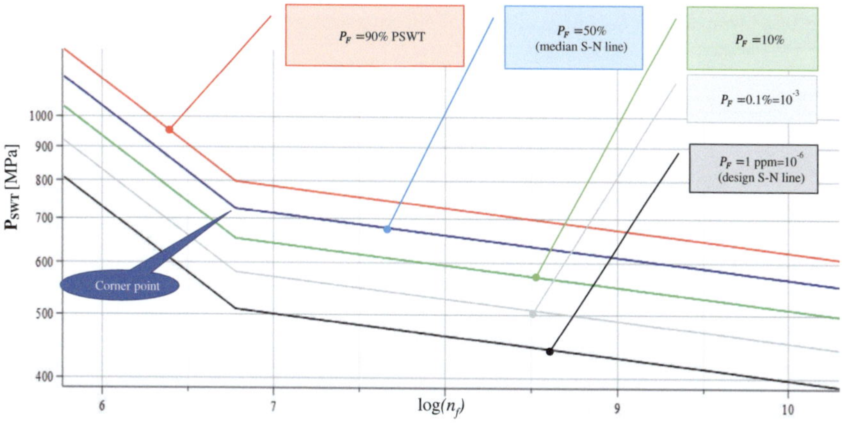

Fig. 14.15 S–N curves for Walker exponent $\tilde{\gamma}$ =1/2, P_{SWT}, (Kaiser & Berger, 2005)

Table 14.5 Formulas for S–N lines in terms of effective stresses τ_{ar} for $T_\sigma = \frac{1}{1.234}$

Walker exponent	$\tilde{\gamma} = \frac{1}{2} \leftrightarrow P_{SWT}$	
Region	UHCF[4] $k_{UHCF} = 30$	HCF[5] $k_{HCF} = 5$
	$2 \times 10^6 \leq n_f \leq 2 \times 10^{10}$	$7 \times 10^5 \leq n_f \leq 2 \times 10^6$
$P_F = 90\%$[6]	$\tau_{ar} = 1342 \cdot n_f^{-1/k_{UHCF}}$	$\tau_{ar} = 18098 \cdot n_f^{-1/k_{HCF}}$
$P_F = 50\% P_F$	$\tau_{ar} = 1220 \cdot n_f^{-1/k_{UHCF}}$	$\tau_{ar} = 16453 \cdot n_f^{-1/k_{HCF}}$
$P_F = 10\% P_F$	$\tau_{ar} = 1098 \cdot n_f^{-1/k_{UHCF}}$	$\tau_{ar} = 14807 \cdot n_f^{-1/k_{HCF}}$
$P_F = 0.1\% = 10^{-3}$	$\tau_{ar} = 976 \cdot n_f^{-1/k_{UHCF}}$	$\tau_{ar} = 13161 \cdot n_f^{-1/k_{HCF}}$
$P_F = 1$ ppm $= 10^{-6}$	$\tau_{ar} = 857 \cdot n_f^{-1/k_{UHCF}}$	$\tau_{ar} = 11560 \cdot n_f^{-1/k_{HCF}}$

The influence of mean stress is accounted using P_{SWT} approach (Exp. Data Kaiser et al. (2011))

curves. As an alternative, the piecewise-linear propagation function is introduced for the explanation of this phenomenon.

For this purpose, we consider the crack, which propagates in a plane plate. The straight line, which connects both tips of the crack, will be referred to as "director vector" of the crack. In the XY coordinates, the director vector of the ideally straight crack lengthens *exactly* towards the X-axis. The tensile stress acts in the direction of the Y-axis. The range of the component tension stress σ_{YY} is $\Delta\sigma$. Thus, the range of stress intensity factor of the planar crack is proportional to the square root of its length a:

[4] Ultra-high cycle fatigue, $2 \cdot 10^6 \leq n_f \leq 2 \cdot 10^{10}$.

[5] High cycle fatigue, $7 \cdot 10^5 \leq n_f \leq 2 \cdot 10^6$.

[6] Failure probability for $T_\sigma = 1/1.234$.

Table 14.6 Formulas for S–N lines in terms of effective stresses τ_{ar} for $T_\sigma = \frac{1}{1.234}$

Walker exponent	$\tilde{\gamma} = \frac{1}{3}$		$\tilde{\gamma} = \frac{1}{2} \leftrightarrow P_{SWT}$		$\tilde{\gamma} = \frac{2}{3}$	
Region	UHCF $k_{UHCF} = 30$	HCF $k_{HCF} = 5$	UHCF $k_{UHCF} = 30$	HCF $k_{HCF} = 5$	UHCF $k_{UHCF} = 30$	HCF $k_{HCF} = 5$
	$2 \times 10^6 \leq n_f \leq 2 \times 10^{10}$	$7 \times 10^5 \leq n_f \leq 2 \times 10^6$	$2 \times 10^6 \leq n_f \leq 2 \times 10^{10}$	$7 \times 10^5 \leq n_f \leq 2 \times 10^6$	$2 \times 10^6 \leq n_f \leq 2 \times 10^{10}$	$7 \times 10^5 \leq n_f \leq 2 \times 10^6$
n_f	$\left(\frac{\tau_{FP}}{\tau_{ar}}\right)^{-k_{UHCF}}$	$\left(\frac{\tau_{FP}}{\tau_{ar}}\right)^{-k_{HCF}}$	$\left(\frac{\tau_{FP}}{\tau_{ar}}\right)^{-k_{UHCF}}$	$\left(\frac{\tau_{FP}}{\tau_{ar}}\right)^{-k_{HCF}}$	$\left(\frac{\tau_{FP}}{\tau_{ar}}\right)^{-k_{UHCF}}$	$\left(\frac{\tau_{FP}}{\tau_{ar}}\right)^{-k_{HCF}}$
$90\% P_F$	$\tau_{FP} = \tau_{90\%}$					
	1519	20,489	1342	18,098	1185	15,986
$50\% P_F$ Median line	$\tau_{FP} = \tau_{50\%}$					
	1381	18,626	1220	16,453	1078	14,531
$10\% P_F$	$\tau_{FP} = \tau_{10\%}$					
	1243			14,807	970	13,080
$0.1\% = 10^{-3} P_F$	$\tau_{FP} = \tau_{0.1\%}$					
	1105	14,900	976	13,161	862	11,626
$1 ppm = 10^{-6} P_F$	$\tau_{FP} = \tau_{1 ppm}$					
	970	13,087	857	11,560	757	10,211

The influence of mean stress is accounted using P_W approach with different Walker exponents $\tilde{\gamma}$ (Exp. Data Kaiser et al. (2011))

$$\Delta K = \Delta\sigma \cdot \sqrt{\pi a}.$$

If the length of crack is tiny, the elongation of crack length pro cycle gives δ_l. This region is the low range fatigue region, $\Delta K \leq \Delta K^*$. Crack tries to follow the outer grain boundaries, because of the high defect density of the crystal lattice and weakness of polycrystalline structure.

When the crack grows, the process of crack extension accelerates and the elongation of crack length pro cycle will be higher δ_s. Crack acquire enough strength to deform the polycrystalline on its tip and could spread in a shortest way across the grain regions. Both processes occur simultaneously, but the tendency of grain-destruction dominates for the higher stresses, $\Delta K \geq \Delta K^*$. The piecewise-linear

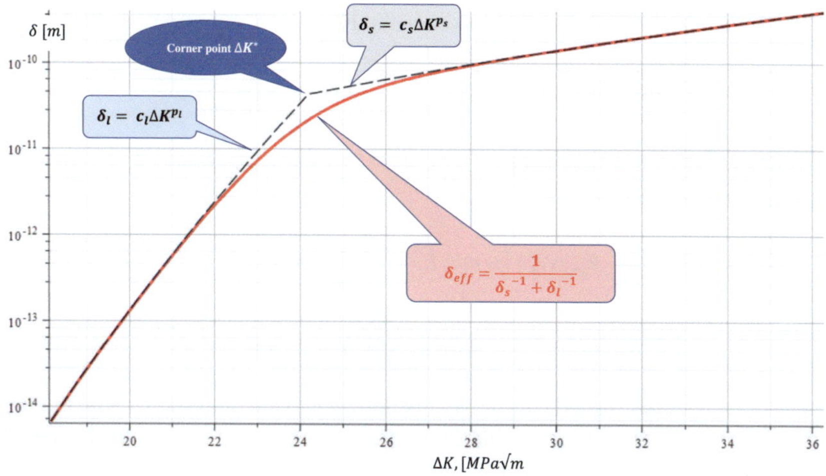

Fig. 14.16 Piecewise-linear propagation function

propagation function is shown on Fig. 14.16. The value ΔK^* matches the ordinate of the "corner point" in coordinates $\Delta K - \delta$ (Fig. 14.16).

The increments of the crack length pro cycle are determined by two Paris-Irvin relations. Planar crack propagates along the *X*-axis. The length increments for both regions pro one load cycle are:

$$\begin{aligned} \delta_s &= c_s \Delta K^{p_s}, \text{ if } \Delta K \geq \Delta K^*; \\ \delta_l &= c_l \Delta K^{p_l}, \text{ if } \Delta K \leq \Delta K^*. \end{aligned} \tag{14.20a}$$

The parameters c_s, c_l, p_s, p_l express the crack growth for both ranges of stress intensity factor.

To describe the transition region between both regimes, the total length increment pro one load cycle is assumed as:

$$\delta_{eff} = \left(\delta_s^{-1} + \delta_l^{-1}\right)^{-1}. \tag{14.20b}$$

The hypothesis (14.20a) leads to the piecewise-linear propagation function and abrupt changing of the propagation character. On the other hand, the expression (14.20b) accounts the parallel contributions of the in-grain (grain-destruction) and the grain-boundary growth. For the setting (14.20b), the smooth transition region arises instead of a solitary cross-point between two regimes (Fig. 14.16).

The value of stress intensity factor at the "corner point" ΔK^* follows from the condition $\delta_s = \delta_l$:

$$\Delta K^* = \left(\frac{c_s}{c_l}\right)^{\frac{1}{p_l - p_s}}. \tag{14.20c}$$

The application of Eq. (14.20) modifies the Paris-Irvin equation Eq. (9.21):

$$\frac{da}{dn} = F(\Delta K),\ F(\Delta K) = \left(\frac{1}{c_s \Delta K^{p_s}} + \frac{1}{c_l \Delta K^{p_l}}\right)^{-1}. \tag{14.21}$$

From (14.21) follow the limit asymptotic S–N lines for both low and high ranges of stress intensity factor:

$$F_s(\Delta K) = c_s \Delta K^{p_s}\text{, if } \Delta K \geq \Delta K^*, \tag{14.22a}$$

$$F_l(\Delta K) = c_l \Delta K^{p_l}\text{, if } \Delta K \leq \Delta K^*. \tag{14.22b}$$

The crack spreads from the initial inclusion size δ to the final size of wire diameter d. The solution of Eq. (14.21) with the initial condition $n(a = \delta) = 0$ delivers the number of cycles to total failure$n_f = n(a = \text{d})$:

$$n_f = n_f(c_s, c_l) = \left[\frac{2}{c_s(p_s-2)\sigma^{p_s} a^{\frac{p_s}{2}-1}\pi^{\frac{p_s}{2}}} + \frac{2}{c_l(p_l-2)\sigma^{p_l} a^{\frac{p_l}{2}-1}\pi^{\frac{p_l}{2}}}\right]_{a=\delta}^{a=\text{d}}. \tag{14.23}$$

The function $n_f(c_s, c_l)$ depends upon physical parameters c_s, c_l and upon geometrical parameters δ and d. Using the reported experimental results (Kaiser, Berger, 2007), it is possible to determine the coefficients and exponents of Eq. (14.22), which lead to the experimental S–N line (blue line, Fig. 14.17). We assume, that the failure caused by the defect of the size $\delta = 30$ [μm]. With this assumption, the reported S–N line delivers for valve spring wire DIN EN 10270 the following physical coefficients of the Eq. (14.20):

$$c_s = 1.064 \cdot 10^{-18}[m] \cdot \left([\text{MPa}] \cdot [m]^{\frac{1}{2}}\right)^{-p_s}, \quad p_s = 5.5,$$

$$c_l = 2.7448 \cdot 10^{-53}[m] \cdot \left([\text{MPa}] \cdot [m]^{\frac{1}{2}}\right)^{-p_l}, \quad p_l = 30.5.$$

According to Eq. (14.20c), the "corner point" of stress intensity factor is equal to:

$$\Delta K^* = 24.185\,[\text{MPa}\sqrt{m}].$$

Equation (14.23) estimate of the influence of inclusion size δ on the fatigue life of springs. For the estimation, we vary the size of the inclusion δ and come back to the different S–N lines with the fixed physical coefficients and wire diameter δ. The results of evaluation of the inclusion influence on the S–N median lines are presented on Fig. 14.17. The calculation was performed for the defects of the sizes $\delta = 40$ [μm] (red line, Fig. 14.17) and $\delta = 50$ [μm] (magenta line, Fig. 14.17).

Further, we can estimate the influence of probabilistic effects on the S–N lines Eq. (14.23). The Eq. (14.23) delivers the median S–N line for the stress, which is

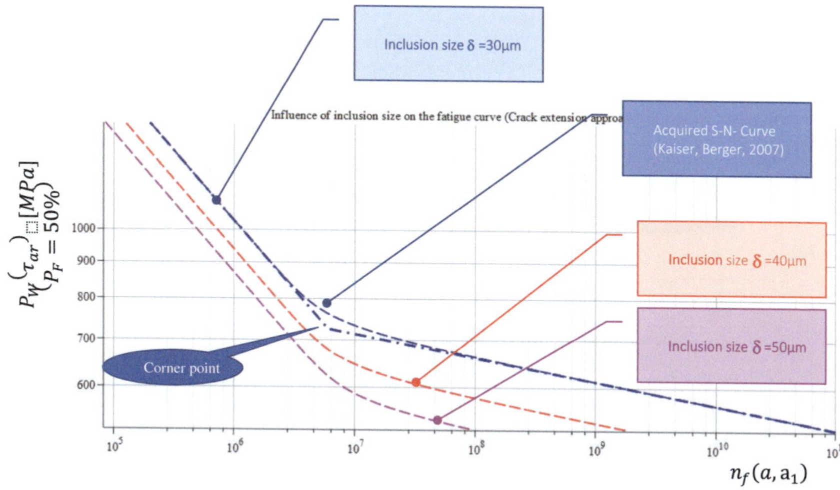

Fig. 14.17 Influence of inclusion size on the fatigue curve (Crack extension approach)

corresponds the mean value of the lognormal distribution $\sigma = \Lambda$. To reduce the failure probability, the cycling stress σ must be reduced, as shown in Table C-4, Appendix C. To draw the S–N line, which corresponds the certain cumulative failure probability, one determines the ratio of stress parameter stress to the median stress Λ (Table C.4). The median stress $\sigma = \Lambda$ should be replaced by the stress parameter to the requisite failure probability.

For example, for $\boldsymbol{P}_F = 10^{-6}$ the Table C.4 gives the value

$$\sigma = \Lambda \cdot T_{\sigma}^{1.845229} = 0.6478\Lambda.$$

Substitution of this value for stress into (14.23) results the S–N line for $\boldsymbol{P}_F = 10^{-6}$.

Based on crack propagation approach, Fig. 14.18 draws S–N lines for different cumulative failure probabilities.

14.4 Scale-Dependent Propagation of Probabilistic Crack

14.4.1 Simulation of Crack Propagation with Random Deviation

In Chap. 9, the method to derive the S–N curves from the micromechanical model of the crack propagation was presented. The method opens the way to somewhat adopt the micromechanics of propagation and derive the new S–N curves.

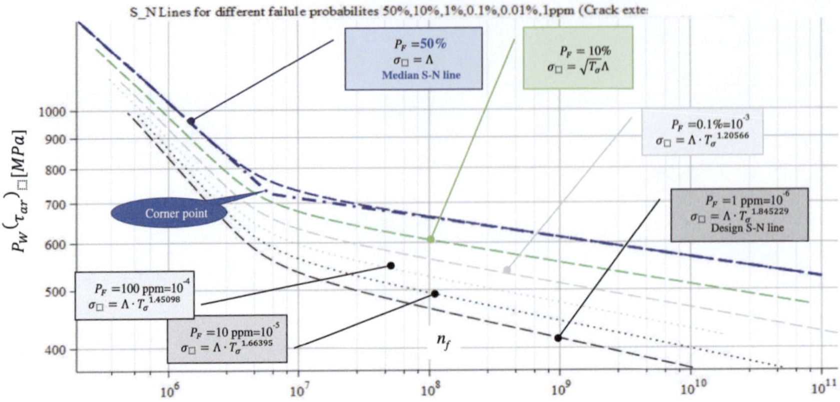

Fig. 14.18 S–N curves for different failure probabilities and for an inclusion size 30 [μm]

We assumed above the planar crack propagation in the energetically preferred direction. Thus, if the material is homogeneous, the crack spreads in the energetically preferred direction. This direction results the maximal possible elongation for the given direction of the greatest principal stress. In the second step, we propose the mechanism of crack extension, which accounts the deviation of crack from its energetically preferred direction. The consideration is the following. Evidently, that the process of fracture is an anisotropic phenomenon. The growth of micro-cracks occurs in the directions, which attempt to maximize the energy release rate and minimize the strain energy density. The crack spreading along the grain boundaries requires less energy in comparison to the spreading across the perfect grain crystals. The grain boundaries are randomly oriented. Consequently, the optimum propagation path deviates randomly from the ideal straight path, normal to the maximum principal stress. The deviation occurs also due to the disordered orientations of anisotropy in grains and the highly warped grain boundaries. Thus, due to an arbitrary anisotropy of single polycrystalline domains and their irregular shapes, the crack does not follow the straight direction and randomly deviates from the energetically preferred direction. The models of disordered crack propagation were reviewed in (Schwalbe, 1980). The damage mechanisms on the VHCF regime are reviewed in (Zimmermann, 2019). We use these considerations for the physical background of the proposed mathematical model.

Our aim is to evaluate quantitatively the influence of steel grain sizes on the extension of fatigue crack. For this purpose, we emphasize the various dimensions of single crystals for diverse steel crystalline structures (Verhoeven, 2007, Chap. 1, Fig. 1.2). For explanation, we show the crystalline structures, which appear in the duration of the quenching of steel alloys (Fig. 14.19). The schematic representation of the temperature- and time-dependent transformation ranges shows the ranges of martensite, bainite and perlite formation):

(1) *Quenching to martensite*;

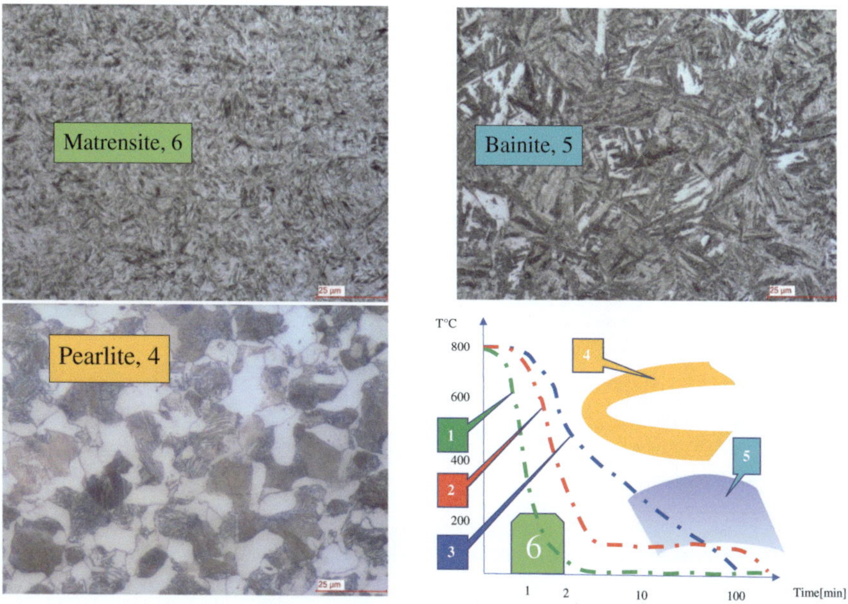

Fig. 14.19 The schematic representation of the temperature- and time-dependent transformation ranges: (1) Quenching to martensite (2) Intermediate grades (3) Cooling to bainite (4) Pearlite area (5) Bainite area (6) Martensite area

(2) *Intermediate grades*;
(3) *Cooling to bainite*;
(4) *Pearlite area*;
(5) *Bainite area.*

The material data of *51CrV4* are given in (DIN EN 10132-4:2003-04).

Consider a zigzag-shaped crack, which propagates in a plane plate (Fig. 14.20). The straight line, which connects both tips of the crack, will be referred to as "director vector" of the crack. In the XY coordinates, the director vector of the crack lengthens *roughly* towards the X-axis. The tensile stress acts in the direction of the Y-axis. The range of the component tension stress σ_{YY} is $\Delta\sigma$.

At low stress intensity factor range a highly faceted fracture surface resulted due to crystallographic fracture along slip bands. We apply a simple model for the description of the crack branching. We assume, that the crack propagates on each load cycle in different, stochastic directions. The crack spreading is analogous to the Brownian motion with drift. The applied stress causes the energy release near the tip of the crack and consequently the drift of the crack path.

As already discussed in Sect. 14.3.2, there are at least two diverse mechanisms of crack elongation and correspondingly two length scales.

The first mechanism governs the propagation of the crack, if the range of stress intensity factor is high. The crack spreads across the grains, but its direction must

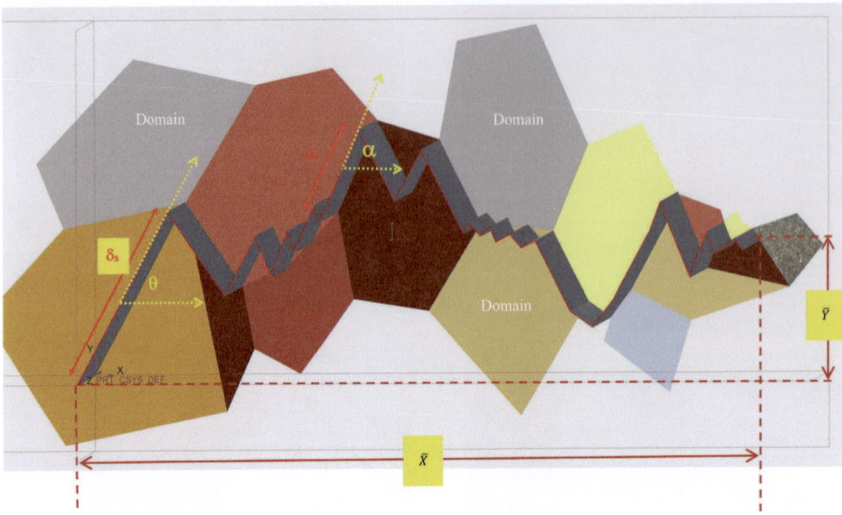

Fig. 14.20 The principal scheme for stochastic crack propagation

not coincide with the X-axis. The random angle between the macroscopically crack director and direction of the microscopical crack spreading is

$$-\pi < \theta < \pi.$$

This angle is a stochastically function on each propagation event. If the elongation of crack, projected on the macroscopically crack plane of crack, is positive, the crack spreads. In this case

$$-\frac{\pi}{2} < \theta < \frac{\pi}{2}.$$

Otherwise, if

$$\theta < -\frac{\pi}{2} \text{ or } \frac{\pi}{2} < \theta,$$

the crack spreads backwards. However, the backward spreading of crack is energetically intolerable, and the crack suspends one elongation cycle.

The second mechanism is responsible for the crack spreading at low ranges of stress intensity factor. In this case a highly faceted fracture surface resulted due to crystallographic fracture along slip bands. The crack spreads over the grain region. The random angle between the macroscopically crack plane and direction of the slip band is:

$$-\pi < \alpha < \pi.$$

The crack propagates forward, if the angle keeps lying between:

$$-\frac{\pi}{2} < \alpha < \frac{\pi}{2}.$$

Otherwise, if

$$-\frac{\pi}{2} > \alpha > -\pi, \quad \pi > \alpha > \frac{\pi}{2}$$

the direction of crack extension is opposite to the director direction. The backward extension of crack is energetically unfavorable and will be jammed.

For each cycle, both angles (θ, α) are the statistically independent random values.

The increments of the crack length pro cycle are determined by two Paris-Irvin relations. In a similar manner as in Eq. (14.20), the length increments correspondingly inside the grains and along slip bands are:

$$\delta_s = c_s[\cos(\theta).\Delta K]^{p_s}, \text{ if } \quad \Delta K \geq \Delta K^*; \tag{14.24a}$$

$$\delta_l = c_l[\cos(\alpha) \cdot \Delta K]^{p_l}, \text{ if } \quad \Delta K \leq \Delta K^*. \tag{14.24b}$$

The components of the director vector $X_p(n), Y_p(n)$ are the functions of number of cycles n. The projected length of the crack is X_p and this value replaces the length of the planar crack in the equation for crack intensity factor for the considered uniaxial stress state. Thus, the range of stress intensity factor of the zigzag crack is proportional to the square root of the projected length X_p:

$$\Delta K = \Delta\sigma \cdot \sqrt{\pi X_p}.$$

The critical projected length separates for the given stress range both regions:

$$X^* = \frac{1}{\pi}\left(\frac{\Delta K^*}{\Delta\sigma}\right)^2.$$

One of two potential mechanisms of crack propagation will dominate. The dominating mechanism is responsible for the crack spreading. The travel of the crack tip in the XY-plane according to (14.24) reads:

$$\begin{cases} \frac{\mathrm{d}X_p}{\mathrm{d}n} = \mathrm{F}_X(\theta, \alpha), \\ \frac{\mathrm{d}Y_p}{\mathrm{d}n} = \mathrm{F}_Y(\theta, \alpha). \end{cases} \tag{14.25}$$

The Eq. (14.25) are the stochastic differential equations. Obviously, that the nonplanar crack deviates from the abscise axis. Two variants of the propagation functions were tested. The first, stiffer variant for components of propagation function in (14.25) is:

$$\begin{cases} \mathrm{F}_X(\theta, \alpha) = \min(\cos(\theta) \cdot \delta_s, \cos(\alpha) \cdot \delta_l), \\ \mathrm{F}_Y(\theta, \alpha) = \min(\sin(\theta) \cdot \delta_s, \sin(\alpha) \cdot \delta_l). \end{cases} \tag{14.26a}$$

The second, smoother variant reads:

$$\begin{cases} \mathrm{F}_X(\theta, \alpha) = \left(\frac{1}{\cos(\theta) \cdot \delta_s} + \frac{1}{\cos(\alpha) \cdot \delta_l}\right)^{-1}, \\ \mathrm{F}_Y(\theta, \alpha) = \left(\frac{1}{\sin(\theta) \cdot \delta_s} + \frac{1}{\sin(\alpha) \cdot \delta_l}\right)^{-1}. \end{cases} \tag{14.26b}$$

14.4.2 Closed-Form Solution of the Stochastic Differential Equation

The closed-form solutions are advantageous for analysis. The closed-form solution of the probabilistic differential equation is based on the averaging of the cyclical crack elongations. For averaging procedure, the angular elongation in Eq. (14.25) must be integrated over the limits:

$$-\frac{\pi}{2} < \theta < \frac{\pi}{2}, \quad -\frac{\pi}{2} < \alpha < \frac{\pi}{2}.$$

The evaluation of integrals is based on the formula from Appendix C:

$$S(p) = \frac{1}{2\pi} \int_{-\pi/2}^{\pi/2} \cos^{2p} \varphi \mathrm{d}\varphi = \frac{1}{2\sqrt{\pi}\Gamma(p+1)} \Gamma\left(p + \frac{1}{2}\right). \tag{14.27}$$

Applying the Eq. (14.27), the averaged propagation functions for short or for long range read:

$$\overline{F_s(\Delta K)} = \frac{c_s}{2\pi} \int_{-\pi/2}^{\pi/2} [\cos(\theta) \cdot \Delta K]^{p_s} \mathrm{d}\theta = S(p_s) c_s \Delta K^{p_s}, \tag{14.28a}$$

$$\overline{F_l(\Delta K)} = \frac{c_l}{2\pi} \int_{-\pi/2}^{\pi/2} [\cos(\alpha) \cdot \Delta K]^{p_l} \mathrm{d}\alpha = S(p_l) c_l \Delta K^{p_l}. \tag{14.28b}$$

Following from (14.26a) or (14.26b), the averaged propagation functions for both short and long ranges correspondingly read:

$$\overline{\mathrm{F}_X(\theta, \alpha)} = \min\big(S(p_s) c_s \Delta K^{p_s}, S(p_l) c_l \Delta K^{p_l}\big) \tag{14.29a}$$

or

$$\overline{\mathrm{F}_X(\theta,\alpha)} = \left(\frac{1}{S(p_s)c_s\Delta K^{p_s}} + \frac{1}{S(p_l)c_l\Delta K^{p_l}}\right)^{-1}. \tag{14.29b}$$

There are two variants of the deterministic differential equations for both propagation functions (14.29a) and (14.29b). The differential equations for the mean values of projected lengths $\overline{X}(n)$, $\overline{Y}(n)$ read according to the stiffer variant Eq. (14.29a):

$$\begin{cases} \frac{\mathrm{d}\overline{X}}{\mathrm{d}n} = \min(S(p_s)c_s\Delta K^{p_s}, S(p_l)c_l\Delta K^{p_l}), \\ \frac{\mathrm{d}\overline{Y}}{\mathrm{d}n} = 0. \end{cases} \tag{14.30a}$$

Similarly, for smoother variant (14.29b) the equations for the mean values of projected lengths $\overline{X}(n)$, $\overline{Y}(n)$ are:

$$\begin{cases} \frac{\mathrm{d}\overline{X}}{\mathrm{d}n} = \frac{1}{(S(p_s)c_s\Delta K^{p_s})^{-1}+(S(p_l)c_l\Delta K^{p_l})^{-1}}, \\ \frac{\mathrm{d}\overline{Y}}{\mathrm{d}n} = 0. \end{cases} \tag{14.30b}$$

In both cases (14.30a) and (14.30b), the mean value of $\overline{X}$ extends randomly with the cycle number and shows the crack elongation. The $\overline{X}$ component is a kind of Brownian motion with the stress-enforced drift. Because the stress $\sigma_{XX} = 0$, the mean value of $\overline{Y}$ is zero. The travelling crack oscillates over abscises axis. The component $\overline{Y}$ of oscillation is the classical Brownian motion without stress-enforcement.

If probabilistic effects are suppressed, both factors $S(p_s)$ and $S(p_l)$ turn to one. This case delivers again the deterministic S–N curves (14.23).

If the crack propagates randomly either in low amplitude or in high amplitude region or in both regions, we derive three S–N curves. Table 14.7 displays the review of possible combinations. The numbers of cycles for both short and long ranges follow from (14.23) after the correction of coefficients c_s, c_l.

The first S–N curve

$$\widetilde{n_f} = n_f(c_s, S(p_l)c_l), \tag{14.30}$$

represents the case of the reduction of crack length increments in the region of high amplitudes of stress intensity factor. The number of cycles to failure $\widetilde{n_f}$ increases in high stress region, because the crack spreads slowly.

The second S–N curve

$$\widetilde{\widetilde{n_f}} = n_f(S(p_s)c_s, c_l), \tag{14.31}$$

characterizes the case of the reduction of crack length increments in the region of low amplitudes of stress intensity factor. The crack propagates slowly, and the number of cycles in low stress region $\widetilde{\widetilde{n_f}}$ growths. The crack spreads for low amplitudes and for low lentgths slowly.

Table 14.7 Propagation patterns (Pattern D-Deterministic planar crack, S-Stochastic Brownian crack)

Stress intensity factor	Crack length	Pattern	Propagation function	Number of cycles to failure
High $\Delta K \geq \Delta K^*$	Short $X \leq X^*$	D	$F(\Delta K) = \left(\frac{1}{c_s \Delta K^{p_s}} + \frac{1}{c_l \Delta K^{p_l}}\right)^{-1}$, $F(\Delta K) = \frac{1}{\frac{9.3914\times10^{17}}{\Delta K^{5.5}} + \frac{3.643\times10^{52}}{\Delta K^{30.5}}}$	$n_f = n_f(c_s, c_l)$, $n_f = \frac{5.4153 \times 10^{92}}{\Delta\sigma^{30.5}} + \frac{3.8188 \times 10^{21}}{\Delta\sigma^{5.5}}$
Low $\Delta K \leq \Delta K^*$	Long $X \geq X^*$	D		
High $\Delta K \geq \Delta K^*$	Short $X \leq X^*$	D	$\tilde{F}(\Delta K) = \left(\frac{1}{c_s \Delta K^{p_s}} + \frac{1}{S(p_l) c_l \Delta K^{p_l}}\right)^{-1}$, $\tilde{F}(\Delta K) = \frac{1}{\frac{9.3914\times10^{17}}{\Delta K^{5.5}} + \frac{7.1617\times10^{53}}{\Delta K^{30.5}}}$	$\widetilde{n_f} = n_f(c_s, S(p_l)c_l)$, $\widetilde{n_f} = \frac{5.4153 \times 10^{92}}{\Delta\sigma^{30.5}} + \frac{3.2477 \times 10^{22}}{\Delta\sigma^{5.5}}$
Low $\Delta K \leq \Delta K^*$	Long $X \geq X^*$	S		
High $\Delta K \geq \Delta K^*$	Short $X \leq X^*$	S	$\tilde{\tilde{F}}(\Delta K) = \left(\frac{1}{S(p_s) c_s \Delta K^{p_s}} + \frac{1}{c_l \Delta K^{p_l}}\right)^{-1}$, $\tilde{\tilde{F}}(\Delta K) = \frac{1}{\frac{7.9688\times10^{18}}{\Delta K^{5.5}} + \frac{3.643\times10^{52}}{\Delta K^{30.5}}}$	$\widetilde{\widetilde{n_f}} = n_f(S(p_s)c_s, c_l)$, $\widetilde{\widetilde{n_f}} = \frac{1.06453 \times 10^{94}}{\Delta\sigma^{30.5}} + \frac{3.8188 \times 10^{21}}{\Delta\sigma^{5.5}}$
Low $\Delta K \leq \Delta K^*$	Long $X \geq X^*$	D		
High $\Delta K \geq \Delta K^*$	Short $X \leq X^*$	S	$\tilde{\tilde{F}}(\Delta K) = \left(\frac{1}{S(p_s) c_s \Delta K^{p_s}} + \frac{1}{S(p_l) c_l \Delta K^{p_l}}\right)^{-1}$, $\tilde{\tilde{F}}(\Delta K) = \frac{1}{\frac{7.9688\times10^{18}}{\Delta K^{5.5}} + \frac{7.1617\times10^{53}}{\Delta K^{30.5}}}$	$\widetilde{\widetilde{\widetilde{n_f}}} = n_f(S(p_s)c_s, S(p_l)c_l)$, $\widetilde{\widetilde{\widetilde{n_f}}} = \frac{1.06453 \times 10^{94}}{\Delta\sigma^{30.5}} + \frac{3.2477 \times 10^{22}}{\Delta\sigma^{5.5}}$
Low, $\Delta K \leq \Delta K^*$	Long $X \geq X^*$			

The third S–N curve

$$\overset{\approx}{n}_f = n_f(S(p_s)c_s, S(p_l)c_l). \tag{14.32}$$

takes both weakenings of elongations in regions of high and low stress intensity factors. The cumulation of both effects leads to the maximal fatigue life $\overset{\approx}{n}_f$.

The results of analytical evaluation of possible the averaged propagation functions are shown on Fig. 14.21. The influence of size for initial defect displays Fig. 14.22. Three solid curves present the S–N curves for the randomly propagated crack for the different with the different initial lengths 30, 40, and 50 [μm]. The dashed curves present the S–N curves for planar, deterministic crack for the different with the different initial lengths 30, 40, and 50 [μm]. The randomly propagated cracks indicate roughly twenty times more cycles to achieve the same crack length in comparison to the planar, deterministically spreading crack.

The plot of fatigue crack growth rate da/dn versus stress intensity factor range for different characters of crack propagation displays Fig. 14.23. At low stress intensity factor range a highly faceted fracture surface resulted due to crystallographic fracture along slip bands and stochastic crack expected (green lines). This effect explains the observed increasing of the fatigue life.

The numerically calculated S–N curves are presented on Fig. 14.24. The numerical results deliver the single points of S–N lines. Green points display the results of numerical simulation for planar elongation of crack. Red points show the results of numerical simulation for Brownian crack spreading. For comparison, the closed form

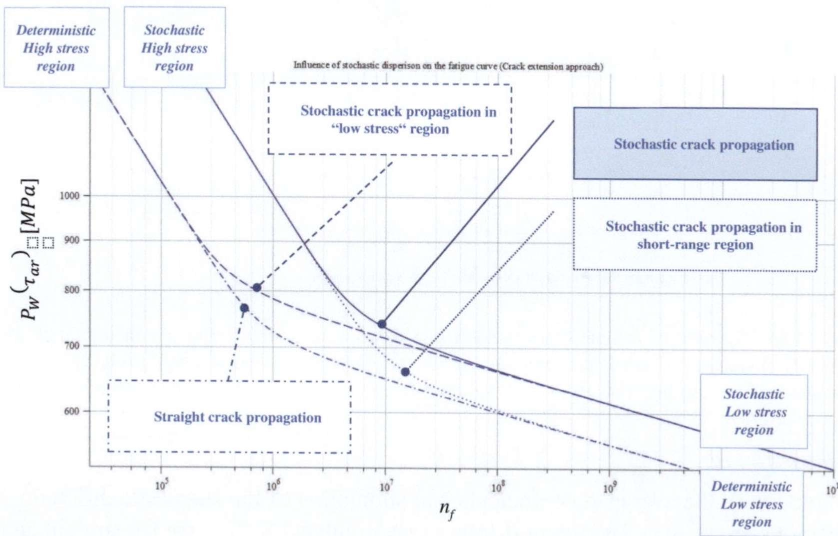

Fig. 14.21 Closed form solutions for S–N curves for different character of crack propagation (straight crack in both regions, straight crack in short and in long cycle regions, stochastic crack propagation in one and an the regions) initial inclusion size 30 [μm]

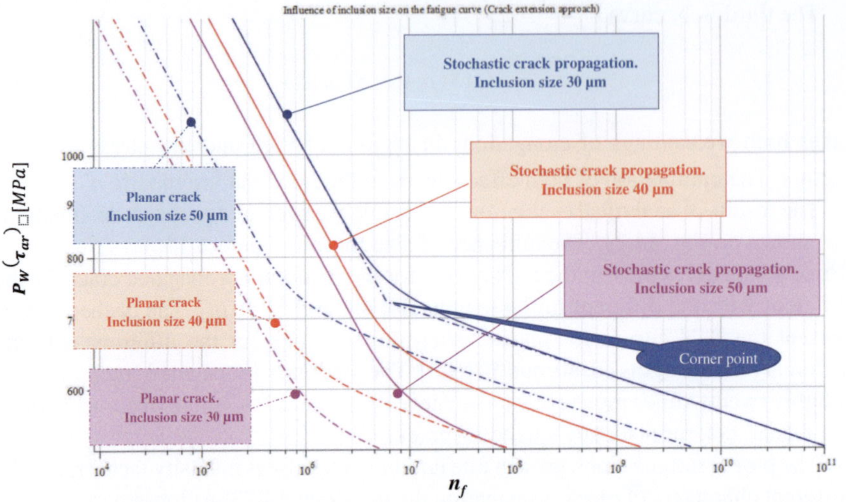

Fig. 14.22 S–N curves for different character of crack propagation (straight crack in both regions and stochastic crack propagation in in both regions) initial inclusion size 30, 40 and 50 [μm]

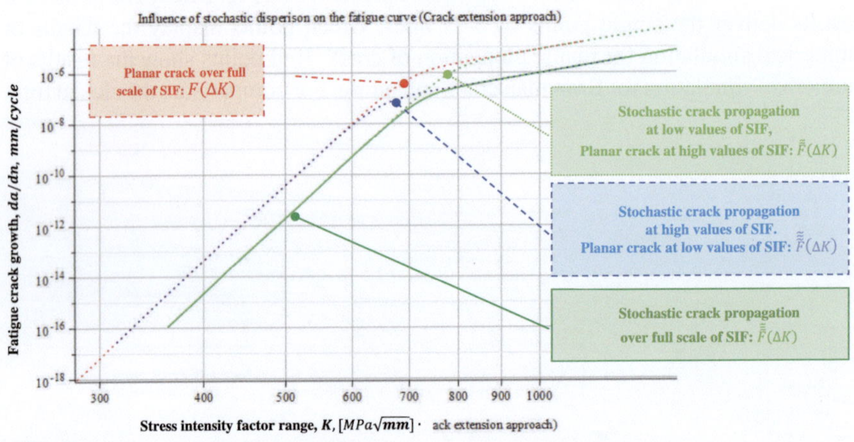

Fig. 14.23 The plot of fatigue crack growth rate da/dn versus stress intensity factor range for different characters of crack propagation. At low stress intensity factor range a highly faceted fracture surface resulted due to crystallographic fracture along slip bands and stochastic crack expected (green lines)

solution with the averaged coefficients and simulation of the stochastic differential equations are shown. The figure displays one simulated S–N curve for straight and one for randomly travelling crack together with closed-form solutions $\overset{\approx}{n}_f, n_f$. For comparison, the S–N curves form closed-form solutions and simulation of stochastic crack growth are portrayed.

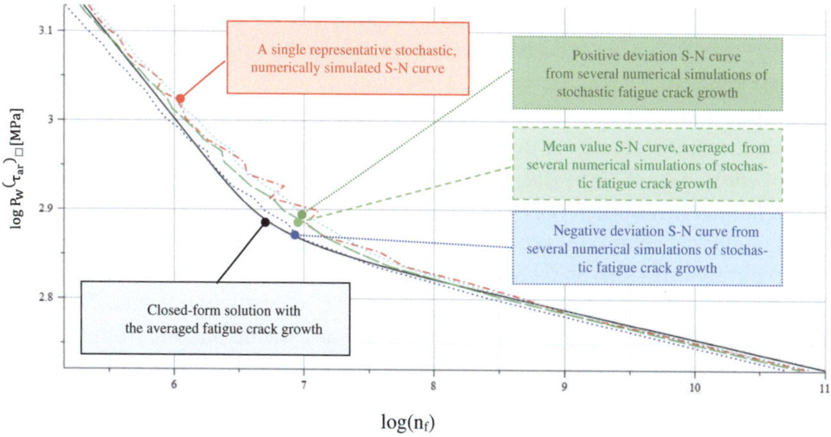

Fig. 14.24 S–N curves for stochastic crack propagation. Comparison of the closed form solution with the averaged coefficients and simulation of the stochastic differential equations

The examples of numerical simulation of drifted Brownian travelling of the cracks are displayed on Fig. 14.25. The above three curves show the jumps of the cracks for high value of stress $\sigma_{YY} = 1346$ MPa. Initially, the crack tip matches origin of coordinates. With each cycle the crack tip travels from the left to the right. The results of three simulations are shown, and the cracks travel on each load cycle considerably long. The fatigue life is the shortest, because the considerable spreading of crack.

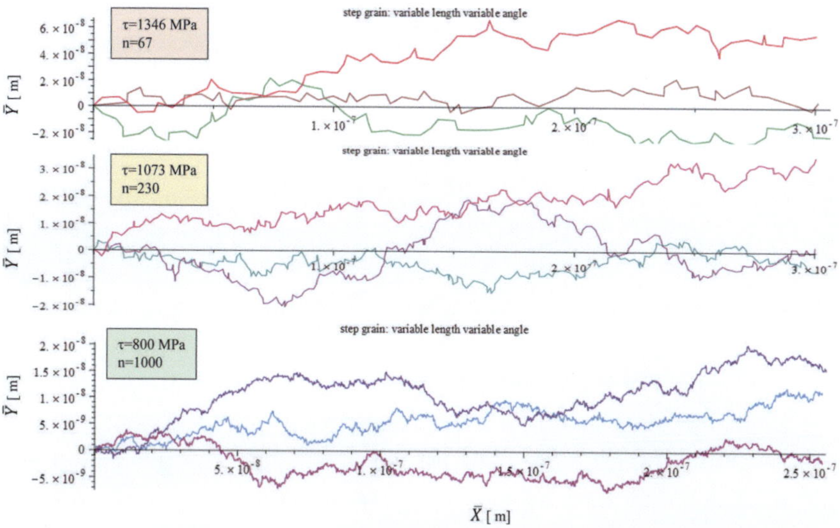

Fig. 14.25 Simulation of stochastic crack growth under different applied stresses and different number of cycles. The initial crack size of 60 [μm]

The central batch of three curves show the jumps of the cracks for mean value of stress $\sigma_{YY} = 1073$ MPa. The cracks start also from the origin of coordinates. The results of three crack paths are shown. The cracks travel on each load cycle evidently shorter, than in the first case. The propagation rate of crack is middling.

The bottommost family of three curves show the jumps of the cracks for lowest value of stress $\sigma_{YY} = 800$ MPa. The cracks travel on each load cycle is the shortest and the fatigue life is longest of all.

The numerical solution is presented on the Fig. 14.26. Several simultaneous evaluations of randomly-driven S–N lines were performed. The single simulated lines oscillate with respect to the averaged S–N line. The several simultaneous S–N lines allows the numerical evaluation of shape factors and the deviations from the averaged S–N line as well. The simulations were performed for 100 loads in the region $\sigma_{YY} = 800..1346$ MPa. Each batch of simulations for a single load delivers a point of the S–N line. The red curve presents an example of stochastic simulation of S–N line. Each point of this curve is based on a single evaluation of crack spreading to failure for each stress level. Each of S–N lines were evaluated eight times. From the generated S–N lines the median fatigue lines and the shape factors of distributions were determined. The green line displays the mean value (median) S–N curve. Two dotted lines show the deviation of the fatigue life based on 8 stochastic simulations. Simulation show, that there are no valuable differences between the numerical results with both variants of propagation functions.

The results of averaging for several simulations are shown on Fig. 14.27. The combinations of stochastic effects, as presented in Table 14.7 and given for closed-form solutions Eq. (14.23), (14.30) to (14.32) were simulated. Evidently, that the

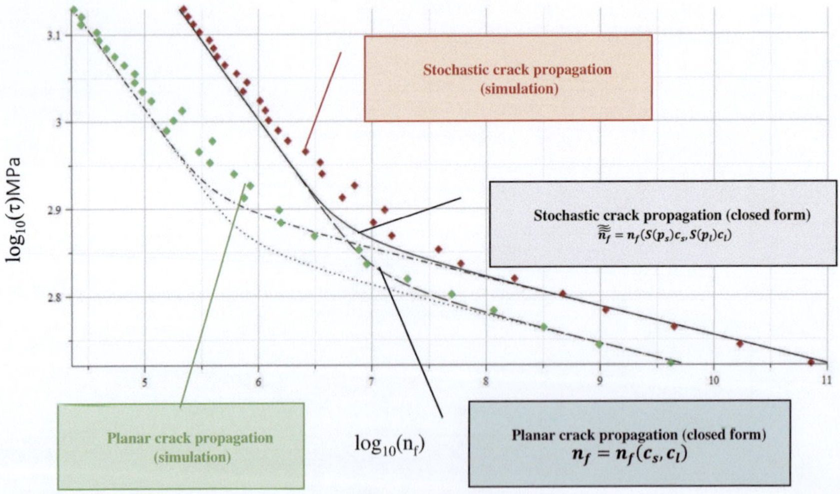

Fig. 14.26 Comparison of S–N curves form closed-form solutions and 'simulation of stochastic crack growth. One simulated S–N curve for straight and one for travelling stochastic crack together with closed-form solutions) $\overset{\approx}{n}_f, n_f$

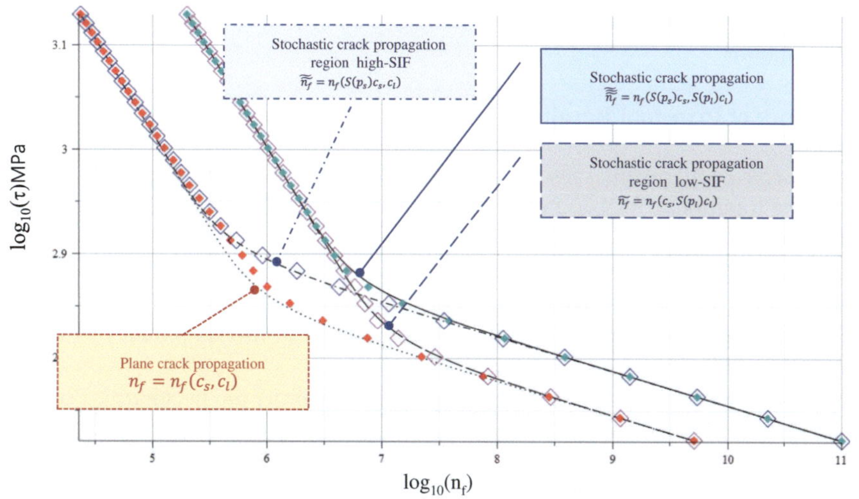

Fig. 14.27 Comparison of S–N curves form closed-form solutions and 'simulation of stochastic crack growth. Mean-values of simulated S–N curves overlapped with closed-form solutions. propagation (straight crack in both regions, straight crack in short and in long cycle regions, stochastic crack propagation in one and an the regions)

numerical and analytical results display the perfect correlation. The numerical simulation confirms the averaging method for the solution of stochastic differential equation, which leads to the closed-form solutions.

14.5 Conclusions

The experimentally acquired data from (AiF. 12 287 N, 2002; AiF 15 064 N, 2010; IGF, 16873N, 2014) provides the source for the statistical modelling of springs' fatigue. Different variants for introduction of effective stresses were compared. Calculations with $\mathbf{P}_{\mathrm{SWT}}$ were performed. Beside this, calculations with Walker stress parameter $\mathbf{P}_{\mathrm{W}}$ were presented. The optimal setting for fitting parameter $\tilde{\gamma}$ was evaluated using variations of stress ratio (Mayer et al., 2015, 2016; Schuller et al., 2013). The fitted value $\tilde{\gamma} = 0.15$ leads to the conformity of the experimentally obtained curves for the range of stress ratio $0 < R_\sigma < 1$.

We also discussed the mechanism of crack extension, which accounts for the deviation of the crack from its energetically preferred direction. The deviation occurs due to the chaotic orientations of the anisotropy in the grains and the stochastic grain boundaries. If the material is homogeneous, the crack will propagate in the energetically preferred direction, resulting in the maximum possible strain. However, due to the arbitrary anisotropy of the microscopic crystal grains and their irregular shapes, the crack does not follow the straight direction and deviates from it randomly.

The shape of a nonplanar crack looks like the path of a Brownian particle in force-driven motion.

Summary of Principal Results

- The accepted experimental results were analyzed from two different perspectives.
- First, the traditional Walker and stress averaging over the wire surface techniques were applied.
- The best fitting of the exponential data provides the Walker parameter with $\tilde{\gamma} \approx 0.15$ and the mean value of basic and corrected stress
- Second, the micromechanics of fatigue crack growth at low amplitudes of cyclic stress was studied.
- The modified micromechanics of crack growth represents the peculiarities of low cyclic amplitudes.
- The fatigue data show the significant extension of the fatigue life of springs in the region of small stress amplitudes.
- This behavior is confirmed by the reduction of damage and the arrest of crack propagation on inclusions and obstacles.
- Remarkable is the influence of microstructural refinement due to heat treatment of spring steels.
- The effect of microstructural refinement was presented from a geometrical point of view.
- The modeling was based on stochastic branching and Brownian traveling of the crack, which leads to the piecewise line-ear terms in the Paris-Irvin equations of crack growth and to the reduction of cyclic strain.
- The closed-form solutions of the stochastic equations are presented.

References

AiF 15 064 N. (2010). Investigation of the endurance stroke strength of coil springs in the range of extremely high oscillating cycles Ed. C. Berger B. Kaiser, Tech. Univ. Darmstadt

Abe, T., Furuya, F., & Matsuoka, S. (2001). Giga-cycle fatigue properties for 1800 MPa-class spring steels. *Transactions of the Japan Society of Mechanical Engineers, 67*(664), 1988–1995.

AiF 12287 N. (2002). Determination of fatigue strength and relaxation diagrams for highly stressed Helical compression springs. In C. Berger & B. Kaiser (Eds.) Tech. Univ. Darmstadt

Bathias, C., & Paris, P. (2005). *Gigacycle fatigue in mechanical practice.*. Marcel Dekker.

Bathias, C. (2014) *Fatigue limit in metals.* Wiley.

Berger, C., Pyttel, B., & Schwerdt, D. (2008), Beyond HCF Is there a fatigue limit? *Materialwissenschaft und Werkstofftechnik, 39*, 769–776. https://doi.org/10.1002/mawe.200800342

Castillo, E., & Fernández-Canteli A. (2009). *A unified statistical methodology for modeling fatigue damage.* Springer Science + Business Media B.V.

Christ, H. -J. (2018). *Fatigue of materials at very high numbers of loading cycles: experimental techniques, mechanisms, modeling and fatigue life assessment*. Springer Spektrum.

FKM. (2018). Rechnerischer Festigkeitsnachweis für Federn und Federelemente. Forschungskuratorium Maschinenbau E.V, Fachkreis Bauteilfestigkeit, FKM-Vorhaben Nr. 600, Heft 332

FKM. (2018). Rechnerischer Festigkeitsnachweis für Federn und Federelemente. FKM-Vorhaben Nr. 600, Heft 332; VDMA: Frankfurt, Germany

Forbes, C., Merran, E., Hastings, N., & Peacock, B. (2011). *Statistical distributions* (4th ed.). Wiley.

Geilen, M. B., Klein, M., & Oechsner, M. (2020). On the influence of ultimate number of cycles on lifetime prediction for compression springs manufactured from VDSiCr class spring wire. *Materials, 13*, 3222. https://doi.org/10.3390/ma13143222

Huhnen, J. (1970). *Final report on preliminary test for the project—effect factor k' for coil springs.* Robert Bosch GmbH.

IGF 16873N. (2014). Investigation of material and manufacturing influences on the VHCF behaviour of coil compression springs. In M. Oechsner (Eds.), TU Darmstadt.

Kaiser, B., & Berger, C. (2005). *Fatigue behaviour of technical springs, Materialwissenschaft und Werkstofftechnik, 36*(11), 685–696.

Kaiser, B., & Berger, C. (2007). Recent findings to the fatigue properties of helical springs. In *Proceedings of Advanced Spring Technology JSSE 60th Anniversary International Symposium*, November 2nd, 2007, Fukiage Hall, Nagoya, Japan.- Tokyo: JSSE, 2007. S. 97–108, Tokyo, JSSE

Kaiser, B., Pyttel, B., & Berger, C. (2011). VHCF-behavior of helical compression springs made of different materials. *International Journal of Fatigue, 33*, 2011.

Kalbfleisch, J., & Prentice, R. (2002). *The statistical analysis of failure time data* (2nd ed.). Wiley.

Karr, U., Penning, B., Tran, D., & Mayer, H. (2017). Development of Nitrided 18NI maraging steel to optimise the very high cycle fatigue properties. In M. Zimmermann & H. -J. Christ (Eds.) *7th International Conference on Very High Cycle Fatigue*, July 3–5, 2017, Dresden, Germany. Siegen, Univ.

Kloos, K. H., & Kaiser, B. (1977). Dauerhaltbarkeitseigenschaften von Schraubenfedern in Abhängigkeit von Wickelverhältnis und Oberflächenzustand. *Draht/Wire, 9*(415–421), 539–545.

Kobelev, V. (2017). Some exact analytical solutions in structural optimization mechanics based design of structures and machines. *An International Journal, 45*(1). https://doi.org/10.1080/15397733.2016.1143374

Lukács, J. (2010). Fatigue crack growth tests on type 321 austenitic stainless steel in corrosive environment and at elevated temperature. *Procedia Engineering, 2*, 1201–1210.

Mayer, H., U. Karr, Y. Sandaiji, & E. Tamura. (2017). Torsional and axial VHCF properties of spring steel at different load ratios. In M. Zimmermann & H. -J. Christ (Eds.) *7th International Conference on Very High Cycle Fatigue*, July 3–5, 2017, Dresden, Germany. Siegen, Univ.

Mayer, H., Schuller, R., Karr, U., Irrasch, D., Fitzka, M., Hahn, M., & Bacher-Höchst, M. (2015). Cyclic torsion very high cycle fatigue of VDSiCr spring steel at different load ratios. *International Journal of Fatigue, 70*(2015), 322–327. https://doi.org/10.1016/j.ijfatigue.2014.10.007

Mayer, H., Schuller, R., Karr, U., Irrasch, D., Fitzka, M., Hahn, M., & Bacher-Höchst, M. (2016). Mean stress sensitivity and crack initiation mechanisms of spring steel for torsional and axial VHCF loading. *International Journal of Fatigue, 93*, 309–317. https://doi.org/10.1016/j.ijfatigue.2016.04.017

Mughrabi, H. (2006). Specific features and mechanisms of fatigue in the ultrahigh-cycle regime. *International Journal of Fatigue, 28*(11), 1501–1508.

Murakami, Y. (2003). *High and ultrahigh cycle fatigue.* In I. Milne, R. O. Ritchie & B. Karihaloo (Eds.) *Comprehensive structural integrity* (Vol. 4). Cyclic Loading and Fatigue, Elsevier.

Nishimura, Y., Endo, M., Yanase, K., Ikeda, Y., Tanaka, Y., Miyamoto, N., & Miyakawa, S. (2017). HCF and VHCF strength of spring steel with small scratches. In M. Zimmermann & H. -J. Christ (Eds.) *7th International Conference on Very High Cycle Fatigue,* July 3–5, 2017, Dresden, Germany. Siegen, Univ.

Pyttel, B., Brunner, I., Kaiser, B., Berger, C., & Mahendran, M. (2014). Fatigue behaviour of helical compression springs at a very high number of cycles– Investigation of various influences. *International Journal of Fatigue, 60*, 101–109.

Pyttel, B., Schwerdt, D., & Berger, C. (2011). Very high cycle fatigue—Is there a fatigue limit? *International Journal of Fatigue, 33*, 49–58p.

SAE AE-22. (1997). *SAE fatigue design handbook*. SAE International.

Sander, M. (2018). *Sicherheit und Betriebsfestigkeit von Maschinen und Anlagen*, (2nd ed.). Springer-Verlag GmbH Deutschland.

Schuller, R., Mayer, H., Fayard, A., Hahn, M., & Bacher-Höchst, M. (2013). Very high cycle fatigue of VDSiCr spring steel under torsional and axial loading. *Materialwissenschaft und Werkstofftechnik, 44*(4). https://doi.org/10.1002/mawe.201300029

Schwalbe, K.-H. (1980). *Bruchmechanik metallischer werkstoffe*. Carl Hanser Verlag.

Schwerdt, D. (2011). *Schwingfestigkeit und Schädigungsmechanismen der Aluminiumlegierungen EN AW-6056 und EN AW-6082 sowie des Vergütungsstahls 42CrMo4 bei sehr hohen Schwingspielzahlen* (Ph.D. Thesis), TU Darmstadt, 210 p.

Verhoeven, J. D. (2007). *Steel metallurgy for the non-metallurgist* (pp. 44073-0001). ASM International, Materials Park.

Zimmermann, M. (2019). Very high cycle fatigue. In C. -H. Hsueh et al. (Eds.), Handbook of mechanics of materials (p. 57). Springer Nature Singapore Pte Ltd. https://doi.org/10.1007/978-981-10-6884-3_43

Appendix A
Models of Spring Materials

A.1 Plastic Deformation

1. The stress $\boldsymbol{\sigma}$ and strain tensors $\boldsymbol{\varepsilon}$ reads (Chen & Han, 1988):

$$\boldsymbol{\sigma} = \begin{bmatrix} \sigma_{xx} & \tau_{xy} & \tau_{xz} \\ \tau_{xy} & \sigma_{yy} & \tau_{yz} \\ \tau_{xz} & \tau_{yz} & \sigma_{zz} \end{bmatrix},$$
$$\boldsymbol{\varepsilon} = \begin{bmatrix} \varepsilon_{xx} & \gamma_{xy}/2 & \gamma_{xz}/2 \\ \gamma_{xy}/2 & \varepsilon_{yy} & \gamma_{yz}/2 \\ \gamma_{xz}/2 & \gamma_{yz}/2 & \varepsilon_{zz} \end{bmatrix} \equiv \begin{bmatrix} \varepsilon_{xx} & \varepsilon_{xy} & \varepsilon_{xz} \\ \varepsilon_{xy} & \varepsilon_{yy} & \varepsilon_{yz} \\ \varepsilon_{xz} & \varepsilon_{yz} & \varepsilon_{zz} \end{bmatrix}. \tag{A.1}$$

The octahedral normal stress σ_{oct} and the octahedral normal strain ε_{oct} are denoted as:

$$\begin{aligned} 3\sigma_{oct} &= \sigma_{xx} + \sigma_{yy} + \sigma_{zz} \equiv \sigma_{kk} = \mathbf{I}(\boldsymbol{\sigma}), \\ 3\varepsilon_{oct} &= \varepsilon_{xx} + \varepsilon_{yy} + \varepsilon_{zz} \equiv \varepsilon_{kk} = \mathbf{I}(\boldsymbol{\varepsilon}). \end{aligned} \tag{A.2}$$

The octahedral normal stress is equal to hydrostatic pressure or spherical stress. The $\mathbf{I}(\boldsymbol{\sigma})$ and $\mathbf{I}(\boldsymbol{\varepsilon})$ are the first invariants of stress and strain tensors.

The tensors $\boldsymbol{s}$ and $\boldsymbol{e}$ are the deviators of tensors (A.1) of with the components respectively:

$$\begin{aligned} s_{ij} &= \sigma_{ij} - \sigma_{oct}\delta_{ij}, \\ e_{ij} &= \varepsilon_{ij} - \varepsilon_{oct}\delta_{ij}, \end{aligned} \tag{A.3}$$

V. Kobelev, *Fundamentals of Springs Mechanics*,
https://doi.org/10.1007/978-3-031-58584-5

$$\boldsymbol{s} = \begin{bmatrix} s_{xx} & \tau_{xy} & \tau_{xz} \\ \tau_{xy} & s_{yy} & \tau_{yz} \\ \tau_{xz} & \tau_{yz} & s_{zz} \end{bmatrix},$$
$$\boldsymbol{e} = \begin{bmatrix} e_{xx} & \gamma_{xy}/2 & \gamma_{xz}/2 \\ \gamma_{xy}/2 & e_{yy} & \gamma_{yz}/2 \\ \gamma_{xz}/2 & \gamma_{yz}/2 & e_{zz} \end{bmatrix} \equiv \begin{bmatrix} e_{xx} & \varepsilon_{xy} & \varepsilon_{xz} \\ \varepsilon_{xy} & e_{yy} & \varepsilon_{yz} \\ \varepsilon_{xz} & \varepsilon_{yz} & e_{zz} \end{bmatrix}. \tag{A.4}$$

The octahedral shear strain is:

$$\gamma_{oct}/2 \overset{\text{def}}{=} \sqrt{\frac{2}{3}\mathbf{II}(\boldsymbol{e})}, \quad \mathbf{II}(\boldsymbol{e}) = \frac{1}{2}e_{ij}e_{ij}. \tag{A.5}$$

$\mathbf{II}(\boldsymbol{e})$ is the second invariant of deviatory strain tensor $\boldsymbol{e}$.
The octahedral shear stress. with reads as:

$$\tau_{oct} \overset{\text{def}}{=} \sqrt{\frac{2}{3}\mathbf{II}(\boldsymbol{s})}, \quad \mathbf{II}(\boldsymbol{s}) = \frac{1}{2}s_{ij}s_{ij}. \tag{A.6}$$

$\mathbf{II}(\boldsymbol{s})$ is the second invariant of deviatory stress $\boldsymbol{s}$.
The equivalent stress is $\sigma_{eq} = 3\tau_{oct}/\sqrt{2}$. .

2. For a linear elastic medium

$$\boldsymbol{s} = 2G\boldsymbol{e}, \quad 3K\varepsilon_{oct} = \sigma_{oct}, \tag{A.7}$$

where G is the shear modulus,K is the modulus of cubic compressibility or bulk modulus of elasticity . From this equation follows, that in case of linear elasticity: $\tau_{oct} = G\gamma_{oct}$.

For a compressible isotropic work-hardening material, without the distinct yielding point the stress could be presented as the function of strain :

$$\boldsymbol{e} = f_\tau(\tau_{oct})\frac{\boldsymbol{s}}{\tau_{oct}}, \quad 3K\varepsilon_{oct} = \sigma_{oct}. \tag{A.8}$$

The dependence of stress in terms of strain reads:

$$\boldsymbol{s} = 2G_p(\gamma_{oct})\boldsymbol{e}, \quad \boldsymbol{s} = f_\tau^{-1}(\gamma_{oct})\frac{\boldsymbol{e}}{\gamma_{oct}}. \tag{A.9}$$

The relation must be satisfied:

$$\frac{\tau_{oct}}{f_\tau(\tau_{oct})} = \frac{f_\tau^{-1}(\gamma_{oct})}{\gamma_{oct}} = 2G_p(\gamma_{oct}). \tag{A.10}$$

From the Eq. (A.9) follows the relation of the octahedral shear stress upon the octahedral shear strain :

$$\tau_{oct} = G_p(\gamma_{oct})\gamma_{oct}. \tag{A.11}$$

3. The Johnson–Cook model (Johnson & Cook, 1983) expresses the stress upon strain . The Johnson–Cook model is appropriate for the large deformation, high strain rate and elevated temperatures. In the case of an uniaxial deformation, the Johnson–Cook stress–strain equation is written in the form:

$$\sigma = F_1(\varepsilon)F_2(T^{\circ})F_3(\dot{\varepsilon}), \tag{A.12}$$

$$F_1(\varepsilon) = A_\sigma + B_\sigma \varepsilon^{1/n}, \tag{A.13}$$

$$F_2(T^{\circ}) = 1 - \left(\frac{T^{\circ} - T^{\circ}_{ref}}{T^{\circ}_m - T^{\circ}_{ref}}\right)^{m^{\circ}}, \tag{A.14}$$

$$F_3(\dot{\varepsilon}) = 1 + C \ln \frac{\dot{\varepsilon}}{\dot{\varepsilon}_{ref}}. \tag{A.15}$$

where A_σ and B_σ are the strain coefficients, T°_m is the melting temperature, T°_{ref} is a reference (ambient) temperature, m° is the temperature exponent, $\dot{\varepsilon}$ is the (plastic) strain rate, $\dot{\varepsilon}_{ref}$ is a reference (quasi-static) strain rate and C is the strain rate sensitivity parameter.

The term $F_1(\varepsilon)$ is called the Ludwik–Hollomon equation. This term reveals the quasi-static yielding and work hardening behavior at a certain temperature.

The term $F_2(T^{\circ})$ determines the temperature dependence.

The third term demonstrates the strain rate sensitivity. The term $F_3(\dot{\varepsilon})$ includes commonly the logarithm of the strain rate, normalized by a reference value. More general approximations are known (El-Magd et al., 1985; Mauerauch & Vöhringer, 1978).

4. Relations between the coefficients of explicit and implicit stress–strain functions are listed in Table A.1. Simplified relations between the coefficients of Ludwik–Hollomon, Johnson–Cook and Ramberg–Osgood stress–strain functions are shown in Table A.2.

Creep Deformation

The isotropic, time dependent stress function (Betten, 2008) for creep laws (Table A.3):

$$\frac{\mathrm{d}e_{ij}}{\mathrm{d}t} = \frac{s_{ij}}{\tau_{oct}} f_c(\tau_{oct}, t), \tag{A.16}$$

Table A.1 Relations between the coefficients of explicit and implicit stress–strain functions

Relations between the coefficients of explicit and implicit stress–strain functions

	Implicit stress–strain function	Explicit stress–strain function
Relation between octahedral shear stress and strain	$\gamma_{oct} = f_\tau(\tau_{oct})$	$\tau_{oct} = f_\tau^{-1}(\gamma_{oct})$
Quasi-static shear-stress-shear-strain relation	$\gamma = f_\tau(\tau)$	$\tau = f_\tau^{-1}(\gamma)$
Quasi-static tensile-stress-tensile-strain relation	$\varepsilon = \frac{1}{\sqrt{3}} f_\tau\left(\frac{\sigma}{\sqrt{3}}\right) \overset{\text{def}}{=} f_\sigma(\sigma)$	$\sigma = \sqrt{3} f_\tau^{-1}\left(\sqrt{3}\varepsilon\right) \overset{\text{def}}{=} f_\sigma^{-1}(\varepsilon)$
Quasi-static stress–strain flow equations	Ramberg–Osgood equation	Ludwik–Hollomon equation
Shear deformation	$\gamma = \frac{\tau}{G} + \left(\frac{\tau}{K_\tau}\right)^n$	$\tau = A_\tau + B_\tau \gamma^{1/n}$
Uniaxial deformation	$\varepsilon = \frac{\sigma}{E} + \left(\frac{\sigma}{K_\sigma}\right)^n$	$\sigma = A_\sigma + B_\sigma \varepsilon^{1/n}$
Stress–strain relation Eq. (6.8), $\tau_w = G\gamma_w$	$\gamma_{oct} = \frac{\tau_{oct}}{G.\sqrt{1-(\tau_{oct}/\tau_w)^2}}$	$\tau_{oct} = \frac{G\gamma_{oct}}{\sqrt{1+(\gamma_{oct}/\gamma_w)^2}}$
Secant modulus in stress–strain relation Eq. (6.8)	$\gamma_{oct} = \tau_{oct}/G_p(\tau_{oct})$, $G_p(\tau_{oct}) = G \cdot \sqrt{1-(\tau_{oct}/\tau_w)^2}$	$\tau_{oct} = G_p(\gamma_{oct})\gamma_{oct}$, $G_p(\gamma_{oct}) = \frac{G}{\sqrt{1+(\gamma_{oct}/\gamma_w)^2}}$
Relations between coefficients for shear and uniaxial plastic flow for incompressible medium	$E = 3G,\ K_\sigma = \sqrt{3}K_\tau$	$A_\tau = \frac{A_\sigma}{\sqrt{3}}$, $B_\tau = \frac{B_\sigma}{\sqrt{3}} \cdot \left(\frac{1}{\sqrt{3}}\right)^{1/n}$
Johnson–Cook equations	$\gamma_{oct} = f_\tau\left(\frac{\tau_{oct}}{F_2(T^\circ)F_3(\dot{\gamma}_{oct})}\right)$	$\frac{\tau_{oct}}{F_2(T^\circ)F_3(\dot{\gamma}_{oct})} = f_\tau^{-1}(\gamma_{oct})$
Dynamic tensile-stress-tensile-strain relation	$\varepsilon = f_\sigma\left(\frac{\sigma}{F_2(T^\circ)F_3(\dot{\varepsilon})}\right)$	$\frac{\sigma}{F_2(T^\circ)F_3(\dot{\varepsilon})} = f_\sigma^{-1}(\varepsilon)$
Dynamic static shear-stress-shear-strain relation	$\gamma = f_\tau\left(\frac{\tau}{F_2(T^\circ)F_3(\dot{\gamma})}\right)$	$\frac{\tau}{F_2(T^\circ)F_3(\dot{\gamma})} = f_\tau^{-1}(\gamma)$

Table A.2 Ludwik–Hollomon, Johnson–Cook and Ramberg–Osgood stress–strain functions

	Pure shear deformation	Uniaxial deformation
Explicit stress–strain functions (A.8): Ludwik–Hollomon and Johnson–Cook equations	$\tau = -\frac{K_\sigma^2}{\sqrt{3}(K_\sigma+nE)} + \frac{K_\sigma}{\sqrt{3}}\left(\frac{\gamma}{\sqrt{3}}\right)^{1/n}$	$\sigma = -\frac{K_\sigma^2}{K_\sigma+nE} + K_\sigma \varepsilon^{1/n}$
Implicit stress–strain function (A.7): Ramberg–Osgood equation	$\gamma = \frac{3\tau}{E} + \left(\frac{\sqrt{3}\tau}{K_\sigma}\right)^n$	$\varepsilon = \frac{\sigma}{E} + \left(\frac{\sigma}{K_\sigma}\right)^n$

Table A.3 Common phenomenological laws for creep

Creep laws	$\overline{\sigma} = \frac{3\overline{\tau}}{\sqrt{2}}; \tilde{\tau} = \frac{\overline{\sigma}}{\sqrt{3}}$
Norton-Bailey law	$g_c = \overline{\varepsilon} \cdot \left(\frac{\tau_{oct}}{\overline{\tau}}\right)^{\xi+1}$
Garofalo creep law	$g_c = \overline{\varepsilon} \cdot \sin\mathrm{h}^{\xi+1}\left(\frac{\tau_{oct}}{\overline{\tau}}\right)$
Exponential creep law	$g_c = \overline{\varepsilon} \cdot \left[\exp\left(\frac{\tau_{oct}}{\overline{\tau}}\right) - 1\right]$
Naumenko-Altenbach-Gorash creep law	$g_c = \overline{\varepsilon} \cdot \left(\frac{\tau_{oct}}{\overline{\tau}}\right) + \overline{\varepsilon} \cdot \left(\frac{\tau_{oct}}{\overline{\tau}}\right)^{\xi+1}$
Time independent stress function $\frac{\sigma_{th}}{\overline{\sigma}} = \frac{\tau_{th}}{\overline{\tau}}, \frac{\sigma_Y}{\overline{\sigma}} = \frac{\tau_Y}{\overline{\tau}}$	$g_c = \overline{\varepsilon} \cdot \left(\frac{\tau_{oct}}{\overline{\tau}}\right)^{\xi+1} \cdot \frac{1}{\left(\frac{\tau_{oct}}{\tau_{th}}\right)^{m_1} - 1} \cdot \frac{1}{1 - \left(\frac{\tau_{oct}}{\tau_Y}\right)^{m_2}}$

$$f_c(\tau_{oct}, t) = h_c(t) g_c(\tau_{oct}). \tag{A.17}$$

Appendix B
Some Special Functions

1. Integrals with polylogarithm.

The weighted integrals of the function

$$f(x) = \ln\bigl(\tanh\bigl(a + \operatorname{arctanh}\bigl(e^{bx}\bigr)\bigr)\bigr) \tag{B.1}$$

are:

$$I_0(a, b; X) \equiv \int_0^X f(x)\mathrm{d}x = X \ln(\tanh(a)) + \frac{1}{b}(\Lambda_2 - \mathrm{M}_2 + \mu_2 - \lambda_2), \tag{B.2}$$

$$\begin{aligned} I_1(a, b; X) &\equiv \int_0^X f(x)x\mathrm{d}x \\ &= \frac{1}{6b^2}\bigl[\pi^2 \ln(\coth(a)) + \ln^3(\coth(a)) + 3b^2X^2 \ln(\tanh(a))\bigr] \\ &\quad + \frac{1}{b^2}(\mathrm{M}_3 - \Lambda_3) - \frac{x}{b}(\mathrm{M}_2 - \Lambda_2), \end{aligned} \tag{B.3}$$

$$\begin{aligned} I_2(a, b; X) &\equiv \int_0^X f(x)x^2\mathrm{d}x \\ &= \frac{X^3}{3} \ln(\tanh(a)) \\ &+ \frac{2}{b^3}(\Lambda_4 - \mathrm{M}_4 + \mu_4 - \lambda_4) + \frac{2X}{b^2}(\mathrm{M}_3 - \Lambda_3) - \frac{x^2}{b}(\mathrm{M}_2 - \Lambda_2). \end{aligned} \tag{B.4}$$

The following abbreviations are used:

V. Kobelev, *Fundamentals of Springs Mechanics*,
https://doi.org/10.1007/978-3-031-58584-5

$$\mathrm{M}_k \stackrel{\text{def}}{=} \mathrm{Li}_k(-\coth(a)\exp(bX)), \quad \Lambda_k \stackrel{\text{def}}{=} \mathrm{Li}_k(-\tanh(a)\exp(bX)),$$

$$\mu_k = \mathrm{M}_k|_{X=0} \equiv \mathrm{Li}_k(-\coth(a)), \quad \lambda_k = \Lambda_k|_{X=0} \equiv \mathrm{Li}_k(-\tanh(a)).$$

In these expressions is $\mathrm{Li}_k(z)$ the polylogarithm (also known as Jonquière's function) of order k and argument z.

2. Integrals with hypergeometric function

The weighted integrals of the function $g = \left(a + x^{-m}\right)^{-1/m}$ are ($p \geq 0$):

$$J_p(a, m; X) \equiv \int_0^X x^p g(x)\mathrm{d}x = {}_2\mathbf{F}_1\left(\frac{1}{m}, \frac{2+p}{m}; \frac{2+p+m}{m}; -aX^m\right) \cdot \frac{X^{2+p}}{2+p}. \tag{B.5}$$

For some cases, the integrals could be expressed in terms of elementary functions:

$$J_1(a, 1; X) = -\frac{X}{a^2} + \frac{X^2}{2a} + \frac{\ln(1+aX)}{a^3},$$
$$J_2(a, 1; X) = \frac{X}{a^3} - \frac{X^2}{2a^2} + \frac{X^3}{3a} - \frac{\ln(1+aX)}{a^4},$$

$$J_1(a, 2; X) = \frac{X\sqrt{1+aX^2}}{2a} - \frac{\arcsin\left(X\sqrt{A}\right)}{2a^{3/2}},$$
$$J_2(a, 2; X) = \frac{2}{3a^2} - \frac{2\sqrt{1+aX^2}}{3a^2} + \frac{X^2\sqrt{1+aX^2}}{3a},$$

$$J_1(a, 3; X) = -\frac{1}{2a} + \frac{\left(1+aX^3\right)^{2/3}}{2a},$$
$$J_2(a, 4; X) = -\frac{1}{3a} + \frac{\left(1+aX^4\right)^{3/4}}{3a}.$$

3. Integrals with incomplete beta function

The weighted integrals of the function $(|c-r|/r)^{1/n}$ are:

$$\begin{aligned} K_n(a, b, c) &\equiv \frac{1}{c}\int_a^b \left(\frac{|c-r|}{r}\right)^{1/n} \mathrm{d}x \\ &= \frac{(-1)^{-1/n}\pi}{n}\left(i + \cot\left(\frac{\pi}{2n}\right)\right) - \mathbf{B}\left(\frac{a}{c}; \frac{n-1}{n}, \frac{n+1}{n}\right) \\ &\quad - (-1)^{1/n}\mathbf{B}\left(\frac{b}{c}; \frac{n-1}{n}, \frac{n+1}{n}\right) \end{aligned} \tag{B.6}$$

and

$$\begin{aligned} L_n(a, b, c) &\equiv \frac{1}{c^2}\left[\int_a^b \left(\frac{|c-r|}{r}\right)^{1/n} (c-r)\mathrm{d}x\right] \\ &= \frac{i\pi}{\exp\left(\frac{i\pi}{n}\right) - 1}\frac{1+n}{n^2} + \mathbf{B}\left(\frac{a}{c}; \frac{2n-1}{n}, \frac{n+1}{n}\right) \end{aligned}$$

$$- \mathbf{B}\left(\frac{a}{c}; \frac{n-1}{n}, \frac{n+1}{n}\right)$$

$$+ (-1)^{-1/n}\left[\mathbf{B}\left(\frac{b}{c}; \frac{2n-1}{n}, \frac{n+1}{n}\right) - \mathbf{B}\left(\frac{b}{c}; \frac{n-1}{n}, \frac{n+1}{n}\right)\right]. \quad \text{(B.7)}$$

In these expressions

$$\mathbf{B}(x; p, q) = \int_0^x z^{p-1}(1-z)^{q-1}\mathrm{d}z \quad \text{(B.8)}$$

is the incomplete beta-function (Pearson, 1968). This function expresses in terms of hypergeometric functions (Abramowitz & Stegun, 1972) by:

$$\mathbf{B}(x; p, q) = {}_2\mathbf{F}_1(p, 1-q; p+1; x)\frac{x^p}{p}. \quad \text{(B.9)}$$

4. Complete elliptic integrals

$$\mathbf{K}(k) = \int_0^1 \frac{1}{\sqrt{(1-k^2t^2)(1-t^2)}}\mathrm{d}t$$

complete elliptic integrals of the first kind.

$$\mathbf{E}(k) = \int_0^1 \sqrt{\frac{1-k^2t^2}{1-t^2}}\mathrm{d}t$$

complete elliptic integrals of the second kind .

$$\mathbf{\Pi}(\kappa, k) = \int_0^1 \frac{1}{(1-\kappa t^2)\sqrt{(1-k^2t^2)(1-t^2)}}\mathrm{d}t$$

complete elliptic integrals of the third kind.

5. Appell hypergeometric function.

Appell hypergeometric function of two variables (Erdélyi, 1950) and (Kampe de Feriet, 1957) reads:

$$\mathbf{F}_1([A, B_1, B_2, C]; x, y)$$
$$= \frac{\Gamma(C)}{\Gamma(A)\Gamma(C-A)}\int_0^1 t^{A-1}(1-t)^{C-A-1}(1-tx)^{-B_1}(1-ty)^{-B_s}\mathrm{d}t. \quad \text{(B.10)}$$

6. Legendre and hypergeometric function.

The expression of the integral for $|R| < 1$ (Gradshteyn & Ryzhik, 2015) in terms of the Legendre function P_p and the hypergeometric function ${}_2\mathbf{F}_1$:

$$\Upsilon(R) = \frac{1}{2\pi}\int_0^{2\pi} \frac{dt}{(1 + R\sin t)^p} = {}_2\mathbf{F}_1\left(\left[\frac{1+p}{2}, \frac{p}{2}\right][1], R^2\right) = \frac{\mathbf{P}_{-p}\left(\frac{1}{\sqrt{1-R^2}}\right)}{\left(1 - R^2\right)^{p/2}}. \tag{B.11}$$

7. Integrals with trigonometric function

The expressions of the integral (Gradshteyn & Ryzhik, 2015, 3.621) in terms of Gamma function are.

$$\int_0^{\pi/2} cos^{2p}\varphi d\varphi = \int_0^{\pi/2} \sin^{2p}\varphi d\varphi = \frac{\sqrt{\pi}}{2}\frac{\Gamma(p)}{\Gamma(p+1/2)}, \tag{B.12}$$

$$\int_0^{\pi/2} cos^{2p+1}\varphi d\varphi = \int_0^{\pi/2} \sin^{2p+1}\varphi d\varphi = \frac{\sqrt{\pi}}{2}\frac{\Gamma(p+1/2)}{\Gamma(p+1)}, \tag{B.13}$$

$$\int_0^{\pi/2} \sin^{p-1}\varphi cos^{q-1}\varphi d\varphi = \frac{\Gamma(p/2)\Gamma(q/2)}{2\Gamma((p+q)/2)}, \quad \mathrm{Re}p > 0, \quad \mathrm{Re}q > 0. \tag{B.14}$$

The following integral evaluates with Legendre function Eq. (3.661) (Gradshteyn & Ryzhik, 2015):

$$\frac{1}{2\pi}\int_0^{2\pi} (a + b\cos t)^q dt = \left(a^2 - b^2\right)^{\frac{q}{2}} \cdot \mathbf{P}_q\left(\frac{a}{\sqrt{a^2 - b^2}}\right). \tag{B.15}$$

Appendix C
Statistical Assessment for Scattering of Fatigue Data

1. Prediction of failure probability under a specific stress level is an important tool for analysis of springs under static and cyclic loading. Several distributions are customary applied for the statistical evaluation of the fatigue data (Forbes et al., 2011).

Scattering analysis of experimental data in fatigue tests was traditionally performed by using:

- Normal (Gaussian) distribution
- Lognormal distribution
- Exponential distribution
- Weibull distribution
- arcsin $\sqrt{P}$—Arcsine distribution.

The variability of a material characteristic is described by using lognormal distribution function, which cover a large scale of applications. The lognormal distribution is recommended in (ASTM E 739-80, 1986; ASTM E 606-80, 1986; DIN 50100, 2016:12) norms. (Schijve, 2005) proposed "log(N-N0)-normal distribution" for fatigue data analysis.

The probability density function is $p_f(\sigma_\delta)$. The symbol σ_δ designates the variate or random variable. In case of fatigue analysis, the variate characterizes the intensity of the harmonically variable stress. The stress amplitude, or stress range, or parameter of Smith, Watson and Topper, of Walker parameter is used for characterization the variable stress in fatigue experiments.

The cumulative failure probability of the argument σ is equal to the probability that the variate δ takes a value less than or equal to σ:

$$P_f(\sigma) = \Pr[\sigma_\delta \le \sigma] = \int_{-\infty}^{\sigma} p_f(\sigma_\delta) \mathrm{d}\sigma_\delta. \tag{C.1}$$

V. Kobelev, *Fundamentals of Springs Mechanics*,
https://doi.org/10.1007/978-3-031-58584-5

Table C.1 Normal distribution

Normal distribution	
Distribution mean	Λ
Scale parameter	μ
Probability density of failures	$p_f(\sigma_\delta) = \frac{1}{\mu\sqrt{2\pi}} exp\left(-\frac{(\Lambda-\sigma_\delta)^2}{2\mu^2}\right)$
Maximum of probability density of failures	$\max p_f(\sigma_\delta) = p_f(\sigma_{\max}) = 1/\sqrt{2\pi}\mu$
Stress at maximum	$\sigma_\delta = \Lambda$
Cumulative failure probability P_F (for given stress value σ	$P_f(\sigma) = \frac{1}{2} - \frac{1}{2}\text{erf}\left(\frac{\Lambda-\sigma}{\mu\sqrt{2}}\right)$
Standard deviation	$S = \mu$
Mean value (median)	$\sigma_{mean} = \Lambda, \quad P_f(\sigma_{mean}) = 1/2$

Table C.2 Lognormal distribution

Lognormal distribution	
Mean log parameter	$\ln\Lambda$
Scale parameter	μ
Probability density of failures	$p_f(\sigma_\delta) = \begin{cases} \frac{1}{\sigma_\delta\mu\sqrt{2\pi}} \exp\left(-\frac{1}{2\mu^2}\left(\ln\frac{\sigma_\delta}{\Lambda}\right)^2\right), & \text{if } \sigma_\delta \geq 0; \\ 0, & \text{if } \sigma_\delta < 0. \end{cases}$
Maximum of probability density	$\max p_f(\sigma_\delta) = p_f(\sigma_\delta) = \frac{\exp(\mu^2/2)}{\Lambda\sqrt{2\pi}\mu}$
Stress at maximum	$\sigma_{\max} = \frac{\Lambda}{\exp(\mu^2/2)}$
Cumulative failure probability P_F (for given stress value σ)	$P_f(\sigma) = \frac{1}{2} - \frac{1}{2}\text{erf}\left(\frac{\ln\Lambda - \ln\sigma}{\mu\sqrt{2}}\right)$
Standard deviation	$S = \Lambda\sqrt{\exp(\mu^2) - 1}\exp(\mu^2/2)$
Mean value (median)	$\sigma_{mean} = \Lambda \exp(\mu^2/2), P_f(\sigma_{mean}) = 1/2$

For a fixed stress parameter, the cumulative failure $P_f(\sigma)$ is equal to the fraction of failures among the large series of samples.

In order to obtain a relationship between stress level and failure probability we use the lognormal distribution. The parameters of probability density functions for normal and lognormal distributions are shown in Tables C.1 and C.2. The normal and lognormal distribution function have two parameters. The scale parameter of the distributions is μ. The location parameter of the distribution is Λ. Each parameter has a specific significance as function of analyzed application. Figures C.1 and C.2 display the graphs of functions $p_f(\sigma_\delta)$, $P_f(\sigma)$. Apparent is the influence of scale parameter μ on the width of scaling region.

For normal distribution, the location parameter Λ is the level of stress for which most fracture events occur. The scale parameter μ is a measure of experimental data scattering. These parameters have the similar meaning for lognormal distribution as well.

2. The method of calculation of failure probability is applicable for the statistical failure analysis of samples under static loading. The same method is applicable cyclic loading, if the number of cycles to failure is fixed. For fatigue tests, the number of cycles is fixed. For example, the simplest "stress parameter" is the amplitude of stress for a symmetric stress cycle. Because of practical reasons, the fatigue tests are performed to failure for several values of fixed stress parameters, such that cycles to failure scatter. To account the scattering of cycles to failure for a fixed stress value, the data must be reorganized, in other words, transposed (Wirsching, 1983). For design purposes, the designer quests the levels of stress σ for the prescribed the fraction of failures P_F^* among the large series of samples.

The cumulative failure probability for the specified intensity of cyclic load $P_f(\sigma)$ must obey the condition:

$$P_f(\sigma) \leq P_F^*. \tag{C.2}$$

Accordingly, we search the fatigue lines, which express the number of cycles to failure as the functions of stress for each fixed failure probability.

3. The graph of the inverse error function is shown on Fig. C.3. The inverse error function approximately evaluates as:

$$\text{ierf}(x) \approx \begin{cases} \text{sign}(x)\sqrt{x^2 - 1}, & \text{if } |x| < 1, \\ \frac{x\sqrt{\pi}}{2} + \frac{(x\sqrt{\pi})^3}{24} + \frac{7(x\sqrt{\pi})^5}{960} + \frac{127(x\sqrt{\pi})^7}{80640} + .., & \text{if } |x| < 1. \end{cases} \tag{C.3}$$

The inverse error function is the odd function of its argument x (Carlitz, 1963):

$$\text{ierf}\,(x) = -\text{ierf}\,(-x). \tag{C.4}$$

Apparently, most samples will fail if the stress parameter is approximately X_0. Thus, the at least 50% of all samples fail, if the stress parameter is less or equal to X_0.

4. For the design of springs, the acceptable level of the stress parameter of certain prearranged cumulative failure probability $50\% - \varepsilon$ after a certain number of cycles must be determined. The value ε is referred to as confidence parameter For example, if $\varepsilon = 50\% - 1$ ppm, the failure probability for corresponding stress level will be less or equal 1 ppm $= 10^{-6}$.

To evaluate the acceptable level of the stress parameter of certain prearranged cumulative probability, the following questions must be answered. Which stress σ_m cause the failure of samples with the probability $50\% - \varepsilon$? For which stress parameter

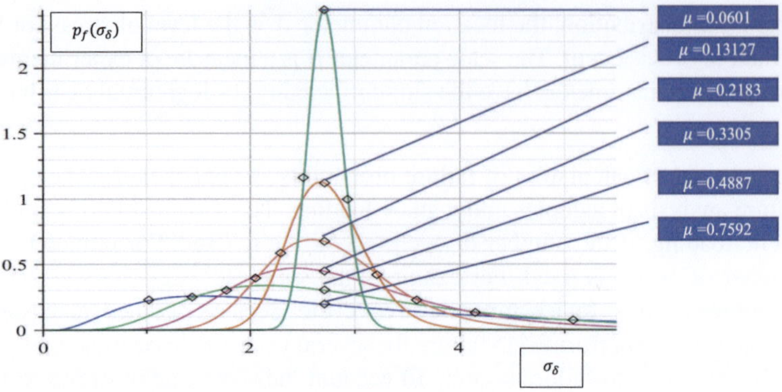

Fig. C.1 Probability density of lognormal distribution for mean value $\ln \Lambda = 1$ and different scaling factors μ

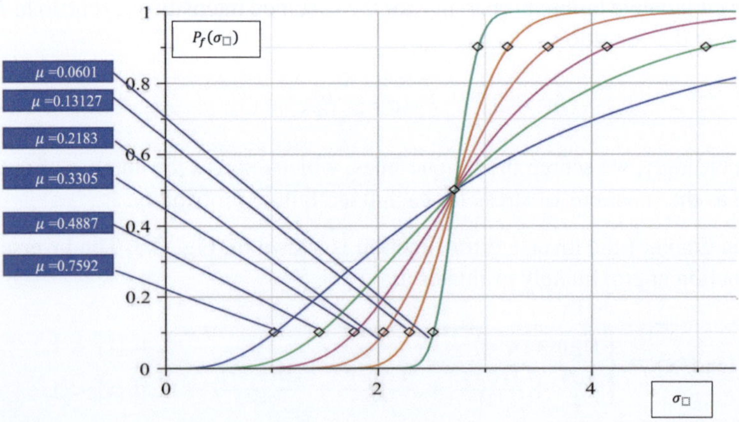

Fig. C.2 Cumulative probability of lognormal distribution for mean value $\ln \Lambda = 1$ and different scaling factors μ

σ_p the spring fails with the definite failure probability $50\% + \varepsilon$? The answers for these questions deliver the admissible stress level for the guaranteed failure probability in series applications.

For a given confidence parameter ε, the stress scatter range $T(\varepsilon)$ is the ratio of both stresses σ_p, σ_m for the given confidence parameter :

$$\frac{\sigma_p(\varepsilon)}{\sigma_m(\varepsilon)} = \frac{1}{T(\varepsilon)}. \tag{C.5}$$

Once the stress scatter range $T(\varepsilon)$ is experimentally evaluated, the scale parameter reads:

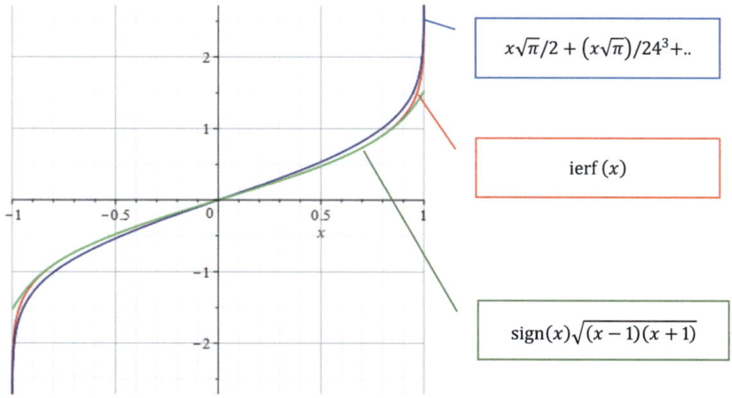

Fig. C.3 Inverse error function

	T_σ	μ
1	0.14285	0.75920
2	0.28571	0.48876
3	0.42857	0.33057
4	0.57142	0.21833
5	0.71428	0.13127
6	0,85714	0,06014

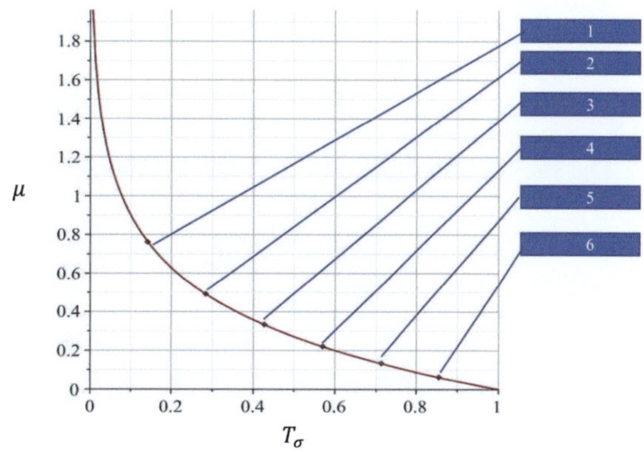

Fig. C.4 Relation between $\boldsymbol{T_\sigma}$ and scaling factor $\boldsymbol{\mu}$

$$\mu = \frac{\ln(T(\varepsilon))}{2\sqrt{2}\,\text{ierf}\,(2\varepsilon)}. \tag{C.6}$$

The scatter range $T(\varepsilon)$ is the acquired quantity and is assumed in this book to be known for the traditional setting of confidence parameter $\varepsilon_0 = 4/10$. The ratio of the stresses, which lead after a definite number of cycles to failure with 90% and 10% probabilities, is referred to as the "standard stress scatter range" T_σ (Haibach, 2006) :. For the acquired curves, the usual assessment of the confidence parameter is $\varepsilon = \varepsilon_0$.

5. The fatigue data arranges data sets in accordance to the number of cycles to failure. The described procedure is applicable directly for these data sets. Baseline fatigue data are obtained by cycling test specimens at constant-amplitude stress (or strain) until failure occurs. For each stress level, the scattering of the

Table C.3 Calculations for Lognormal distribution for given failure confidence parameter ε

Failure probability is 50% for stress X_0	$P_f(\sigma_0) = \frac{1}{2}$	$\sigma_0 = \Lambda$
Failure probability is 50% + ε for stress X_p	$P_f(\sigma_p) = \frac{1}{2} + \varepsilon$	$X_p(\varepsilon) = \Lambda \exp\left(\sqrt{2}\mu \text{ ierf}(2\varepsilon)\right)$
Failure probability is 50% − ε for stress X_m	$P_f(\sigma_m) = \frac{1}{2} - \varepsilon$	$X_m(\varepsilon) = \Lambda \exp\left(-\sqrt{2}\mu \text{ ierf}(2\varepsilon)\right)$
Scatter range of failure probabilities,$0 < T(\delta) < 1$	$\frac{\sigma_p(\varepsilon)}{\sigma_m(\varepsilon)} = \frac{1}{T(\delta)}$	$\mu = \frac{\ln(T(\varepsilon))}{2\sqrt{2}\text{ierf }(2\varepsilon)}$

number cycles to failure is performed. The "cycle scatter range" T_N for a fixed stress is evaluated. The data could be rearranged, that "standard stress scatter range" T_σ evaluates (Wirsching, 1983). For this purpose, two S–N lines are evaluated. These S–N lines display the levels of failure probabilities 90% and 10% correspondingly:

$$\sigma_{p0} = \sigma_p(\varepsilon_0), \ \sigma_{m0} = \sigma_m(\varepsilon_0), \varepsilon_0 = \frac{4}{10}. \tag{C.7}$$

The stress scatter range of the lognormal distribution is:

$$\frac{\sigma_{p0}}{\sigma_{m0}} = \frac{1}{T_\sigma}, \ T_\sigma = T(\varepsilon_0), 0 < T_\sigma < 1. \tag{C.8}$$

The scale parameter μ of lognormal distribution relates to the standard stress scatter range as follows:

$$\mu = \frac{\ln(T_\sigma)}{2\sqrt{2}\text{ ierf }(2\varepsilon_0)} = -0.3901520 \ln(T_\sigma). \tag{C.9}$$

Generally, the stress scatter range and the scale parameters depend on the cyclic stress or on number of cycles. The acquired data habitually assumes the constant scatter range over the total interval of stress parameter. In this case, the σ_{p0} and σ_{m0} are parallel to the baseline S–N line.

6. Once the scale parameter μ is determined, the cumulative failure probability distribution could be evaluated. For this purpose, of the formula for the cumulative failure probability distribution from the Table C.3 is applicable in the fatigue experiments, the value σ represents the measure of cyclic stress, for example, their amplitude or stroke. For practical purposes, the inverse problem must be solved. The engineer must determine the amplitude of the cyclic stress, which leads to the certain cumulative failure probability. Accordingly, the acceptable value σ stress parameter is the function of the required failure probability P_F^*. The design of spring is performed for this acceptable value of stress parameter.

Table C.4 Acceptable stress parameters for required cumulative failure probability P_F

Cumulative failure probability $\boldsymbol{P_F}$	Stress parameter to requisite failure probability	Ratio of stress parameter stress to the median stress $\boldsymbol{\Lambda}$
$P_f(\sigma)$	σ	σ/Λ
0.9	$\Lambda/\sqrt{T_\sigma}$	$1/\sqrt{T_\sigma}$
0.5 median	Λ	1
0.1	$\Lambda\sqrt{T_\sigma}$	$\sqrt{T_\sigma}$
10^{-2}	$\Lambda \cdot T_\sigma^{0.90762}$	$T_\sigma^{0.90762}$
10^{-3}	$\Lambda \cdot T_\sigma^{1.20566}$	$T_\sigma^{1.20566}$
10^{-4}	$\Lambda \cdot T_\sigma^{1.45098}$	$T_\sigma^{1.45098}$
10^{-5}	$\Lambda \cdot T_\sigma^{1.66395}$	$T_\sigma^{1.66395}$
10^{-6}	$\Lambda \cdot T_\sigma^{1.845229}$	$T_\sigma^{1.845229}$

For example, assume for standard stress scatter range the following value $T_\sigma = 1/1.25 = 0.8$. If the obligatory failure probability is equal to $P_F^* = 0.1\% = 10^{-3}$, the reduction of stress parameter relative to its median is:

$$T_\sigma^{1.20566..} = 0.7641\ldots \tag{C.10}$$

This means, that the acceptable value of stress is 76% of the median stress parameter. If the required failure probability is equal to $P_F^* = 1ppm = 10^{-6}$, we have

$$T_\sigma^{1.8455..} = 0.6478\ldots \approx 65\%. \tag{C.11}$$

The applicability of the lognormal distribution is restricted to the range of the cumulative failure probability between 1 and 99%. The estimations of cyclic stress for the lower values of the cumulative failure probability are displayed for recommendation. Figure C.4 displays the relation between T_σ and scaling factor μ of the lognormal distribution (Table C.4).

Appendix D
Influence of Mean Stress on Fatigue Life

D.1 Fatigue Strength Diagrams

Appendix D summarizes the different approaches to mean stress sensitivity of structural materials in general, and particularly of spring alloys. For estimating fatigue life in engineering applications, Goodman and Haigh diagrams are commonly used (Chaps. 11 and 12). For numerical, and particularly the finite-element computation, the analytical formulas are much more preferable. The analytical formulas could be immediately condensed to program languages scripts. For example, ANSYS uses the ANSYS Parametrical Design Language (ANSYS, 2024; Thompson & Thompson, 2017). APDL is the scripting language of the Ansys solver and interpreter, which was renamed to MAPDL (Mechanical APDL) around 2010. The APDL macros reads the nodal and finite element information and calls the solution and postprocessing. Scripting is a powerful tool that allows you to combine the functionality of the Graphical User Interface of ABAQUS (ABAQUS, 2020) and the power of the programming language Python (ABAQUS, 2024).

The conventional hypotheses by Smith–Watson–Topper, Walker and Bergmann have been successfully used to describe uniaxial cyclic fatigue with non-zero mean value over the whole range of the fatigue life . However, the physical characteristics of the mean stress sensitivities in these hypotheses are different. The mean stress sensitivity according to Smith–Watson–Topper is identical for all materials and stress levels. This weakness reduces the applicability of the Smith–Watson–Topper parameter. At first glance, the mean stress sensitivities according to Walker and Bergmann are diverse. The mean stress sensitivities depend upon two different additional correction parameters, namely the Bergmann parameter and the Walker exponent. The possibility of fitting the mean stress sensitivity in these hypotheses overcomes the significant drawback of the Smith–Watson–Topper schema. Appendix D displays the dependence between the Bergmann parameter and the Walker exponent, which leads to a certain mean stress sensitivity.

V. Kobelev, *Fundamentals of Springs Mechanics*,
https://doi.org/10.1007/978-3-031-58584-5

The strength hypotheses lead to associations for a reasonable curve expressed in stress or strain. From this curve, the definitive admissible amplitudes of the stresses or strains are obtained. The definitive admissible amplitudes correspond to a certain number of cycles to failure. The dominant stresses in leaf and disc springs are the normal stresses. For shear compression or helical tension springs, the shear stresses should be used instead of the normal stresses. Fatigue characteristics are determined by analyzing the failure parameters and the fatigue characteristics of the materials. Among the equivalents, stress hypotheses have the greatest potential. For a loading with a fully occupied stress tensor, the application of the magnitudes for the equivalent scalar stress is appropriate. Alternatively, multiaxial stress can be reduced to a certain scalar energy parameter (Łagoda, 2022; Gadouini, 2008). The equivalent scalar stress for isotopic materials can be presented as a function of the stress invariants. If only the second invariant of the stress tensor plays a role, and the corresponding scalar is referred to as the Mises equivalent stress.

Based on the same equivalent stress, higher fatigue strengths typically result for springs made of the same material subjected to normal stress than for springs subjected to shear stress. For this reason, numerous alternative equivalent stress criteria have been proposed, such as the multiaxial fatigue criteria embrace stress criteria, strain criteria and energy criteria (stress–strain). The proper choice depends on the physical nature of the failure parameter. The fatigue criteria describe the failure after a large number of cycles. Accordingly, the plastic deformation is very small, and can be neglected. Alternatively, for large plastic deformations, stress criteria have been proposed that condense the tensor stress state to the uniaxial scalar value. The corresponding equivalent stress σ is used for calculations instead of tensor stress functions. For the subsequent fatigue calculations, the equivalent stress σ is used.

Appendix explains the correct choice between the traditional Bergmann and Walker criteria. Namely, if the fitting parameters of Bergmann and Walker correspond to the same mean stress sensitivity due to Schütz, the curves of the two criteria will be slightly different. With the additional experimental data, it is possible to determine whether Bergmann or Walker is preferable.

Fatigue strength is the stress range or stress amplitude at which a component can endure an infinite number of load cycles. The fatigue strength σ_w denotes the stress amplitude that can be sustained under alternating stress. On the Wöhler line, this range is marked by the horizontal curve to the right of the inflection point. As soon as a component is subjected to higher stresses, the fatigue check is referred to as the fatigue strength (FKM, 2018). In addition to fatigue strength σ_w, the analysis of a fatigue strength diagram is based on several static strength parameters. The static strength parameters include tensile strength R_m and yield strength R_e.

The Wöhler line for the spring elements depends on various parameters for the following influencing factors:

- Spring material, incl. thermal treatment;
- Notch effect, corrosion, roughness;
- Stress (normal or shear stress, mixed stress);
- Load spectrum;

- Component size and specifically diameter of spring wire;
- Operating temperature;
- Residual stress (shot peening, roller burnishing, laser peening, tapping);
- Mean stress;
- Stress gradient.

The correlation of fatigue strength with certain material properties, such as ductility and strength, has been reported in the literature. Experience shows that fatigue strength is almost proportional to the tensile strength of the material. The higher the tensile strength, the higher the fatigue strength. Other influences of the material condition on the fatigue strength arise from the ductility, grain size, heat treatment, presence of defects and inclusions in the material.

The notch effect on the spring surface is a critical factor in fatigue strength. Surface defects such as blowholes, hard inclusions, corrosion cavities, hairline cracks, decarburization or grooves act as notches. The sharper the notch, the lower the fatigue strength of the component with respect to its rated stress.

Experimental studies show that fatigue strength decreases significantly with environmental degradation and corrosion of springs. In fatigue-stressed components, corrosion leads to the expansion of incipient cracks and also to progressive crack growth. The roughness profile on the component surface created by the manufacturing process acts as a notch in conjunction with the surface microstructure. The influence of roughness on the reduction of fatigue strength increases with increasing brittleness (and indirectly with increasing strength), according to VDI. 2226 (VDI, 1965).

The influence of the component size is caused by different mechanisms, namely:

- Technological influences (edge layer thickness, edge strength, surface hardening);
- The size of the highly stressed volume or surface of a spring. Depending on how far the stressed area extends into the volume of the spring, there may be an increased influence of statistically distributed defects and inclusions;
- The greater the stress gradient, the greater the support provided by the surrounding material. A higher support effect positively influences the fatigue behavior. For springs, the support effect does not play a significant role because the spatial gradients for the stresses are small and the stress distributions are identical for similar springs;
- For geometrically similar springs with equally high stress maxima, the smaller spring will have a higher stress gradient and thus more favorable fatigue behavior;
- The smaller diameter wires generally have the higher fatigue strength.

Both the fatigue and static strengths of most materials generally increase at low temperatures. However, notch sensitivity and brittle fracture susceptibility are increased. Accordingly, higher temperatures generally result in a decrease in fatigue strength.

There are some special characteristics of spring materials that are important in fatigue design. Residual stresses occur in springs as a result of almost all treatments in the manufacturing process. These can be up to the yield point. The effect on fatigue strength depends on the value of the residual stresses. There are predominantly compressive or predominantly tensile residual stresses. Compressive residual stresses increase the fatigue strength, while tensile residual stresses usually have a negative effect on the fatigue strength. Tensile residual stresses are added to the average stresses in each cycle and cause a higher level of damage in each load cycle.

Fatigue strength diagrams represent the dependence of fatigue strength on the mean stress. The fatigue strength diagrams according to Smith and Haigh are commonly used in spring industry. In the Smith diagram, the admissible upper stresses $\sigma_{\max}$ and lower stresses $\sigma_{\min}$ are the functions of the mean stress (σ_m). The mean stress is the average value of the upper stress ($\sigma_{\max} = \sigma_m + \sigma_a$) and the lower stress ($\sigma_{\min}$) during one load cycle:

$$\sigma_m = \frac{\sigma_{\max} + \sigma_{\min}}{2}, \quad \sigma_a = \frac{\sigma_{\max} - \sigma_{\min}}{2}. \tag{D.1}$$

The ratio of lower stress to upper stress in a cycle is referred to as the stress ratio:

$$R_\sigma = \frac{\sigma_{\min}}{\sigma_{\max}}, \quad \sigma_m = \frac{1 + R_\sigma}{1 - R_\sigma}\sigma_a. \tag{D.2}$$

The compression and shear tests are performed in the stress–positive region: $0 < \sigma_{\min} < \sigma_{\max}$. The tension test is executed in the stress–negative region: $\sigma_{\min} < \sigma_{\max} < 0$. The tensile strength is plotted as the top stress on the ordinate and the average stress on the abscissa. The resulting point is on a 45° line with the coordinate origin. From this point, the Goodman line for the top stress is generated in the simplified Smith diagram by connecting it to the alternating stress on the ordinate. The same procedure is used to generate the Goodman line for the lower stress. The resulting fatigue range is bounded at the top by the yield strength. Consequently, the fatigue range below the 45° line is adjusted due to the symmetry between the upper and lower stresses. From the Smith diagram, it is possible to read the upper and lower stresses that can be sustained and the stress amplitudes σ_a as a function of the mean stress σ_m. In general, alloys with increasing tensile strength exhibit greater sensitivity to mean stress. Another correlation is found in the ductility of the materials. Brittle alloys tend to have higher mean stress sensitivity.

The damage $\mathbf{P}(\sigma_m, \sigma_a)$ of the structural element is the function of the σ_m and σ_a. The isolines, which correspond to the constant damage, are $\mathbf{P}(\sigma_m, \sigma_a) = \mathbf{P}_0$. In the fatigue strength diagrams the damage along the isoline corresponds to the ultimate damage for the given number of load cycles. If the damage is greater as $\mathbf{P}_0$, the structural element is unacceptable. Therefore, the isoline with the prescribed ultimate damage is essential for the design of the structural elements. This isoline is known as the fatigue diagram. In the Haigh diagram, the steady-state stress amplitude over the mean stress can be read directly. In addition, the stress ratios can be assigned to the pairs of stress amplitude and mean stress values using the origin lines. Similar to the Smith diagram, the Goodman line is plotted as a simplification, starting from the alternating strength on the ordinate. In the Haigh diagram, however, this point is

Table D.1 Stress ratio

Stress ratio	Position of the origin straight line	Stress range
$R_\sigma = 1$	Positive range of mean stress axis	Constant stress
$R_\sigma = 0$	45° line in the first quadrant (tension area)	Oscillating tension stress $\sigma_{\min} = 0$
$R_\sigma = -1$	Positive range of the amplitude axis	Alternating stress $\sigma_{\min} = -\sigma_{\max}$
$R_\sigma = -\infty$	45° line in the second quadrant (compression area)	Oscillating compression stress $\sigma_{\max} = 0$

connected to the point of tensile strength on the mean stress axis (abscissa). Below the Goodman line is the fatigue region, which is bounded on the right by a straight line between the yield strengths on the amplitude axis and the mean stress axis. In the compression region, the Goodman line is extended to the yield point or conservatively replaced by a horizontal line at the level of the alternating fatigue strength. (σ_w). Again, the fatigue limit is bounded by a straight line between the fatigue strength and the yield strength (or, in the case of compression, the crush limit).

D.2 Mean Stress Sensitivity According to Schütz

As shown above, the inclination of the Goodman straight line determines the mean stress sensitivity. An equivalent slope in the straight line has a similar meaning in the Haigh diagram . The number of load cycles to failure (fatigue life) does not change along the straight lines. In other words, both diagrams use the linear dependence of the ultimate stress amplitude on the mean stress. This approach is traditional in industrial engineering.

The formal definition of mean stress sensitivity is outlined below. The mean stress sensitivity according to Schütz uses only two experimental values $\sigma_a(R_\sigma = -1), \sigma_a(R_\sigma = 0) = \sigma_m(R_\sigma = 0)$ (see Table D.1).

The first value in (D.3a) is the mean stress sensitivity M_r, according to (Schütz, 1967). According to Schütz, the damage is represented as the straight line connecting the points $\mathbf{P}_S(R_\sigma = 0)$ and $\mathbf{P}_S(R_\sigma = -1)$. The mean stress sensitivity according to Schütz is assumed to be constant. The positive constant M_r designates the tangent of the angle between the horizontal and the straight line of constant damage on the Haigh diagram. The higher damage leads to the lower cycles to failure. This means, that the positive mean stress σ_m increases the damage, which leads to lower fatigue life. The calculation of the mean stress sensitivity is shown on Fig. D.1. There are three points a, b, c. The stress amplitude in point a is equal to $\sigma_a(R_\sigma = -1)$. The stress amplitude in point c is equal to $\sigma_a(R_\sigma = 0)$. The vertical distance between the points a and c is equal to $bc = \sigma_a(R_\sigma = -1) - \sigma_a(R_\sigma = -0)$. The positive value bc corresponds to the length of the opposite side of the triangle abc. The horizontal distance between the points a and b is equal to

$$ab = \sigma_m(R_\sigma = 0) - \sigma_m(R_\sigma = -1) \equiv \sigma_m(R_\sigma = 0).$$

The positive value ab corresponds to the length of the adjacent side of the triangle abc. Note, that:

$$\sigma_m(R_\sigma = -1) = 0, \quad \sigma_m(R_\sigma = 0) = \sigma_a(R_\sigma = 0).$$

The hypothenuse is ac. The angle α is assumed traditionally to be positive, and is equal to $\alpha = \operatorname{atan}(bc/ab) > 0$. The left image on Fig. D.1 displays the Haigh diagram. The slope of the straight line ac is negative. The tangent of the declination angle in Haigh diagram is equal to the positive mean stress sensitivity:

$$\tan\alpha = \frac{bc}{ab} = \frac{\sigma_a(R_\sigma = -1) - \sigma_a(R_\sigma = 0)}{\sigma_m(R_\sigma = 0)} = \frac{\sigma_a(R_\sigma = -1)}{\sigma_m(R_\sigma = 0)} - 1 \overset{\text{def}}{=} M_r. \tag{D.3a}$$

The *positive* mean stress sensitivity M_r leads to the *negative slope* of the curve, i.e. the fatigue line on Haigh diagram declines with the higher mean stress.

The right picture on Fig. D.1 shows the Goodman diagram. The slope of the straight line $a'c'$ is positive. For the evaluation of the angles $\alpha = \measuredangle cab$ and $\beta = \measuredangle c'a'b'$, take into account the geometric relations between the sides of the abc and$a'b'c'$:

$$aO = a'O, \; cO = c'O, \; ac = a'c',$$

$$Ob' = a'b' = \frac{Oa'}{\sqrt{2}}, \; ab = \frac{Oc}{\sqrt{2}}, \; bc = Oa - \frac{Oc}{\sqrt{2}}.$$

From the comparison of triangles abc and $a'b'c'$ we get the relation between the tangents of the angles:

$$\frac{1 + \tan\alpha}{\sqrt{2}} = \frac{\sqrt{2}}{1 + \tan\beta}.$$

According to Eq. (D.3a), the tangent of the angle α is the mean stress sensitivity M_r. Finally, the tangent of the inclination angle β of the fatigue line in Goodman diagrams depends upon the mean stress sensitivity:

$$\mu_r \overset{\text{def}}{=} \tan\beta = \frac{b'c'}{a'b'} = \frac{1 - M_r}{1 + M_r}. \tag{D.3b}$$

The inversion of Eq. (D.3b) reads:

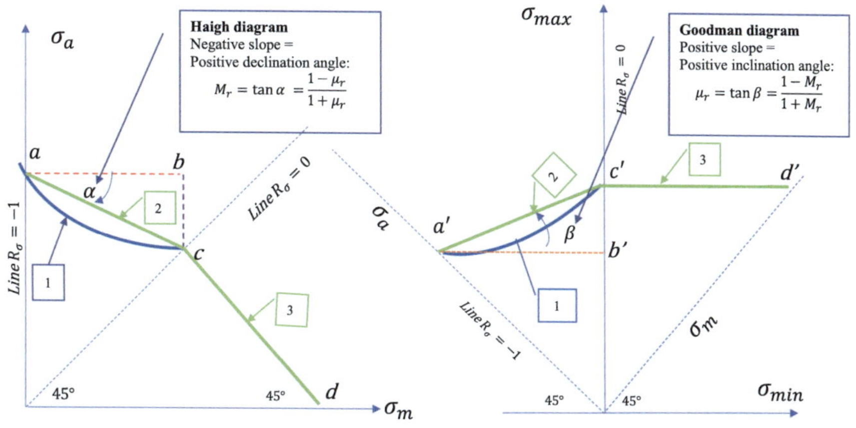

Fig. D.1 Definition of the mean stress sensitivity: (1) Curved isoline of the equal fatigue life, (2) Idealized straight isoline of equal fatigue life for calculation of mean stress influence and (3) Relaxation region

Table D.2 Damage parameter $\mathbf{P_S}$ and the mean stress sensitivity according to Schütz

Stress ratio	Lower stress	Mean stress	Upper stress	$\mathbf{P}_S \overset{\text{def}}{=} \sigma_a + \sigma_m.M_r$
$R_\sigma = 0$	$\sigma_{\min} = 0$	$\sigma_m = \sigma_a$	$\sigma_{\max} = 2\sigma_a$	$\mathbf{P_S}(R_\sigma = 0) = (1 + M_r)\sigma_a$
$R_\sigma = -1$	$\sigma_{\min} \equiv \sigma_a = -\sigma_{\max}$	$\sigma_m = 0$	$\sigma_{\max} = \sigma_a$	$\mathbf{P_S}(R_\sigma = -1) = \sigma_a$

$$M_r \equiv \tan\alpha = \frac{1-\mu_r}{1+\mu_r} \tag{D.3c}$$

From Eq. (D.3b) follows, that for $M_r = 0, \alpha = 0°$, $\beta = 45°$. Similarly, for $M_r = 1$ is valid $\alpha = 45°$, $\beta = 0°$. For all admissible $0 < M_r < 1$, the angle $0 < \beta < 45°.0 < \beta < 45°$.

The damage parameters are defined in Tables D.2, D.3, D.4 and D.5. The evaluation of the values $\sigma_a(R_\sigma = -1)$ and $\sigma_a(R_\sigma = 0)$ for the certain number of cycles up to damage are required. The method is explained in Chap. 12.

However, real materials rarely exhibit a linear dependence of stress amplitude on mean stress. The simple linear dependence of ultimate stress amplitude on mean stress, which leads to the same damage and the same fatigue life, is inappropriate. If the line of constant damage on the Haigh diagramm differs from the straight one, the stress amplitude for the certain damage σ_a depends nonliearly upon the mean value σ_m, $\sigma_a = \sigma_a(\sigma_m)$. According to the experimental findings, the amplitude σ_a approximately follows a hyperbola with the increasing mean stress σ_m (Fig. D.1). One possible correction was proposed by Guideline FKM (2020). This standard uses the piecewise-linear approximation. The piecewise-linear function, according to FKM, consists of three regions. The programming of scripts for finite-element analysis is somewhat tricky.

Table D.3 Determination of the damage parameter $\mathbf{P_{SWT}}$ and the mean stress sensitivity according to Smith–Watson–Topper

Stress ratio	Lower stress	Mean stress	Upper stress	$\mathbf{P_{SWT}} \overset{def}{=} \sqrt{\sigma_a \cdot \sigma_{max}}$ $\mathbf{P_{SWT}} = \sqrt{\sigma_a \cdot (\sigma_a + \sigma_m)}$
$R_\sigma = 0$	$\sigma_{min} = 0$	$\sigma_m = \sigma_a$	$\sigma_{max} = 2\sigma_a$	$\mathbf{P_{SWT}}(R_\sigma = 0) = \sqrt{2}.\sigma_a$
$R_\sigma = -1$	$\sigma_{min} = -\sigma_{max}$	$\sigma_m = 0$	$\sigma_{max} = \sigma_a$	$\mathbf{P_{SWT}}(R_\sigma = -1) = \sigma_a$
The average, constant mean stress sensitivity according to Smith–Watson–Topper	$M_r \overbrace{=}^{Eq(D.3)} \dfrac{\mathbf{P_{SWT}}(R_\sigma = -1) - \frac{\mathbf{P_{SWT}}(R_\sigma = 0)}{\sqrt{2}}}{\frac{\mathbf{P_{SWT}}(R_\sigma = 0)}{\sqrt{2}}} = \sqrt{2} - 1 = 0.41 \ldots \overset{def}{=} M_{r_{SWT}}$			

Table D.4 Determination of the damage parameter $\mathbf{P_W}$ and the mean stress sensitivity according to Walker

Stress ratio	Lower stress	Mean stress	Upper stress	$\mathbf{P_W} \overset{def}{=} \sigma_a^{\tilde{\gamma}} . \sigma_{max}^{1-\tilde{\gamma}}$ $\mathbf{P_W} = \sigma_a^{\tilde{\gamma}} . (\sigma_a + \sigma_m)^{1-\tilde{\gamma}}$
$R_\sigma = 0$	$\sigma_{min} = 0$	$\sigma_m = \sigma_a$	$\sigma_{max} = 2\sigma_a$	$\mathbf{P_W}(R_\sigma = 0) = 2^{1-\tilde{\gamma}} \sigma_a$
$R_\sigma = -1$	$\sigma_{min} = -\sigma_{max}$	$\sigma_m = 0$	$\sigma_{max} = \sigma_a$	$\mathbf{P_W}(R_\sigma = -1) = \sigma_a$
The average, constant mean stress sensitivity according to Walker	$M_r \overbrace{=}^{Eq.(D.3)} \dfrac{\mathbf{P_W}(R_\sigma = -1) - \frac{\mathbf{P_W}(R_\sigma = 0)}{2^{1-\tilde{\gamma}}}}{\frac{\mathbf{P_W}(R_\sigma = 0)}{2^{1-\tilde{\gamma}}}} = \dfrac{\sigma_a - \frac{\sigma_a}{2^{1-\tilde{\gamma}}}}{\frac{\sigma_a}{2^{1-\tilde{\gamma}}}} = 2^{1-\tilde{\gamma}} - 1 > 0.$			

Table D.5 Determination of the damage parameter $\mathbf{P_B}$ and the mean stress sensitivity according to Bergmann

Stress ratio	Lower Stress	Mean stress	Upper Stress	$\mathbf{P_B} \overset{def}{=} \sqrt{\sigma_a . (\sigma_{max} - \zeta \sigma_m)}$ $\mathbf{P_B} = \sqrt{\sigma_a . [\sigma_a + (1 - \zeta)\sigma_m]}$
$R_\sigma = 0$	$\sigma_{min} = 0$	$\sigma_m = \sigma_a$	$\sigma_{max} = 2\sigma_a$	$\mathbf{P_B}(R_\sigma = 0) = \sigma_a \sqrt{2 - \zeta}$
$R_\sigma = -1$	$\sigma_{min} = -\sigma_{max}$	$\sigma_m = 0$	$\sigma_{max} = \sigma_a$	$\mathbf{P_B}(R_\sigma = -1) = \sigma_a$
The average, constant mean stress sensitivity according to Bergmann	$M_r \overbrace{=}^{Eq.(D.3)} \dfrac{\mathbf{P_B}(R_\sigma = -1) - \frac{\mathbf{P_B}(R_\sigma = 0)}{\sqrt{2-\zeta}}}{\frac{\mathbf{P_B}(R_\sigma = 0)}{\sqrt{2-\zeta}}} = \sqrt{2 - \zeta} - 1 > 0.$			

Three alternative criteria are widely used in industry and in finite element codes. These criteria, namely the Smith–Watson–Topper, Walker, and Bergmann criteria, exhibit the analogous hyperbolic dependence between amplitude and mean stress. In each of these traditional criteria, additional fitting parameters are used to describe the continuous nonlinear dependence of stress amplitude on mean stress for a given fatigue life. Among them, the Smith–Watson–Topper criteria lack the additional parameter. Any other criterion (Walker or Bergmann) uses a fitting parameter. The methodology for unified evaluation of mean stress sensitivity is explained below (Table D.3).

D.3 Mean Stress Sensitivity According to Smith–Watson–Topper

The Smith–Watson–Topper parameter ($\mathbf{P_{SWT}}$) (Smith et al., 1970) was proposed for the evaluation of the fatigue life of metallic materials in uniaxial loading conditions. This fatigue parameter characterizes the crack initiation with non-zero mean values. In the above cited paper, the numerous modifications to the original $\mathbf{P_{SWT}}$ are summarized. The Smith–Watson–Topper parameters have been widely used to describe uniaxial cyclic fatigue with non-zero mean values over the whole range of the fatigue life. The results from the cyclic tests were presented using the $\mathbf{P_{SWT}}$ parameter (Boeller & Seeger, 1987). This parameter was used for the description of the fatigue life in a complex loading.

The determination of the damage parameter $\mathbf{P_{SWT}}$ and mean stress sensitivity, according to Smith–Watson–Topper, uses the following schema. Table D.3 displays the steps to link the mean stress sensitivity, according to Schütz, to the mean stress sensitivity, according to Smith–Watson–Topper. In the actual study, the mean stress sensitivity uses three values $\sigma_a(R_\sigma = -1), \sigma_a(R_\sigma = 0) = \sigma_m(R_\sigma = 0)$, in accordance with Schütz [Schütz, 1967]. Accordingly, two distinct stress ratios $R_\sigma = 0$ and $R_\sigma = 1$ are used. The calculation requires only two experimentally acquired values, namely $\sigma_a(R_\sigma = -1)$, M_{rSWT}. With these values, the values

$$\mathbf{P_{SWT}}(R_\sigma = 0) = \sqrt{2} \cdot \sigma_a(R_\sigma = 0) \equiv \sqrt{2} \cdot \sigma_m(R_\sigma = 0),$$

$$\mathbf{P_{SWT}}(R_\sigma = -1) = \sigma_a(R_\sigma = -1).$$

are calculated for a certain fixed fatigue life. After substituting these damage parameters, equation (D.3a) gives the mean stress sensitivity according to Smith–Watson–Topper. Obviously, this value is the same for all materials and stress levels, namely:

$$M_{rSWT} = 0.41$$

This situation reveals a significant disadvantage of the Smith–Watson–Topper hypothesis, since this criterion has no additional fitting parameters. Thus, the application of the Smith–Watson–Topper criterion to materials with significantly different mean stress sensitivity is rather doubtful.

D.4 Mean Stress Sensitivity According to Walker

An alternative empirical relationship was developed by Walker (1970). This relationship provides more freedom in the description of the mean stress effects and satisfactorily represents the effect of the mean stress. In the relationship by Walker, an additional correction parameter $\tilde{\gamma}$ is introduced . The parameter $\tilde{\gamma}$ characterizes the exponent of the stress amplitude. Walker's parameters were used to fit the experimental data and predict the fatigue life behavior at different mean strain values during cyclic loading. Based on the Walker approach, the authors (Geilen et al., 2020; Karr et al., 2022) evaluated the influence of the mean strain on the fatigue life of the spring material.

The determination of the damage parameter $\mathbf{P_W}$ and mean stress sensitivity, according to Walker, is performed analogously (Table D.4). Two distinct stress ratios are used, namely the ratios according to Schütz $R_\sigma = 0$ and $R_\sigma = 1$. The same values as above, are required for the calculations:

$$\sigma_a(R_\sigma = -1), \quad \sigma_a(R_\sigma = 0), \quad \sigma_m(R_\sigma = 0).$$

With these values, the damage parameters are calculated with the Walker formula as:

$$\mathbf{P_W}(R_\sigma = 0) = 2^{1-\tilde{\gamma}} \cdot \sigma_a(R_\sigma = 0) \equiv 2^{1-\tilde{\gamma}} \cdot \sigma_m(R_\sigma = 0),$$

$$\mathbf{P_W}(R_\sigma = -1) = \sigma_a(R_\sigma = -1).$$

The substitution of these values in Equation (D.3a) results in the mean stress sensitivity, according to Walker:

$$M_r = 2^{\tilde{\gamma}-1} - 1.$$

Importantly, this value is not constant anymore and depends on parameter γ. This feature of the Walker hypothesis overcomes the significant drawback to the Smith–Watson–Topper schema.

It is possible to express the Walker exponent $\tilde{\gamma}$ in terms of the Schütz mean stress sensitivity. Important to mark, that the Walker exponent $\tilde{\gamma}$ depends directly on M_r:

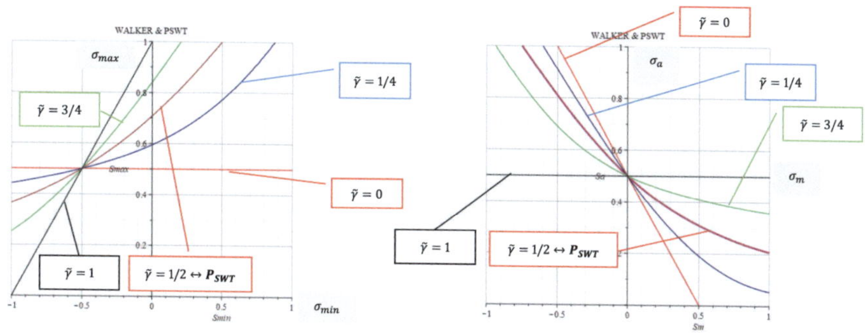

Fig. D.2 Visualization of the mean stress sensitivity for Goodman (left) and Haigh (right) diagrams with the Walker parameter

$$\tilde{\gamma}(M_r) = 1 - \frac{\ln(1 + M_r)}{\ln 2}. \tag{D.4}$$

Figure D.2 shows of Haigh and Goodman diagrams for Walker criterion. Figure D.2 demonstrates the dependence of the cycle amplitude on the mean stress according to Walker, which leads to a certain fixed fatigue life. Left picture on the Fig. D.2 shows the curves of the equal fatigue life in the coordinates $\sigma_{\max} = \sigma_{\max}(\sigma_{\min})$ according to Goodman method. Right picture on the Fig. D.2 displays the curves of the equal fatigue life in the coordinates $\sigma_a = \sigma_a(\sigma_m)$ according to Haight method.

D.5 Mean Stress Sensitivity According to Bergmann

Similar nonlinear approximations were proposed by Bergmann and Seeger (1979). The factor used in this paper slightly differs from the original parameter by Bergmann, $k = 1 - \zeta$. The sense of the new parameter is the following: the parameter ζ clearly reveals the similarity with the criterion by Walker. Figure D.3 shows of Haigh and Goodman diagrams for Bergmann criterion. Bergmann's adaptation assumes the non-zero value of the ζ-factor. For example, for carbon steel $k = 0.45$, $\zeta = 0.55$. A better agreement with the experiment was obtained for $k = 0.4$, $\zeta = 0.6$ (Nihei et al., 1986). For the 1045 steel grade, the best fitting for the experimental data was obtained for $k = 1.2$, $\zeta = -0.2$ (Wehner & Fatemi, 1991). Notably, for the following value in the Bergmann parameter $\zeta = 1, k = 0$, the mean stress sensitivity fades.

The determination of the damage parameter $\mathbf{P_B}$ and mean stress sensitivity, according to Bergmann, is assessed similarly. Table D.5 uses the stress ratios $R_\sigma = 0$ and $R_\sigma = 1$. Once again, the values $\sigma_a(R_\sigma = -1)$, $\sigma_a(R_\sigma = 0) = \sigma_m(R_\sigma = 0)$ are evaluated. From these values, the parameters are determined as:

$$\mathbf{P_B}(R_\sigma = 0) = \sqrt{2 - \zeta} \cdot \sigma_a(R_\sigma = 0) \equiv \sqrt{2 - \zeta} \cdot \sigma_m(R_\sigma = 0),$$

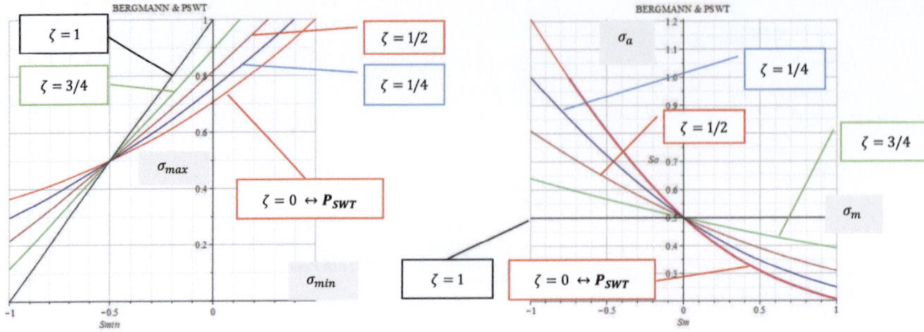

Fig. D.3 Visualization of the mean stress sensitivity for Goodman (left) and Haigh (right) diagrams according to Bergmann

$$\mathbf{P_B}(R_\sigma = -1) = \sigma_a(R_\sigma = -1).$$

Figure D.3 displays the Bergmann dependence of cycle amplitude on the mean stress, which leads to the equal fatigue life. Left picture on the Fig. D.3 shows the curves of the equal fatigue life in the coordinates $\sigma_{\max} = \sigma_{\max}(\sigma_{\min})$ according to Goodman and method. Right picture on the Fig. D.3 shows the curves of the equal fatigue life in the coordinates $\sigma_a = \sigma_a(\sigma_m)$ according to Haight method. According to Equation (D.5), each curve corresponds to a certain value in the parameter ζ.

In conclusion, Equation (D.3a) results in the mean stress sensitivity, which follows from the Bergmann criterion:

$$M_r = \sqrt{2 - \zeta} - 1.$$

Significantly, this value is the function of another parameter $\boldsymbol{\zeta}$. This property of the Bergmann hypothesis overcomes the aforementioned disadvantage of the Smith–Watson–Topper scheme.

Analogously to Equation (D.4), the Bergmann parameters are definite functions of the mean stress sensitivity as well:

$$\zeta(M_r) = 2 - (M_r + 1)^2, \tag{D.5}$$

$$k(M_r) = (M_r + 1)^2 - 1 \rightarrow k = 1 - \zeta$$

Figure D.4 displays the Goodman diagrams for Walker and Bergmann criteria of cycle amplitude on the mean stress, which leads to the equal fatigue life. Both intervals of positive and negative stresses are calculated from the formulas for $\mathbf{P_W}.\mathbf{P_B}$ values. The relaxation regions are displayed for clarity

Figure D.5.demonstrates the graphs of Haigh diagrams for the functions $\zeta(M_r), \widetilde{\gamma}(M_r)$. The green vertical line shows the determined mean stress sensitivity $M_r = M_{rSWT} \equiv \sqrt{2} - 1 = 0.41$, which results from the calculation of the damage

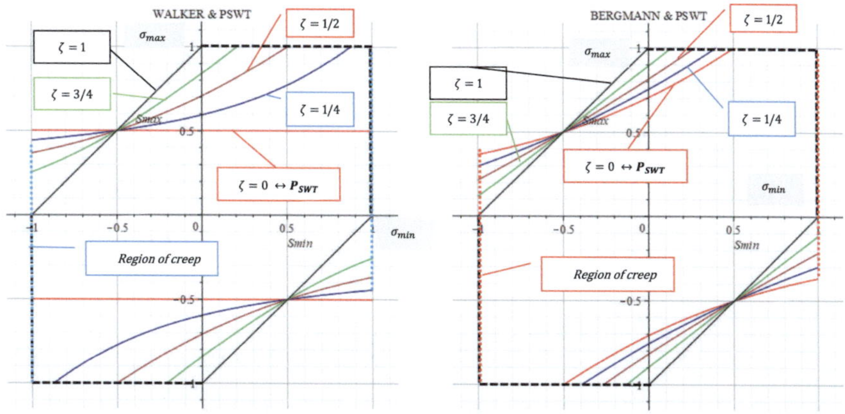

Fig. D.4 Visualization of the mean stress sensitivity for Goodman diagrams according to Walker and Bergmann criteria

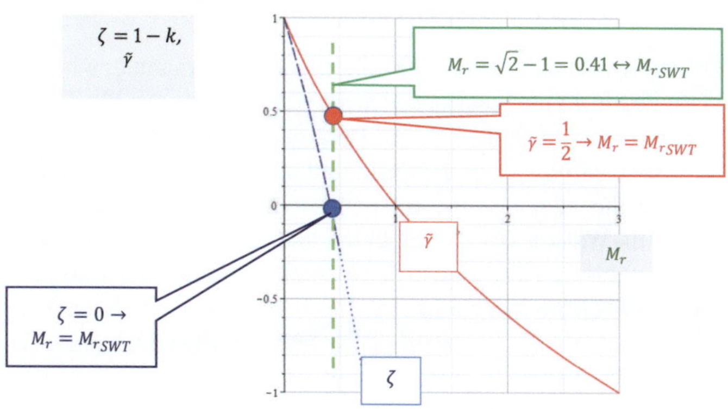

Fig. D.5 Parameters of Walker $\tilde{\gamma}$ and Bergmann ζ as functions of the mean stress sensitivity M_r.

parameter $\mathbf{P_{SWT}}$, according to Smith–Watson–Topper. From the graphs (Fig. D.5), the reduction of both parameters with the increasing mean stress sensitivity is clearly visible

D.6 Recalculation Between Bergmann and Walker Criteria

From Equations (D.4) and (D.5), follows the interrelations between the Bergmann parameter ζ and the Walker exponent $\tilde{\gamma}$. The parameters of both criteria match to the identical mean stress sensitivity μ_r:

Table D.6 Damage parameters, $\mathbf{P_S}, \mathbf{P_{SWT}}, \mathbf{P_W}, \mathbf{P_B}$, for different stress ratios R_σ

Damage parameter	$R_\sigma = 0$	$R_\sigma = -1$	$R_\sigma = -1/2$
$\mathbf{P_{SWT}} = \frac{\sqrt{1-R_\sigma}}{\sqrt{2}}\sigma_{\max}$	$\frac{\sigma_{\max}}{\sqrt{2}}$	$\sigma_{\max}$	$\frac{\sqrt{3}}{2}\sigma_{\max}$
$\mathbf{P_S} = \left(\frac{1-R_\sigma}{2} + \frac{1+R_\sigma}{2}M_r\right)\sigma_{\max}$	$(1+M_r)\frac{\sigma_{\max}}{2}$	$\sigma_{\max}$	$\left(\frac{3}{4} + \frac{M_r}{4}\right)\sigma_{\max}$
$\mathbf{P_W} = \left(\frac{1-R_\sigma}{2}\right)^{\tilde{\gamma}}\sigma_{\max}$	$2^{-\tilde{\gamma}}\sigma_{\max}$	$\sigma_{\max}$	$\left(\frac{3}{4}\right)^{\tilde{\gamma}}\sigma_{\max}$
$\mathbf{P_B} = \frac{\sqrt{(R_\sigma-1)\cdot(\zeta R_\sigma+\zeta-2)}}{2}\sigma_{\max}$	$\frac{\sqrt{2-\zeta}}{2}\sigma_{\max}$	$\sigma_{\max}$	$\frac{\sqrt{3\cdot(4-\zeta)}}{4}\sigma_{\max}$

$$M_r = 2^{1-\tilde{\gamma}} - 1 = \sqrt{2-\zeta} - 1. \tag{D.6}$$

A resolution of Eq. (D.6) with respect to ζ, gives the looked-for dependence on the Walker exponent $\tilde{\gamma}$ as:

$$\zeta = 2 - 2^{2-2\tilde{\gamma}}, \quad \tilde{\gamma} = 1 - \frac{\ln(2-\zeta)}{2\ln 2}. \tag{D.7}$$

Figure D.5 shows the parameters of Walker $\tilde{\gamma}$ and Bergmann ζ, which correspond to the same fatigue life, as the functions of mean stress sensitivity M_r. The relationship between the parameters of Walker $\tilde{\gamma}$ and Bergmann ζ for equal fatigue life and for equal mean stress sensitivity μ_r is demonstrated in Fig. D.6. The calculation uses Eq. (D.7). Meanwhile, the value $\zeta = 0$ corresponds to the mean stress sensitivity, M_{rSWT} according to Smith–Watson–Topper. Both the Walker and Bergmann parameters are capable of adequately describing the experimentally obtained mean stress sensitivity. The difference between the two parameters lies in the slightly different dependencies of the cycle amplitude on the mean stress, as shown in Figs. D.2 and D.3.

The present study shows the possibility of a simple unification of the nonlinear dependence of the stress amplitude on the stress mean value with a single parameter. The above results illustrate the difference and similarity between the traditional Walker and Bergmann criteria.

The Smith–Watson–Topper rule has no fitting constant and demonstrates the fixed values of damage. The damage parameters $\mathbf{P_{SWT}}, \mathbf{P_B}$,$\mathbf{P_W}$ for different ratios of lower stress to upper stress in the cycle and the positive σ_o are shown in Table D.6. The damage values are always equal for pure alternating stress $R_\sigma = -1$. The fitting constants can be chosen to equalize the damage for $R_\sigma = 0$. As shown above, if the Walker exponent $\tilde{\gamma}$ relates to the Bergmann parameter according to Eq. (6), the damage parameters for $R_\sigma = 0$ for the three criteria, $\mathbf{P_S}, \mathbf{P_B}$,$\mathbf{P_W}$, will all be equal. However, for $R_\sigma = -1/2$ the damage parameters $\mathbf{P_S}, \mathbf{P_B}$,$\mathbf{P_W}$ result in different values. The parameter, which shows the minimal deviation between the acquired and predicted damage in the third case $R_\sigma = -1/2$, will be preferrable for the fatigue estimation in the full range $0 \leq R_\sigma \leq -1$.

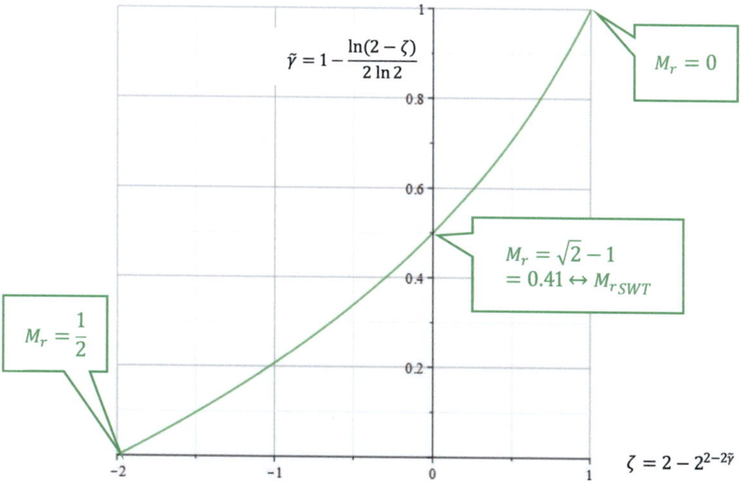

Fig. D.6 Walker parameter $\tilde{\gamma}$ as a function of Bergmann constant ζ

References

ABAQUS Scripting User's Manual. (2024). https://classes.engineering.wustl.edu/2009/spring/mase5513/abaqus/docs/v6.5/books/cmd/default.htm?startat=pt02ch06s01.html. Accessed February 15, 2024

Abdallah, Z., Gray, V., Whittaker, M., & Perkins, K. A. (2014). Critical analysis of the conventionally employed creep lifting methods. *Materials, 7*, 3371–3398.

Abramowitz, M., & Stegun, I. A. (1972). *Handbook of mathematical functions with formulas, graphs, and mathematical tables*. Dover Publications.

Abu-Haiba, M. S., Fatemi, A., & Zoroufi, M. (2002). Creep deformation and monotonic stress-strain behavior of Haynes alloy 556 at elevated temperatures. *Journal of Materials Science, 37*, 2899–2907.

Akaki Y., Matsuo, T., Nishimura, Y., Miyakawa, S., & Endo, M. (2017). Microscopic observation of shear-mode fatigue crack growth behavior under the condition of continuous hydrogen-charging, 6th International conference on fracture fatigue and wear. *Journal of Physics: Conference Series, 843*, 012051.

Akiniwa, Y., Stanzl-Tschegg, S., Mayer, H., Wakita, M., & Tanaka, K. (2008). Fatigue strength of spring steel under axial and torsional loading in the very high cycle regime. *International Journal of Fatigue, 30*, 2057–2063.

Akiniwa, Y., & Tanaka, K. (2004). Evaluation of fatigue strength of high strength steels in very long life regime. In: *Third international conference on very high cycle fatigue (VHCF-3)* (Shiga, Japan, September 16–19, 2004, pp. 464–471).

Analytical Strength Assessment of Components in Mechanical Engineering: Made of Steel, Cast Iron and Aluminium Materials. (2020). FKM Guideline 7th Edition 2020 VDMA (Verlag). ISBN 978-3-8163-0745-7.

ANSYS. (2022). ANSYS, Inc. Southpointe, 2600 ANSYS Drive, Canonsburg, PA 15317c. http://www.ansys.com/

ANSYS. (2024). ANSYS, Inc. Southpointe, 2600 ANSYS Drive, Canonsburg, PA 15317c, http://www.ansys.com/

Askey, R. A., Daalhuis, A., & Olde, B. (2010). Meijer G-function. In F. W. J. Olver, D. M. Lozier, R. F. Boisvert, & W. C. Charles (Eds.), *NIST handbook of mathematical functions*. Cambridge University Press, ISBN 978-0521192255, MR 2723248.

ASM. (1986). *Atlas of fatigue curves* (Ed. Howard E. Boyer). ASM International, ISBN: 978-0-87170-214-2.

ASTM DS67A. (2002). Handbook of comparative world steel standards (J. E. Bringas (ed.), 2nd edn.) (ASTM Data Series; DS 67A). ASTM International.

ASTM E 606-80 (1986) *Constant-amplitude low-cycle fatigue testing, annual book of ASTM standards, Section 3* (pp. 656–673).

V. Kobelev, *Fundamentals of Springs Mechanics*,
https://doi.org/10.1007/978-3-031-58584-5

ASTM E 739-80. (1986). Statistical analysis of linear or linearized stress-life (S-N) and strain-life (ε-N). In Fatigue data annual book of ASTM standards, Section 3 (pp. 737–745).

AVIF A 210. (2006). Investigations on the corrosion behaviour of disc springs and disc spring columns. In C. Berger & B. Kaiser (Eds.), Untersuchungen zum Korrosionsverhalten von Tellerfedern und Tellerfedersäulen. Tech. Univ. Darmstadt

Barrois, W. (1979). Repeated plastic deformation as a cause of mechanical surface damage in fatigue, wear, fretting-fatigue, and rolling fatigue: A review. *International Journal of Fatigue, 1*(4), 167–189.

Bateman, H., & Erdélyi, A. (1953). *Higher transcendental functions* (Vol. I). McGraw–Hill. (See § 7.3, "Definition of the G-Function", p. 206)

Bazant, Z. P., & Cedolin, L. (1991). *Stability of structures.* Oxford University Press.

Bergmann, J., & Seeger, T. (1979). On the influence of cyclic stress-strain curves, damage parameters, and various evaluation concepts on the life prediction by the local approach. Corpus ID: 137185840.

Betten, J. (2008). *Creep mechanics* (3rd ed.). Springer.

Beden, S. M., Abdullah, S., & Ariffin, A. K. (2009). Review of fatigue crack propagation models for metallic components. *European Journal of Scientific Research, 28*(3), 364–397. ISSN 1450-216X.

Bhadeshia, H. K. D. H. (2003). Mechanisms and models for creep deformation and rupture, In: I. Milne, R. O. Ritchie, & B. Karihaloo (Eds.), *Comprehensive structural integrity, creep and high-temperature failure* (Vol. 5). Elsevier.

Blasón, S., Correia, J. A. F. O., Apetre, N., Arcari, A., De Jesus, A. M. P., Moreira, P., & Fernández-Canteli, A. (2016). Proposal of a fatigue crack propagation model taking into account crack closure effects using a modified CCS crack growth model. *Procedia Structural Integrity, 1*, 110–117. XV Portuguese Conference on Fracture, 10–12 February 2016, Paço de Arcos, Portugal.

Boardman, F. D. (1965). Derivation of creep constants from measurements of relaxation creep in springs. *Philosophical Magazine, 11*(109), 185–187. https://doi.org/10.1080/14786436508211935

Boardman, B. (1990). Fatigue resistance of steels. In *ASM handbook, Volume 1: Properties and selection: Irons, steels, and high-performance alloys* (pp. 673–688). ASM Handbook Committee, ASM International.

Boeller, C., & Seeger, T. (1987). Materials data for cyclic loading, part A: Unalloyed steels. In *Materials data for cyclic loading.* Materials Science Monographs, 42A. Elsevier. Available online: https://www.elsevier.com/books/materials-data-for-cyclic-loading/boller/978-0-444-42871-4. Accessed on February 15, 2024.

Boresi, A. P., & Schmidt, R. J. (2003). *Advance mechanics of materials* (6th ed.). Wiley.

Brown, M. W. (1989). Analysis and design methods in multiaxial fatigue. In C. Moura Branco, L. Guerra Rosa (Eds.), *Advances in fatigue science and technology.* Kluwer Academic Publishers.

Bussoloti, R., Luigi, L. L. M., Canale, L. C. F., & Totten, C. E. (2015). Delta ferrite: cracking of steel fasteners. *Encyclopedia of Iron, Steel, and Their Alloys,* 1070–1081. https://doi.org/10.1081/E-EISA-120049491

Carlitz, L. (1963). The inverse of the error function. *Pacific Journal of Mathematics, 13*, 459–470.

Carpinteri, A. (Ed.). (1994). *Handbook of fatigue crack propagation in metallic structures.* Elsevier Science B.V.

Castillo, E., Fernández-Canteli, A., & Siegele, D. (2014). Obtaining S-N curves from crack growth curves: An alternative to self-similarity. *International Journal of Fracture, 187*, 159–172. https://doi.org/10.1007/s10704-014-9929.3

Castillo, E., & Fernández-Canteli, A. (2009). *A unified statistical methodology for modeling fatigue damage.* Springer Science + Business Media B.V.

Chen, W. F., & Han, D. J. (1988). *Plasticity for structural engineers.* Springer-Verlag.

Chen, F., Wang, F., & Cui, W. (2011). Fatigue life prediction of engineering structures subjected to variable amplitude loading using the improved crack growth rate model. *Fatigue & Fracture of*

Engineering Materials & Structures, 35, 278–290. https://doi.org/10.1111/j.1460-2695.2011.01619.x

Cherniy, V. P. (2001). Effect of curved bar properties on bending of curved pipes. *Transactions of the ASME, Journal of Applied Mechanics,* 68. https://doi.org/10.1115/1.1357518

Chilver, A. H. (1967). *Thin-walled structures.* Wiley.

Choi, B. L., et al. (2009). Torsion beam axle system design with a multidisciplinary approach. *International Journal of Automotive Technology, 10*, 1.

Christ, H.-J. (1991). *Wechselverformung von Metallen.* Springer-Verlag.

Clark, R. A., & Reissner, E. (1951). *Bending of curved tubes, advances in applied mechanics* (Vol. II, pp. 93–122). Academic Press.

Collins, J. A., Busby, H. R., & Staab, G. H. (2010). *Mechanical design of machine elements and machines: A failure prevention perspective.* Wiley.

Correia, J. A. F. O., Blasón, S., Arcari, A., Calvente, M., Apetre, N., Moreira, P. M. G. P., de Jesus, A. M. P., & Canteli, A. F. (2016a). Modified CCS fatigue crack growth model for the AA2019-T851 based on plasticity-induced crack-closure. In *XV Portuguese conference on fracture and fatigue, theoretical and applied fracture mechanics* (Vol. 85, Part A, pp. 26–36).

Correia, J. A. F. O., Blasón, S., de Jesus, A. M. P., Canteli, A. F., Moreira, P. M. G. P., & Tavares, P. J. (2016b). Fatigue life prediction based on an equivalent initial flaw size approach and a new normalized fatigue crack growth model. *Engineering Failure Analysis, 69*, 15–28.

Davison, M., & Essex, G. C. (1998). Fractional differential equations and initial value problems. *The Mathematical Scientist,* 108–116.

de Castro, J. T. P., Meggiolaro, M. A., Miranda, A. C. (2009). On the estimation of fatigue crack propagation lives under variable amplitude loads using strain-life data. In *Proceedings of COBEM 2009 20th International congress of mechanical engineering (ABCM)*, November 15–20, 2009, Gramado, RS, Brazil.

DIN 50100:2016-12. (2016). *Load controlled fatigue testing—Execution and evaluation of cyclic tests at constant load amplitudes on metallic specimens and components.* Beuth Verlag.

DIN EN 10132-4:2003-04. (2003). *Cold-rolled narrow steel strip for heat-treatment—Technical delivery conditions—Part 4: Spring steels and other applications.* Beuth Verlag.

DIN EN 10270-2:2012-01. (2012). *Steel wire for mechanical springs.* Beuth Verlag.

DIN HANDBOOK 349 (2013) Technical springs, Beuth Verlag, Berlin

DIN-TASCHENBUCH 349. (2012). *Standards for basic materials and semi-finished products.* Federn 2, Werkstoffe, Halbzeuge, Beuth Verlag.

DIN-TASCHENBUCH 29. (2015). *Berechnungs- und Konstruktionsgrundlagen, Qualitätsanforderungen, Bestellangaben, Begriffe, Formelzeichen und Darstellungen.* Federn 1 Beuth Verlag.

Dominguez Almaraz, G. M. (2008). Prediction of very high cycle fatigue failure for high strength steels, based on the inclusion geometrical properties. *Mechanics of Materials, 40*, 636–640. https://doi.org/10.1016/j.mechmat.2008.03.001

Donahue, R. J., Clark, H. M., Atanmo, P., Kumble, R., & McEvily, A. J. (1972). Crack opening displacement and the rate of fatigue crack growth. *International Journal of Fracture Mechanics, 8*, 209–219. https://doi.org/10.1007/BF0070388

Dowling, N. E. (2004). *Mean stress effects in stress-life and strain-life fatigue, F2004/51, 2nd SAE Brasil International Conference on Fatigue.* SAE International.

Dumir, P. C. (1986). Nonlinear axisymmetric response of orthotropic thin truncated conical and spherical caps. *Acta Mechanica, 60*, 121–132.

Ellyin, F. (1986). Stochastic modelling of crack growth based on damage accumulation. *Theoretical and Applied Fracture Mechanics, 6*, 95–101.

El-Magd, E., Abdel-Ghany, W., & Homayun, M. (1985). Application of the FEM to elastoplastic wave propagation in metals. *Engineering Computations, 2*(2), 114–120. https://doi.org/10.1108/eb023609

Erdélyi, A. (1950). Hypergeometric functions of two variables. *Acta Mathematica, 83*(1), 131–164.

Evans, H. E. (1984). *Mechanisms of creep fracture.* Elsevier Applied Science Publishing.

Fatemi, A., & Yang, L. (1998). Cumulative fatigue damage and life prediction theories: A survey of the stat of the art for homogeneous materials. *International Journal of Fatigue, 20*(1), 9–34.

FEM for Springs. (2003). Masayoshi Shimoseki, Toshio Hamano, Toshiyuki Imaizumi, Springer-Verlag Berlin Heidelberg.

Fleck, N. A., Kang, K. J., & Ashby, M. F. (1994). The cyclic properties of engineering materials. *Ada Metallurgy Materials, 42*(2), 365–381.

Flügge, W., & Marguerre, K. (1950). Wölbkräfte in dünnwandigen Profilstäben. *Ing.-Arch., 18*, 23–38.

Forbes, C., Merran, E., Hastings, N., & Peacock, B. (2011). *Statistical distributions* (4th ed.). Wiley.

Freudenthal, A. M. (1973). Fatigue and fracture mechanics. *Engineering Fracture Mechanics, 5*(2), 403–414. https://doi.org/10.1016/0013-7944(73)90030-1

Furuya, Y., Hirukawa, H., & Takeuchi, E. (2019). Gigacycle fatigue in high strength steels. *Science and Technology of Advanced Materials, 20*(1), 643–656. https://doi.org/10.1080/14686996.2019.1610904

Gadouini, H., Nadot, Y., Rebours, C. (2008). Influence of mean stress on the multiaxial fatigue behaviour of defective materials. *International Journal of Fatigue, 30*, 1623–1633.

Garud, Y. (1981). Multiaxial fatigue: A survey of the state of the art. *Journal of Testing and Evaluation, 9*(3), 165–178.

Geilen, M.B., Klein, M., Oechsner, M. (2020), A novel algorithm for the determination of walker damage in loaded disc springs. *Materials, 13*, 1661. https://doi.org/10.3390/ma13071661

Göhner, O. (1931). Schubspannungsverteilung im Querschnitt eines gedrillten Ringstabs mit Anwendung auf Schraubenfedern, *Ing.-Archiv., 2*(1), 1–19.

Gooch, D. J. (2012). Techniques for Multiaxial Creep Testing, Springer Science & Business Media, 364 p.

Graham, J. A., Millan, J. F., & Appl, F. J. (1968). *Fatigue design handbook*. SAE.

Graße, W. (1965). Wölbkrafttorsion dünnwandiger prismatischer Stäbe beliebigen Querschnitts. *Ing.-Arch., 24*, 330–338.

Gradshteyn, I. S., & Ryzhik, I. M. (2015). *Table of integrals, series, and products* (8th ed.). ISBN 978-0-12-384933-5. Academic Press, Elsevier Inc., Waltham, MA 02451, USA

Gross, D., & Seelig, T. (2018). *Fracture mechanics. With an introduction to micromechanics* (3rd edn.). Springer-Verlag Berlin Heidelberg.

Guerra Rosa, L., & Radon (1989) Fatigue threshold behaviour part II: Theoretical aspects and open questions, In K. C. Moura Branco & L. Guerra Rosa (Eds.), *Advances in fatigue science and technology* (pp. 141–156). Kluwer Academic Publishers.

Gumbel, E. J. (1954). Statistical theory of extreme values and some practical applications. In *Applied mathematics series* (Vol. 33, 1st edn.). U.S. Department of Commerce, National Bureau of Standards.

Hagiwara, T., Kawami, A., Ueno, A., & Kido, A. (1991). Super-clean steel for valve spring quality. *Wire Journal International*, 29–34.

Heilig, R. (1961). Beitrag zur Theorie der Kastenträger beliebiger Querschnittsform. *Der Stahlbau, 30*, 333–349.

Helical Springs. (1974). *Engineering design guides.* The United Kingdom Atomic Energy Authority and Oxford University Press, ISBN 019:859142X

Hesselmann B., & Kobelev, V. (2002) *Disk spring*. European Patent Application EP1331416. https://worldwide.espacenet.com/patent/search/family/026010978/publication/EP1331416A1?q=pn%3DEP1331416A1

Honeykomb, R. W. K. (1968). *The plastic deformation of metals.* Edward Arnold (Publishers).

Hutchings, M. T., & Krawitz, A. D. (1992). *Measurement of residual and applied stress using neutron diffraction.* Springer Science Business Media.

Hutchinson, J. (1968). Singular behaviour at the end of a tensile crack in a hardening material. *Journal of the Mechanics and Physics of Solids, 16*, 13–31.

Illgner, K. H. (1995). Die modernen Wärmebehandlungsanlagen für Federn. *Draht, 46*(11), 573–578.

Ince, A., & Glinka, G. (2011). A modification of Morrow and Smith–Watson–Topper mean stress correction models. *Fatigue & Fracture of Engineering Materials & Structures, 34*, 854–867. https://doi.org/10.1111/j.1460-2695.2011.01577.x

ISO. (2010). *Road vehicles—Vehicle dynamics and road-holding ability—Vocabulary, ISO Standard 8855, Rev. 2010*. International Organization for Standardization.

James, M. N., Hughes, D. J., Chen, Z., Lombard, H., Hattingh, D. G., Asquith, D., Yates, J. R., & Webster, P. J. (2007). Residual stresses and fatigue performance. *Engineering Failure Analysis, 14*(2007), 384–395.

Johnson, G. R., Cook, W. H. (1983). A constitutive model and data for metals subjected to large strains, high strain rates and high temperatures. In *Proceedings of the 7th International Symposium on Ballistics, 21*, 541–7.

Jönsson, J., & Andreassen, M. J. (2009). Distortional modes of thin-walled beams. In: J. Ambrosio, et al. (Eds.) *7th EUROMECH solid mechanics conference*. Lisbon, Portugal.

Juvinall, R. C., & Marshek, K. M. (2017). *Fundamentals of machine component design* (6th edn.). Wiley.

Kalbfleisch, J., & Prentice, R. (2002). *The statistical analysis of failure time data* (2nd edn.). Wiley.

Kampe de Feriet J. (1957) *Fonctions de la Physique Mathématique, Formulaire de Mathématiques à l'usage des Physiciens et des Ingénieurs*. Centre National de la Recherche Scientifique.

Kanninen, M. F., & Popelar, C. H. (1985). *Advanced fracture mechanics*. Oxford University Press.

Karman, Th. (1911). Über die Formänderung dünnwandiger Rohre, insbesondere federnder Ausgleichrohre. *Z. Ver. Deut. Ing., 55*, 1889–1895.

Karr, U., Sandaiji, Y., Tanegashima, R., Murakami, S., Schönbauer, B., & Fitzka Mayer, M. I. H. (2020). Inclusion initiated fracture in spring steel under axial and torsion very high cycle fatigue loading at different load ratios. *International Journal of Fatigue, 134*(2020), 105525. https://doi.org/10.1016/j.ijfatigue.2020.105525

Karr, U., Schönbauer, B. M., Sandaiji, Y., & Mayer, H. (2022). Effects of non-metallic inclusions and mean stress on axial and torsion very high cycle fatigue of SWOSC-V Spring steel. *Metals, 12*, 1113. https://doi.org/10.3390/met12071113

Kawahara, J., Tanabe, K., Banno, T., & Yoshida, M. (1992). Advance of valve spring steel. *Wire Journal International*, pp. 55–61.

Kennedy, A. J. (1967). *Processes of creep and fatigue in metals*. Oliver & Boyd.

Kessler, D. A., & Rabin, Y. (2003). Stretching instability of helical springs. *Physical Review Letters, 90*, 024301.

Kobelev, V. (2012). Thin-walled rods with semi-open profile for semi-solid automotive suspension. *International Journal of Automotive Technology, 13*(2), 231–245.

Kobelev, V. (2013). Thin-walled rods with semiopened profiles. *ASME Journal of Applied Mechanics, 80*. https://doi.org/10.1115/1.4006935

Kobelev, V. (2017c) A proposal for unification of fatigue crack growth law. 6th International conference on fracture fatigue and wear. *IOP Conference Series: Journal of Physics: Conference Series, 843*, 012022. https://doi.org/10.1088/1742-6596/843/1/012022

Kobelev, V. (2017b). Unification proposals for fatigue crack propagation laws. *Multidiscipline Modeling in Materials and Structures, 13*(2), 262–283. https://doi.org/10.1108/MMMS-10-2016-0052

Kobelev, V. (2018) *Durability of springs* (1st edn). Springer International Publishing AG. ISBN: 978-3-319-58477-5.

Kobelev, V. (2021). *Durability of springs* (2nd ed.). Springer.

Kobelev, V. (2023a). Exact shell solutions for conical springs. II. Radial cylindric curb. *Mechanics Based Design of Structures and Machines, An International Journal*. https://doi.org/10.1080/15397734.2023.2189942

Kobelev, V. (2023b). Exact shell solutions for conical springs. III. disk springs with variable thickness. *Mendeley Data, V1*. https://doi.org/10.17632/gtk67w9kp7.1. https://data.mendeley.com/datasets/gtk67w9kp7

Kobelev, V., & Larichev, A. D. (1988). Model of thin-walled anisotropic rods. *Mechanics of Composite Materials, 24*, 97–104.

Kobelev, V., Hesselmann, B., & Rinsdorf, A. (2003). Corrugated spring with gradual progressive spring characteristic. US 7334784 B2, US Patent Office.

Kobelev, V., Klaus, U., Scheffe, U., & Ivo, J. (2009). Querträger für eine Verbundlenkerachse. European Patent EP2281701, European Patent Office.

Kollar, L. P., & Pluzsik, A. (2002). Analysis of thin-walled composite beams with arbitrary layup. *Journal of Reinforced Plastics and Composites, 21*, 1423.

Kong, Y. S., Abdullah, S., Schramm, D., Omar, M. Z., & Haris, S. M. (2019). Correlation of uniaxial and multiaxial fatigue models for automobile Spring life assessment. *Experimental Techniques, 44*, 197–215.

Krupp, U. (2007). *Fatigue crack propagation in metals and alloys: microstructural aspects and modelling concepts*. Wiley-VCH, ISBN: 979.3-527-31537-6.

Kujawski, D., & Ellyin, F. (1984). A fatigue crack propagation model. *Engineering Fracture Mechanics, 20*, 695–704.

La Rosa, G., Messina, M., & Risitano, A. (2001). Stiffness of variable thickness disk springs. *Journal of Mechanical Design, 123*(2), 294–299. https://doi.org/10.1115/1.1357162

Łagoda, T., Vantadori, S., Głowacka, K., Kurek, M., & Kluger, K. (2022). Using the Smith-Watson-Topper parameter and its modifications to calculate the fatigue life of metals: The state-of-the-art. *Materials, 15*, 3481. https://doi.org/10.3390/ma15103481

Lamacq, V., Baïetto-Dubourg, M.-C., & Vincent, L. (1996). Crack path prediction under fretting fatigue - a theoretical and experimental approach. *Journal of Tribology, American Society of Mechanical Engineers, 118*(4), 711–720.

Laudiero, F., & Savoia, M. (1990). Shear strain effects in flexure and torsion of thin-walled beams with open or closed cross-section. *Journal of Thin-Walled Structures, 10*, 87–119.

Lehnert, W. (1997). Umformung - Gefüge - mechanische Eigenschaften hochgekohlter Stahldrähte für Federn. *Draht, 44*(4), 44–48.

LibreOffice. (2023). LibreOffice Calc. https://www.libreoffice.org/diskover/calc/

Librescu, L., & Song, O. (2006). *Thin-walled composite beams, theory and application*. Springer Berlin Heidelberg.

Linnig, W., et al. (2009). The twist-beam rear axle—Design, materials, processes and concepts. ATZ Worldwide eMagazines Edition, 2.

Mainardi, F., & Gorenflo, R. (2007). Time-fractional derivatives in relaxation processes: A tutorial survey. *Fractional Calculus and Applied Analysis, 10*, 269–308. http://arxiv.org/abs/0801.4914

Mainardi, F., & Spada, G. (2011). Creep, relaxation and viscosity properties for basic fractional models in rheology. *The European Physical Journal, 193*, 133–160.

MAPLE. (2022). Maplesoft, a division of Waterloo Maple Inc. 2022, 615 Kumpf Drive Waterloo, ON Canada, N2V 1K8. https://www.maplesoft.com/

Mauerauch, E., Vöhringer, O. (1978). Z. Werkstofftechnik 9(11), 370–391.

Mayer, H., Schuller, R., Karr, U., Irrasch, D., Fitzka, M., Hahn, M., & Bacher-Höchst, M. (2015). Cyclic torsion very high cycle fatigue of VDSiCr spring steel at different load ratios. *International Journal of Fatigue, 70*(2015), 322–327. https://doi.org/10.1016/j.ijfatigue.2014.10.007

Mazari, M., Bouchouicha, B., Zemri, M., Benguediab, M., & Ranganathan, N. (2008). Fatigue crack propagation analyses based on plastic energy approach. *Computational Materials Science, 41*(2008), 344–349.

McEvily, A. J., & Groeger, J. (1977). On the threshold for fatigue-crack growth. In *Fourth International conference on fracture* (Vol. 2, pp. 1293–1298). University of Waterloo Press.

McLean, D. (1966). (1966) The physics of high temperature creep in metals Rep. *Progr. Phys., 29*, 1.

Microsoft EXCEL. (2023). Microsoft Corporation, One Microsoft Way, Redmond, WA 98052–6399, USA. https://www.microsoft.com

Miller, M. S., & Gallagher, J. P. (1981). An analysis of several fatigue crack growth rate (FCGR) descriptions. In S. J. Hudak, Jr. and R. J. Bucci (Eds.), *Fatigue crack growth measurement and data analysis* (pp. 205–251). ASTM STP 738, American Society for Testing and Materials.

Miller, K. J., & Brown, M. W. (Eds.). (1985). *Multiaxial fatigue.* ASTM STP 853.

Milne, R., Ritchie, O., & Karihaloo, B. (2003) *Comprehensive structural integrity, Vol. 4, cyclic loading and fatigue.* Elsevier.

Mishnaevsky, L., Jr., & Brøndsted, P. (2007). Modeling of fatigue damage evolution on the basis of the kinetic concept of strength. *International Journal of Fracture, 144*, 149–158. https://doi.org/10.1007/s10704-007-9086-1

Müller, E., Yapi, A., & Rhönisch, B. (2006). Die Ausprägung von Druckeigenspannungen durch Spannungsstrahlen an Minibloc-Federn. *VDI-Berichte, 1972*, 285.

Müller, E. (1994). Plastizieren, Kugel- und insbesondere Spannungsstrahlen zur Lebensdauersteigerung von Federelementen. In: Bauteil ´94—Die Feder. Tagungsband des DVM - Tages 1994, Deutscher Verband für Material-forschung und -prüfung e.V. Berlin, pp. 339–349.

Mun, K.-J., Kim, T.-J., & Kim, Y.-S. (2010). Analysis of the roll properties of a tubular-type torsion beam suspension. *Proceedings of IMechE, Part D: Journal of Automobile Engineering, 224*(D1), 1–13.

Murakami, Y. (1985). Analysis of stress intensity factors of modes I, II and III for inclined surface racks of arbitrary shape. *Engineering Fracture Mechanics, 22*, 101–114.

Murakami, Y., Nomoto, T., & Ueda, T. (1999). *Fatigue & Fracture Engineering & Material Structures, 22*, 581–590.

Nadot, Y., & Billaudeaub, T. (2006). Multiaxial fatigue limit criterion for defective materials. *Engineering Fracture Mechanics, 73*(1), 112.

Nihei, M., Heuler, P., Boller, Ch., & Seeger, T. (1986). Evaluation of mean stress effect on fatigue life by use of damage parameters. *International Journal of Fatigue, 8*(3), 119–126. ISSN 0142-1123, https://doi.org/10.1016/0142-1123(86)90002-2.

Noroozi, A. H., Glinka, G., & Lambert, S. (2008). Prediction of fatigue crack growth under constant amplitude loading and a single overload based on elasto-plastic crack tip stresses and strains. *Engineering Fracture Mechanics, 75*(2), 188–206.

Noroozi, A. H., Glinka, G., & Lambert, S. (2005). A two parameter driving force for fatigue crack growth analysis. *International Journal of Fatigue, 27*(10–12), 1277–1296.

O´Malley, M., & Hayes, M. P. (1990). Der Einfluß der Oberflächenqualität auf das Ermüdungsvermögen von austenitischen Edelstahlfedern. In *Proceedings of European Spring Federation Conference*, Düsseldorf.

OpenOffice. (2023). Apache OpenOffice Calc. www.OpenOffice.org/calc

Paris, P., & Erdogan, F. (1963) A critical analysis of crack propagation laws. *Journal of Basic Engineering, Transactions of ASME,* 528–534.

Pearson, K. (1968). *Tables of incomplete beta functions* (2nd ed.). Cambridge University Press.

Pineau, A. (1989). Mechanisms of creep-fatigue interactions. In: C. M. Branco, & L. G. Rosa (Eds.), *Advances in fatigue science and technology.* NATO ASI Series, Vol. 159. Springer. https://doi.org/10.1007/978-94-009-2277-8_12

Polak, J. (2003). Cyclic deformation, crack initiation, and low-cycle fatigue. In: I. Milne et al. (Eds.), *Comprehensive structural integrity.*

Postma, T. (1993). Hohe Anforderungen. Neue Entwicklungen bei Ventilfederstahl. *Drahtwelt, 6*, 22–24.

Prokic, A. (2003). Stiffness method of thin-walled beams with closed cross-section. *Computers and Structures, 81*, 39–51.

Pugno, N., Ciavarella, M., Cornetti, P., & Carpinteri, A. (2006). A generalized Paris' law for fatigue crack growth. *Journal of the Mechanics and Physics of Solids., 54*, 1333–1349.

Pugno, N., Cornetti, P., & Carpinteri, A. (2007). New unified laws in fatigue: From the Wöhler's to the Paris' regime. *Engineering Fracture Mechanics, 74*, 595–601.

Rajamani, R. (2006). *Vehicle dynamics and control.* Springer.

Reich, R., & Kletzin, U. (2011). Fatigue damage parameters and their use in estimating lifetime of helical compression springs. In *56th International Scientific Colloquium*. Ilmenau University of Technology, 12–16 September 2011.

Reimpell, J., Stoll, H., & Betzler, W. (2001). *The automotive chassis: Engineering principles*. SAE.

Reissner, E. (1946). Analysis of shear lag in box beams by the principle of minimum potential energy. *Quarterly of Applied Mathematics, 4*, 268–278.

Rice, J., & Rosengren, G. (1968). Plane strain deformation near a crack tip in a power-law hardening material. *Journal of the Mechanics and Physics of Solids, 16*, 1–12.

Richard, H. A., & Sander, M. (2012). Ermüdungsrisse. Springer-Vieweg. ISBN 979.3-8349.1594-1. https://doi.org/10.1007/979.3-8349.8663-7

Ritchie, R. O., Boyce, B. L., Campbell, J. P., Roder, O., Thompson, A. W., & Milligan, W. W. (1999). Thresholds for high-cycle fatigue in a turbine engine Ti–6Al–4V Alloy. *International Journal of Fatigue, 21*, 653–662. https://doi.org/10.1016/S0142-1123(99)00024-9

SAE HS 788. (1980). *Manual on design and application of leaf springs*. SAE International.

SAE. (2008). Vehicle dynamics terminology. SAE Standard J670, Rev. 2008-01-24.

Salmon, C. G., & Johnson, J. E. (1996). *Steel structures* (4th ed.). HarperCollins College Publishers.

Sangid, M. D. (2013). The physics of fatigue crack initiation. *International Journal of Fatigue, 57*(2013), 58–72.

Schijve, J. (2005). Statistical distribution functions and fatigue of structures. *International Journal of Fatigue, 27*, 1031–1039.

Schmitz-Cohnen, K. (1994) Präzisionshalbzeuge für die Federnindustrie aus nichtrosten-den Werkstoffen. *Draht, 45*(3), 228–237.

Schütz, W. (1967). Über Eine Beziehung zwischen der Lebensdauer bei Konstanter und bei Veränderlicher Beanspruchungsamplitude und Ihre Anwendbarkeit auf Die Bemessung von Flugzeugbauteilen; Zeitschrift für Flugwissenschaften und Weltraumforschung/Journal of flight sciences and space research/hrsg. von Deutschen Gesellschaft für Luft- und Raumfahrt e.V. u. d. Deutschen Forschungs- und Versuchsanstalt für Luft- und Raumfahrt e.V, 15 H. 11, S. 407/419, 1967.

Schwalbe, K.-H. (1980). *Bruchmechanik metallischer Werkstoffe*. Carl Hanser Verlag.

Shi, K., Cai, L., & Bao, C. (2014). Crack growth rate model under constant cyclic loading and effect of different singularity fields. *Procedia Materials Science, 3*, 1566–1572.

Shigley, J. E., & Mischke, C. R. (1986). *Standard handbook of machine design* (p. 1632). McGraw-Hill, Inc. ISBN 0-07-056892-8.

Siev, A. (1966). Torsion in closed sections. *Engineering Journal, AISC, 3*(1), 46–54.

Slivker, V. (2007). *Mechanics of structural elements, theory and applications*. Springer Berlin Heidelberg.

Smith, K. N., Watson, P., Topper, T. H. (1970). A stress-strain function for the fatigue of metals. *Journal of Materials, ASTM, 5*(4), 767–778.

Socie, D. F., & Morrow, J. D. (1980). Review of contemporary approaches to fatigue damage analysis. In J. J. Burke & V. Weiss (Eds.), *Risk and failure analysis for improved performance and reliability* (pp. 141–194). Plenum Publication Corporation.

Stachowiak, G. W., & Batchelor, A. W. (2014). Engineering tribology. ISBN: 978-0-12-397047-3. Butterworth-Heinemann.

Steele, M. C. (1954) The Plastic Bending And Twisting of Square Section Members, Journal of the Mechanics And Physics Of Solids, 8188–166.

Suresh, S. (1998). *Fatigue of materials*. Cambridge University Press.

Tamberg, K. G., Mikluchin, P. T. (1973). *Torsional phenomena analysis and concrete structure design, analysis of structural systems for torsion, SP-35* (pp. 1–102). American Concrete Institute.

Tanaka, K. (2003). Fatigue crack propagation. In I. Milne, O. Ritchie, & B. Karihaloo (Eds.), *Comprehensive structural integrity, Vol. 4, cyclic loading and fatigue*. Elsevier.

Thompson, M. K., & Thompson, J. M. (2017) *ANSYS mechanical APDL for finite element analysis*. Butterworth-Heinemann, Oxford. ISBN: 978-0-12-812981-4. https://doi.org/10.1016/B978-0-12-812981-4.00011-3

Timoshenko, S., & Gere, J. M. (1961). *Theory of elastic stability*. McGraw-Hill.

Timoshenko, S., & Woinowsky-Krieger, S. (1957). *Theory of plate and shell* (2nd ed.). McGraw Hill.

Timoshenko, S. (1945). Theory of bending, torsion, and buckling of thin-walled members of open cross-section. *Journal of Franklin Institute, 239*, 201–219, 249–268, 343–361.

Torres, M. A. S., & Voorwald, H. J. C. (2002). An evaluation of shot peening, residual stress and stress relaxation on the fatigue life of AISI 4340 steel. *International Journal of Fatigue, 24*(2002), 877–886.

Totten, G. (2008). Fatigue crack propagation. *Advanced materials & processes* (pp. 39–41).

VDI 2226:1965-07. (1965). *Empfehlung für Die Festigkeitsberechnung Metallischer Bauteile*. VDI.

Venter, A. M., Luzin, V., & Hattingh, D. G. (2014). Residual stresses associated with the production of coiled automotive springs. *Materials Science Forum, 777*, 78–83. https://doi.org/10.4028/www.scientific.net/MSF.777.78

Verhoeven, J. D. (2007). *Steel metallurgy for the non-metallurgist*. ASM International, Materials Park, Ohio 44073-0002.

VHCF7. (2017). *7th International conference on very high cycle fatigue*, July 3–5, 2017 (M. Zimmermann, & H.-J. Christ (Editors).), Dresden, Germany, Siegen, Univ.

Vlasov, V. Z. (1961). *Thin-walled elastic beams*. Office of Technical Services, U.S. Department of Commerce, Washington 25, DC, TT-61–11400.

Walker, K. (1970). The effect of stress ratio during crack propagation and fatigue for 2024-T3 and 7075-T6 Aluminum. In *Effects of environment and complex load history on fatigue life, ASTM STP 462*. Ameriacan Society for Testing and Materials, West Conshohocken, PA, pp. 1–14.

Wang, Q. Y., Berard, Y., Dubarre, A., Baudry, G., Rathery, S., & Bathias, C. (1999). Gigacycle fatigue of ferrous alloys. *Fatigue & Fracture Engineering and Materials Structure, 22*, 667–672.

Wang, Q. Y., Bathias, C., Kawagoishi, N., & Chen, Q. (2000). Effect of inclusion on subsurface crack initiation and gigacycle fatigue strength. *International Journal of Fatigue, 24*(12), 1269–1274. https://doi.org/10.1016/S0142-1123(02)00037-3

Weertman, J. (1991). Fatigue. In: Rita G. Lerner, & George L. Trigg (Eds.), *Encyclopedia of physics* (2nd ed.). VCH Publishers Inc.

Wehner, T., Fatemi, A. (1991). Effects of mean stress on fatigue behaviour of a hardened carbon steel. *International Journal of Fatigue, 13*(3), 241–248. ISSN 0142-1123, https://doi.org/10.1016/0142-1123(91)90248-W

Whatham, J. F. (1981). Thin shell analysis of non-circular pipe bends. *Nuclear Engineering and Design, 67*, 287–296.

Wiemer, H.-E. (1998). Übersicht über Verfahren und Anlagen der Sekundärmetallurgie. *Stahl und Eisen, 118*(7), 27–29.

Wirsching, P.H. (1983). Statistical summaries of fatigue data for design purposes. NASA Contractor Report 3697, Levis Research Center.

WO 2008125076 A1. (2007). Leaf spring made of a fiber-plastic composite and force transmission element therefore. European Patent Office, 2008-10-23, Espacenet.

Wormsen, A., Sjödin, B., Hrkegrd, G., & Fjeldstad, A. (2004). Non-local stress approach for fatigue assessment based on weakest-link theory and statistics of extremes. *Fatigue & Fracture Engineering and Material Structures, 30*, 1214–1227.

Xue, L. (2008). A unified expression for low cycle fatigue and extremely low cycle fatigue and its implication for monotonic loading. *International Journal of Fatigue, 30*, 1691–1698.

Yoshimoto, T., & Matsuo, T. (2017). An investigation into the frequency dependence upon the fatigue crack growth rate conducted by a novel fatigue testing method with in-situ hydrogen-charging. *Journal of Physics: Conference Series, 843*, 012050. 6th International Conference on Fracture Fatigue and Wear.

Zimmermann, M. (2019). Very high cycle fatigue, 57. In C.-H. Hsueh et al. (Eds.), *Handbook of mechanics of materials*. Springer Nature Singapore Pte Ltd. https://doi.org/10.1007/978-981-10-6884-3_43

Index

V. Kobelev, *Fundamentals of Springs Mechanics*,
https://doi.org/10.1007/978-3-031-58584-5